Principles of Bioenergetics

Vladimir P. Skulachev · Alexander V. Bogachev
Felix O. Kasparinsky

Principles of Bioenergetics

 Springer

Vladimir P. Skulachev
Belozersky Institute of Physico-Chemical
  Biology
Moscow State University
Moscow
Russia

Alexander V. Bogachev
Belozersky Institute of Physico-Chemical
  Biology
Moscow State University
Moscow
Russia

Felix O. Kasparinsky
Faculty of Biology
Moscow State University
Moscow
Russia

ISBN 978-3-642-33429-0   ISBN 978-3-642-33430-6 (eBook)
DOI 10.1007/978-3-642-33430-6
Springer Heidelberg New York Dordrecht London
Library of Congress Control Number: 2012950861

© Springer-Verlag Berlin Heidelberg 2013
This work is subject to copyright. All rights are reserved by the Publisher, whether the whole or part of the material is concerned, specifically the rights of translation, reprinting, reuse of illustrations, recitation, broadcasting, reproduction on microfilms or in any other physical way, and transmission or information storage and retrieval, electronic adaptation, computer software, or by similar or dissimilar methodology now known or hereafter developed. Exempted from this legal reservation are brief excerpts in connection with reviews or scholarly analysis or material supplied specifically for the purpose of being entered and executed on a computer system, for exclusive use by the purchaser of the work. Duplication of this publication or parts thereof is permitted only under the provisions of the Copyright Law of the Publisher's location, in its current version, and permission for use must always be obtained from Springer. Permissions for use may be obtained through RightsLink at the Copyright Clearance Center. Violations are liable to prosecution under the respective Copyright Law.
The use of general descriptive names, registered names, trademarks, service marks, etc. in this publication does not imply, even in the absence of a specific statement, that such names are exempt from the relevant protective laws and regulations and therefore free for general use.
While the advice and information in this book are believed to be true and accurate at the date of publication, neither the authors nor the editors nor the publisher can accept any legal responsibility for any errors or omissions that may be made. The publisher makes no warranty, express or implied, with respect to the material contained herein.

Printed on acid-free paper

Springer is part of Springer Science+Business Media (www.springer.com)

Preface

Energy and life. These are the phenomena of objective reality, subjective notions, and simply the words of our language. It has been already over 2000 years that philosophers have been arguing about the essence of these terms. Scientists have been trying to comprehend them for centuries. It has been already for many decades that the positive emotional impact of these words has been used in the advertising industry to attract consumers' attention. Millions of people use the words "energy" and "life" every day while being satisfied with only an illusion of intuitive understanding of their meaning.

What is the secret of such popularity? We would like to suggest that evolution of human language has been determined by the success of those individuals who could attract wide attention to the notions that proved to be the most essential for the very existence of the society as a whole and its individual representatives. Advantages of civilization allow modern people to avoid many problems that previous generations had to face. This makes it possible to change the system of priorities. "Energy and life" seem to be not of such a crucial importance today—both for an individual and human society in general. Emotional attractiveness encoded in the language can be viewed today as a historical step on the path that was taken by people so as to understand Nature around them as well as themselves. The years 1961–1995 witnessed achievement of unbelievable progress in understanding of the molecular mechanisms of energy supply. This fact could create the illusion that practically everything has already been discovered in , and the next generations of biologists would better choose some other areas of scientific research.

But with the arrival of year 1996 there appeared the first publications on a completely new role of mitochondria—energy transforming organelles of animal, plant, and fungal cells—in the destiny of these cells. Mitochondria were found to play a key role in the processes leading to programmed cell death. They are not only the main energy providers for this process (and this fact leads to a new approach to the problem of "energy and death" that was previously viewed mainly in connection to nuclear weapons), but also serve as extremely powerful facilitators of lethal signals. Hence there appeared a new area of research connected to the role of mitochondria in the programmed elimination of tissue parts, organs, and even the whole organisms. This discovery led to a substantial increase in the number of scientific publications on mitochondria-related issues. The number of such articles in the first decade of the twenty-first century was about 2.5 times higher than in the last decade of the previous century.

We hope this book will become the contemporary textbook on bioenergetics. Most of it is based on the course of lectures on that has been presented by one of the book's authors (V. P. S.) to Moscow State University students over the last 40 years. This course embraces fundamental data on molecular mechanisms of accumulation of energy and its usage in mitochondrial, chloroplast, and bacterial membranes. These issues have already been reviewed in the previous textbook by the first author (Skulachev VP (1988) *Membrane Bioenergetics*. Springer, Berlin), but the present book includes also a number of new aspects, e.g. evolution of bioenergetic mechanisms, toxicology and physiology of reactive oxygen species, their role in programmed death phenomena, such as apoptosis, mitoptosis and phenoptosis (aging), as well as some other topics.

Contents

Part I General Aspects of Bioenergetics

1 Introduction 3
   1.1 Definition of the Term "Bioenergetics" and Some
         Milestones of its History 3
   1.2 Bioenergetics in the System of Biological Sciences 5
   1.3 Laws of Bioenergetics. 9
   1.4 Evolution of Bioenergetic Mechanisms 13
        1.4.1 Adenosine Triphosphate 14
        1.4.2 Hypothesis of Adenine-Based Photosynthesis. 15
        1.4.3 Reserve Energy Sources and Glycolysis 19
        1.4.4 Proton Channels and H^+-ATPase as Means
               to Prevent Glycolysis-Induced Acidification
               of the Cell. 21
        1.4.5 Bacteriorhodopsin-Based Photosynthesis
               as the Primordial Mechanism of Visible
               Light Energy Transduction 22
        1.4.6 Chlorophyll-Based Photosynthesis 23
        1.4.7 Respiratory Mechanism of Energy Supply 25
   References 27

Part II Generators of Proton Potential

2 Chlorophyll-Based Generators of Proton Potential. 31
   2.1 Light-Dependent Cyclic Redox Chain of Purple Bacteria 32
        2.1.1 Main Components of Redox Chain and Principle
               of Their Functioning. 33
        2.1.2 Reaction Center Complex 36
        2.1.3 $CoQH_2$: Cytochrome c-Oxidoreductase 49

		2.1.4	Ways to Use $\Delta\bar{\mu}_{H^+}$ Generated by the Cyclic Photoredox Chain	52
	2.2		Noncyclic Photoredox Chain of Green Bacteria	53
	2.3		Noncyclic Photoredox Chain of Chloroplasts and Cyanobacteria	55
		2.3.1	Principle of Functioning	55
		2.3.2	Photosystem 1	58
		2.3.3	Photosystem 2	61
		2.3.4	Cytochrome b_6f Complex	63
		2.3.5	Fate of $\Delta\bar{\mu}_{H^+}$ Generated by the Chloroplast Photosynthetic Redox Chain	66
	References			68
3	**Organotrophic Energetics**			71
	3.1		Substrates of Organotrophic Energetics	71
	3.2		Short Review of Carbohydrate Metabolism	71
	3.3		Mechanism of Substrate Phosphorylation	75
	3.4		Energetic Efficiency of Fermentation	79
	3.5		Carnosine	82
	References			85
4	**The Respiratory Chain**			87
	4.1		Principle of Functioning	87
	4.2		NADH:CoQ-Oxidoreductase (Complex I)	92
		4.2.1	Protein Composition of Complex I	93
		4.2.2	Cofactor Composition of Complex I	94
		4.2.3	Subfragments of Complex I	95
		4.2.4	Inhibitors of Complex I	96
		4.2.5	Possible Mechanisms of $\Delta\bar{\mu}_{H^+}$ Generation by Complex I	97
	4.3		CoQH$_2$:Cytochrome c-Oxidoreductase (Complex III)	102
		4.3.1	Structural Aspects of Complex III	102
		4.3.2	X-Ray Analysis of Complex III	104
		4.3.3	Functional Model of Complex III	106
		4.3.4	Inhibitors of Complex III	108
	4.4		Cytochrome c Oxidase (Complex IV)	108
		4.4.1	Cytochrome c	109
		4.4.2	Cytochrome c Oxidase: General Characteristics	110
		4.4.3	X-Ray Analysis of Complex IV	112
		4.4.4	Electron Transfer Pathway in Complex IV	113
		4.4.5	Mechanism of $\Delta\bar{\mu}_{H^+}$ Generation by Cytochrome c Oxidase	115
		4.4.6	Inhibitors of Cytochrome Oxidase	116
	References			117

Contents

5 Structure of Respiratory Chains of Prokaryotes and Mitochondria of Protozoa, Plants, and Fungi 119
   5.1   Mitochondrial Respiratory Chain of Protozoa, Plants, and Fungi 120
   5.2   Structure of Prokaryotic Respiratory Chains............... 122
        5.2.1   Respiratory Chain of *Paracoccus denitrificans*...... 123
        5.2.2   Respiratory Chain of *Escherichia coli*............ 124
        5.2.3   Redox Chain of *Ascaris* Mitochondria: Adaptation to Anaerobiosis 127
        5.2.4   Respiratory Chain of *Azotobacter vinelandii* 128
        5.2.5   Oxidation of Substrates with Positive Redox Potentials by Bacterial Respiratory Chains 129
        5.2.6   Respiratory Chain of Cyanobacteria 131
        5.2.7   Respiratory Chain of Chloroplasts 132
   5.3   Electron Transport Chain of Methanogenic Archaea......... 132
        5.3.1   Oxidative Phase of Methanogenesis 134
        5.3.2   Reducing Phase of Methanogenesis 135
   References 137

6 Bacteriorhodopsin 139
   6.1   Principle of Functioning 139
   6.2   Structure of Bacteriorhodopsin........................ 141
   6.3   Bacteriorhodopsin Photocycle 144
   6.4   Light-Dependent Proton Transport by Bacteriorhodopsin..... 145
   6.5   Other Retinal-Containing Proteins 149
        6.5.1   Halorhodopsin.......................... 149
        6.5.2   Distribution of Bacteriorhodopsin and its Analogs in Various Microorganisms 151
        6.5.3   Sensory Rhodopsin and Phoborhodopsin 151
        6.5.4   Animal Rhodopsin 153
   References 155

Part III $\Delta\bar{\mu}_{H^+}$ Consumers

7 $\Delta\bar{\mu}_{H^+}$-Driven Chemical Work 159
   7.1 H^+-ATP Synthase................................. 159
        7.1.1   Subunit Composition of H^+-ATP Synthase 159
        7.1.2   Three-Dimensional Structure and Arrangement in the Membrane 161
        7.1.3   ATP hydrolysis by Isolated Factor F_1 167
        7.1.4   Synthesis of Bound ATP by Isolated Factor F_1 169
        7.1.5 F_o-Mediated H^+Conductance 169

- 7.1.6 Possible Mechanism of Energy Transduction by F_oF_1-ATP Synthase 172
- 7.1.7 H^+/ATP Stoichiometry 174
- 7.2 H^+-ATPases as Secondary $\Delta\bar{\mu}_{H^+}$ Generators 176
  - 7.2.1 F_oF_1-Type H^+-ATPases 177
  - 7.2.2 V_oV_1-Type H^+-ATPases 179
  - 7.2.3 E_1E_2-Type H^+-ATPases 180
  - 7.2.4 Interrelations of Various Functions of H^+-ATPases 182
- 7.3 H^+-Pyrophosphate Synthase (H^+-Pyrophosphatase) 183
- 7.4 H^+-Transhydrogenase 186
- 7.5 Other Systems of Reverse Transfer of Reducing Equivalents 190
- References 191

8 $\Delta\bar{\mu}_{H^+}$-Driven Mechanical Work: Bacterial Motility 195
- 8.1 $\Delta\bar{\mu}_{H^+}$ Powers the Flagellar Motor 196
- 8.2 Structure of the Bacterial Flagellar Motor 197
- 8.3 A Possible Mechanism of the H^+-motor 200
- 8.4 $\Delta\bar{\mu}_{H^+}$-Driven Movement of Non-Flagellar Motile Prokaryotes and Intracellular Organelles of Eukaryotes 202
- 8.5 Motile Eukaryote: Prokaryote Symbionts 204
- References 205

9 $\Delta\bar{\mu}_{H^+}$-Driven Osmotic Work 207
- 9.1 Definition and Classification 207
- 9.2 $\Delta\Psi$ As Driving Force 208
- 9.3 ΔpH As Driving Force 210
- 9.4 Total $\Delta\bar{\mu}_{H^+}$ as Driving Force 211
- 9.5 $\Delta\bar{\mu}_{H^+}$-Driven Transport Cascades 213
- 9.6 Carnitine: An Example of a Transmembrane Group Carrier 214
- 9.7 Some Examples of $\Delta\bar{\mu}_{H^+}$-Driven Carriers 217
  - 9.7.1 *Escherichia coli* Lactose, H^+-Symporter 218
  - 9.7.2 Mitochondrial ATP/ADP-Antiporter 221
- 9.8 Role of $\Delta\bar{\mu}_{H^+}$ in Transport of Macromolecules 224
  - 9.8.1 Transport of Mitochondrial Proteins: Biogenesis of Mitochondria 225
  - 9.8.2 Transport of Bacterial Proteins 226
  - 9.8.3 Role of $\Delta\Psi$ in Protein Arrangement in the Membrane 227
  - 9.8.4 Bacterial DNA Transport 227
- References 228

Contents   xi

10 $\Delta\bar{\mu}_{H^+}$ as Energy Source for Heat Production 231
    10.1 Three Ways of Converting Metabolic Energy into Heat 231
    10.2 Thermoregulatory Activation of Free
         Respiration in Animals 232
         10.2.1 Brown Fat 232
         10.2.2 Skeletal Muscles 236
    10.3 Thermoregulatory Activation of Free Respiration in Plants . . . 240
    References 241

Part IV Interaction and Regulation of Proton Potential Generators and Consumers

11 Regulation, Transmission, and Buffering of Proton Potential 245
    11.1 Regulation of $\Delta\bar{\mu}_{H^+}$ 245
         11.1.1 Alternative Functions of Respiration 245
         11.1.2 Regulation of Flows of Reducing Equivalents
                 Between Cytosol and Mitochondria 248
         11.1.3 Interconversion of $\Delta\Psi$ and ΔpH 249
         11.1.4 Relation of $\Delta\bar{\mu}_{H^+}$ Control to the Main Regulatory
                 Systems of Eukaryotic Cells 250
         11.1.5 Control of $\Delta\bar{\mu}_{H^+}$ in Bacteria 251
    11.2 Energy Transmission Along Membranes
         in the Form of $\Delta\bar{\mu}_{H^+}$ 252
         11.2.1 General Remarks 252
         11.2.2 Lateral Transmission of $\Delta\bar{\mu}_{H^+}$ Produced by
                 Light-Dependent Generators in Halobacteria
                 and Chloroplasts 253
         11.2.3 Trans-Cellular Power Transmission
                 Along Cyanobacterial Trichomes 253
         11.2.4 Structure and Functions of Filamentous Mitochondria
                 and Mitochondrial Reticulum 254
    11.3 Buffering of $\Delta\bar{\mu}_{H^+}$ 265
         11.3.1 Na^+/K^+ Gradients as a $\Delta\bar{\mu}_{H^+}$ Buffer in Bacteria 265
         11.3.2 Other $\Delta\bar{\mu}_{H^+}$-Buffering Systems 268
    References 269

Part V The Sodium World

12 $\Delta\bar{\mu}_{Na^+}$ Generators 275
    12.1 Na^+-Motive Decarboxylases 275
    12.2 Na^+-Translocating NADH:Quinone-Oxidoreductase 277
         12.2.1 Primary Structure of Subunits of Na^+-Translocating
                 NADH:Quinone Oxidoreductase 277

		12.2.2 Na^+-NQR Prosthetic Groups	279
	12.3	Na^+-Motive Methyltransferase Complex from Methanogenic Archaea	280
	12.4	Na^+-Motive Formylmethanofuran Dehydrogenase from Methanogenic Archaea	281
	12.5	Secondary $\Delta\bar{\mu}_{Na^+}$ Generators: Na^+-Motive ATPases and Na^+-Pyrophosphatase	282
		12.5.1 Bacterial Na^+-ATPases	282
		12.5.2 Animal Na^+/K^+-ATPase and Na^+-ATPase	283
		12.5.3 Na^+-Motive Pyrophosphatase	284
	References		285
13	Utilization of $\Delta\bar{\mu}_{Na^+}$ Produced by Primary $\Delta\bar{\mu}_{Na^+}$ Generators		287
	13.1	Osmotic Work Supported by $\Delta\bar{\mu}_{Na^+}$	287
		13.1.1 Na^+, Metabolite-Symporters	287
		13.1.2 Na^+ Ions and Regulation of Cytoplasmic pH	288
	13.2	Mechanical Work	289
	13.3	Chemical Work	291
		13.3.1 $\Delta\bar{\mu}_{Na^+}$-Driven ATP Synthesis in Anaerobic Bacteria	291
		13.3.2 $\Delta\bar{\mu}_{Na^+}$ Consumers Performing Chemical Work in Methanogenic Archaea	293
	References		294
14	Relations Between the Proton and Sodium Worlds		297
	14.1	How Often is the Na^+ Cycle Used by Living Cells?	297
	14.2	Probable Evolutionary Relationships of the Proton and Sodium Worlds	298
	14.3	Membrane-Linked Energy Transductions Involving Neither H^+ Nor Na^+	300
	References		302

Part VI Mitochondrial Reactive Oxygen Species and Mechanisms of Aging

15	Concept of Aging as a Result of Slow Programmed Poisoning of an Organism with Mitochondrial Reactive Oxygen Species		305
	15.1	Nature of ROS and Paths of their Formation in the Cell	306
	15.2	How Do Living Systems Protect Themselves from ROS?	309
		15.2.1 Antioxidant Compounds	309
		15.2.2 Decrease in Intracellular Oxygen Concentration	309
		15.2.3 Decrease in ROS Production by the Respiratory Chain	312
		15.2.4 Mitoptosis	315

		15.2.5	Apoptosis	318
		15.2.6	Necrosis	320
		15.2.7	Phenoptosis	322
	15.3	Biological Function of ROS		323
	15.4	Aging as Slow Phenoptosis Caused by Increase in mROS Level		326
		15.4.1	Definition of the Term "Aging" and a Short Historical Overview of the Problem	326
		15.4.2	Phenoptosis of Organisms that Reproduce Only Once	329
		15.4.3	Can Aging be a Slow Form of Phenoptosis?	333
		15.4.4	Mutations that Prolong Lifespan	336
		15.4.5	ROS and Aging	339
		15.4.6	Naked Mole-Rat	340
		15.4.7	Aging Program: Working Hypothesis	342
		15.4.8	Paradox of Protein p53	343
		15.4.9	Arrest of Age-Dependent Increase of Mitochondrial ROS as a Possible Way to Slow the Aging Program	344
	References			346
16	Possible Medical Applications of Membrane Bioenergetics: Mitochondria-Targeted Antioxidants as Geroprotectors			355
	16.1	SkQ Decelerates the Aging Program		355
	16.2	Comparison of Effects of Food Restriction and SkQ		372
	16.3	From *Homo sapiens* to *Homo sapiens liberatus*		376
	16.4	Conclusions		377
	References			378

Appendix 1: Energy, Work, and Laws of Thermodynamics 383

Appendix 2: Prosthetic Groups and Cofactors 393

Appendix 3: Inhibitors of Oxidative Phosphorylation 403

Appendix 4: Plant Hormones 407

Appendix 5: Mitochondria-Targeted Antioxidants and Related Penetrating Compounds 409

Appendix 6: Mitochondria-Targeted Natural Rechargeable Antioxidant 413

Appendix 7: Key Participants of the Project "Practical Application of Penetrating Cations" 417

References 421

Author Index 423

Subject Index 427

Index of Organisms.................................... 435

Abbreviations

$\Delta\Psi$	Transmembrane difference of electric potentials
$\Delta\bar{\mu}_{H^+}$	Transmembrane difference of electrochemical potentials of hydrogen ions
$\Delta\bar{\mu}_{Na^+}$	Transmembrane difference of electrochemical potentials of sodium ions
ΔpH	Transmembrane difference of pH values
ADP	Adenosine 5′-diphosphate
AMP	Adenosine 5′-monophosphate
AOX	Alternative cyanide-resistant oxidase
ATP	Adenosine 5′-triphosphate
BLM	Bilayer planar phospholipid membrane
BChl	Bacteriochlorophyll
$(BChl)_2$	Bacteriochlorophyll dimer
$(BChl)_2^*$	Excited bacteriochlorophyll dimer
$(BChl)_2^{\bullet+}$	Cation radical of bacteriochlorophyll dimer
BPheo	Bacteriopheophytin
cAMP	Adenosine 3′,5′-cyclic monophosphate
Capsaicin	8-Methyl-N-vanillyl-6-nonenamide
CCCP	m-Chlorocarbonyl cyanide phenylhydrazone
Chl	Chlorophyll
$(Chl)_2$	Chlorophyll dimer
$(Chl)_2^*$	Excited chlorophyll dimer
$(Chl)_2^{\bullet+}$	Cation radical of chlorophyll dimer
CoQ	Ubiquinone
$CoQH_2$	Ubiquinol
$CoQH^\bullet$	Neutral ubisemiquinone
$CoQ^{\bullet-}$	Ubisemiquinone anion
E_m	Midpoint redox potential
EPR	Electron paramagnetic resonance
FAD	Flavin adenine dinucleotide
FMN	Flavin mononucleotide

Fd	Ferredoxin
GDP	Guanosine 5′-diphosphate
GTP	Guanosine 5′-triphosphate
HQNO	2-Heptyl-4-hydroxyquinoline N-oxide
$H^+/\bar{e}$	Ratio of protons to electrons transferred across the membrane by respiratory or photosynthetic chain enzyme complexes
MitoQ	10-(6′-ubiquinonyl)decyltriphenylphosphonium
MQ	Menaquinone
NAC	N-Acetylcysteine
NAD^+	Nicotinamide adenine dinucleotide
NADH	Reduced nicotinamide adenine dinucleotide
$NADP^+$	Nicotinamide adenine dinucleotide phosphate
NADPH	Reduced nicotinamide adenine dinucleotide phosphate
Na^+-NQR	Na^+-translocating NADH:ubiquinone-oxidoreductase
$Na^+/\bar{e}$	Ratio of sodium ions to electrons transferred across the membrane by a respiratory chain enzyme
NDH-1	Bacterial H^+-translocating NADH:quinone-oxidoreductase, homologue of mitochondrial complex I
NDH-2	Bacterial noncoupled NADH:quinone-oxidoreductase
NMR	Nuclear magnetic resonance
PC	Plastocyanin
Pheo	Pheophytin
P_i	Inorganic phosphate
PQ	Plastoquinone
PQH_2	Plastoquinol
$PQ^{\bullet-}$	Plastosemiquinone anion
PP_i	Inorganic pyrophosphate
ROS	Reactive oxygen species
SkQ	Plastoquinone derivative bound to delocalized cation through decyl or amyl linker
SkQ1	10-(6′-Plastoquinonyl)decyltriphenylphosphonium
SkQR1	10-(6′-Plastoquinonyl)decylrhodamine
TNF	Tumor necrosis factor

Part I
General Aspects of Bioenergetics

Chapter 1
Introduction

1.1 Definition of the Term "Bioenergetics" and Some Milestones of its History

What are the connections between energy and life? At the beginning of the twentieth century, J. B. S. Haldane defined life as *the mode of existence of self-reproducing structures at the cost of external energy sources.* According to this definition, the ability to access energy is one of two main properties of life. But how does energy promote the existence of life? What are the mechanisms involved in this process? We will try to answer these questions while not using complicated mathematical equations.

First let us define the theme of our course. *Bioenergetics (biological energetics) is the sum of the processes of transformation of external sources of energy into biologically useful work of living systems, and also a division of biology that studies these processes.*

At the source of bioenergetics, one can find discussions of ancient thinkers about the nature of fermentation and the role of air in the use of food by living organisms. Leonardo da Vinci was one of the first to compare animal nutrition to a candle burning (beginning of the sixteenth century). This idea was developed by the Dutch naturalist Jan Baptiste van Helmont in his experiments with plants in 1648. Until the beginning of the twentieth century, bioenergetics was mainly based on studies of overall balances of the processes of energy supply of living organisms (respiration and fermentation) and the effect of different conditions (such as transfer from state of rest to state of work, change in external temperature, etc.) on the energy balance of the organism. Rapid development of biochemistry in the first half of the twentieth century facilitated formulation of the main laws of energy transformation in living cells. Key roles in this were played by the outstanding Russian scientists Vladimir Engelhardt (Engelhardt 1930, 1931, 1932) and Vladimir Belitser (Belitser and Tsibakova 1939) who discovered phosphorylating respiration and found that ATP formation is somehow coupled to electron transfer in the respiratory chain. But it was not until the key principle of energy

Fig. 1.1 King Carl XVI Gustaf of Sweden (*right*) awarding the Nobel Prize to Peter Mitchell (1978)

transformation in living cells was formulated by Peter Mitchell that bioenergetics came into existence as a separate science.

Every branch of science has its own birth date. For instance, molecular genetics started with the publication of an article by James Watson and Francis Crick (who later received the Nobel Prize in Physiology or Medicine) in the prestigious scientific journal *Nature* in March of 1956. The article described the helical structure of DNA found in viruses. Bioenergetics also had its own birth date, and it is also connected to a publication in *Nature*, but in this case by another Nobel Prize winner, Peter Mitchell (Fig. 1.1). In 1961, Mitchell published an article offering a new explanation for the mechanism of oxidative and photosynthetic phosphorylation (the "chemiosmotic hypothesis"). But the word "bioenergetics" was first used 4 years earlier, in 1957, by yet another Nobel Prize winner, Albert Szent-Györgyi (Fig. 1.2). *Bioenergetics* was the title Szent-Györgyi gave to his small book—a rather strange book, full of unrestrained fantasies about the nature of life (Szent-Gyorgyi 1957). Many thoughts expressed in that book proved to be wrong, but its title heralded a new science.

In 1968, in Polignano a Mare, a tiny town in the south of Italy, a conference was held on oxidative phosphorylation. Discussions on the name for a new branch of biology dedicated to research on molecular mechanisms of energy transduction in living cells became the main topic of one of the sessions. One of the authors of this book (V.P.S.) was fortunate to be a participant of that conference. His English language skills were quite limited then, and when he needed to express personal opinions it sometimes turned out to be too blunt. When the discussion became too cumbersome and discordant (from his point of view), he stood up and said that he had recently organized a "Department of Bioenergetics" at Moscow State University, and hence suggested giving this name to the new science. During the break one of the participants of the conference, Karel van Dam from The Netherlands, scolded V.P.S. for such an abrupt expression of his position in the presence of such pillars of science as Hans Krebs, Lars Ernster, and Edward Charles Slater, but surprisingly enough the discussion led to general acceptance of the name. And

1.1 Definition of the Term "Bioenergetics" and Some Milestones of its History

Fig. 1.2 Nobel Laureate Albert Szent-Györgyi

after a few years journals on bioenergetics, conferences on bioenergetics, and an international organization of bioenergetics with national sections in different countries appeared. Unfortunately, psychics later decided to usurp the new term. V.P.S. tried hard to protest in the mass media, but he was defeated. Now there are two terms—"scientific bioenergetics" and "pseudoscientific bioenergetics". This phenomenon started in Russia and then spread abroad. Language is like an ocean—a power that one cannot fight with alone.

Having broken away into a separate science, bioenergetics quickly gained its unique place among the sciences. In 1978, its leading founder Peter Mitchell received the Nobel Prize for his chemiosmotic theory. In 1988, Hartmut Michel, Johann Deisenhofer, and Robert Huber became Nobel Prize winners for their success in determining the structure of the photosynthetic reaction centers using X-ray crystallography. In 1997, Paul Boyer and John Walker received the Nobel Prize for their research on the proton-driven ATP-synthase, and Jens Skou received the Nobel Prize for the discovery of Na^+, K^+-ATPase. In 2003, Peter Agre and Roderick MacKinnon received the Nobel Prize for disclosing and X-ray studies of aquaporin and the K^+-channel, respectively. And the discovery of the central role of mitochondria in programmed cell death (made by Guido Kroemer, Xin Wang, and Donald Newmeyer) is apparently waiting for its prize. The number of publications on this theme has been increasing exponentially over the last decade.

1.2 Bioenergetics in the System of Biological Sciences

To classify biological sciences, one can use the analogy of an eight storey building where each storey corresponds to a new level of structural complexity of living matter (Fig. 1.3). Coming down from floor to floor, the researcher moves from

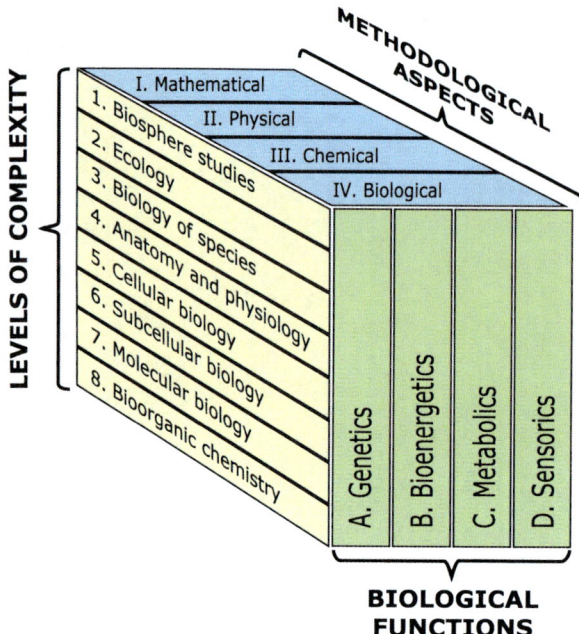

Fig. 1.3 The "biology building"—interrelationships of the biological sciences (Skulachev 1988)

more general to more specific matters. The top storey is for *biosphere studies*. The second storey from the top is occupied by *ecology*. It deals with the communities of individual organisms that form ecosystems that in turn constitute the biosphere when brought together. The next storey deals with the group of sciences that can be united under the title "*biology of species*". These are the classical descriptive biological sciences such as zoology, botany, mycology, microbiology, and virology.

The species-composing populations are in turn composed of individuals who are studied at the fourth storey from the top. Here specialists in *anatomy and physiology* study the structure and functioning of individual organisms and their organs and tissues. Tissues are composed of cells, and biological phenomena studied on this level from the realm of *cellular biology* (the fifth storey from the top). The next level of simplification is *subcellular biology*, which studies intracellular organelles and supramolecular systems. This realm includes studies on structure and functioning of plasmalemma, cytosol, nucleus, mitochondria, endoplasmic reticulum, lysosomes, peroxisomes, chloroplasts, and some other subcellular organelles.

The seventh storey from the top is of crucial importance. It is occupied by *molecular biology*. Individual molecules that compose living beings—especially macromolecules and their complexes—are the domain of molecular biology. It is on this storey that the transition to quantitative description of biological objects occurs. The biology of molecules is a very complex, but at the same time it is a science that operates on a level of precision similar to that of chemistry and

1.2 Bioenergetics in the System of Biological Sciences

physics. Molecules and their complexes that are the objects of research here must be kept in their native condition, i.e., they need to maintain their biological function when being studied. When we move to the lowest storey (eighth from the top) this condition is not necessarily observed. *Bioorganic chemistry* studies the structural and physicochemical properties of pure individual macromolecules as well as their constituents and low molecular mass compounds present in living organisms. Some of these compounds are inorganic, but it makes little sense to consider "bioinorganic chemistry" as a special science and make a separate storey for it, since the number of its objects is very much smaller than that of bioorganic chemistry.

The classification of the biological sciences is not limited to the levels of complexity. A biological object can be studied by a biologist, chemist, physicist, or mathematician. Each of these specialists uses different *methodological approaches* and *research philosophy*. Sometimes even the conclusions of the concrete researcher depend on the methodological "porch" of the "biology building" he is working in. The *mathematician* strives to find equations that describe the object of research or applies informatics to analyze "biological texts". The *biophysicist*, while studying the reactions flow over time, wants his observations to have high time resolution over the whole period of observation.[1] In the case of the *chemist*, it is the purity of molecules and system homogeneity that will be of primary importance. The *biologist* is not likely to strive for femtosecond definition, mathematical accuracy, or chemical purity. He is especially interested in the living system's behavior (regulation, birth, and death), and also in its history—phylogenesis and ontogenesis. The authors of this book are biologists by education, and that is why the book to a substantial degree reflects the biological approach.

There is yet other criterion for the classification of biological studies—*biological functions* of the systems which are the objects of research. There are four main functions of biological objects. One of them is genetic, connected with the transfer of individual properties to the next generation. The science dealing with this function is *genetics*. The second vital condition of a living system's existence is energy supply, and it is *bioenergetics* that studies these phenomena. There is also an important biological function consisting in interconversion of different chemical substances. The compounds we consist of are basically different from those that can be found in the environment. The great majority of them we synthesize ourselves—that is why transformation of substances can be viewed as a special biological function. By analogy with genetics and bioenergetics, the science investigating metabolism of substances can be called *metabolics* (Skulachev 1988). This is an example of a neologism. We basically speak here about classical biochemistry. But at a certain point biochemistry gave birth to the sciences that

---

[1] Russian biophysicists have been quite successful in this area. For instance, the laboratories of the Belozersky Institute of Physico-Chemical biology at Moscow State University have measured kinetics of the primary processes of photosynthesis in living nature to the resolution of 2×10^{-15} s, which is the highest achieved in biological studies) (Shuvalov 2007).

occupy the three lowest stories in the biology building—subcellular biology, molecular biology, and bioorganic chemistry. This is why biochemistry can be seen as a branch of biology that uses a chemical approach in its studies. And finally there is *sensorics* (Skulachev 1988). This is also a neologism similar to the previous term. It defines the science that studies reception, transmission, and processing of signals coming into the living system from its inner or outer sources.

We should stress that such a functional approach crowns studies of any biological system. Finally, it is not so important what specific molecule participates in a reaction, and it is even less important whether it was discovered by biologists, physicists, chemists, or mathematicians. But it is important to understand what the processes this molecule is involved in are and what the role of its functioning in the general scheme of vital functions of the object of research is. Therefore, four functional sciences constitute the goal of biology. When we have fulfilled the task in its functional aspect, we can pause and then move forward to new challenges. And then the question is which direction to take. As Alexander Pushkin said—"But whither do we sail?"

It can be suggested that, having studied the world around us and human nature as part of it, we as humans will turn toward the task of creation of new entities and improvement of our own nature. Some time ago, such aspects were considered to belong to the realm of science fiction, even though plant and animal selection has been known for many centuries. People have been successful in creating some new animal breeds and plant varieties. But this process has been extremely slow—and success accidental—for it was a special branch of industry rather than fundamental science that has been dealing with selection. Deepening of our knowledge in all three dimensions depicted in Fig. 1.3 has opened the possibility of creating new biological molecules, new processes, and finally new, previously nonexistent living forms on the basis of a plan prepared in advance by a researcher. These types of questions are being addressed by the science that received the name "*bioengineering*". Bioengineering does not fit into the system depicted in Fig. 1.3 because the "inhabitants" of the "biology building" are trying to gain knowledge about the living beings that have appeared in the process of evolution. As to bioengineering, it represents the human attempt to create a new living form or its components. This is a principally new, innovative approach that represents the future of biology, even though we may spend hours discussing the ethical problems of the creation of new biological objects—the consequences and dangers of these practices for the future of humanity. Biologists have already entered this path, and it is not fully logical that bioengineering has given rise to such uncompromising debates, for selection is also an attempt to create new, previously nonexistent beings, and nobody condemns the selectionists. On the other hand, scientists need to be extremely careful when dealing with bioengineering research—otherwise humanity might pay a high price for a mistake. For instance, it might be possible to create superpathogenic bacteria that will be able to cause simultaneous symptoms of plague, cholera, anthrax, and some other lethal diseases. It would not be possible to fight such pathogens using contemporary antibiotics. Such a monster might be produced accidentally or purposefully by terrorists so as to threaten humanity. One should keep in mind this extremely dangerous aspect of bioengineering research, but at the same time it should not become a reason to ban bioengineering. Instead, such research should continue under strict public and governmental control.

1.3 Laws of Bioenergetics

Today bioenergetics has reached such progress in its development that we can already formulate its main laws.

First law of bioenergetics. Living beings avoid direct utilization of the energy of external sources when performing useful work. They first transform this energy into a convertible form—ATP, $\Delta\bar{\mu}_{H^+}$, or $\Delta\bar{\mu}_{Na^+}$—and then use it in various energy-consuming processes (Eq. 1.1). Comparing this situation to our everyday life, one can say that organisms prefer money to barter.

$$\text{Energy sources} \rightarrow \text{ATP}, \Delta\bar{\mu}_{H^+}, \text{ or } \Delta\bar{\mu}_{Na^+} \rightarrow \text{work} \quad (1.1)$$

This principle was first stated in 1941 by Fritz Lipmann (Fig. 1.4), who at that time knew of only one form of biological "currency"—ATP (Lipmann 1941). It is true that ATP is a universal sign of life—there are no living cells without ATP. But later two additional kinds of biological "currency" were found. These are electrochemical potential differences of hydrogen ions and sodium ions—$\Delta\bar{\mu}_{H^+}$ and $\Delta\bar{\mu}_{Na^+}$, respectively.

The electrochemical potential difference in H^+ ions ($\Delta\bar{\mu}_{H^+}$) can exist in two forms—electrical and chemical. The first is a transmembrane difference in electric potential ($\Delta\Psi$), and the second is the transmembrane difference in H^+ concentration ($\Delta p H$). Figure 1.5 illustrates these two notions. If there is an electric potential difference across the membrane (generated, e.g. by a battery), H^+ tends to move from the positively charged to the negatively charged compartment (Fig. 1.5a). One can reach the same situation by adding, for instance, hydrochloric acid to the left compartment, thus increasing the concentration of H^+ in the left compartment in comparison to the right one. In this case, a proton current will also take place "downhill", but it is $\Delta p H$, and not $\Delta\Psi$, that will be the driving force for the process (Fig. 1.5b).

Potential energy accumulated in the form of $\Delta\Psi$ or $\Delta p H$ can be utilized if the membrane has a device capable of coupling downhill H^+ movement to the performance of useful work.

The energy stored in $\Delta\bar{\mu}_{H^+}$ can be calculated from Eq. 1.2.

$$\Delta\bar{\mu}_{H^+} = F\Delta\Psi + RT \cdot \ln\frac{[H^+]_p}{[H^+]_n} \quad (1.2)$$

where $\Delta\Psi$ is the transmembrane electric potential difference, R is the gas constant, T is the absolute temperature, F is the Faraday constant, and $[H^+]_p$ and $[H^+]_n$ are

Fig. 1.4 Fritz Lipmann

Fig. 1.5 Two forms of H$^+$ potential between compartments separated by a membrane—electric field gradient $\Delta\Psi$ (**a**) and acidity gradient ΔpH (**b**). The direction of proton current is shown by the *dotted arrows*

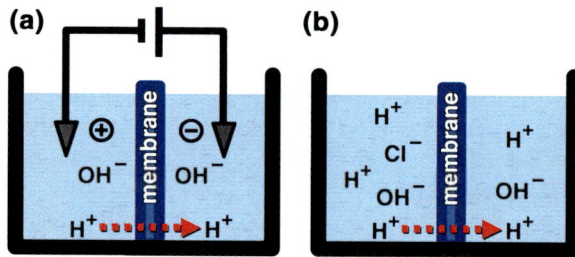

the molar concentrations of H$^+$ in the positively charged (or more acidic) and negatively charged (or more alkaline) compartments, respectively (Mitchell 1973).

The units of $\Delta\bar{\mu}_{H^+}$ are J mol^{-1}. To express $\Delta\bar{\mu}_{H^+}$ in volts (V), one should divide it by Faraday's constant (*F*). For this quantity, Mitchell introduced the term "proton motive force" (Δp) (Mitchell 1973), which at 25 °C can be calculated according to Eq. 1.3.

$$\Delta p = \Delta\bar{\mu}_{H^+}/F = \Delta\Psi - 0.06 \cdot \Delta p\text{H}. \tag{1.3}$$

The *difference* between $\Delta\Psi$ and ΔpH is due to the fact that the pH is a *negative* logarithm of the H$^+$ concentration. Indeed, $\Delta\bar{\mu}_{H^+}$ increases when the left compartment in Fig. 1.5 becomes more positively charged or its pH is lowered.

According to Eq. 1.3, ΔpH = 1 is equivalent to $\Delta\Psi = 0.06$ V (or 60 mV). The same value expressed in kJ mol^{-1} is 5.7, and that in kcal mol^{-1} is 1.37.

Similar equations can be applied to sodium energetics. In this case, in Eq. 1.3 $\Delta\bar{\mu}_{H^+}$ is replaced by $\Delta\bar{\mu}_{Na^+}$, and the proton motive force by the sodium ion motive force (Eq. 1.4). The latter can be abbreviated as Δs (from Latin *sodium*) (Skulachev 1984, 1988).

1.3 Laws of Bioenergetics

$$\Delta s = \Delta \bar{\mu}_{Na^+}/F = \Delta \Psi - 0.06 \cdot \Delta p\text{Na} \qquad (1.4)$$

Figure 1.6a shows a scheme that describes the energetics of living cells using $\Delta \bar{\mu}_{H^+}$ as the convertible membrane-linked energy currency. According to the scheme, the energy of light or respiratory substrates can be utilized by enzymes of the photosynthetic or respiratory redox chains or by bacteriorhodopsin to form $\Delta \bar{\mu}_{H^+}$. The latter can support various types of work in the "protonic" membrane, with ATP synthesis being the most important. Substrate-level phosphorylations serve as an alternative mechanism of ATP formation that operates with no $\Delta \bar{\mu}_{H^+}$ involved. Such phosphorylations occur in the glycolytic chain and in oxidative decarboxylation of α-ketoglutarate.

The $\Delta \bar{\mu}_{H^+}$-linked formation of ATP is a major but not the only process of transformation of $\Delta \bar{\mu}_{H^+}$ into chemical work. The $\Delta \bar{\mu}_{H^+}$-supported synthesis of inorganic pyrophosphate and transfer of reducing equivalents in the direction of more negative redox potentials (e.g., reverse electron transfer in the respiratory chain and the transhydrogenase reaction) are also of this type of energy transduction.

The $\Delta \bar{\mu}_{H^+}$-driven uphill transport of various substances across the coupling membrane can be described as $\Delta \bar{\mu}_{H^+} \rightarrow$ osmotic work transduction, while the rotation of the protonic motor of motile bacteria represents $\Delta \bar{\mu}_{H^+}$-driven mechanical work. Heat production by mitochondria of cold-exposed animals is an example of $\Delta \bar{\mu}_{H^+} \rightarrow$ heat production.

All the above types of energy transduction have also been described for the nonmembranous parts of cells. Here they are supported by the energy of ATP or other high-energy compounds.

There are systems specializing in the buffering of $\Delta \bar{\mu}_{H^+}$ or ATP levels. For $\Delta \bar{\mu}_{H^+}$, this function is performed by gradients of Na^+ and K^+; for ATP, by creatine phosphate (Skulachev 1988).

In certain bacteria, $\Delta \bar{\mu}_{Na^+}$ instead of $\Delta \bar{\mu}_{H^+}$ is formed at the expense of energy released by respiration or by nonoxidative decarboxylation of some organic acids. Then the $\Delta \bar{\mu}_{Na^+}$ can be used to support chemical, osmotic, or mechanical work. The K^+/H^+ gradient can serve as a buffering system for $\Delta \bar{\mu}_{Na^+}$ (Skulachev 1988) (Fig. 1.6b).

The most complicated pattern of energy transduction is inherent in animal cells, where there are three different interconvertible energy currencies—$\Delta \bar{\mu}_{H^+}$ for mitochondria and some other intracellular vesicles, $\Delta \bar{\mu}_{Na^+}$ for the outer cell membrane, and ATP for nonmembranous cell constituents.

Coming back to the first law of bioenergetics, we should state that its three components (ATP, $\Delta \bar{\mu}_{H^+}$, and $\Delta \bar{\mu}_{Na^+}$) do not have equal value. ATP is always present, while the two other components are interchangeable, and hence some organisms have only one or the other. But one of the latter two should always be present. Thus we can formulate the second law.

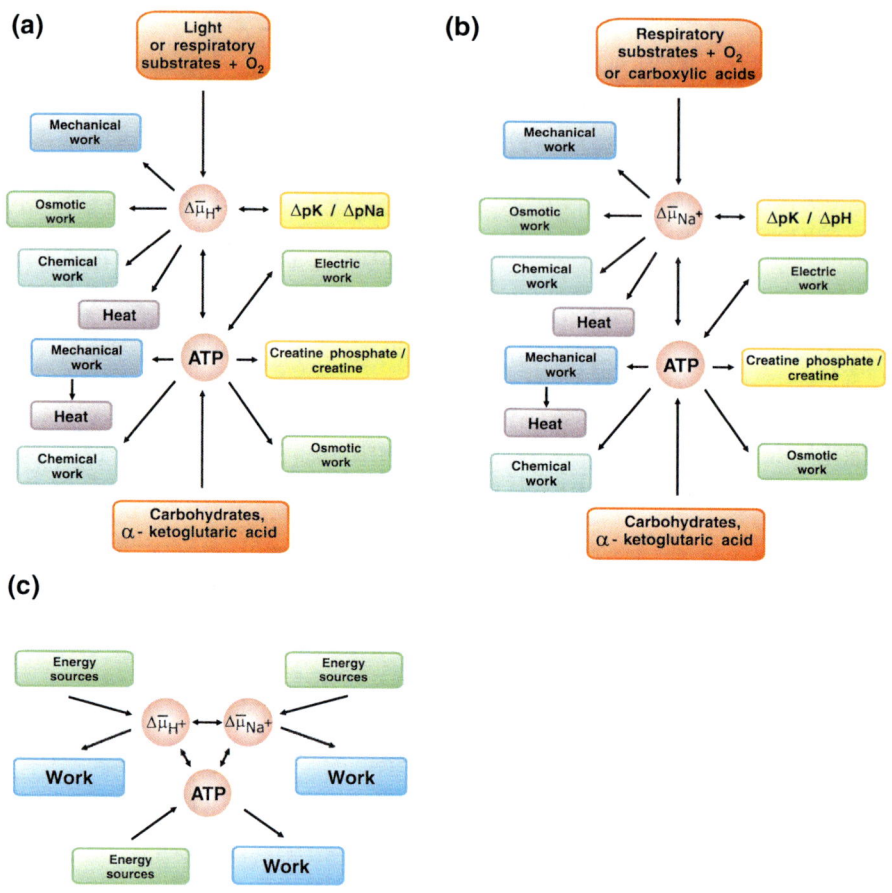

Fig. 1.6 General scheme of energetics of a living cell with $\Delta\bar{\mu}_{H^+}$ and ATP (**a**) or $\Delta\bar{\mu}_{Na^+}$ and ATP (**b**) as interconvertible energy forms. (**c**) Interconversion of the three forms of energy

Second law of bioenergetics. Every living cell has at least two forms of interconvertible energy—ATP, and $\Delta\bar{\mu}_{H^+}$ or $\Delta\bar{\mu}_{Na^+}$. If we continue our analogy to money, the living cell always has both cash and a credit card in its wallet.

Third law of bioenergetics. This law states that there is interconversion of three forms of energy stored in a cell. That is why *a cell can satisfy all its energy needs if it can obtain at least one of the three interconvertible energy forms from external energy sources* (Fig. 1.6c). In other words, it makes no difference for the cell whether it pays with cash or with a credit card.

There are some kinds of cells that use only one type of energy source. Some of them live by respiration or photosynthesis and produce only $\Delta\bar{\mu}_{H^+}$; they have no glycolysis, hence they have no possibility for direct synthesis of ATP. For some others, vice versa, glycolysis (which forms ATP without mediation of $\Delta\bar{\mu}_{H^+}$) is the only energy source. Certain bacteria have neither glycolysis nor respiration nor photosynthesis. They accumulate $\Delta\bar{\mu}_{Na^+}$ via one of the decarboxylation reactions and use $\Delta\bar{\mu}_{Na^+}$ for production of ATP and $\Delta\bar{\mu}_{H^+}$ (Skulachev 1988).

Most people are not aware of the scale of events that accompany our body's energy supply. Here are some of the numbers. The average human being consumes about 140 l of oxygen per day and synthesizes about 40 kg of ATP. To obtain such values, about 0.5 kg of protons needs to be transported across mitochondrial membranes, while $\Delta\bar{\mu}_{H^+}$ generators maintain the voltage of the electric field across the mitochondrial membrane corresponding to hundreds of kilovolts per centimeter of the membrane thickness.

The science that this book describes still contains many blank spots. It is unclear what the molecular mechanisms of the interconversion of the majority of different energy forms are. Even more, the schemes in Fig. 1.6 do not describe all the possible types of biological energy transformation. For instance, they lack the generation of light. Light is not just an energy source. All living beings generate photons that are produced in the course of free radical reactions that take place in living cells. It is not clear whether this is just a byproduct of such reactions, or whether cells use such light as a kind of signaling.

A very vivid illustration of the depth of our ignorance is the history of the discovery of the role of mitochondria in programmed cell death—apoptosis. At the very end of the twentieth century, "death proteins" were discovered in the mitochondrial intermembrane space—they trigger apoptosis when leaving the mitochondria into the cytosol. Hence mitochondria are not only the electric stations of cells, but also destiny-defining organelles. And one of the simplest respiratory chain electron carriers—cytochrome c, which has been studied by so many biochemists starting from David Keilin who discovered it in 1925—proved to be a key player in the cell suicide scenario. It is important to note that this property is a completely different function of cytochrome c that has nothing to do with its role as an electron carrier (see Chap. 15).

1.4 Evolution of Bioenergetic Mechanisms

How did those three interconvertible energy forms we spoke about in the previous section come into existence? The search for an answer to this question is not easy. We cannot conduct a direct experiment, and paleontologists can provide us only with an opinion based on indirect data that certain traces of cyanobacteria were found in the oldest known sedimentary formations that are about 3.5 billion years old. But we certainly cannot omit discussion of this problem. One cannot understand functioning of contemporary living systems without having at least some general idea of their past, i.e., their origin and evolution. So we will present below our hypothesis on the origin of bioenergetic mechanisms.

1.4.1 Adenosine Triphosphate

Why was this substance chosen in the course of evolution as a universal biological energy currency? Let us look at each of the three components of ATP. The first is the triphosphate residue where the energy of ATP is stored. So, this is why ATP a phosphate-containing compound (A**T**P). It is known that when the third phosphorus atom in ATP is changed to arsenic (the element closest in its properties to phosphorus), the resulting ADP-arsenate is unstable, and it spontaneously decomposes to ADP and free arsenate at room temperature and neutral pH. The arsenic atom is a little larger than the phosphorus atom, and this makes it easier for water molecules to attack it. Also, arsenic is a far rarer substance than phosphorus, so it would be much more difficult for a living being to gather arsenic from the environment.

Many biologists share a myth that ATP is a unique biological energy form, for this substance has high-energy (or macroergic) bonds. Strictly speaking, not only ATP, but also all the substances capable of forming ATP in group-transfer enzymatic reactions can be defined as *high-energy compounds*. The free energy of hydrolysis of these compounds is not lower than that of ATP, i.e., about 10 kcal mol^{-1} under physiological conditions (under standard conditions, this value for ATP is about 7 kcal mol^{-1}).

It should be stressed that the term "high-energy compound" is of biological rather than chemical significance. For example, according to the above-given definition, creatine phosphate should be regarded as a high-energy compound in vertebrates (having the enzyme creatine phosphate kinase to phosphorylate ADP), but not in some invertebrates where arginine phosphate kinase substitutes for the mentioned enzyme.

The false idea of the unique character of pyrophosphate bonds in ATP led to the next mistake—the belief that it is ATP *hydrolysis* that releases energy. But in many cases, energy was shown to be released on ATP *sorption* on a protein, and not by hydrolysis of a pyrophosphate bond. When ATP is bound to the active site of an enzyme, it often loses its "macroergic character", and then such a bound ATP molecule already cannot be turned into a normal (free) "macroergic" one, for the energy of ATP has been invested into a strained conformation of the enzyme molecule. In such cases, free ATP can be obtained only by denaturing the protein. Often, the conformational energy of the strained protein is used to perform some useful work, and ATP hydrolysis is needed for obtaining the end products of the reaction (ADP and phosphate). The latter, after being cleaved off the protein, leaves the space for a new ATP molecule to attach to the protein. One should not be surprised by ATP deenergization in the process of its sorption, for *free* ATP has little in common with bound ATP, which interacts simultaneously with many functional groups in the active site of the protein. As the result of these interactions, electron densities in the *bound* ATP can be distributed very differently than in a free ATP molecule.

Now we turn to the question—why ATP? This is not a simple question, for ADP is also macroergic. There is an enzyme called adenylate kinase that can turn

1.4 Evolution of Bioenergetic Mechanisms

two ADP molecules into one molecule of ATP and one of AMP. The standard energy price of the pyrophosphate bond in the terminal phosphate of ADP is similar to that in ATP. So it seems that it would have been possible to base all of cell energetics on the hydrolysis of ADP to AMP. But for some reason evolution chose the more complicated ATP, and not the simpler ADP. This paradox can be explained if we look at *two ways of ATP hydrolysis* (Eqs. 1.5 and 1.6).

$$ATP + H_2O \rightarrow ADP + P_i \tag{1.5}$$

$$ATP + H_2O \rightarrow AMP + PP_i \tag{1.6}$$

Terminal phosphate hydrolysis of ATP is shown to take place when the energy-consuming process driven by the ATP hydrolysis requires energy *less* than or *equal* to 10 kcal mol^{-1}. This is also the case when the energy requirement is *much larger* than 10 kcal mol^{-1}. In the latter case, a special mechanism (e.g., actomyosin filament) is at work making it possible to use the energy of *many* ATP molecules simultaneously to perform a functional act.

If the energy requirement is only *slightly higher* than 10 kcal mol^{-1}, i.e., by several kcal mol^{-1}, ATP is hydrolyzed to AMP and inorganic pyrophosphate (see below, Sect. 7.3). In the living cell, the energy release is usually 4–5 kcal mol^{-1} higher when AMP and PP$_i$ are formed instead of ADP and P$_i$. This is because of a much lower cytosolic concentration of PP$_i$ than of P$_i$. This effect is the result of hydrolysis of PP$_i$ by soluble pyrophosphatases.

So, we can state that the use of adenosine *tri*phosphate rather than adenosine *di*phosphate (also a high-energy compound) as a convertible energy currency adds some flexibility to the biological energy supply system (Skulachev 1988).

1.4.2 Hypothesis of Adenine-Based Photosynthesis

But perhaps the most intriguing question is why it is adenosine that became the carrier of high-energy phosphates, i.e., why **ATP**? All the above-discussed facts are also true for inorganic triphosphate. Below, we will try to demonstrate that today adenosine in ATP is but a birthmark of evolution.

There are different ideas about the source of energy necessary for the birth of life on Earth or elsewhere in the universe. Some consider it to be the energy of electric discharges of thunder storms—others support theories of thermal energy or energy of oxidation of ancient sedimentary rocks. But we believe the development of life to have been so long and difficult that the primordial energy source could not have been limited by any earthly circumstances. There is only one such energy source, the one that up until now has been serving as the primary energy source for life on Earth—the light of the Sun. Sunlight includes both visible and ultraviolet quanta. Ultraviolet is thought to have passed freely through the Earth's atmosphere at the time of the origin of life. It could reach the surface of primordial ocean because the amount of oxygen was still too small to produce the ultraviolet-impermeable ozone layer. The amount of colored substances (i.e., those that absorb visible light) is very small, while ultraviolet light is absorbed by almost all chemical compounds. Ultraviolet photons could initiate many chemical reactions between simple substances that were present in the primordial ocean and ancient Earth's atmosphere. The American biochemist Cyril Ponnamperuma conducted the following experiment. He put a sterile solution of hydrocyanic acid into an ampule, sealed it, and then irradiated it with ultraviolet light. Adenine and other purines and pyrimidines were formed in the solution as a result of the irradiation (Ponnamperuma et al. 1963; Ponnamperuma 1966).

According to Ponnamperuma, ultraviolet also caused synthesis of adenosine under the same conditions, and if ethyl metaphosphate was added to hydrocyanic acid, then AMP, ADP, ATP, and adenosine tetraphosphate were synthesized. In the latter case the ADP → ATP transformation was much more effective than AMP and adenosine synthesis or AMP → ADP transformation.

When modeling the absorption of the ancient Earth's atmosphere, Carl Sagan came to the conclusion that it had a "gap" in the 240–290 nm region and was thus transparent for ultraviolet light. This can be explained by the fact that the main simple compounds of that atmosphere (H_2O, CH_4, NH_3, CO_2, CO, and HCN) absorb light of wavelengths shorter than 240 nm, and the absorption maximum of formaldehyde (which is also supposed to have been part of that original atmosphere) occurs at wavelengths longer than 290 nm. It is to this "gap" that the spectral maxima of purines and pyrimidines belong (Ponnamperuma et al. 1963; Sagan 1966; Ponnamperuma 1966).

Here are the reasons why, according to Sagan and Ponnamperuma, that adenine has certain advantages in serving as an ultraviolet light antenna when compared to other purines and pyrimidines: (1) the strongest light absorption in the mentioned spectral "gap"; (2) higher resistance to the destructive effect of ultraviolet light, and (3) longer duration of excited state caused by absorption of an ultraviolet quantum (Ponnamperuma et al. 1963; Sagan 1966).

Lev Blumenfeld and Mikhail Temkin (1962) calculated that the magnitude of free energy changes accompanying the disruption of the aromatic structure of adenine are similar to the energy required to form ATP from ADP and inorganic phosphate.

On the basis of all the above-mentioned data, we suggested the following mechanism of ultraviolet light-based phosphorylation in primordial living cells (Skulachev 1969, 1994, 1996).

(1) The adenine part of ATP absorbs an ultraviolet quantum, which leads to an excited state with a disturbed system of double bonds. At the same time, the amino group of adenine, which normally corresponds to an aromatic amine, adopts the properties of an aliphatic group, which makes it easier for a phosphorus atom of inorganic phosphate to attack it.[2]

---

[2] There is also another possibility—phosphate being connected to N1 of adenine and not to its amino group. For instance, natural compounds where NAD^+ is connected to ribose in N1 position have been described (Lee et al. 1989).

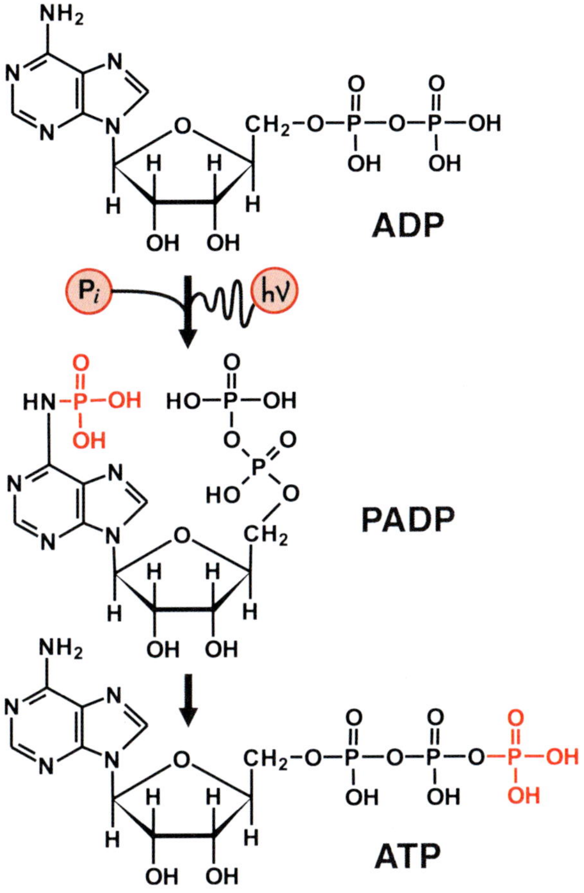

1.4 Evolution of Bioenergetic Mechanisms

Fig. 1.7 Scheme of "adenine photosynthesis", a suggested primordial mechanism of energy storage by living cells. A quantum of ultraviolet light is absorbed by the adenine part of adenosine diphosphate (ADP), thus transforming it into its excited state. Excitation facilitates the addition of inorganic phosphate (P_i) to the adenine amino group of ADP. PADP is formed as a result—an ATP isomer where the third phosphate group is connected not to the pyrophosphate "tail", but to the adenine "head" of ADP. This phosphate is later transferred from the "head" to the "tail", a process leading to the formation of the canonical ATP (Skulachev 1969, 1994)

(2) Excited adenine of ADP is phosphorylated, thus producing a PADP, the ATP isomer with the third phosphoryl group at the adenine amino group.

(3) A phosphoryl group is transferred from adenine to the terminal (second) ADP phosphate. This transfer is enhanced by the fact that the distance between the adenine and the second phosphate ADP is exactly equal to the size of another (the third) phosphate residue (it was Szent-Györgyi who was the first to take note of this fact (Szent-Gyorgyi 1957)). The transfer of the phosphoryl group from the adenine "head" of the nucleotide to the phosphate "tail" is coupled with its stabilization, because a very labile phosphoamide is being replaced with a less labile phosphoanhydride (Fig. 1.7).

Stages (2) and (3) are hypothetical; they are supposed to explain the mechanism of ultraviolet-induced ATP synthesis in Ponnamperuma's experiments (Ponnamperuma et al. 1963; Ponnamperuma 1966).

Adenine and less often other purines and pyrimidines are known to be part of key coenzyme and enzyme prosthetic groups, such as nicotinamide adenine dinucleotide (NAD$^+$), nicotinamide adenine dinucleotide phosphate (NADP$^+$), flavin adenine dinucleotide (FAD), coenzyme A (CoA), thiamine pyrophosphate (vitamin B$_1$ derivative), and vitamin B$_{12}$ (for formulas, see Appendix 2). The structures of all these compounds are based on the same principle. They include: (1) a functional group directly involved in catalysis; (2) purine (or sometimes pyrimidine); and (3) a flexible linker that allows two other parts of the molecule to interact. It can be seen especially clearly in dinucleotide structure—in this case, flat nicotinamide residues (in NAD$^+$ and NADP$^+$) or isoalloxazine (in FAD) lie on the similarly flat adenine residue. Energy transfer from the adenine residue to the nicotinamide or isoalloxazine residue is induced by absorption of an ultraviolet quantum by adenine. It seems reasonable to suggest the following scheme—adenine is converted to its excited state by ultraviolet light, then it transfers energy to the coenzyme functional group, which in turn uses this energy for energy-consuming chemical reactions (for instance, reduction of simple compounds of the external medium to more complicated compounds of primordial cells).

Later this rather nonspecific and quite uncontrollable catalysis carried out by low molecular mass coenzymes was replaced by processes driven by other catalysts—high molecular mass enzymes that are characterized by very high substrate specificity and the possibility for regulation of catalysis. Ribonucleic acids—polymers that are formed from nucleotide monomers—are likely to have been the first enzymes. Adenine-based photosynthesis might be catalyzed by complexes of RNA and magnesium salts of ADP and phosphate. RNA could also have served as an antenna that collected ultraviolet light and transferred the excitation to ADP.

Two properties of RNA are likely to have determined further evolution—the ability of the molecule to create its own copy (replicate itself) and the ability of two different RNA molecules to exchange pieces of their nucleotide sequences (recombination). According to Alexander Spirin (1976), the primordial ocean might have been a giant reactor containing an endless variety of constantly recombining RNA molecules (similar to the novel *Solaris* by Stanislaw Lem).

It is interesting to note that even today certain biochemical reactions can be catalyzed by ribonucleic acids (so-called *ribozymes*). The most important example is the ribosome, which catalyzes protein synthesis due to participation of functional groups especially of ribosomal RNA, while the protein functional groups are not as important. But in general it is beyond doubt that the catalytic functions of modern organisms are usually fulfilled by proteins that have a far greater variety of chemical groups and their combinations than does RNA.

Coding of protein structure was probably originally carried out by the same ribonucleic acids. Later this coding function was transferred to the more stable deoxyribonucleic acids (DNA)—the absence of the OH group in deoxyribose stabilizes the ester bond of DNA (Spirin 1976).

Lipids and lipid-like substances, especially phospholipids, were probably the next important invention of biological evolution. Phospholipids have an amazing property—they can spontaneously, without any kind of outside support, form a very thin membrane that is impermeable for hydrophilic substances such as nucleotide coenzymes, RNA, DNA, proteins, and carbohydrates. This membrane has the structure of a bilayer of two phospholipid molecule leaflets that contact each other with their hydrophobic (hydrocarbon) "tails", while the hydrophilic "heads" (phosphate residues) of the phospholipids are on the two opposite membrane surfaces. The thickness of such membranes is about 50 Å (Skulachev 1988, 1994).

Formation of the first membranes meant the creation of the first cells, the inner content of which was separated from the outer environment with a sufficiently reliable barrier. Appearance of these tiny separate vesicles could have played an important role in protection from unfavorable effects of ultraviolet radiation.

1.4.3 Reserve Energy Sources and Glycolysis

Ultraviolet light is a kind of double-edged sword. On the positive side, it can initiate a variety of chemical reactions, including useful ones, e.g., phosphorylation of the adenine group of ADP. But at the same time (and for the same reason!) ultraviolet light is dangerous—it can destroy in living cells the molecules that have been already synthesized. Reserve substances—energy sources that are synthesized in the light and then are utilized in darkness—might have become one of the ways to reduce "ultraviolet danger".

Ultraviolet light that reaches the surface of the ocean cannot penetrate it to any substantial depth because of turbidity and the presence of dissolved substances that absorb ultraviolet quanta. As a result, only a very thin surface layer is bombarded by ultraviolet quanta. This fact suggests the following mechanism of the energy supply of the first living cells.

Because of the movement of liquid layers in the ocean, cells constantly circulated between the thin upper water layer that was irradiated by ultraviolet light, and deeper layers that the light could not reach. The synthesis of ATP occurred close to the surface of the water, and then the ATP was used for the formation of reserve substances that were later decomposed in the depth of the ocean, thus supporting resynthesis from ATP of ADP and P_i, which were in turn formed in the course of endergonic processes associated with life activity. As a result, short ultraviolet light exposures were alternated with much longer periods with no ultraviolet danger (Fig. 1.8). The reserve substances also helped living cells to survive throughout the night (Skulachev 1994).

Inorganic pyrophosphate and polyphosphates might have been good candidates for the role of primordial energy reserves of cells. Even today they fulfill this function in some organisms. For instance, polyphosphates are formed from ATP in fungal cells when energy is available in excess, and they are decomposed leading to ATP production when energy resources are limited. But it is carbohydrates, and not polyphosphates, that serve as easily mobilized energy reserves in the majority of organisms. Their synthesis requires the energy of ATP (glyconeogenesis), and it is a long sequence of reactions far more complicated than ATP-based polyphosphate synthesis. In contrast to polyphosphates, carbohydrates accumulate not only

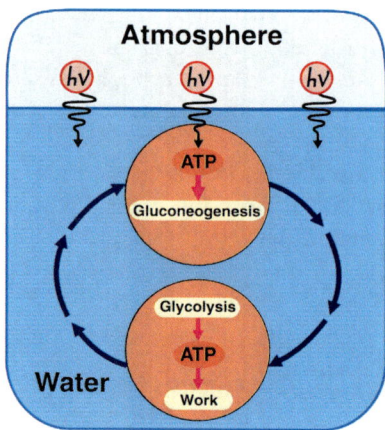

Fig. 1.8 "Adenine photosynthesis"-based energetics of a primordial living cell. According to this hypothesis, cells used ultraviolet quanta that reached the surface of the primordial ocean for synthesis of ATP which in turn is used as the energy source for carbohydrate synthesis (glyconeogenesis). A cell, when swept by the ocean water flow from the surface to some depth, was shielded from ultraviolet radiation. That is where ATP synthesis-coupled cleavage of accumulated carbohydrates (glycolysis) took place. This ATP could later be used to perform different types of work in the cell (Skulachev 1994)

energy, but also "construction material". Cleavage of carbohydrates (glycolysis) produces not just ATP, but also different carboxylic acids, such as pyruvic acid, which can be used later for synthesis of very different compounds.

Two main types of glycolysis have been described. In the first case (alcoholic fermentation) ethanol and carbon dioxide are the final products of the cleavage of carbohydrates. These compounds can easily penetrate the cell membrane, a fact that has both advantages and drawbacks. On the positive side, the final products of glycolysis do not overflow the cell, and on the negative side it is difficult to reverse the process and go back to carbohydrate if the final products have already left the cell and have been diluted in the ocean of the external environment (see Chap. 3).

This drawback is absent from the second type of glycolysis, which is far more widespread today. In this case it is lactic acid or some other carboxylic acid that is the final product of glycolysis. The lactate anion does not penetrate cell membranes, does not leave the cell, and because of this it can be used for resynthesis of carbohydrates when such a possibility appears. It is unfortunate though that lactic acid molecules can easily dissociate, thus forming lactate and hydrogen ions. The latter also cannot penetrate the membrane. They stay in the cell and acidify its content. When not prevented, this acidification can lead to cell death because of acid denaturation of proteins. A possible solution to this problem will be described in the next section.

1.4.4 Proton Channels and H^+-ATPase as Means to Prevent Glycolysis-Induced Acidification of the Cell

Living cells today have many proteins that are embedded in the membrane and act as carriers. It is with their help that cells solve the problem of transmembrane transport of the compounds that are not capable of penetrating the membrane by themselves. For instance, there are proteins that carry protons. The so-called F_o factor, a protein that is part of H^+-ATP-synthase, acts as a proton carrier or channel (see Chap. 7).

One can suggest that during glycolysis in primordial cells the F_o factor was active in the absence of the F_1 factor, the second (catalytic) component of H^+-ATP-synthase. Thus, protons that were released by the reactions of glycolysis could leave the cell. This would be a way to prevent acidification of the interior of the cell, and as a result H^+ concentrations in the inner and outer environment would be in balance. In this situation two processes became the main factors limiting glycolysis—acidification of the external environment and generation of transmembrane electric potential ($\Delta\Psi$) caused by outward diffusion of H^+ leading to excess negative charge inside the cell. It is noteworthy that all known living cells today have negatively charged inner space, i.e., $\Delta\Psi$ on the plasma membrane is always inside negative.

It would have been possible to diminish the influence of these limitations by adding to the protein H^+ carrier (F_o factor) another protein (F_1 factor), which could use the energy of ATP for active H^+ transport through the F_o factor across the cell membrane. H^+-ATP-synthase (a complex of factors F_o and F_1) is known to have the ability to catalyze ATP hydrolysis instead of ATP synthesis when working in the opposite direction. This process is coupled to outward transport of protons and is called the H^+-ATPase reaction (Fig. 1.9) (for details, see Chap. 3). The appearance of H^+-ATPase is supposed to have completed formation of the primordial cell that used ultraviolet light as the main energy source.

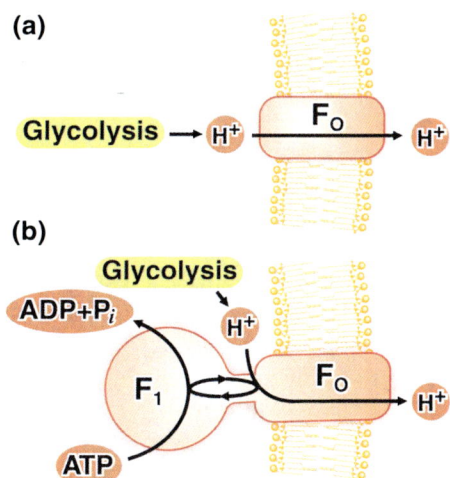

Fig. 1.9 Possible ways for primordial cells to export protons released by glycolysis. **a** facilitated diffusion of H^+ ions via protein (F_o factor) that creates a H^+-conducting path across the cell membrane; **b** F_oF_1 complex (H^+-ATPase) carries out active outward transport of H^+ from the cell at the expense of ATP hydrolysis

1.4.5 Bacteriorhodopsin-Based Photosynthesis as the Primordial Mechanism of Visible Light Energy Transduction

With the passing time fewer ultraviolet quanta were reaching the Earth's surface. The reason for this was the formation of the ozone layer in the atmosphere caused by the increase in oxygen concentration. Originally, oxygen was probably formed because of ultraviolet radiation-induced photolysis of water vapor. To survive under those new conditions, ancient cells had to switch from ultraviolet light to some other energy source that was still available. Visible light was probably this source.

But another scenario of evolution also seems possible. Visible light-based photosynthesis might have appeared before the atmosphere became strongly UV-absorbing, when living cells penetrated the deeper ocean layers where no ultraviolet light was available. Replacement of dangerous ultraviolet radiation with safe visible light could have become the basis of natural selection at that stage of evolution. Within this concept, the creation of the ozone layer might have been of biogenic nature, as it is the result of photolysis of water by the chlorophyll-based photosynthetic system of green bacteria and cyanobacteria (Skulachev 1994).

A new type of photosynthesis should have still produced ATP, which by that time had firmly occupied a position in the very center of the metabolic map as the interconvertible energy currency of cells. But adenine could no longer have served as the light-harvesting antenna because its absorption was in the ultraviolet and not in the visible spectral range. Today there are two types of photosynthetic structures that use visible light. In the first type it is chlorophyll that serves as an antenna, while in the second this function belongs to a derivative of vitamin A (retinal), which is connected to a special protein, bacteriorhodopsin. Chlorophyll is found in green plants and almost all photosynthetic bacteria. The group of halotolerant and thermotolerant archaea and certain bacteriorhodopsin-containing bacteria are the only exceptions. Nevertheless, it was probably bacteriorhodopsin that was the primordial evolutionary mechanism of visible light energy storage by the living cell.

Bacteriorhodopsin is a light-driven proton pump (see Chap. 6). It can actively transport protons out of the cell at the expense of the energy of visible light that is absorbed by the retinal moiety in its molecule. Light energy is transformed into $\Delta\bar{\mu}_{H^+}$ as a result. Protons transported out of the cell by bacteriorhodopsin can return into the cell via an F_oF_1 complex so that the energy released through the "downhill" proton movement is used for ATP synthesis. It does not seem so difficult to imagine the way this type of photosynthesis—the process catalyzed by bacteriorhodopsin and F_oF_1 complex—came into being. With the appearance of bacteriorhodopsin, it became possible to create $\Delta\bar{\mu}_{H^+}$ at the expense of visible light. Once generated, this $\Delta\bar{\mu}_{H^+}$ could reverse the H^+-ATPase reaction that had been used earlier for outward transport of protons produced by glycolysis. In this way, the F_oF_1 complex might have been transformed from ATPase to ATP-synthase (Fig. 1.10).

The structure of bacteriorhodopsin is much simpler than that of the system of chlorophyll photosynthesis. The protein part of bacteriorhodopsin consists of one

1.4 Evolution of Bioenergetic Mechanisms

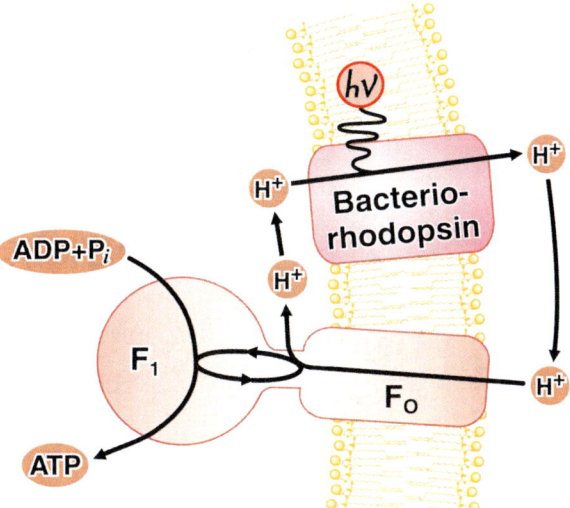

Fig. 1.10 Bacteriorhodopsin photosynthesis. Protons are pumped from the cell by bacteriorhodopsin, a retinal-containing protein, where retinal acts as a chromatophore which absorbs visible light. Protons reenter the cell ("downhill" transport) via the F_oF_1 H^+-ATPase complex which catalyzes synthesis (rather than hydrolysis) of ATP under such conditions

middle-sized polypeptide chain that has no coenzymes or prosthetic groups besides retinal. Bacteriorhodopsin is characterized by extremely high stability—it manifests the same activity even after having been boiled in an autoclave at +130 °C, its activity remains the same when NaCl concentration in the external solution is changed from zero to saturation, and it can operate over a wide range of pH. Even more so, one can remove the terminal parts of the polypeptide chain that protrude from the membrane and even cleave the chain in the middle without any apparent effect on the pump activity. However, the activity of bacteriorhodopsin as an energy transformation system is relatively low—only about 20 % of the energy of the light quantum is transformed into $\Delta\bar{\mu}_{H^+}$. And there is only one proton transported across the membrane per photon absorbed.

1.4.6 Chlorophyll-Based Photosynthesis

Chlorophyll-based photosynthesis utilizes the photon energy more efficiently than its bacteriorhodopsin-based analog. It transports two protons across the membrane for each quantum (versus only one for bacteriorhodopsin). And in addition to H^+ transport, energy is stored in the form of carbohydrates synthesized from CO_2 and H_2O (see Chap. 2). This could be the reason for evolution to largely substitute chlorophyll-based for bacteriorhodopsin-based photosynthesis.

Chlorophyll photosynthesis is catalyzed by an enzyme system composed of several proteins. A light quantum is first absorbed by chlorophyll, initiating oxidation of excited chlorophyll by the next component of the photosynthetic electron transfer chain. This chain consists of a number of redox enzymes and coenzymes that are located in the bacterial inner membrane, chromatophores, or in the

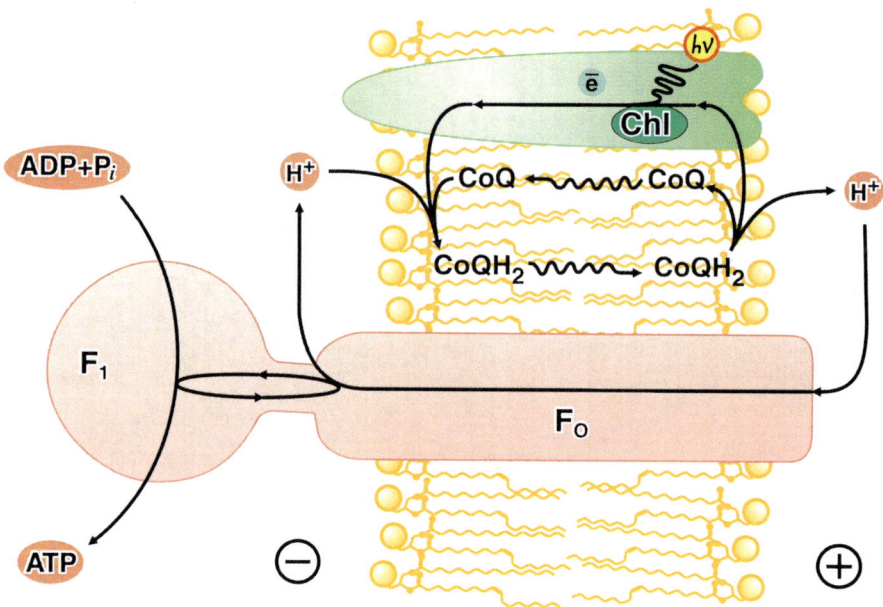

Fig. 1.11 Chlorophyll-based photosynthesis of purple bacteria (simplified scheme). Explanations are given in the text

membrane of thylakoids of plant chloroplasts. The components of the electron transfer chain usually contain some metal ions with variable valency (iron, copper, or manganese). Iron can be part of a heme, and such proteins are called cytochromes. Nonheme-iron proteins, where an iron ion is connected to a protein molecule via the sulfur atom of cysteine or (less often) via the nitrogen atom of histidine, also play an important role. Not only metal ions can be used as electron carriers. This function can also be fulfilled by quinone derivatives, such as ubiquinone, plastoquinone, and vitamins of the K group (all the structural formulas of mentioned prosthetic groups and enzyme cofactors of electron transfer chains can be found in Appendix 2).

The final stages of electron transfer from the excited chlorophyll are different in different types of photosynthesis.

In purple bacteria (Fig. 1.11), after light induces the transition of chlorophyll to the excited state, an electron is transferred to the opposite side of the membrane, where it is used to reduce quinone of CoQ (less frequently, MQ). Then reduced CoQ binds two protons, thus forming $CoQH_2$. The latter diffuses to the other membrane side so as to return the electron to the chlorophyll molecule. At the same time, the protons of $CoQH_2$ are released to the aqueous phase outside the cell. The process is completed by the return of H^+ into the cell via H^+-ATP-synthase, the latter transport being coupled to ATP synthesis.

Gloeobacter-like cyanobacteria were probably the next step in the evolution of photosynthesis. In this case chlorophyll together with some special enzymes

1.4 Evolution of Bioenergetic Mechanisms

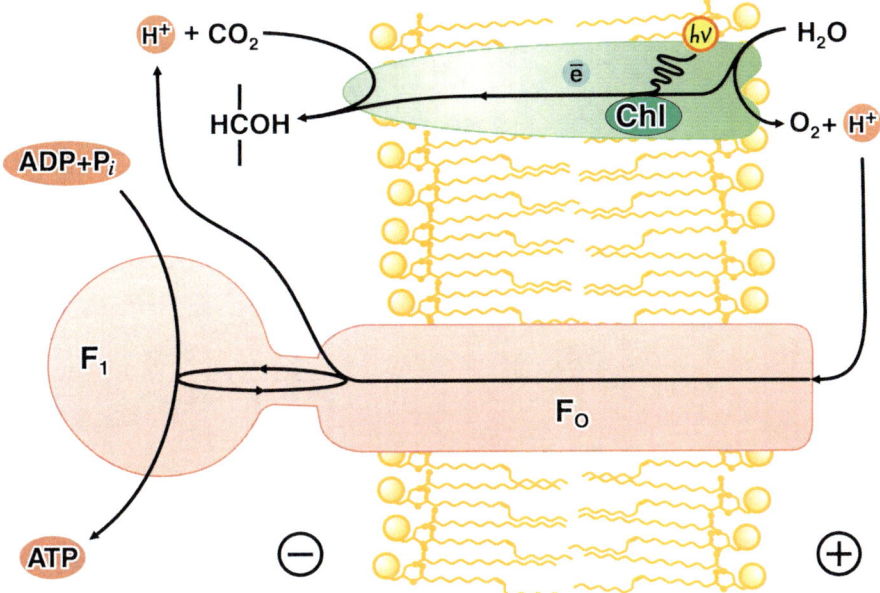

Fig. 1.12 Chlorophyll-based photosynthesis of cyanobacteria of genus *Gloeobacter* (simplified scheme). Explanations are given in the text. $H\overset{|}{C}OH$ represents a part of the glucose molecule

reduces intracellular carbon dioxide to glucose. Protons are taken from the cytosol during this process. Extracellular water that is cleaved to O_2 and H^+ serves as a reductant of the photooxidized chlorophyll. The inward H^+ flow via F_oF_1 is coupled to generation of ATP (Fig. 1.12). So, cyanobacterial photosynthesis produces not only ATP, but also carbohydrate, the main reserve substance of present-day living cells. Cyanobacteria are the most probable evolutionary predecessors of chloroplasts, the green plant organelles that function on the basis of the same scheme shown in Fig. 1.12.

1.4.7 Respiratory Mechanism of Energy Supply

Molecular oxygen is a byproduct of cyanobacterial and plant photosynthesis. Increase in its concentration in the atmosphere led to the development of enzymes capable of eliminating this dangerous powerful oxidant. It was probably decreasing intracellular oxygen concentration that was the original function of the enzymes that reduced O_2 to H_2O (see Chap. 15). But later aerobic bacteria learned to benefit from this process by creating a respiratory electron transporting chain coupled to outward H^+ transfer (see Chap. 4).

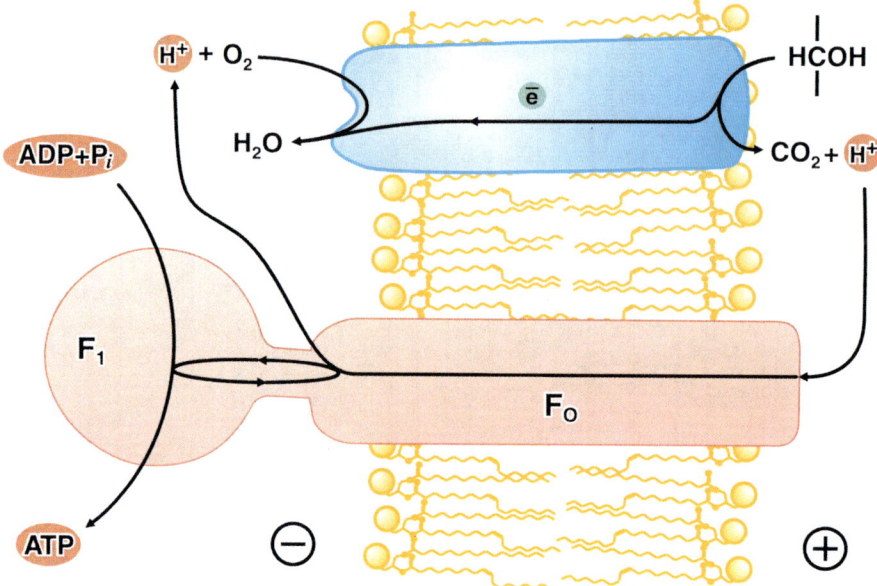

Fig. 1.13 Mechanism of oxidative phosphorylation in aerobic bacteria and mitochondria (simplified scheme). Explanations are given in the text. $\mathrm{H\overset{|}{C}OH}$ represents a part of the glucose molecule

The respiratory chain provides a mechanism for oxidation of respiratory substrates. These substrates can be obtained, for example, from glucose, which is in turn produced by photosynthetic organisms. We can say that respiratory chain enzymes fulfill functions that are opposite to those of cyanobacterial and chloroplast photosynthesis. It is not synthesis, but oxidation of organic substances that occurs in the course of respiration. As a result, O_2 is reduced to water instead of H_2O being oxidized to O_2. However, in both cases ATP synthesis is coupled to "downhill" H^+ transfer (Fig. 1.13).

Like chloroplasts being developed from cyanobacteria—animal, plant, and fungal mitochondria probably originated from aerobic bacteria. That is why it is not surprising that the functions of the mitochondrial respiratory chain can be described by the scheme of bacterial respiration presented in Fig. 1.13.

Replacement of H^+ by Na^+ in a number of bioenergetic structures was probably the final step in the creation of the energy scheme of modern living beings. Sodium ions have two advantages when compared to protons. First, sodium is far more widely available (compare 5×10^{-1} M Na^+ to 1×10^{-7} M H^+ in the ocean's water). Second, membranes are more permeable for H^+ than for Na^+. Proton acceptor groups of membrane proteins increase the H^+ permeability of membranes, while this does not occur in the case of Na^+. It is very difficult to create a membrane which is not permeable for H^+—it is much easier to do this for Na^+.

And protons are especially difficult to use as coupling ions under alkaline conditions when their concentration is very low. Protons are also inconvenient for life at high temperatures, for this factor increases H^+-permeability of membranes much more than their Na^+-permeability. If we apply a principle well-known to enzymologists—"protein has unlimited abilities"—then we can imagine a protein that facilitates outward transport of Na^+ at the expense of some energy source as well as another protein that allows Na^+ to enter the cell, the latter process being coupled to ATP synthesis or performance of some other useful work. In fact, such "sodium" membranes use Na^+ instead of H^+ as the coupling ion. For instance, there are membranes of some marine bacteria that have a Na^+-transporting respiratory chain (or Na^+-decarboxylase) and Na^+-ATP-synthase. There are also external membranes of animal cells with Na^+/K^+-ATPase and Na^+-dependent proteins that function as carriers of different substances (see Chaps. 12 and 13).

Appearance of "calcium" energetics might have been one more step of evolution. Bivalent calcium ions fulfill two functions in the majority of organisms—structural (comprising skeletons) and regulatory (facilitating reversible change in properties of a variety of extracellular and intracellular components). Different systems of calcium ion transport (Ca^{2+}-ATPases, Na^+/Ca^{2+}-exchangers, Ca^{2+}-channels) are present in the membranes of all the main organelles of cells (plasmalemma, sarcoplasmic reticulum, endoplasmic reticulum, Golgi apparatus, and mitochondria). Part of the energy production in animal cells is always spent to support Ca^{2+} flows that regulate such important processes as nerve impulse transfer and intracellular hormone signaling, initiation of muscle contraction and cell growth regulation, stimulation of anaerobic and aerobic catabolism, secretory cell exocytosis and intermembrane contact formation, control of cell aging and death, etc. One of this book's authors (F.O.K.) obtained some data suggesting that calcium ion transmembrane gradients can stabilize for some time the $\Delta\Psi$ level on cell membranes when $\Delta\bar{\mu}_{H^+}$-generators are inhibited (Kasparinsky and Vinogradov 1996). Current data on the spacial organization of close contacts of potential-dependent plasmalemma Ca^{2+} channels, endoplasmic reticulum, and mitochondria suggest the hypothesis of direct delivery of calcium gradient free energy to many cell compartments.

References

Belitser VN, Tsibakova ET (1939) On the coupling mechanism of the respiration and phosphorilation. Biokhimiya (Russ) 4:516–521

Blumenfeld LA, Temkin MI (1962) On possible mechanism of formation of adenosine triphosphoric acid in oxidative phosphorylation. Biophysika (Russ) 7:731–734

Engelhardt WA (1930) Ortho- und Pyrophosphat im aeroben und anaeroben Stoffwechsel der Blutzellen. Biochem Z 251:16–21

Engelhardt WA (1931) Anaerobic decomposition and aerobic resynthesis of pyrophospate in erythrocytes of birds. Kazan Med Zhurn (Russ) 27:496–504

Engelhardt WA (1932) Die Beziehungen zwischen Atmung und Pyrophosphatumsatz in Vogelerythrocyten. Biochem Z 251:343–345

Kasparinsky FO, Vinogradov AD (1996) Slow Ca^{2+}-induced inactive/active transition of the energy-dependent Ca^{2+} transporting system of rat liver mitochondria: clue for Ca^{2+} influx cooperativity. FEBS Lett 389:293–296

Lee HC, Walseth TF, Bratt GT, Hayes RN, Clapper DL (1989) Structural determination of a cyclic metabolite of NAD^+ with intracellular Ca^{2+}-mobilizing activity. J Biol Chem 264:1608–1615

Lipmann F (1941) Metabolic generation and utilization of phosphate bond energy. Adv Enzymol 1:99–107

Mitchell P (1961) Coupling of phosphorylation to electron and hydrogen transfer by a chemiosmotic type of mechanism. Nature 191:144–148

Mitchell P (1973) Hypothesis: cation-translocating adenosine triphosphatase models: how direct is the participation of adenosine triphosphate and its hydrolysis products in cation translocation? FEBS Lett 33:267–274

Ponnamperuma C (1966) Non-biological synthesis of some components of nucleic acids. In: Oparin AI (ed) The origin of pre-biological systems. Mir, Moscow, pp 224–238

Ponnamperuma C, Sagan C, Mariner R (1963) Synthesis of adenosine triphosphate under possible primitive earth conditions. Nature 199:222–226

Sagan C (1966) The u.v.-supported primary synthesis of nucleoside phosphates. In: Oparin AI (ed) The origin of pre-biological systems. Mir, Moscow, pp 211–223

Shuvalov VA (2007) Electron and nuclear dynamics in many-electron atoms, molecules and chlorophyll-protein complexes: a review. Biochim Biophys Acta 1767(6):422–433

Skulachev VP (1969) Energy accumulation processes in the cell. Nauka, Moscow

Skulachev VP (1984) Sodium bioenergetics. TIBS 9:483–485

Skulachev VP (1988) Membrane bioenergetics. Springer, Berlin

Skulachev VP (1994) Bioenergetics: the evolution of molecular mechanisms and the development of bioenergetic concepts. Antonie Van Leeuwenhoek 65(4):271–284

Skulachev VP (1996) Evolution of convertible energy currencies of the living cell: from ATP to $\Delta\mu_H^+$ and $\Delta\mu_{Na}^+$. In: Balltschettsky H (ed) Origin and evolution of biological energy conversation. VCH Publishers, New York, pp 11–35

Spirin AS (1976) Cell-free systems of polypeptide biosynthesis and approaches to the evolution of translation apparatus. Orig Life 7(2):109–118

Szent-Gyorgyi A (1957) Bioenergetics. Academic Press, New York

Part II
Generators of Proton Potential

Chapter 2
Chlorophyll-Based Generators of Proton Potential

Generators of $\Delta\bar{\mu}_{H^+}$ play a leading role in the conversion of external energy sources into forms that can be used by living cells. We begin by discussing light-dependent (photosynthetic) generators. First of all, we consider them to have been the primordial systems for proton potential generation driven by external energy sources during the evolution of life. Second, photosynthesis still plays the key role in energy supply to the biosphere. The biosphere is known to consist mainly of photosynthetic organisms and organotrophs that directly or indirectly consume products of photosynthesis—organic substances and oxygen. Oxygenic photosynthesis, i.e. photosynthesis that produces oxygen, takes place in green plants and cyanobacteria. It is this type of photosynthesis that provides oxygen for the planet. One can point to the precise geographic coordinates of the areas of the Earth that are responsible for this function. Most of the Earth's oxygen is produced in the taiga forests in Siberia. Canada, with its temperate zone forests, especially coniferous ones, provides a smaller contribution. The famous tropical forests (African and South American jungles) as well as the World's ocean do not play an important role in this process due to the presence of vast numbers of heterotrophic bacteria that absorb that very oxygen that is produced by photosynthetic organisms. So, Siberia can be compared to the planet's lungs, and if Siberian forests are destroyed it will mean the end of aerobic life on Earth.

There is one other technical but nevertheless important point that explains our decision to start our presentation with photosynthesis. This process is initiated by a quantum of light, and this fact makes it possible to use modern physical instrumentation for the analysis of its mechanism. For instance, when using an ultrafast laser one can generate a 2×10^{-14} s long flash of light so as to study kinetics and mechanisms of the primary processes of light energy storage.

2.1 Light-Dependent Cyclic Redox Chain of Purple Bacteria

Let us start not from that main (oxygenic) photosynthesis that has just been mentioned above, but from the more primitive photosynthetic mechanisms that have been described in phototrophic bacteria (they have already been mentioned in the Chap. 1). The photosynthetic apparatus of purple bacteria is located in the internal (i.e. cytoplasmic) membrane and in chromatophores—intracellular vesicles that split off that membrane.

The biological membrane is composed of lipids and proteins. It is essentially a lipid bilayer with transmembrane proteins floating in it. There are also other proteins that are attached to the surface of the bilayer. Biomembranes usually contain more protein than lipid, even though there are some where lipid prevails. Usually it is phospholipids that form the lipid component of biomembranes, but sometimes this function is fulfilled by sulfolipids or glycolipids.

One of the varieties of chlorophyll—bacteriochlorophyll—plays the key role in photosynthesis in the purple bacteria (see structure in Appendix 2). In the second half of the 1940s, Alexander Krasnovsky (Fig. 2.1) discovered reactions of photoinduced chlorophyll oxidation and reduction (Krasnovsky 1948), this discovery being the first step in the understanding of the mechanism of functioning of the photosynthetic apparatus.

Absorption of visible light by a chromophore molecule is known to induce transfer of an electron from the main orbital (S_0) to one of the singlet (S_1^*, S_2^* ...) or triplet (T) excited orbitals (Fig. 2.2). Photon energy is spent during this process to transfer of the electron to an orbital that is further away from the nucleus. When in the excited orbital, the electron has a relatively weak connection to the nucleus, and thus it can be easily torn away from the chromophore molecule. So, this molecule in an excited state becomes a good *reductant*. Absorption of a light quantum produces at the same time a "vacancy" for an electron in the main orbital ("a hole"). This "hole" has substantial affinity for an electron, i.e. it is a good *oxidant*.

The lifetime of a chromophore molecule in an excited state (especially in a singlet state) is extremely short. The electron is inclined to return from an excited

Fig. 2.1 Alexander Krasnovsky

2.1 Light-Dependent Cyclic Redox Chain of Purple Bacteria

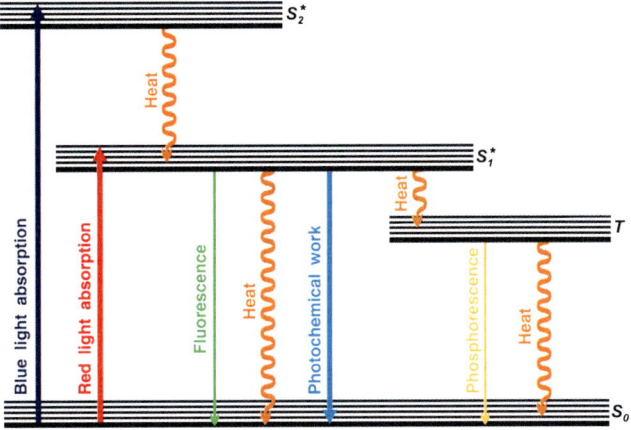

Fig. 2.2 Scheme of electron transfer induced by photon absorption by a chromophore

to the main orbital. Such a return is accompanied with dissipation of the absorbed photon energy in the form of heat or in the form of a light of longer wavelength (fluorescence or phosphorescence). But photosynthesizing organisms have learned to use the effect of chromophore transition into an excited state for the storage of the energy of the quantum ($h\nu$) in a form which is useful for a cell. For instance, if during the excited state lifetime an electron is transferred from the excited orbital to some acceptor X, and the vacancy on the main chromophore orbital is filled from some donor Y, then part of the energy of the photon can be stored as a difference of redox potentials of substances X and Y under the condition that the redox potential of the $Y_{oxidized}/Y_{reduced}$ couple is more positive than that of the $X_{oxidized}/X_{reduced}$ couple. Even more so, if this redox reaction is organized in such a way that an electron is transferred from Y to X *across* the membrane, then a certain part of the energy of the light quantum can also be stored as $\Delta\bar{\mu}_{H^+}$. Just in this way the most important photosynthetic process, i.e. oxygenic photosynthesis, is organized.

In an alternative and simplified version of photosynthesis, an electron from an excited orbital is returned to the main orbital, but not directly. It has to cross the membrane first, which leads to generation of $\Delta\bar{\mu}_{H^+}$. This type of photosynthesis is found in purple bacteria.

2.1.1 Main Components of Redox Chain and Principle of Their Functioning

Light is absorbed by a magnesium porphyrin moiety of bacteriochlorophyll, which is bound to a special protein (the *light-harvesting antenna complex*) or by additional pigments—*carotenoids*—located in the same membrane and transferring the

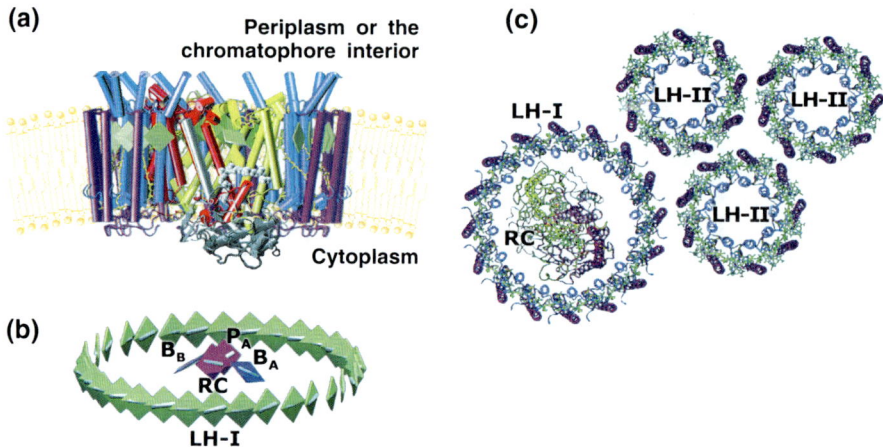

Fig. 2.3 a, b Structure of the reaction center complex (RC) and of antenna I (LH-I) of purple bacteria. Bacteriochlorophylls are represented as rhombs. Bacteriochlorophylls of antenna I are shown in green color; (BChl)$_2$ (it is labeled P$_A$); and bacteriochlorophyll monomers (B$_A$ and B$_B$) of the reaction center are *red* and *blue*, respectively. The α-helical columns of the reaction center subunits are presented as colored *rods—yellow* (*L*-subunit), red (*M*-subunit), and *gray* (*H*-subunit). Antenna I proteins are shown as *blue* and *pink* rods. Bacteriopheophytins are not shown. **c** Structural arrangement of the reaction center complex, antenna I, and antenna II in purple bacteria. The polypeptide chain of the proteins and bacteriochlorophyll molecules are shown. (Hu et al. 1998)

excitation energy to the antenna bacteriochlorophyll. One should note that in the case of acidophilic bacteria Mg^{2+} ions are replaced by Zn^{2+} ions in bacteriochlorophyll, so that the Mg^{2+} of the porphyrin complex would not be replaced by H$^+$ ions, which would otherwise occur under acidic conditions (Wakao et al. 1996). The excitation migrates through antenna bacteriochlorophyll until it reaches what is called a "special pair", a *bacteriochlorophyll dimer* bound to other proteins forming a *photosynthetic reaction center complex* (Fig. 2.3).

Besides the dimer, the complex contains two *bacteriochlorophyll monomers*, two *bacteriopheophytins* (i.e. bacteriochlorophyll analogs where the Mg^{2+} ion is replaced by two H$^+$ ions), and two ubiquinone molecules (CoQ).[1] The amount of reaction center pigment complex is usually about two orders of magnitude smaller than that of the antenna pigment complex.

The excited bacteriochlorophyll dimer serves as an electron donor to ubiquinone (CoQ). This process is directed across the membrane to its cytoplasmic side. It includes a number of intermediate stages with consecutive reduction and oxidation of bacteriochlorophyll monomer, bacteriopheophytin, and bound MQ and/or CoQ (reactions 1–4 in the scheme of the cyclic redox chain, Fig. 2.4).

---

[1] In certain bacteria, one of the ubiquinone molecules is replaced by a menaquinone (MQ); see structure in Appendix 2.

2.1 Light-Dependent Cyclic Redox Chain of Purple Bacteria

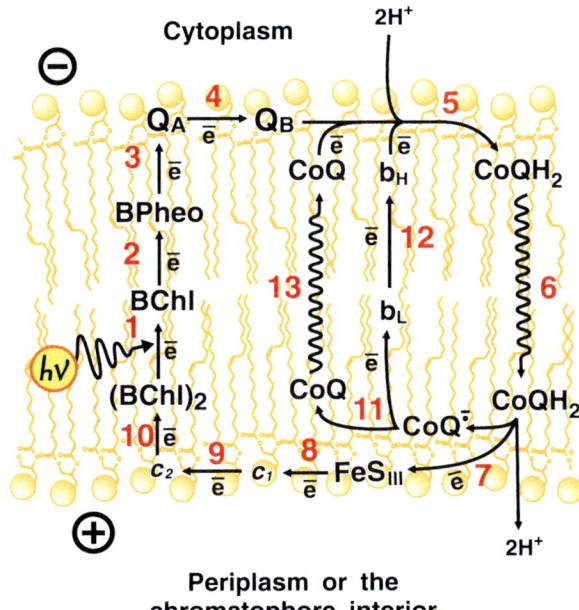

Fig. 2.4 Cyclic redox chain of the purple photosynthetic bacterium *Rhodospirillum rubrum*: (BChl)$_2$ and BChl—dimer and monomer of reaction center bacteriochlorophylls; BPheo—bacteriopheophytin; Q$_A$ and Q$_B$—two ubiquinone molecules bound to reaction center proteins; b$_H$ and b$_L$—high- and low-potential hemes of ; FeS$_{III}$—nonheme iron protein; c_1 and c_2—respective cytochromes (Skulachev 1988)

For reduction, CoQ needs two electrons and two protons (Fig. 2.5). One of these electrons is donated by the reaction center and the other by heme b_H of cytochrome b.[2] Having accepted two electrons, CoQ combines with two protons from the nearest water phase (above the membrane, i.e. in the cytosol in Fig. 2.4; CoQH$_2$ formation is described by reaction 5 in Fig. 2.4 and in more detail in Fig. 2.5).

When formed, CoQH$_2$ diffuses across the membrane in the direction of its opposite (periplasmic) side (Fig. 2.4, reaction 6). Close to this membrane surface, CoQH$_2$ is oxidized to the semiquinone form (reaction 7) by the *nonheme iron–sulfur (FeS) protein* abbreviated as FeS$_{III}$. When CoQH$_2$ is oxidized, two H$^+$ ions are released to the periplasm or inside the chromatophore (Fig. 2.4).

The electron accepted by FeS$_{III}$ is later used for the reduction of the bacteriochlorophyll dimer that was previously oxidized during the first stage of the cycle. Cytochromes of c type participate in the electron transfer from FeS$_{III}$ to (BChl)$_2^{\bullet+}$ (reactions 8–10; in the case of *Rhodospirillum rubrum* these are cytochromes c_1 and c_2).

The radical ion CoQ$^{\bullet-}$ that appears after CoQH$_2$ has lost one electron and two protons reduces heme b_L (reaction 11). The latter transfers the electron to the

---

[2] Cytochromes are heme-containing proteins catalyzing redox processes (for heme structures, see Appendix 2). Cytochromes having the same heme group and differing in the protein moiety are usually indicated by the same letter but different numerical indexes.

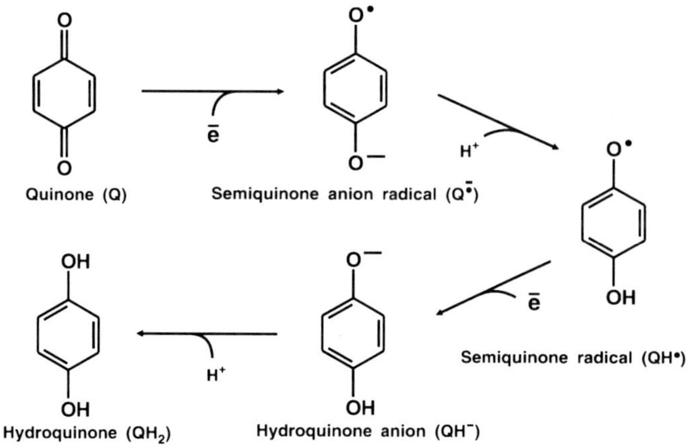

Fig. 2.5 Steps of conversion of quinone to hydroquinone (quinol) (Rich 1984)

previously oxidized heme b_H (reaction 12). The cycle is completed by diffusion of the formed CoQ to the cytoplasmic surface of the membrane (reaction 13).

The net result of the entire cycle is $\Delta\bar{\mu}_{H^+}$ *generation due to the transmembrane uphill transport of two* H^+ *ions per photon absorbed by BChl.*

The energy for the cycle is provided by photoexcitation of bacteriochlorophyll. The latter, being excited by a photon, greatly changes its redox potential, which becomes about −950 mV instead of +440 mV in the nonexcited state. All the further electron transfers go energetically downhill, i.e. from negative to more positive redox potentials (Fig. 2.6a).

The proteins catalyzing the cyclic electron transfer can be separated into two distinct complexes—the reaction center complex and the bc_1 complex.

2.1.2 Reaction Center Complex

Protein Composition. The reaction center complex isolated from *Rhodospirillum rubrum* is formed by *three protein subunits*—the so-called heavy (H), medium (M), and light (L) subunits. The subunit stoichiometry in the complex is 1:1:1 (Okamura et al. 1982).

In *Blastochloris* (previously called *Rhodopseudomonas*) *viridis, four* protein constituents were shown to compose the reaction center complex. The additional subunit was identified with a peculiar *c-type cytochrome containing tetraheme c groups*. The molecular masses of this cytochrome and subunits H, M, and L, obtained using SDS electrophoresis, were shown to be 38, 35, 28, and 24 kDa, respectively, this fact being the reason for the names given to non-cytochrome subunits H, M, and L. However, determination of the primary structure (which happened some years later) gave values of 40.5, 28.3, 35.9, and 30.6 kDa (336,

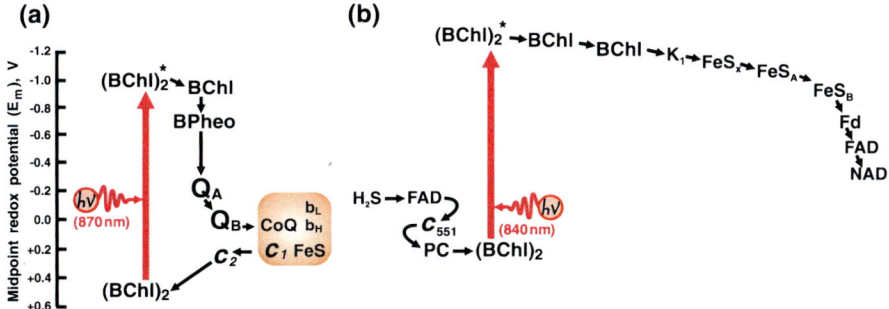

Fig. 2.6 Redox potentials of electron carrier involved in the photosynthetic systems of purple bacteria (**a**) and green sulfur bacteria (**b**). For explanations, see Sects. 2.1.1 and 2.2.2, respectively

258, 323, and 273 amino acid residues, respectively), which is in obvious contrast to the previously accepted nomenclature. The old subunits names are nevertheless still used in the literature.

The hydropathy profile of the M-subunit shows five very hydrophobic columns composed of at least 20 amino acid residues each. Each column is long enough to span the membrane in an α-helix. This conclusion was directly confirmed by X-ray analysis of *Blastochloris viridis* reaction center complexes (Hartmut Michel, Johann Deisenhofer, and Robert Huber, Nobel Prize in Chemistry, 1988, Fig. 2.7). They obtained 0.5–1-mm crystals of the complexes using $(NH_4)_2SO_4$ treatment of the reaction center complex solution supplemented with a detergent.

An X-ray study of these crystals revealed for the first time the *three-dimensional structure of a membrane energy transducer* at 0.3-nm resolution (Deisenhofer et al. 1984, 1985a, b). It was found that the 13-nm-long complex is composed of two peripheral domains (tetraheme cytochrome *c* and the H-subunit) and the central part (M- and L-subunits). It is the M- and L-subunits that bear bacteriochlorophylls and bacteriopheophytins of *b*-type, quinones, and nonheme iron that is connected to imidazoles of the four histidines of the protein. Comparing these data with earlier observations, one can assume that cytochrome *c* faces the periplasm, the H-subunit the cytoplasm, the M- and L-subunits being integrated in the hydrophobic membrane layer.

The X-ray picture of the reaction center complex reveals eleven transmembrane α-helical columns that are oriented perpendicular to the plane of the membrane and parallel to the long axis of the complex, being located in its middle (membrane) part. The H-subunit has only one α-helical column, while the M- and L-subunits, being similar to each other in their amino acid sequences and spatial structures, contain five such columns each. These columns are oriented symmetrically about the longitudinal axis of the complex (Fig. 2.8).

The single hydrophobic α-helix of the H-subunit serves as an anchor binding this hydrophilic polypeptide to the membrane. Another peripheral subunit, tetraheme cytochrome *c*, it is attached to the membrane by contacts with other subunits

Fig. 2.7 Nobel Laureates Hartmut Michel, Johann Deisenhofer, and Robert Huber

and by two fatty acids covalently bound to a glycerol residue forming a thioether bond with the SH group of the N-terminal cysteine of this cytochrome (Fig. 2.9).

When looking at the packing of polypeptide chains of the H-, M-, and L-subunits in the membrane, one should note complete absence of charged amino acids residues (Fig. 2.10a) and almost complete absence of bound water (Fig. 2.10b) in the transmembrane α-helical columns. Such amino acids and water are located on the membrane surfaces; the majority of the negatively charged residues are found on the outer membrane surface, whereas positively charged amino acids are exposed on the inner membrane surface. This fact suggests that the assembly of the reaction center complex polypeptides inside the membrane is directed by transmembrane $\Delta\Psi$ (the cell interior is negative). Electrophoresis of the protein domains is apparently involved in this assembly.

It should be noted that crystallization of the membrane proteins is extremely difficult to perform. The success in attempts to crystallize the *Bl. viridis* reaction center complex seems to be due to the fact that its central hydrophobic part (the M- and L-subunits) is to some extent protected from water by large hydrophilic polypeptides—the H-subunit and tetraheme cytochrome *c*.

Arrangement of Redox Groups. The most important result of the X-ray study of the *Bl. viridis* reaction center complex is that the arrangement of redox prosthetic groups is revealed (Deisenhofer et al. 1984). In Fig. 2.11, the results of the X-ray analysis of the complex is shown in such a way that only prosthetic groups are seen. According to these data, four cytochrome *c* hemes are located above the bacteriochlorophyll dimer. The Fe–Fe distances between the hemes are 1.4–1.6 nm. The Fe atom of the heme closest to $(BChl)_2$ is situated 2.1 nm above two Mg atoms of the bacteriochlorophyll dimer. The dimer is oriented parallel to the long axis of the reaction center complex. It is fixed between two α-helical D segments belonging to the M- and L-subunits (D_M and D_L, respectively). In each α-helical column there is a histidine residue, the imidazole group of which serves as an axial ligand for Mg^{2+} in $(BChl)_2$.

Two bacteriochlorophyll monomers were found to be arranged slightly below the dimer (the distance between Mg^{2+} ions in $(BChl)_2$ and BChl is 1.3 nm, and the

2.1 Light-Dependent Cyclic Redox Chain of Purple Bacteria

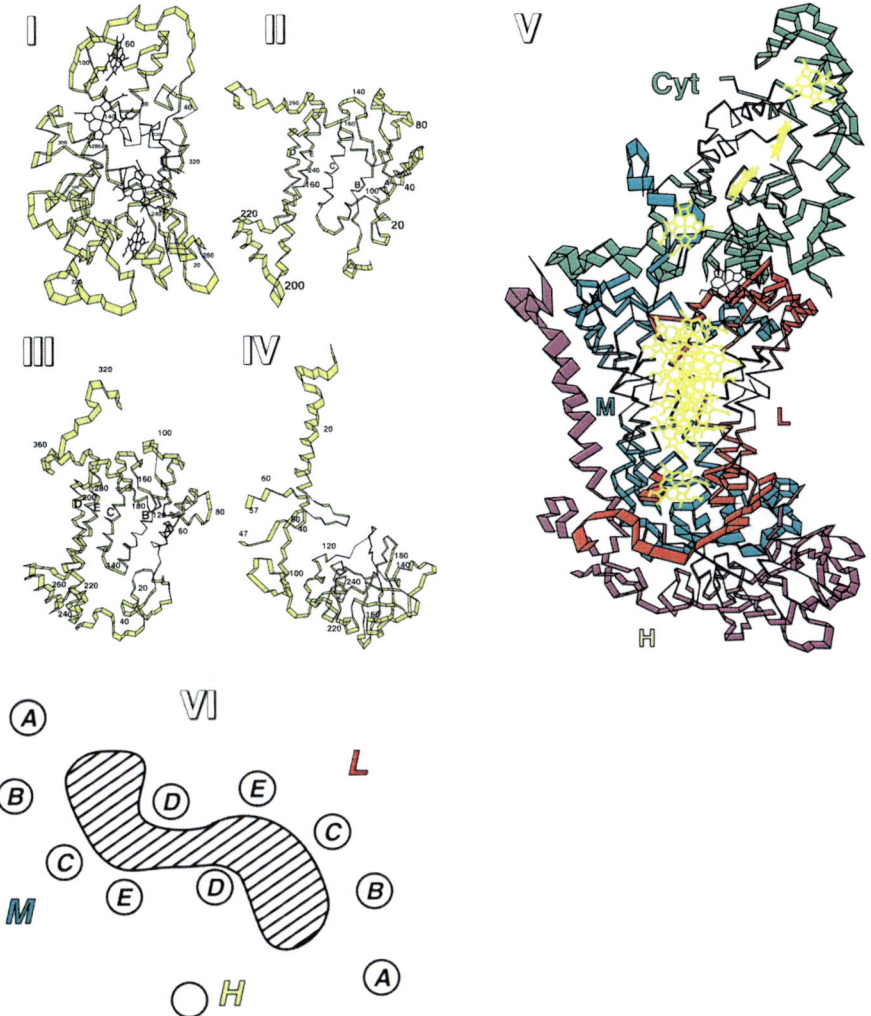

Fig. 2.8 Structure of protein subunits of the *Blastochloris viridis* reaction center: *I, II, III*, and *IV* stand for tetraheme cytochrome *c*, *L*-, *M*-, and *H*-subunits, respectively; *V*—general view of the reaction center complex; yellow electron-transporting prosthetic groups; *VI*—cross-section of the complex showing relationship between 11 α-helical segments (*top view*; A, B, C, D, E, five transmembrane α-helical segments of *M*- and *L*- subunits). The region occupied by the chromophores is shaded. The segments are tilted by an angle of 38° (segments *D*) and < 25° (other segments) to the membrane plane. Figure section *VI* shows a section at a level close to the center of the hydrophobic membrane layer (Deisenhofer et al. 1984; Henderson 1985)

angles between the porphyrin ring planes are ~ 70°). Imidazoles of histidine residues located in C_M and D_M or C_L and D_L segment connections are ligands for Mg^{2+} of the monomers. Bacteriochlorophyll monomers are in contact with

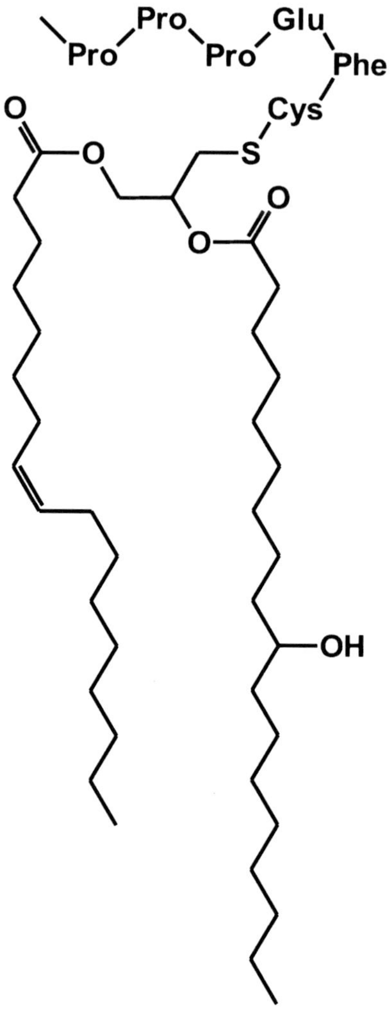

Fig. 2.9 Hydrophobic anchor of the tetraheme cytochrome *c* from *Blastochloris viridis*—fatty acid residues bound via glycerol to the SH-group of the N-terminal cysteine of the protein

bacteriopheophytins (BPheo). The corresponding distances and angles are 1.1 nm and 64°, respectively. Phytol residues participate in the formation of the contact. Menaquinone is located 1.8 nm below the BPheo molecule that is bound to the L-subunit. It lies on the *Trp-252* residue of the M-subunit. The symmetric position below another BPheo molecule is most probably occupied by the quinone "head" of CoQ, which is less tightly bound to the reaction center complex than menaquinone (it is released when the complexes are isolated and crystallized). This ubiquinone seems to lie on the *Phe-216* residue of the L-subunit.

Halfway between the menaquinone and the ubiquinone pocket is a nonheme iron. It is connected with the protein in a manner other than that in other nonheme iron proteins. Four iron ligands were identified as imidazole groups of histidines of segments D_M, E_M, D_L, and E_L.

2.1 Light-Dependent Cyclic Redox Chain of Purple Bacteria

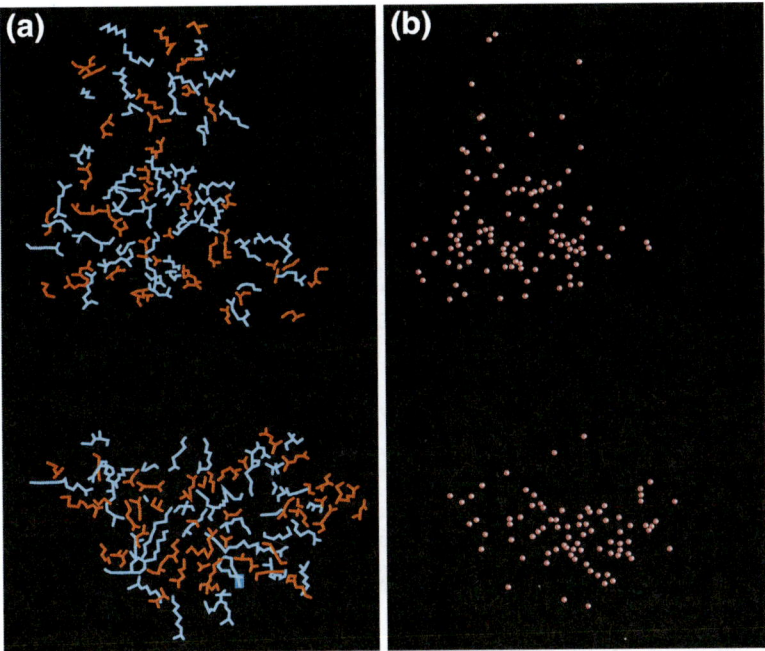

Fig. 2.10 Structure of reaction center complex of *Blastochloris viridis*. **a**—charged amino acid residues in the complex. Positively charged amino acids are orange, and negatively charged ones are blue. **b**—bound water molecules in the reaction center complex of *Bl. viridis*. The central part of the complex, which is immersed in the membrane, contains neither charged amino acid residues nor water, this fact being responsible for the sharp decrease in dielectric constant and, hence, small dissipation of energy occurring during the transmembrane electron transfer in this part of the complex (Deisenhofer et al. 1984)

Connections between segments A_M and B_M, C_M and D_M, A_L and B_L, and C_L and D_L as well as four of the five initial amino acid residues from the N-terminus of the H-subunit polypeptide are in contact with tetraheme cytochrome c. The *Tyr-L162* residue is located between (BChl)$_2$ and the nearest heme. It is supposed to participate in the electron transfer from this heme to the oxidized (BChl)$_2$. The *Glu-H177* residue, which has access to the CoQ pocket, is presumably involved in H$^+$ transfer from the cytoplasm to CoQ$^{\bullet-}$. Removal of the H-subunit was shown to exert a strong effect on the functioning of the reaction center at the level of CoQ.

According to spectral studies, the bacteriochlorophyll dimer is arranged perpendicularly to the plane of the *Bl. viridis* cytoplasmic membrane. Therefore, it seems highly probable that the long axis of the reaction center complex is arranged across the membrane. The distance between the Mg atoms of the bacteriochlorophyll dimer and the menaquinone is about 3.0 nm, so that an electron, running from the dimer to the quinone, crosses a major part of the hydrophobic barrier of the membrane.

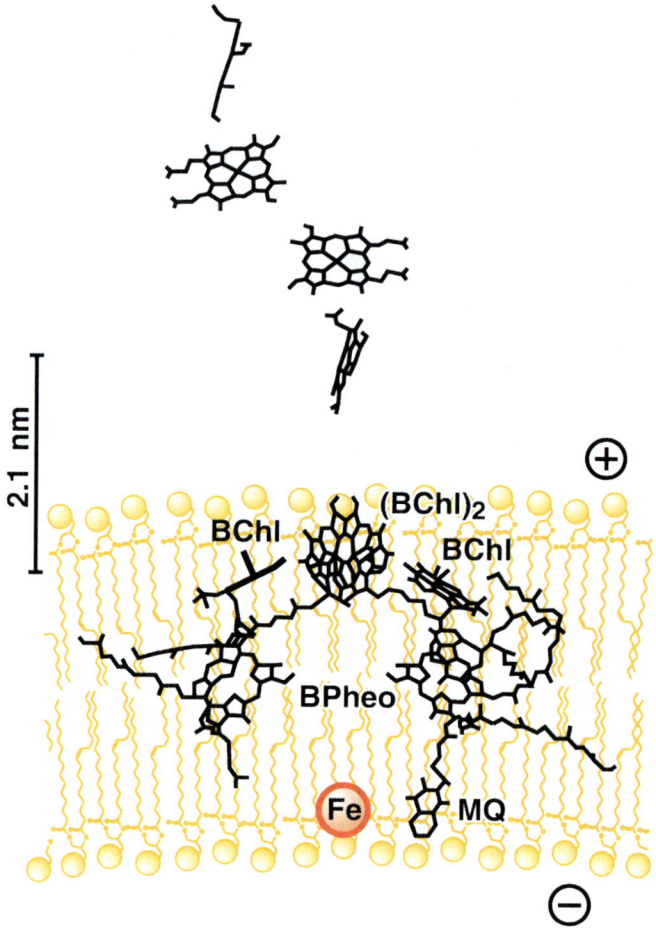

Fig. 2.11 Location of prosthetic groups in the reaction center complex of *Bl. viridis* (X-ray analysis data). Four hemes (above the membrane), bacteriochlorophyll dimer, two bacteriochlorophyll monomers, two bacteriopheophytins, menaquinone (MQ), and nonheme iron (Fe) are shown (Deisenhofer et al. 1984)

In general, the system composed of bacteriochlorophylls, bacteriopheophytins, and quinones resembles two parallel electron-transfer pathways. One can speculate that an electron reducing the bound ubiquinone (CoQ) to semiubiquinone (CoQ$^{•-}$) goes along the right-hand pathway via the menaquinone, whereas the other electron reducing the semiquinone (CoQ$^{•-}$) to ubiquinol passes via the left-hand pathway with no menaquinone involved. Experiments showed, however, that this is not the case—one of the pathways (the left one) does not participate in CoQ reduction (Parson 1982; Robert et al. 1985). When considering the possible function of this pathway, one should take into account the presence of a carotenoid molecule not far from the monomer BChl of the left branch. It is to this molecule

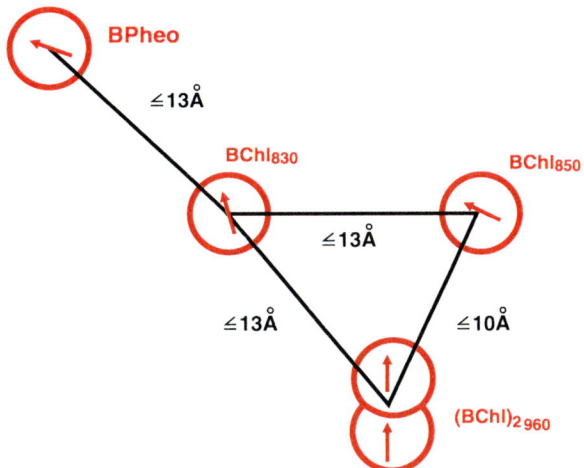

Fig. 2.12 Arrangement of chromophore components of *Bl. viridis* reaction center complex predicted on the basis of spectroscopic observations (Shuvalov and Asadov 1979)

that excitation of (BChl)$_2$ is transferred when (BChl)$_2$ is not oxidized by the right branch and is transformed instead from a singlet to the more stable triplet state. The excitation of the carotenoid is in turn dissipated through heat or emission of a light quantum of a longer wavelength (fluorescence). If oxygen manages to attack and oxidize the excited carotenoid, it leaves the reaction complex so as to be exchanged for a new (intact) carotenoid molecule. In such way the carotenoid molecule "sacrifices itself" to save (BChl)$_2$, BChl, or BPheo, the oxidation of which would lead to irreversible inactivation of the whole complex.

To conclude this section, we note that the X-ray analysis data confirmed the chromophore arrangement scheme (Fig. 2.12) postulated by Shuvalov and Asadov in 1979 on the basis of optical spectroscopy (Shuvalov and Asadov 1979).

Sequence of Electron Transfer Reactions. The X-ray study shows the electron pathway through the reaction center complex to be determined by the spatial arrangement of redox groups:

$$(\text{BChl})_2 \xrightarrow{h\nu} (\text{BChl})_2^* \rightarrow \text{BChl} \rightarrow \text{BPheo} \rightarrow Q_A \rightarrow Q_B \qquad (2.1)$$

where in the case of *Bl. viridis* Q_A is MQ and Q_B is CoQ.

In fact, exactly the same sequence was suggested when picosecond-laser flash-induced changes in the light absorption of (BChl)$_2$, BChl, and BPheo were measured. The flash-induced excitation of (BChl)$_2$ was shown to be followed by oxidation of (BChl)$_2$ and reduction of BChl and BPheo. All these events are completed within 20 ps (Shuvalov and Duysens 1986; Shuvalov et al. 1986). Reduction of menaquinone by BPheo is the next step. It occurs with $t_{1/2}$ of about 230 ps. In the case of *Bl. viridis*, the $t_{1/2}$ for (BChl)$_2^{\bullet+}$ reduction to (BChl)$_2$ is about 270 ns (Fig. 2.13) (Matveetz et al. 1987).

It is crucial that the redox potential of a bacteriochlorophyll dimer in the nonexcited state (+440 mV) is substantially more positive when compared to the monomer potential (about −900 mV). The excitation of the dimer levels this

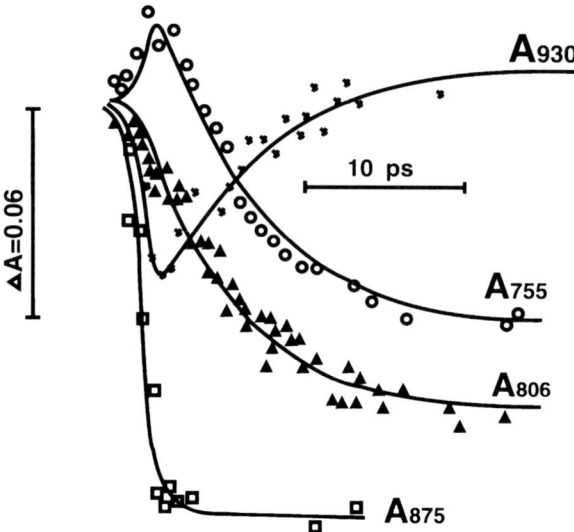

Fig. 2.13 Fast kinetics of light absorption changes (ΔA) in modified *Rhodobacter sphaeroides* reaction center complexes lacking one of two bacteriochlorophyll monomers ($BChl_M$). Absorption decrease at 930 nm (A_{930}) indicates formation of the excited bacteriochlorophyll dimer $(BChl)_2^*$; A_{875} decrease indicates formation of $(BChl)_2^+$. A_{805} decrease, reduction of bacteriochlorophyll monomer, $(BChl)_L$. A_{755} decrease, reduction of bacteriopheophytin, BPheo. Initial A_{775} increase is a result of the $(BChl)_2 \rightarrow (BChl)_2^*$ transition. Excitation was induced by a laser flash ($t_{1/2} = 0.2$ ps). (Matveetz et al. 1986)

difference. Its potential is about −950 mV, i.e. even slightly more negative than in the case of the nonexcited monomer, and even more so when compared to non-excited bacteriopheophytin (about −750 mV).

As shown in the group of one of this book's authors (V.P.S.), the heme of the tetraheme cytochrome *c*, serving as a $(BChl)_2^{\bullet+}$ reductant, is characterized by an α-band at 559 nm and a redox potential of about +380 mV (Dracheva et al. 1986). One can speculate that this heme is located just above $(BChl)_2$ in the reaction center complex. $Heme_{559}$ is, in turn, reduced by $heme_{556}$ (whose redox potential is about 310 mV). The electron donor for $heme_{556}$ is most probably water-soluble cytochrome c_2. The role of the two other hemes of the tetraheme cytochrome *c* (redox potentials +30 mV and −50 mV) remains obscure.

The most remarkable feature of the primary processes of photosynthetic electron transfer is their extremely fast rates. These processes represent *the fastest chemical reactions known in biological systems thus far*. As a result, the first intermediates of the photoredox chain are very short-lived, which is needed to prevent their oxidation by oxygen.

In *Bl. viridis*, anion-radical $MQ^{\bullet-}$ is the first intermediate that has a turnover rate comparable to that of the intermediates of the majority of enzymatic processes. Oxidation of $MQ^{\bullet-}$ by bound CoQ takes about 0.1 ms. Therefore, MQ is often described as a *primary electron acceptor*. In some bacteria, this role is played not by MQ, but also by CoQ. In these cases the reaction center complex contains two molecules of CoQ—the primary acceptor CoQ_A and the secondary acceptor CoQ_B.

Electron transfer between the two quinones is facilitated by the presence of a nonheme iron that does not undergo oxidoreduction and persists in the Fe^{2+} form. In fact, Mn^{2+} can effectively substitute for Fe^{2+} in the reaction center complex (Okamura et al. 1982).

From one point of view, when looking at the mechanism of electron transfer in the reaction center complex, $CoQ_B^{\bullet-}$ can be considered the final product of the operation of the complex:

$$Q_A^{\bullet-} + Q_B \rightarrow Q_A + Q_B^{\bullet-} \tag{2.2}$$

More probable, however, is that not only Q_B, but also $Q_B^{\bullet-}$ is reduced by $Q_A^{\bullet-}$ (see Eq. 2.3). Supporting this version, one should note that the affinity of $Q_B^{\bullet-}$ for the L-subunit of the photosynthetic reaction center complex is much higher than that of Q_B and Q_BH_2.

Reduction of Q_B is accompanied by addition of two protons:

$$Q_A^{\bullet-} + Q_B^{\bullet-} + 2\ H^+ \rightarrow Q_A + Q_BH_2 \tag{2.3}$$

Mechanism of $\Delta\bar{\mu}_{H^+}$ Generation. Because *cytochrome c_2 and quinone are located on the opposite sides of the membrane* (Okamura et al. 1982; Deisenhofer et al. 1984), the transfer of reducing equivalents is directed through the membrane to its cytoplasmic surface. Assuming that it is an electron that moves from cytochrome c_2 to quinone, one can predict that the cytoplasm should be charged negatively relative to the periplasm or the interior of the chromatophore.

This prediction was confirmed by a series of observations concerning $\Delta\Psi$ generation by the reaction center complexes. To monitor $\Delta\Psi$, L. Drachev, A. Kaulen, and A. Semenov in our group developed a method that made it possible to *carry out direct voltmeter measurements of $\Delta\Psi$ generated by enzymes built into proteoliposomes* (Drachev et al. 1974, 1979). The term "proteoliposome" was suggested by one of the authors of this book (V.P.S.) in 1972 for closed membrane vesicles that could be obtained as a result of self-assembly from phospholipids and proteins (Skulachev 1972; Kayushin and Skulachev 1974). In these experiments, proteoliposomes were sorbed on one of the surfaces of a collodion film that was soaked with a solution of phospholipids in decane.

The experiments showed that proteoliposomes are sorbed in such a way that their inner solution does not mix with the outer one (Severina 1982). It is likely that during sorption a part of the proteoliposome membrane that is in contact with the film is dissolved by the decane the film is soaked with, while other parts of the membrane remain intact, thus forming a barrier between the inner solution of the film-attached proteoliposome and the outer solution surrounding the film.

The generation of $\Delta\Psi$ on the membrane of sorbed proteoliposomes can be registered by a voltmeter by using two electrodes that are placed on the two sides of the collodion film. The time resolution of this system is about 50 ns, which is

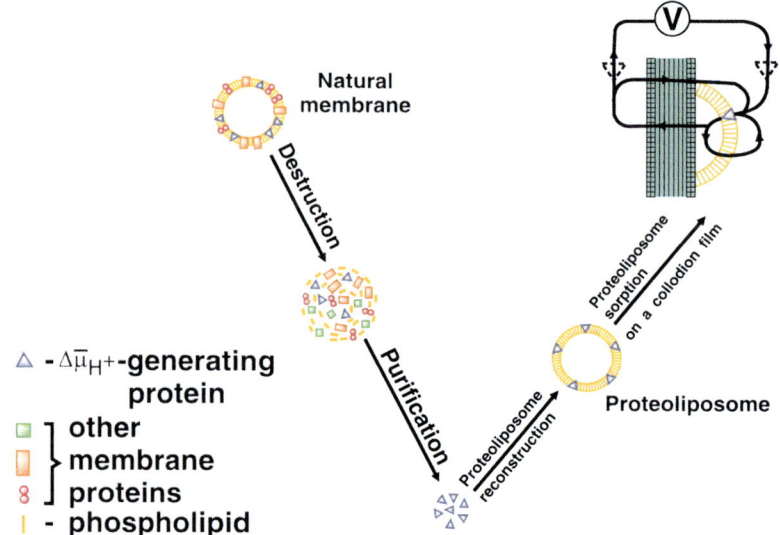

Fig. 2.14 General scheme of studies on $\Delta\bar{\mu}_{H^+}$-generating proteins: purification, self-assembly, and direct measurement of their electrogenic activity (Skulachev 1988)

much shorter than a single turnover rate of any $\Delta\bar{\mu}_{H^+}$. So this method can be used not only to register very fast $\Delta\Psi$ generation, but also to resolve its separate stages. In other words, it is possible to register transfer of a charge (e.g. an electron or a proton) inside the molecule of a protein generating $\Delta\bar{\mu}_{H^+}$.

Figure 2.14 presents a general scheme of the study of $\Delta\bar{\mu}_{H^+}$—generating proteins. It includes their purification, self-assembly, and finally direct measurement of the function of these proteins—transformation of light or chemical energy into electrical form.

Our method made it possible to measure $\Delta\Psi$ generation by reaction centers on nanosecond and microsecond time scales. The main contribution to $\Delta\Psi$ generation was found to be a process that is faster than 50 ns (the time resolution of the measuring system). In the reaction sequence, such fast kinetics is inherent in steps between $(BChl)_2$ and Q_A. Combining this observation with the transmembrane arrangement of the , one can infer that the transfer of reducing equivalents between the $(BChl)_2$ and the quinone is electrogenic, i.e. the electron transfer is not compensated by the movement of other charged component(s) (Dracheva et al. 1986).

The superfast monitoring of $\Delta\Psi$ generation by the *Bl. viridis* cells by electric measurements of the light-gradient type was done by H. Trissl and coworkers (Deprez et al. 1986). The $\Delta\Psi$ measured by this indirect method is lower in magnitude than in our experiments by a factor of at least 100, but the time resolution appears to be as good as 40 ps. Two electrogenic phases ($t_{1/2} \leq 40$ ps, $t_{1/2} = 125$ ps) of almost equal contributions were observed. The first phase corresponded to BPheo reduction by $(BChl)_2$ and the second to BPheo oxidation by menaquinone.

In addition to the fast electrogenic phases, we discovered three slower phases (Dracheva et al. 1986). One appeared when the decane solution of phospholipids used to impregnate the film was supplemented with CoQ_{10}. The magnitude of the phase was about 5 % of the overall $\Delta\Psi$ generated after a laser flash. The time constant of this process was 400 µs, i.e. somewhat slower than the Q_B reduction rate. It was sensitive to o-phenanthroline, an inhibitor of Q_B reduction. We concluded that the slow electrogenic phase is somehow associated with reduction of Q_B. Decane apparently extracts Q_B from the chromatophores or proteoliposomes attached to the collodion film saturated with this hydrocarbon, the effect being prevented by added CoQ_{10}.

There are two possible mechanisms of this slow electrogenic phase that requires CoQ_{10}. (1) Electron transfer from Q_A to Q_B is electrogenic. (2) There is a proton-conducting path from the membrane surface to Q_B—protons move along this path to the reduced anion Q_B and combine with this anion, thus forming QH_2.

To discriminate between these two possibilities, we studied the effect of two consecutive 15-ns laser flashes. If the first version were correct, both flashes would be effective in slow electrogenesis. If the second mechanism were operative, only the second flash would be effective since the $CoQ^{\bullet-}$ formed after the first flash cannot bind H^+ at physiological pH. It is only after the addition of the second electron (which can happen after the second laser flash) that protonation can occur.

As shown in Fig. 2.15, the second flash, but not the first, induced the slow electrogenic phase. This means that version (2) is valid, i.e. the slow electrogenic phase is due to H^+ transfer from water to the place inside the reaction center complex where Q_B is located. If we presume that the dielectric properties of the membrane hydrophobic barrier are the same throughout its entire depth, it becomes possible to define what fraction of this barrier is crossed by H^+. Since the slow stage constitutes 5 % of the entire electric response caused by transmembrane charges transfer, one might assume that the proton is transferred by 1/20 part of the membrane depth. Further research, however, proved the real situation to be more complicated (we will discuss it later).

To study the contribution of the tetraheme cytochrome c, conditions were used when two hemes (c_{556} and c_{559}) were reduced. No CoQ was added to avoid electrogenic effects accompanying formation of $CoQH_2$. A laser flash that photooxidized $(BChl)_2$ was found to induce temporary oxidation of these hemes ($t_{1/2}$ of about 300 ns for heme c_{559} and about 3 µs for heme c_{556}). This process was accompanied by an additional biphasic electrogenesis. The kinetics of each of the phases corresponded to the kinetics of heme oxidation. The contributions of these fast and slow cytochrome electrogenic phases to the overall $\Delta\Psi$ generation by the reaction center complex were about 15 and 5 %, respectively (Fig. 2.16).

A scheme illustrating the structure–function relationships in the *Bl. viridis* reaction centers is shown in Fig. 2.17. The distances between redox groups are given according to the X-ray data. It is interesting that H^+ transfer from the cytoplasmic membrane surface to CoQ constitutes only 5 % of the overall $\Delta\Psi$. There are two possible explanations for this fact—either the proton path (from water to CoQ) is shorter than the path of an electron (from cytochrome c to CoQ),

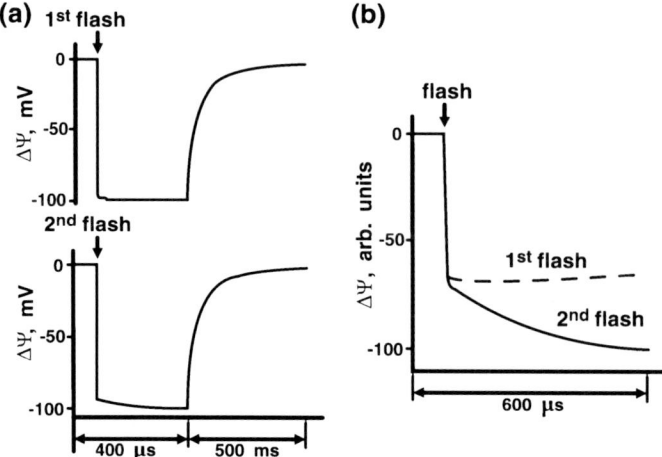

Fig. 2.15 Generation of $\Delta\Psi$ by *Bl. viridis* reaction center complexes in response to two consecutive 15-ns laser flashes. The time between flashes was 0.5 s. Only the second flash generates a "slow" electrogenic phase developing on the microsecond time scale (**a**). This phase is shown at higher resolution in (**b**), where dashed and solid lines represent the first and second flashes, respectively. $\Delta\Psi$ was measured with the proteoliposome-collodion film system. CoQ_{10} was added to the solution used to impregnate the film. The redox potential of the medium was +240 mV. (Dracheva et al. 1986)

or H^+ ions are transferred through a less hydrophobic part of the complex. Most likely both factors contribute to this phenomenon. It seems quite probable that a similar situation occurs during the stages of electron transfer.

The distances between Q_A and $(BChl)_2$, $(BChl)_2$ and c_{559}, and c_{559} and c_{556} along the axis vertical to the membrane plane are 2.9, 2.1, and 1.2 nm, respectively. If the electrogenesis were a linear function of the distance between the redox groups involved, the relative contributions of these electron transfer steps would be 1:0.7:0.4. In fact, they were found to be 1:0.2:0.7. This means that the contribution is greater when the electron transfer stage is immersed deeper into the membrane. This pattern becomes quite understandable when we take into account the fact that the membrane dielectric constant value increases approaching the water phase (Skulachev 1988).

In the 1960s, when Peter Mitchell formulated his chemiosmotic hypothesis, he suggested that an electron moving from cytochrome *c* to CoQ via bacteriochlorophyll crosses the hydrophobic membrane barrier (*the electron transfer half-loop*) (Mitchell 1966). This assumption, quite speculative at that time, has now been directly proved. The only new point that was detected in the result of experimental trial of the original Mitchell scheme is that besides the electron transfer across the membrane, there is a small but measurable electrogenic phase arising because of the movement of the proton in the opposite direction (this movement being necessary for formation of $CoQH_2$ from CoQ) (Skulachev 1988).

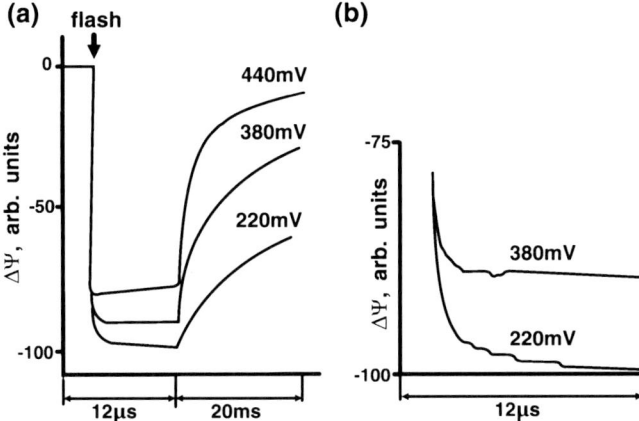

Fig. 2.16 Generation of $\Delta\Psi$ by *Bl. viridis* reaction center complexes—the role of high-potential hemes of cytochrome *c*. Measurements were carried out at redox potential of the medium equal to +220 mV (two high-potential hemes of the tetraheme cytochrome *c* were reduced and two low-potential hemes were oxidized), +380 mV (one of high-potential hemes was 50 % reduced), and +440 mV (all four hemes were oxidized). The figure to the right (b) shows part of figure (a) in more detail. (Dracheva et al. 1986)

2.1.3 CoQH$_2$: Cytochrome c-Oxidoreductase

CoQH$_2$: cytochrome *c*-oxidoreductase (other names, Complex III and bc_1 complex) was discovered and studied in detail when the mitochondrial respiratory chain was investigated (see Sect. 4.3). Later, it was described in photoredox chains. There are still scanty information on the molecular properties of the constituents of the bc_1 complex in photosynthetic bacteria. However, almost all that is known about it is in agreement with that previously reported for its mitochondrial analog. That is why in this section only a short description of the bacterial bc_1 complex will be given.

The bc_1 complex comprises cytochrome *b* containing two hemes—the low-potential heme b_L (another name b_{556}, since its α-peak in the absorption spectrum is located at 556 nm) and the high potential heme b_H (or b_{560}), one iron–sulfur center (FeS$_{III}$), one cytochrome c_1, and several colorless protein subunits without any prosthetic groups. CoQH$_2$ and cytochrome c_2 serve as the reductant and the oxidant of the bc_1 complex, respectively. Cytochrome c_2 is a water-soluble cytochrome similar to mitochondrial cytochrome *c*. In reconstituted systems, mitochondrial cytochrome *c* can substitute for cytochrome c_2 (see Fig. 2.6 for the potentials of these redox carriers).

The CoQH$_2$:cytochrome *c*-oxidoreductase complex is *the slowest component of the photosynthetic redox chain*. Its maximal turnover in situ is generally around once per 10 ms (Rich 1984).

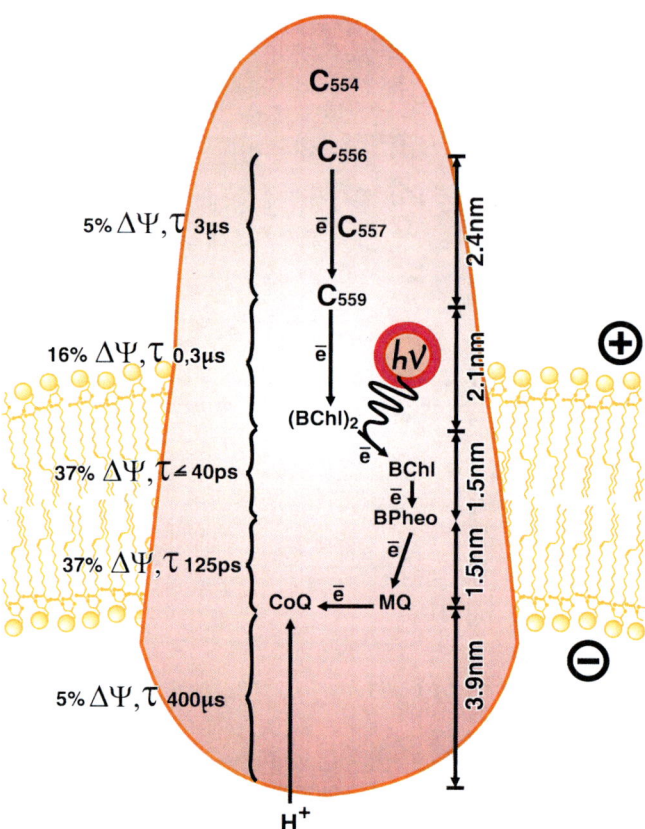

Fig. 2.17 Mechanism of electrogenic phases in *Bl. viridis* reaction center complex (Dracheva et al. 1986; Skulachev 1988)

By analogy with the more elaborate mitochondrial complex, one can suppose that $\Delta\bar{\mu}_{H^+}$ generation by this system is partially due to the electron transfer from heme b_L to b_H, directed perpendicularly to the membrane plane, as shown in Fig. 2.4. This process is oriented in the direction of the cytoplasmic surface of a bacterial membrane. An electron was shown to cross 35–40 % of the membrane thickness while moving from heme b_L to b_H (Glaser and Crofts 1984). The rest is assumed to be crossed by H^+ ions moving in the direction opposite to that of an electron (see Fig. 2.4).

It remains unclear whether CoQ, CoQ$^{\bullet-}$, or both serve as electron acceptor(s) for cytochrome b_H. If this function is specific only for one of these components, it means that formation of a completely reduced form of ubihydroquinone (CoQH$_2$) must involve the dismutation reaction (2 CoQ$^{\bullet-}$ + 2 H$^+$ ↔ CoQ + CoQH$_2$). The reaction center complex is another source of electrons for CoQ reduction to CoQH$_2$. Steady-state functioning of the system requires that for each two CoQ

2.1 Light-Dependent Cyclic Redox Chain of Purple Bacteria

molecules reduced to $CoQH_2$, one would be reduced by the reaction center and the other by b_H heme. Protons required to convert CoQ to $CoQH_2$ are in both cases transported from the cytoplasm.

Having been formed, $CoQH_2$ crosses the greater part of the hydrophobic membrane barrier to be oxidized by FeS_{III}. As a result, $CoQ^{\bullet-}$, reduced FeS_{III}, and 2 H^+ are formed, the latter being transported through the remainder of the hydrophobic barrier to be released to the periplasm or chromatophore interior. In the case of *Rh. rubrum* and *Rhodobacter sphaeroides*, electron transfer from FeS_{III} to $(BChl)_2$ via cytochrome c_1 and cytochrome c_2 completes the cycle, as shown in Fig. 2.4. In *Bl. viridis*, tetraheme cytochrome c participates in the electron transfer between cytochrome c_2 and $(BChl)_2^{\bullet+}$.

The reaction sequence catalyzed by the $CoQH_2$:cytochrome c-oxidoreductase, as shown in Fig. 2.4, represents, in fact, a version of the Q-cycle scheme postulated by Mitchell (Mitchell 1975). The key postulate of this scheme was that $CoQH_2$ oxidation and CoQ reduction occur in such a manner that the fate of each of the two electrons removed from $CoQH_2$ or added to CoQ appears to be different. Now this suggestion is firmly established at least for $CoQH_2$ oxidation. *FeS_{III} was shown to act as a $CoQH_2$ oxidase, whereas heme b_L of cytochrome b was shown to serve as an oxidase of $CoQ^{\bullet-}$*, the latter being formed from $CoQH_2$. There is a chemical reason for the fact that $CoQH_2$ and $CoQ^{\bullet-}$ are oxidized by different enzymes—the redox potentials of hydroquinone and semiquinone are quite different (Rich 1984).

It is noteworthy that the binding of $CoQH_2$ to the FeS_{III} protein proceeds in such a way that the hydroquinone "head" of $CoQH_2$ is bound near a positively charged group (presumably Fe^{3+}). This induces an acidic shift of the pK value of $CoQH_2$, which makes its deprotonation easier. Having lost a proton, $CoQH_2$ changes into $CoQH^-$, which is oxidized by FeS_{III}. Deprotonation seems to be necessary for this oxidation reaction since the redox potential of *the $CoQH_2$/ $CoQH_2^{\bullet+}$* couple is higher than +850 mV, while that of $CoQH^-/CoQH^{\bullet}$ couple is +190 mV (Rich 1984), which is more negative than that of FeS_{III}.

In mitochondria, electron transfer from ubiquinol to *FeS_{III}* and then via cytochrome c_1 to cytochrome c is catalyzed by respiratory chain complex III, which is also known as bc_1 complex or ubiquinol:cytochrome c-oxidoreductase. This complex has been found in a number of bacteria; in many respects it is similar to plastoquinol:plastocyanin-oxidoreductase of thylakoids (*bf* complex). Redox groups of the bc_1 complex include an [2Fe − 2S] cluster, which is part of a so-called Rieske protein, two hemes of a b-type cytochrome and a cytochrome c_1 heme. The [2Fe–2S] cluster of the Rieske protein is connected to the polypeptide by the bonds of one iron atom to two cysteine residues and of the other iron atom to two histidine residues. Thus, the FeS cluster is part of the inner globular structure of the polypeptide chain that is anchored in the membrane with its hydrophobic N-terminal helix and is protruded from the lipid bilayer into the water phase.

Cytochrome c_1 has a globular domain similar to the Rieske protein and a hydrophobic anchor that is located on the C-terminus of the protein. The cytochrome b subunit has eight transmembrane α-helices. Four conserved histidine residues are arranged in pairs on each of two helices so as to serve as axial ligands to the heme (Crofts and Wraight 1983; Rich 1984).

2.1.4 Ways to Use $\Delta\bar{\mu}_{H^+}$ Generated by the Cyclic Photoredox Chain

The proton potential generated in chromatophores by the cyclic redox chain can be used to perform *four types of chemical work*: (1) *ATP formation* from ADP and inorganic phosphate by H^+- ATP synthase; (2) *formation of inorganic pyrophosphate* from two molecules of inorganic phosphate by H^+-pyrophosphate synthase; (3) *reverse transfer of reducing equivalents* via the NADH:CoQ-oxidoreductase complex from the hydrogen donors of about a zero redox potential to NAD^+; and (4) reduction of $NADP^+$ by NADH catalyzed by *transhydrogenase*.

NADH:CoQ-oxidoreductase is the first (of three) $\Delta\bar{\mu}_{H^+}$ generators in the respiratory chain. When operating in the opposite direction ($CoQH_2 \rightarrow NAD^+$), this enzyme consumes $\Delta\bar{\mu}_{H^+}$ produced by the photosynthetic redox chain in chromatophores. This makes it possible to reduce NAD^+ (redox potential − 320 mV) by such hydrogen donors as H_2S (redox potential about 0 mV). The formed NADH is used later in reductive biosyntheses.

One of the ways to utilize NADH is to reduce $NADP^+$ in a transhydrogenase reaction. The $\Delta\bar{\mu}_{H^+}$-consuming enzyme transhydrogenase is also located in the chromatophore membrane. When consuming $\Delta\bar{\mu}_{H^+}$, transhydrogenase strongly increases the redox potential of the $NADPH/NADP^+$ couple, an effect favorable for those reductive syntheses that use NADPH as a hydrogen donor.

In the dark, NADH:CoQ-oxidoreductase can operate in the forward direction oxidizing NADH by CoQ, a process coupled to $\Delta\bar{\mu}_{H^+}$ generation. The formed $CoQH_2$ is further oxidized by the bc_1 complex which reduces cytochrome c_2 and produces $\Delta\bar{\mu}_{H^+}$. Cytochrome c_2 in turn reduces cytochrome o, which transfers electrons to O_2 and generates $\Delta\bar{\mu}_{H^+}$. Thus, purple photosynthetic bacteria have a respiratory chain with three $\Delta\bar{\mu}_{H^+}$ generators, and it starts to operate when light is unavailable.

The cytoplasmic membrane of photosynthetic bacteria has some additional $\Delta\bar{\mu}_{H^+}$ consumers that are absent from chromatophores, i.e.: (1) H^+ motors that rotate bacterial flagella; (2) H^+ that allow accumulation of metabolites in the bacterial cell; (3) Na^+, K^+-transport systems responsible for the asymmetrical distribution of these ions across the cytoplasmic membrane. Gradients of Na^+ and K^+ ions function as a $\Delta\bar{\mu}_{H^+}$-buffer. A similar function seems to be served by pyrophosphate.

A general scheme of light energy transduction to various types of work in the cytoplasmic membrane of purple bacteria is shown in Fig. 2.18.

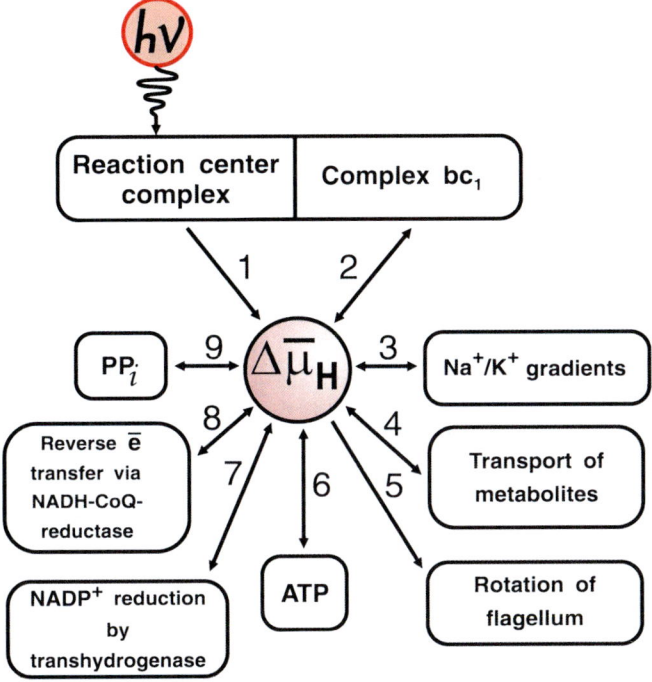

Fig. 2.18 Light-dependent energy transductions in the cytoplasmic membrane of purple photosynthetic bacteria (1–9). In the chromatophore membrane of the same bacteria, a simplified energy transduction pattern occurs when pathways 4, 5, and possibly 3 are absent (Skulachev 1988)

2.2 Noncyclic Photoredox Chain of Green Bacteria

In the preceding section it was mentioned that purple photosynthetic bacteria can reduce NAD^+ by H_2S at the expense of light energy converted to $\Delta\bar{\mu}_{H^+}$ by the cyclic redox chain. Another mechanism is employed by green anaerobic photosynthetic bacteria. It is organized such that *the light simultaneously causes $\Delta\bar{\mu}_{H^+}$ generation and NAD^+ reduction by hydrogen sulfide.*

This mechanism is the simplest version of the so-called *noncyclic photosynthetic redox chain*. Its tentative scheme is shown in Fig. 2.19. In this case the reaction center complex is similar to the photosystem 1 complex of chloroplasts and cyanobacteria (this chapter, Sect. 2.3.2), but the core of the reaction center consists of a homodimer of one subunit instead of a heterodimer of two homologous subunits in the case of photosystem 1. Bacteriochlorophyll dimer, two bacteriochlorophyll monomers, vitamin K_1 (phylloquinone or menaquinone), and three FeS clusters (FeS_X, FeS_A, and FeS_B) that are included sequentially in the redox chain that constitutes the reaction center complex. The water-soluble FeS protein ferredoxin (Fd), which is located on the cytoplasmic surface of the

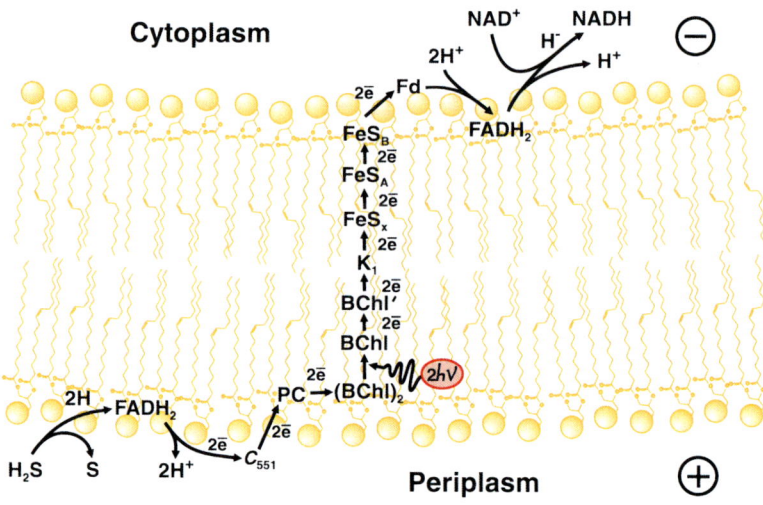

Fig. 2.19 Noncyclic photosynthetic redox chain of green sulfur bacteria (Hauska et al. 2001). Explanations are provided in the text

membrane, serves as an electron acceptor. Electrons are transferred from Fd to a flavoprotein (with FAD prosthetic group) and then to NAD$^+$.

Photooxidized $(BChl)_2^{\bullet +}$ is reduced by the donor segment of the redox chain starting from H$_2$S. The segment contains flavoprotein with FAD as a prosthetic group, cytochrome c_{551}, and Cu-containing protein plastocyanin (PC). The redox potentials of the main components of this redox chain have already been shown in Fig. 2.6b.

The photoelectrogenic activity of the redox chain of green bacteria was registered in experiments with membrane fragments associated with a planar phospholipid membrane. The mechanism of this effect consists, as in other chlorophyll-containing $\Delta\bar{\mu}_{H^+}$ generators, of a transmembrane electron transfer. The electron transfer shown in Fig. 2.19 must be accompanied by H$^+$ release to the outer medium due to the oxidation of H$_2$S to sulfur by the electron donor segment of the redox chain, which is located close to the outer surface of the membrane. On the opposite (inner, i.e. cytoplasmic) side of the membrane, H$^+$ uptake should occur, since NAD$^+$ is reduced by the bacteriochlorophyll system that can donate electrons—not protons. To form NADH from NAD$^+$, one H$^+$ per two electrons accepted by NAD$^+$ must be consumed.

Thus, on the *outer membrane surface* of green bacteria the following reaction takes place:

$$H_2S \rightarrow S + 2\,H^+ + 2\bar{e}, \quad (2.4)$$

while the process occurring on the *inner membrane surface* can be described as:

$$NAD^+ + H^+ + 2\bar{e} \rightarrow NADH. \quad (2.5)$$

2.2 Noncyclic Photoredox Chain of Green Bacteria

It should be noted that *both* $\Delta\bar{\mu}_{H^+}$ *generation and NAD^+ reduction are carried out by one and the same light-driven redox chain* (for review, see (Hauska et al. 2001)).

It should be mentioned that the described metabolic pattern (see Fig. 2.19) is characteristic of only *anaerobic* green sulfur bacteria, such as *Chlorobium sps*. In contrast, *facultatively aerobic* green bacteria (e.g. of the *Chloroflexus* genus) have a redox chain more similar to that found in the purple bacteria (for review see (Merchant and Sawaya 2005)).

2.3 Noncyclic Photoredox Chain of Chloroplasts and Cyanobacteria

The noncyclic photosynthetic redox chain of cyanobacteria and chloroplasts resembles that of the green bacteria. It also utilizes light to form $\Delta\bar{\mu}_{H^+}$ and to reduce $NAD(P)^+$ by an electron donor of a more positive redox potential. The main specific feature of cyanobacteria and chloroplasts compared to green bacteria is that water (instead of hydrogen sulfide) is utilized as the electron donor. The redox potential difference between $NADP^+$ and H_2O is about 1.2 V, i.e. much larger than that between NAD^+ and H_2S used by the green bacteria (0.3 V). Therefore, the transport of one electron along the chloroplast (cyanobacterial) redox chain requires two photons to be consumed, instead of one photon as in the green bacteria.

In plant cells the photosynthetic redox chain is located in the intrachloroplast membranes, mainly in *thylakoids*, i.e. closed, flattened membrane vesicles stacked in batches called *grana* that can be connected by extended membrane structures (lamellas of the chloroplast stroma) (Fig. 2.20). Thylakoids are also found in many cyanobacteria.

2.3.1 Principle of Functioning

A simplified version of the light-dependent redox chain in the thylakoid membrane of chloroplasts is given in Fig. 2.21. The chain includes two types of reaction centers (they are called photosystems 1 and photosystems 2) and several enzymes catalyzing the dark reactions of water oxidation, Q-cycle, and $NADP^+$ reduction.

The process is initiated by absorption of a photon by antenna chlorophyll. Excitation migrates from the antenna chlorophyll to the chlorophyll of a photosystem. If this is the *chlorophyll dimer of photosystem 2* (absorption maximum at 680 nm), it is oxidized by a monomer of chlorophyll II; an electron then moves to *pheophytin*. From pheophytin the electron is transferred to the bound plastoquinone (PQ_A). Then the electron reaches another bound plastoquinone molecule (PQ_B). Similarly to bacterial photosynthesis, this process is catalyzed by nonheme

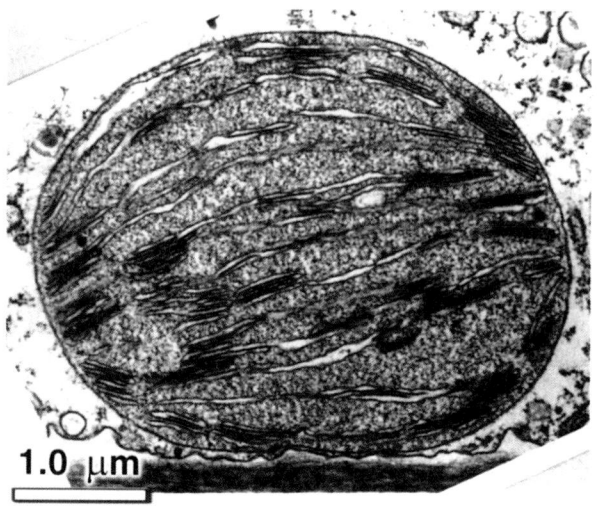

Fig. 2.20 Structure of chloroplast

iron connected to four imidazoles of histidines residues of the protein. The chlorophyll cation radical $(ChlII)_2^{\bullet+}$ accepts an electron from one of the tyrosine residues of the protein, which, in turn, is reduced by an electron removed from a H_2O molecule by means of a rather complicated *manganese-containing water-oxidizing complex* (WOC). Electrons, molecular oxygen, and protons are released when the water molecule is oxidized. The released protons appear in the intrathylakoid space. The reduced PQ is oxidized by the PQH_2:plastocyanin-oxidoreductase (also called b_6f complex, which is a chloroplast analog of the bacterial and mitochondrial bc_1 complex). The b_6f complex comprises a three-heme cytochrome b (hemes b_L, b_H, and c_i), FeS_{III}, and cytochrome f, which is functionally similar to cytochrome c_1.

The b_6f complex catalyzes a Q-cycle similar to that described in purple bacteria and mitochondria. As a result, the transport of one electron from PQ to cytochrome f is coupled to the translocation of one charge across the thylakoid membrane, the absorption of two H^+ ions from the chloroplast stroma, and the release of two H^+ ions into the thylakoid interior. These protons are taken up on the outer surface of the thylakoid membrane and are released on its inner surface.

From cytochrome f the electron moves to the copper-containing protein plastocyanin (Pc), which reduces the cation radical of the chlorophyll dimer of photosystem 1 (in the ground state, the absorption maximum for this chlorophyll is near 700 nm). The $(ChlI)_2^{\bullet+}$ radical cation is formed as a result of the oxidation of the excited $(ChlI)_2^*$. The electron taken from $(ChlI)_2^*$ is transferred via the chlorophyll monomers (ChlI and ChlI′) and vitamin K_1 (phylloquinone) to the *iron–sulfur center* FeS_X, which acts as the primary stable electron acceptor in photosystem 1.

During the next stages the electron is transported to other iron–sulfur clusters (FeS_B, FeS_A, and ferredoxin) and further to a flavoprotein containing FAD as the

2.3 Noncyclic Photoredox Chain of Chloroplasts and Cyanobacteria

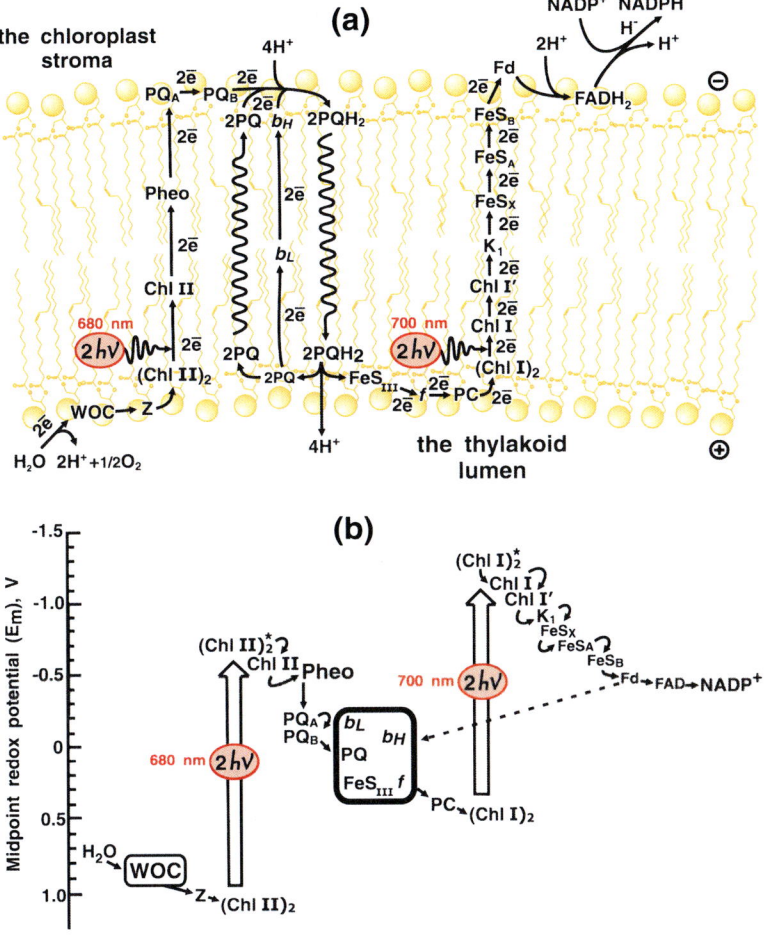

Fig. 2.21 Noncyclic light-dependent redox chain of chloroplasts (Blankenship and Prince 1985): WOC—water-oxidizing complex; *yellow*, electron-transporting prosthetic groups; Z—tyrosine residue in a protein of photosystem 2; (ChlII)$_2$—dimer of photosystem 2 chlorophyll; ChlII— monomer of photosystem 2 chlorophyll; PQ$_A$ and PQ$_B$—plastoquinones bound to photosystem 2; PQ—free plastoquinone; b_H and b_L—high- and low-potential hemes of cytochrome b_6; f—cytochrome f; PC—plastocyanin; (ChlI)$_2$—dimer of photosystem 1 chlorophyll; ChlI and ChlII- monomers of photosystem 1 chlorophylls; K$_I$—vitamin K$_1$; FeS$_X$—primary stable electron acceptor of photosystem 1; FeS$_A$ and FeS$_B$—two iron-sulfur clusters tightly bound to photosystem 1; Fd—ferredoxin. **a** Diagram illustrating the most probable arrangement of redox centers in the thylakoid membrane. **b** Reducing equivalent carriers are arranged according to their redox potentials; the dotted line shows the electron transfer pathway from ferredoxin to b_6f complex that provides cyclic electron transfer in the photoredox chain

prosthetic group. From FAD, the reducing equivalents come to NADP$^+$. The formed NADPH is oxidized by a water-soluble enzyme system reducing CO$_2$ to carbohydrates.

Within the framework of the scheme in Fig. 2.21, the noncyclic redox chain of chloroplasts and cyanobacteria carries out transmembrane movement of 6 H^+ per 4 quanta. This corresponds to the $H^+/\bar{e}$ ratio of 1.5 (the ratio of the number of H^+ ions transported across the membrane to the number of electrons transferred from H_2O to CO_2). *Thus, there are three $\Delta\bar{\mu}_{H^+}$ generators in the thylakoid redox chain—photosystem 1, photosystem 2, and the b_6f complex.*

2.3.2 Photosystem 1

Chloroplast photosystem 1 contains 14 different polypeptides. Two of them with molecular masses of 83.2 and 82.5 kDa show 45 % similarity (percentage of identical amino acid residues in the sequences). Other subunits are of substantially lower mass (for review see (Okamura et al. 1982; Crofts and Wraight 1983; Nelson and Ben-Shem 2004; Nelson 2011)).

(ChlI)$_2$, ChlI, ChlI', vitamin K_1 (phylloquinone), and FeS_X are connected to large subunits, while FeS_A and FeS_B are connected to one of the small hydrophobic subunits. In contrast to photosystem 1 of cyanobacteria, four proteins carrying antenna chlorophylls (Lhca1-4) are integrated into chloroplast photosystem 1. When looking at the thylakoid membrane from the normal to the membrane plane, one will note that antenna proteins form a half-circle around the other subunits that constitute the reaction center complex of photosystem 1 (Fig. 2.22). The transfer of excitation energy from the antenna chlorophyll to (ChlI)$_2$ is extremely effective—almost every photon absorbed by the antenna leads to oxidation of (ChlI)$_2$. The copper-containing protein plastocyanin (see below), being part of photosystem 1, is located on the inner side of the thylakoid membrane, being absorbed by membrane surface. Another surface protein, ferredoxin (11 kDa), is located on its outer side. This protein contains a FeS cluster and transfers electrons from photosystem 1 to the flavoprotein that reduces $NADP^+$ (34 kDa).

The photosystem 1 reaction center complexes form trimers in the thylakoid membranes of cyanobacteria. In the case of higher plant chloroplasts, photosystem 1 is in the form of monomers. Both the trimers of cyanobacteria and the monomers of chloroplasts have been purified, crystallized, and studied by X-ray structural analysis. The results of X-ray studies of chloroplasts are shown in Figs. 2.22 and 2.23. Figure 2.23 shows two pathways of electron transfer through (ChlI)$_2^*$ to FeS_X, each containing the same redox groups (ChlI, ChlI', and vitamin K_1). Both pathways are capable of electron transfer, but one of them is ten times faster than the other.

The primary photoprocess in photosystem 1 is the electron shift from (ChlI)$_2^*$ (redox potential -1.2 V) to ChlI (redox potential about -1 V). The electron is then transferred to ChlI' and later to vitamin K_1 (redox potential -0.8 V). FeS_X (-0.7 V) is the next electron carrier. It is a cluster of four iron ions and four sulfide ions. In addition to this component, there are two more [4Fe − 4S] clusters (FeS_A

2.3 Noncyclic Photoredox Chain of Chloroplasts and Cyanobacteria

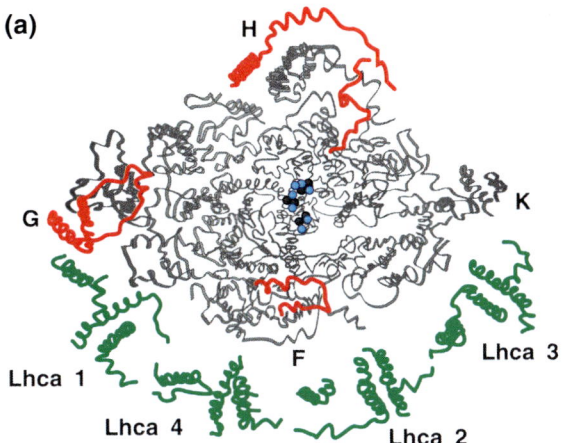

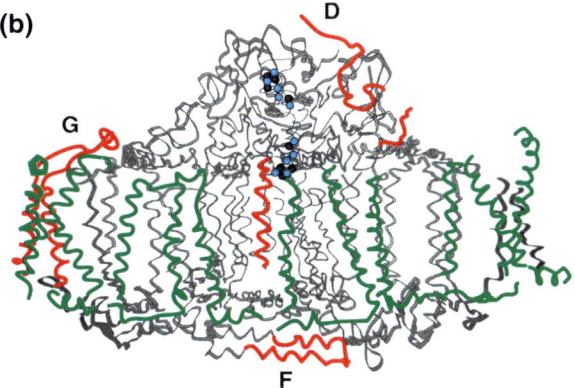

Fig. 2.22 Polypeptide body of photosystem 1 from pea chloroplasts. Four antenna proteins (Lhca 1–4) are shown in green color. Structural elements of chloroplast photosystem 1 that are different from cyanobacterial photosystem 1 are red, and those similar in chloroplasts and cyanobacteria are gray. Iron–sulfur clusters FeS_X, FeS_A, and FeS_B that all share the composition [4Fe-4S] are shown as black (Fe) and blue (S) bold dots. **a** *Top view* from chloroplast stroma side. Subunits F, G, H, and K are shown. **b** *Side view* from antenna protein side. N-terminal domains of F and D subunits that are specific for chloroplasts are in red color. X-ray analysis (Nelson and Ben-Shem 2004)

and FeS_B) in photosystem 1. Their redox potentials are -0.58 and -0.53 V, respectively. The electron transfer sequence in this segment of the redox chain is $FeS_X \rightarrow FeS_A \rightarrow FeS_B$. The latter component serves as the reductant of the water-soluble protein ferredoxin that contains a [2Fe − 2S] cluster.

Another water-soluble electron carrier is the reductant of photooxidized $(ChlI)_2^{\bullet+}$. This is plastocyanin, a copper-containing protein. The primary structure of plastocyanin has sequences homologous to those in subunit II of cytochrome oxidase, the terminal enzyme of the respiratory chain (see Sect. 4.4).

Plastocyanin is reduced by cytochrome f, which is a component of the $b_6 f$ complex. The latter transfers electrons from photosystem 2 to photosystem 1 (see below, Sect. 2.3.3).

There is *an alternative function of the photosynthetic redox chain in chloroplasts, namely cyclic electron transport*, with photosystem 1 and $b_6 f$ complex but not photosystem 2 being involved. This system can be demonstrated in vitro when photosystem 2 is blocked by the herbicide dichlorophenyl dimethylthiourea

Fig. 2.23 Arrangement of prosthetic groups in chloroplast photosystem 1 on the basis of X-ray analysis (Nelson and Ben-Shem 2004)

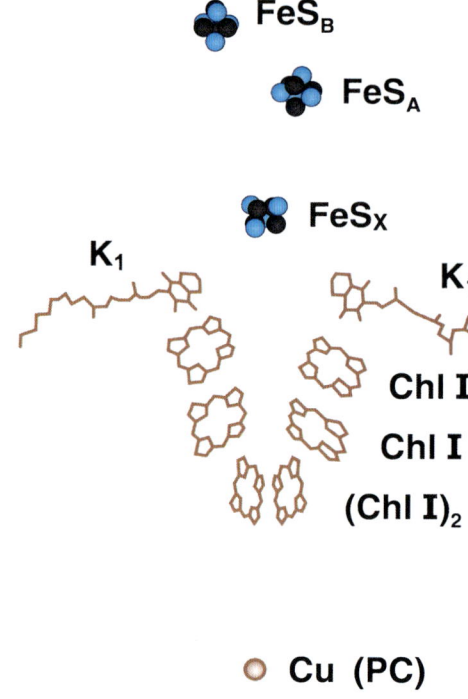

(DCMU, also called Diuron) and an artificial electron carrier is added. Under certain natural conditions, the cyclic electron transfer seems to be operative in vivo. It has been proposed that in such case reducing equivalents are transported from ferredoxin back to $(ChlI)_2^{\bullet+}$ instead of being accepted by NADP$^+$. This process involves plastoquinone and the b_6f complex. It is suggested that ferredoxin reduces PQ, so photosystem 2 does not appear to be necessary to regenerate $(ChlI)_2$ from $(ChlI)_2^{\bullet+}$. This mechanism is similar to the cyclic redox chain of purple bacteria generating $\Delta\bar{\mu}_{H^+}$ without photolysis of water and reduction of NADP$^+$ (Crofts and Wraight 1983).

Photosystem 1 is clearly competent in generating $\Delta\bar{\mu}_{H^+}$. The reduction of Photosystem 1 by plastocyanin was shown to occur on the inner side of the thylakoid membrane, while ferredoxin and ferredoxin-NADP$^+$ reductase, which oxidize photosystem 1, are located on its outer side. Thus, the electron transport via photosystem 1 must be directed across the membrane. The photosystem 1 complex has been reconstituted into proteoliposomes, which were then found to generate $\Delta\bar{\mu}_{H^+}$ under continuous illumination. To mediate cyclic electron transfer around photosystem 1, the electron donor ascorbate and an artificial electron carrier phenazine methosulfate were added to the incubation mixture. Incorporation of a purified H$^+$-ATP synthase complex into the same proteoliposomes made photophosphorylation possible (Nelson and Hauska 1979; Hauska et al. 1980).

The mechanism of $\Delta\bar{\mu}_{H^+}$ generation mediated by photosystem 1 seems to be similar to that mediated by bacterial reaction centers. A nanosecond laser flash induces very fast ($\tau < 100$ ns) $\Delta\Psi$ generation, this process being much faster than any reaction in photosystem 1 other than electron transfer from (ChlI)* to FeS$_X$. The generated $\Delta\Psi$ composes a major part of the total electrogenesis linked with the operation of photosystem 1, so all the other stages of electron transfer make just a relatively small contribution to the energy transformation in this segment of the photosynthetic redox chain (Witt 1979; Lopez and Tien 1984).

2.3.3 Photosystem 2

Photosystem 2 was found to consist (depending on the object and purification method used) of about 30 different types of polypeptides, some of which are coded by nuclear and some by chloroplast DNA. It forms dimers. The X-ray analysis-based structure of photosystem 2 and the location of redox centers are shown in Fig. 2.24.

The photosystem 2 chlorophylls (similar to those from photosystem 1) belong to the group of a-type chlorophylls and have an α-absorption band at 680 nm (700 nm in case of photosystem 1). Subunits D1 and D2 (39.5 and 39 kDa, respectively) construct the central part of photosystem 2. These subunits resemble the M- and L-subunits of the purple bacteria reaction center complex (Hearst and Sauer 1984; Cramer et al. 1985). They bind (ChlII)$_2$, two ChlII, two pheophytins, primary and secondary quinones (PQ$_A$ and PQ$_B$), nonheme iron connected to four histidines, and also two antenna chlorophylls (Chl$_Z$). The 39-kDa subunit specifically binds DCMU.

The presence of 12 very small subunits (3–5 kDa) is characteristic of photosystem 2. Subunits of 9.5 and 4.5 kDa form the *b*-type cytochrome with α-band maximum at 559 nm (cytochrome b_{559}). Each of the mentioned subunits contains a histidine residue with imidazole forming a coordinating bond with heme. The role of cytochrome b_{559} remains unclear. The same can be said about the *c*-type cytochrome (c_{550}), which is located on the inner (directed toward the intrathylakoid space of the thylakoid) side of the membrane. This cytochrome is present in cyanobacteria, but it is absent from chloroplasts. Cytochrome b_{559} is located closer to the opposite membrane side.

A very high redox potential of the basic state of the chlorophyll dimer (the (ChlII)$_2$/(ChlII)$_2^{\bullet+}$ couple) is a distinctive feature of photosystem 2 (Nelson and Ben-Shem 2004; Nelson 2011). It is equal to +1.1 V. This explains why (ChlII)$_2^{\bullet+}$ can serve as the electron acceptor for water (the redox potential of the H$_2$O/O$_2$ couple at neutral pH is + 0.82 V). On the other hand, the redox potential of the excited (ChlII)$_2^*$ is in negative (less than −0.6 V).

Plastoquinone (PQ$_A$) was shown to be the primary stable electron acceptor. Another bound quinone molecule (PQ$_B$) participates in the further electron transfer along the photosynthetic redox chain. Similar to the purple bacteria, this process is

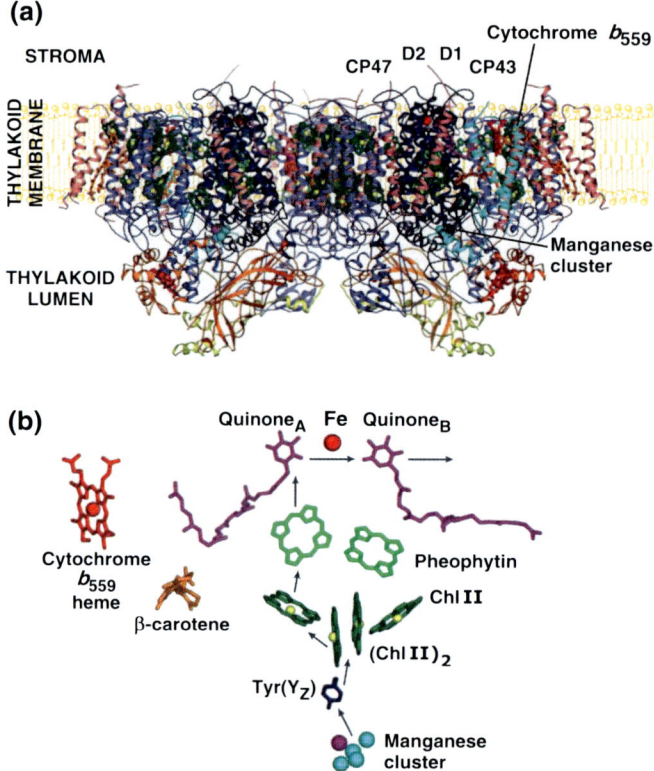

Fig. 2.24 Chloroplast photosystem 2 structure. **a** *Side view* of photosystem 2 dimer: subunits D1 and D2 are in dark blue color, chlorophyll-binding antenna proteins CP43 and CP47 are red, cytochrome b_{559} is light-blue, peripheral subunits orange and yellow, and other subunits pink. **b** spatial arrangement of photosystem 2 prosthetic groups: chlorophylls are dark green, manganese cluster light-blue, a water molecule bound to the manganese cluster pink, heme and nonheme iron red, and quinones purple. The electron transfer pathway is indicated by arrows. (Nelson and Ben-Shem 2004)

accelerated by nonheme iron, which is chelated by four histidines, and it is inhibited by *o*-phenanthroline.

A tyrosine residue of a protein acts as the primary electron donor for $(ChlII)_2^{\bullet+}$. It reduces $(ChlII)_2^{\bullet+}$ within 50–500 ns.

The specific *water-oxidizing complex* (WOC) is the reductant for the tyrosine radical that is formed when this amino acid residue is oxidized by $(ChlII)_2^{\bullet+}$. This system operates such that a splitting of two H_2O molecules results in (i) the reduction of four $(ChlII)_2^{\bullet+}$ molecules and (ii) the formation of one O_2 molecule and four H^+ ions.

The WOC is composed of four polypeptides, one of which contains four Mn^{2+} ions (Cramer et al. 1985). The latter participate directly in electron transfer from water to the tyrosine radical.

The mechanism of $\Delta\bar{\mu}_{H^+}$ generation by photosystem 2 can be pictured by analogy with the reaction center complex of purple bacteria that has already been discussed. Apparently, here there is also a light-dependent transmembrane movement of an electron from the excited chlorophyll dimer to Q_A located on the opposite side of the hydrophobic barrier, and only one of two ChlII- and pheophytin-containing chains is operative.

2.3.4 Cytochrome b_6f Complex

The Q-cycle reaction system is catalyzed by the b_6f complex in chloroplasts and cyanobacteria. This complex contains four large (18–32 kDa) and four small subunits. Three large subunits contain redox centers. These are: *cytochrome f* (32 kDa), *three-heme cytochrome b_6* (23 kDa), and an *FeS protein* (20 kDa). The fourth large subunit (18 kDa) does not directly participate in electron transfer. The complex forms a dimer of 217 kDa. Cytochrome b_6 is unique as it contains two b hemes and a c_i heme (Fig. 2.25) (Stroebel et al. 2003). Both b hemes of cytochrome b_6 have α-maxima at 563 nm. They differ in their redox potentials. The b_L and b_H hemes have midpoint potentials (E_m) of −30 and +150 mV, respectively (Clark and Hind 1983). In chloroplast cytochrome b_6, heme c_i is covalently bound to only one cysteine, and not to two as in all other c-type cytochromes. The other specific feature of c_i heme is the absence of axial ligands represented by functional groups of the protein. The c_i heme is located next to the b_H heme, thus forming a kind of screen dividing it from PQ. PQ probably serves as one of the axial ligands of heme c_i. One can speculate that such a configuration of the PQ-reductase center of the b_6f complex facilitates the two-electron reduction of PQ by electrons of hemes c_i and b_H; it is also possible that this configuration reduces the lifetime of the semiquinone form of plastoquinone (PQ$^{\bullet-}$), which is dangerous due to its ability to reduce O_2 into $O_2^{\bullet-}$. The danger of such a transformation is especially high in cells with oxygenic photosynthesis, where intracellular O_2 concentration is always high.

A comparison of the amino acid sequence of cytochrome b_6 from spinach chloroplasts with those of mitochondrial cytochrome b from animals, plants, and fungi revealed similarity in the first 200 residues from the N-terminus. However, there is an obvious difference—cytochrome b_6 is substantially shorter (211 residues) than mitochondrial cytochromes b (330–385 residues). On the other hand, a considerable sequence homology was found between the fourth (18-kDa) subunit of the b_6f complex and the C-terminal part (residues 269–353) of the mitochondrial cytochrome b. One can assume that a function carried out by the C-terminal part of the mitochondrial cytochrome b is performed in the chloroplasts by the 18-kDa subunit (Widger et al. 1984; Saraste 1984).

Cytochrome b_6 contains five histidine residues, four of which participate in the coordination of the b_L and b_H hemes. These residues at positions 82, 96, 183, and 198 (197 in mitochondrial cytochrome b) are completely conserved, i.e. they are

Fig. 2.25 Redox centers of the b_6f complex dimer from chloroplasts of the green alga *Chlamydomonas reinhardtii*. **a** Iron atoms of c_i, b_H and b_L hemes of cytochrome b, [2Fe-2S] cluster of FeS$_{III}$ protein, and cytochrome f. **b** Surroundings of c_i and b_H hemes in cytochrome b_6. Potential axial ligand of heme c_i iron (most likely plastoquinone) and two water molecules that are bound close to c_i and b_H hemes are shown in green color (Stroebel et al. 2003)

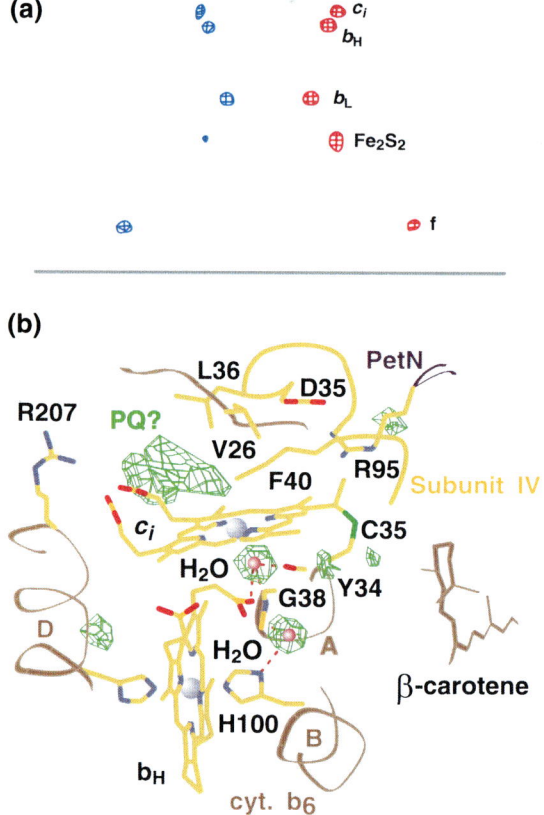

found in cytochrome b_6 as well as in all the mitochondrial cytochromes b from different kingdoms of living organisms. All these histidines are always located in hydrophobic segments of the polypeptide chain. They participate in the binding of hemes b_L and b_H, thus acting as axial ligands and determining the location of the hemes perpendicular to the membrane surface. Each heme b contains two negatively charged propionate groups. These groups oppose conserved, positively charged arginines (Fig. 2.26) (Cramer et al. 1985).

The distance between the heme edges, according to the model, is 1.2 nm, and that between the iron atoms is 2.0 nm. This means that an electron moving from heme b_L to heme b_H needs to cross about half of the hydrophobic membrane core.

A partner of cytochrome b_6 in PQH$_2$ oxidation is the FeS protein of the b_6f complex that is functionally similar to the FeS$_{III}$ component of the respiratory chain. This FeS$_{III}$ contains two iron atoms and two sulfur atoms (Cramer et al. 1985). The next electron carrier is cytochrome f. The amino acid sequence of cytochrome f (it contains 285 residues) shows the presence of the Cys-X–Y-Cys-His sequence, which is characteristic of covalent heme binding by c-type cytochromes. It occurs at residues 21–25 in the N-terminal segment. His-25 and

2.3 Noncyclic Photoredox Chain of Chloroplasts and Cyanobacteria

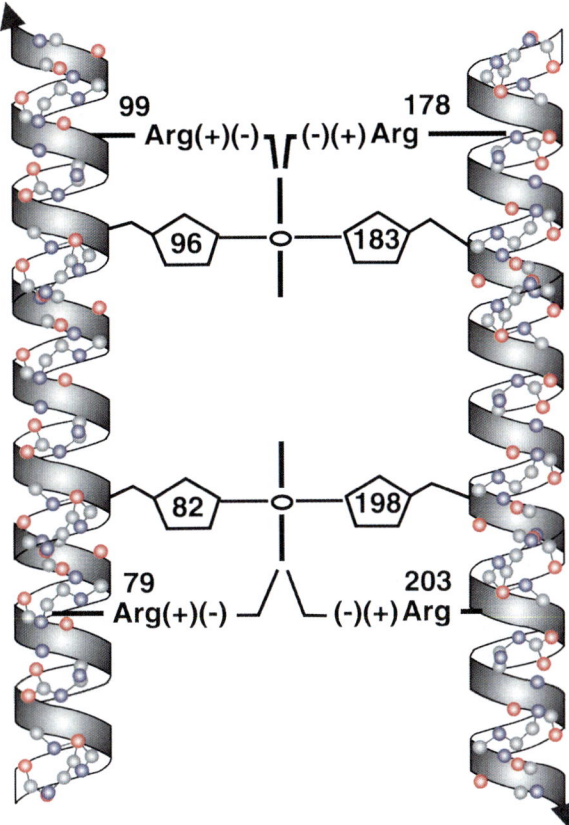

Fig. 2.26 Transmembrane arrangement of two hemes (thick vertical lines) of cytochrome b_6. Upper and lower hemes are b_L and b_H, respectively; 96, 183, 83, and 198, imidazoles of corresponding histidine residues; possible stabilization of negatively charged (–) heme propionate residues by mostly conserved arginine residues (+) is shown (Cramer et al. 1985)

Lys-145 or Lys-222 serve as the ligands for the heme iron. The heme is located in the large (residues 1–250) N-terminal domain exposed to the aqueous phase in the thylakoid lumen. Residues 251–270 are hydrophobic and form an α-helical segment that functions as a membrane-spanning anchor. The remaining 15 C-terminal amino acids are located on the outer surface of the thylakoid membrane and can be removed by proteases. Segment 190–249 includes many acidic amino acid residues facing the thylakoid interior and is probably involved in interaction with a positively charged segment of cytochrome b_6. Ten conserved basic amino acids in the region 58–154 are assumed to participate in binding the next electron carrier—plastocyanin. The cytochrome f spectrum shows an α-band maximum at 555 nm. The redox potential of this cytochrome is +365 mV (Cramer et al. 1985).

The average diameter of the b_6f complex is about 8.5 nm. In natural membranes the complexes tend to form dimers (Cramer et al. 1985).

It should be stressed that b_6f complex is the slowest and most vulnerable part of the photosynthetic redox chain. The entrance to the Q-cycle from photosystem 2 is blocked by the potent herbicide DCMU. Certain steps of the Q-cycle are very

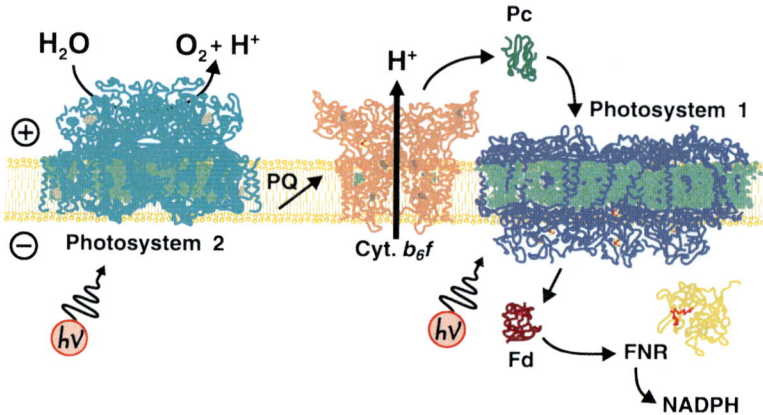

Fig. 2.27 Three $\Delta\bar{\mu}_{H^+}$ generators in the thylakoid membrane of cyanobacteria. To simplify the picture, proteins that are bound to the membrane surface, i.e. plastocyanin (PC), ferredoxin (Fd), and ferredoxin:NADP$^+$ reductase (FNR), are shown in the aqueous phases inside or outside the thylakoids

sensitive to dibromothymoquinone and 2-heptyl-4-hydroxyquinoline N-oxide. The latter two poisons are inhibitory also in mitochondrial and bacterial Q-cycles.

An inhibitor analysis of the thylakoid redox chain is complicated by the existence of electron transfer pathways alternative to the main noncyclic reaction sequence. This is the cyclic electron flow around photosystem 1 (and probably around photosystem 2 as well).

Figure 2.27 shows three $\Delta\bar{\mu}_{H^+}$ generators of oxygenic photosynthesis. These data were obtained by X-ray analysis of photosystem 1 and photosystem 2 and b_6f complex of cyanobacteria. The system of oxygenic photosynthesis looks very similar to the one presented in Fig. 2.27.

2.3.5 Fate of $\Delta\bar{\mu}_{H^+}$ Generated by the Chloroplast Photosynthetic Redox Chain

Thylakoids of chloroplasts perform only two functions—reduction of NADP$^+$ and synthesis of ATP mediated by $\Delta\bar{\mu}_{H^+}$ generation and consumption. The energy of $\Delta\bar{\mu}_{H^+}$ is quantitatively transformed into energy of ATP by the H$^+$-ATP synthase.

The thylakoid does not have to perform osmotic work to transport metabolites, for the aqueous intrathylakoid space is in fact enzymatically empty. The only major process that takes place on the inner surface of the thylakoid membrane is photolysis of water, resulting in formation of molecular oxygen. For both the substrate (H$_2$O) and product (O$_2$) of this reaction, crossing the membrane poses no problem. The conversion of ADP and P$_i$ to ATP and of NADP$^+$ to NADPH both occur on the outer surface of thylakoids.

Fig. 2.28 General scheme of energy transduction in thylakoid membrane

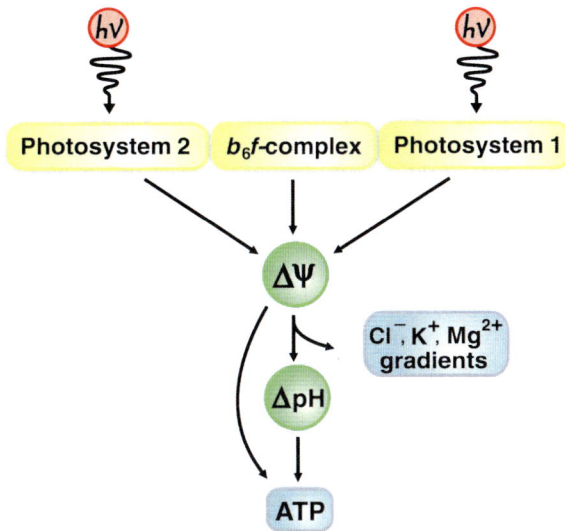

Reverse electron transfer against the redox potential gradient is absent from thylakoids, which solve the reducing power generation problem by formation of NADPH in the noncyclic photoredox chain. Such $\Delta\bar{\mu}_{H^+}$ consumers as H⁺-motors rotating bacterial flagella are also absent from thylakoids.

Since there is only one $\Delta\bar{\mu}_{H^+}$-linked function in the thylakoid membrane, there is no need to buffer $\Delta\bar{\mu}_{H^+}$ as strongly as in bacteria. Nevertheless, thylakoids have a system that can buffer $\Delta\bar{\mu}_{H^+}$ to some degree. The thylakoid membrane is rather permeable to Cl⁻, K⁺, and Mg²⁺. The transport of these ions down the electric gradient formed by $\Delta\bar{\mu}_{H^+}$ generators results in the transduction of $\Delta\Psi$ to ΔpH. Since the amount of energy equivalent stored in the form of ΔpH is much larger than that in $\Delta\Psi$, the $\Delta\Psi \to \Delta$pH transition increases the total reserve of membrane-linked energy in the system. This effect should stabilize the rate of ATP synthesis in spite of light intensity fluctuations occurring in vivo. A general scheme of energy transduction in the thylakoid membrane is shown in Fig. 2.28.

There is one more essential point that should be mentioned when discussing chloroplast energetics. It was found that H⁺-ATP synthase can be absent from those parts of thylakoids in which the membranes of two adjacent thylakoids adjoin each other. This occurs when thylakoids are not swollen (i.e. have a discoid shape) and a very narrow cavity separates one thylakoid from another. In such a case, all the H⁺-ATP synthase molecules were shown to be concentrated at the very periphery of the thylakoid disks exposed to the chloroplast stroma and (mainly) in membranous lamellas stretched through the stroma (see above, Fig. 2.20). These lamellas are in fact the continuation of some thylakoids. The lamellas were shown to contain mainly photosystem 1, whereas thylakoids have both photosystems. It appears that lamellas specialize in ATP synthesis coupled to cyclic electron transport around photosystem 1, and they are not capable of photolysis of water and reduction of NADP⁺ (these two processes occur in thylakoids). Apparently, to form ATP the thylakoid-generated $\Delta \bar{\mu}_{H^+}$ must be first transmitted along the membrane to the lamellas, where a major pool of H^+-ATP synthase is located.

When thylakoids are in a swollen state, H^+-ATP synthase molecules diffuse from the lamellas to thylakoids so as to be equally distributed between all the areas on the intrachloroplast membranes.

All the above-mentioned considerations also apply to the thylakoids of cyanobacteria, which apparently were the evolutionary precursors of chloroplasts. In this case, however, we must supplement the scheme shown in Fig. 2.27 by a respiratory chain arranged in the same thylakoid membrane. (In plant cells, the respiratory chain is located in mitochondria.) The interaction of the photosynthetic and respiratory chains that share the same components of the Q cycle in their middle stages will be discussed later (see Sect. 4.3.3).

References

Blankenship RE, Prince RC (1985) Excited state redox potentials and the Z scheme of photosynthesis. TIBS 10:382–383

Clark RD, Hind G (1983) Spectrally distinct cytochrome b-563 components in a chloroplast cytochrome b-f complex: Interaction with a hydroxyquinoline N-oxide. PNAS 80:6249–6253

Cramer WA, Widger WR, Herrmann RG, Trebst A (1985) Topography and function of thylakoid membrane proteins. TIBS 10:125–129

Crofts AR, Wraight CA (1983) The electrochemical domain of photosynthesis. Biochim Biophys Acta 426:149–185

Deisenhofer J, Epp O, Miki K, Huber R, Michel H (1984) X-ray structure analysis of a membrane protein complex. Electron density map at 3 Å resolution and a model of the chromophores of the photosynthetic reaction center from *Rhodopseudomonas viridis*. J Mol Biol 180:385–398

Deisenhofer J, Epp O, Miki K, Huber R, Michel H (1985a) Structure of the protein subunits in the photosynthetic reaction centre of *Rhodopseudomonas viridis* at 3Å resolution. Nature 318:618–624

Deisenhofer J, Michel H, Huber R (1985b) The structural basis of photosynthetic light reaction in bacteria. TIBS 10:243–248

Deprez J, Trissl HW, Breton J (1986) Excitation trapping and primary charge stabilization in *Rhodopseudomonas viridis* cells, measured electrically with picosecond resolution. PNAS 83:1699–1703

Drachev LA, Kaulen AD, Ostroumov SA, Skulachev VP (1974) Electrogenesis by bacteriorhodopsin incorporated in a planar phospholipid membrane. FEBS Lett 39:43–45

Drachev LA, Kaulen AD, Semenov AY, Severina II, Skulachev VP (1979) Lipid-impregnated filters as a tool for studying the electric current-generating proteins. Anal Biochem 96:250–262

Dracheva SM, Drachev LA, Zaberezhnaya SM, Konstantinov AA, Semenov A, Skulachev VP (1986) Spectral, redox and kinetic characteristics of high-potential cytochrome c hemes in *Rhodopseudomonas viridis* reaction center. FEBS Lett 205:41–46

Glaser EG, Crofts AR (1984) A new electrogenic step in the ubiquinol:cytochrome c_2 oxidoreductase complex of *Rhodopseudomonas sphaeroides*. Biochim Biophys Acta 766: 322–333

References

Hauska G, Samoray D, Orlich G, Nelson N (1980) Reconstitution of photosynthetic energy conservation. II. Photophosphorylation in liposomes containing photosystem-I reaction center and chloroplast coupling-factor complex. Eur J Biochem 111:535–543

Hauska G, Schoedl T, Remigy H, Tsiotis G (2001) The reaction center of green sulfur bacteria (1). Biochim Biophys Acta 1507:260–277

Hearst LE, Sauer K (1984) Protein sequence homologies between portions of the L and M subunit of reaction centers of *Rhodopseudomonas capsulata* and the Q_B-protein of chloroplast thylakoid membranes; a proposed relation to quinone-binding sites. Z Naturforsch 85:515–521

Henderson R (1985) Membrane proteins: structure of a bacterial photosynthetic reaction centre. Nature 318:598–599

Hu X, Damjanovic A, Ritz T, Schulten K (1998) Architecture and mechanism of the light-harvesting apparatus of purple bacteria. PNAS 95:5935–5941

Kayushin LP, Skulachev VP (1974) Bacteriorhodopsin as an electrogenic proton pump: reconstitution of bacteriorhodopsin proteoliposomes generating $\Delta\Psi$ and ΔpH. FEBS Lett 39:39–42

Krasnovsky AA (1948) Reversible photochemical reduction of chlorophylls by ascorbic acid. Dokl Akad Nauk SSSR (Russ) 60:421–424

Lopez JR, Tien HT (1984) Reconstitution of photosystem I reaction center into bilayer lipid membranes. Photobiochem Photobiophys 7:25–39

Matveetz YA, Chekalin SV, Yartsev AA (1987) Femtosecond spectroscopy of the primary photoprocesses in *Rhodopseudomonas sphaeroides* reaction centers. Dokl Akad Nauk SSSR (Russ) 292:724–728

Merchant S, Sawaya MR (2005) The light reactions: a guide to recent acquisitions for the picture gallery. Plant Cell 17:648–663

Mitchell P (1966) Chemiosmotic coupling in oxidative and photosynthetic phosphorylation. Biol Reviews 41:445–502

Mitchell P (1975) Protonmotive redox mechanism of the cytochrome bc_1 complex in the respiratory chain: protonmotive ubiquinone cycle. FEBS Lett 56:1–6

Nelson N (2011) Photosystems and global effects of oxygenic photosynthesis. Biochim Biophys Acta 1807:856–863

Nelson N, Ben-Shem A (2004) The complex architecture of oxygenic photosynthesis. Nat Rev Molec Cell Biol 5:971–982

Nelson N, Hauska G (1979) Topography, resolution and reconstitution of the chloroplast membrane Membrane Bioenergetics. Addison-Wesley, London

Okamura MY, Feher G, Nelson N (1982) Reaction centers. In: Govindjee (ed) Photosynthesis: energy conservation by plants and bacteria, vol 1. Academic Press, New York, pp 1394–1403

Parson WW (1982) Photosynthetic bacterial reaction centers: interactions among the bacteriochlorophylls and bacteriopheophytins. Ann Rev Biophys Bioeng 11:57–80

Rich PR (1984) Electron and proton transfers through quinones and cytochrome *bc* complexes. Biochim Biophys Acta 768:53–79

Robert B, Lutz M, Tiede DM (1985) Selective photochemical reduction of either of the two bacteriopheophytins in reaction centers of *Rps. sphaeroides*. FEBS Lett 183:326–330

Saraste M (1984) Location of haem-binding sites in the mitochondrial cytochrome *b*. FEBS Lett 166:367–372

Severina II (1982) Nystatin: induced increase in photocurrent in the system "bacteriorhodopsin proteliposome/bilayer planar membrane". Biochim Biophys Acta 681:311–317

Shuvalov VA, Asadov AA (1979) Arrangement and interaction of pigment molecules in reaction centers of *Rhodopseudomonas viridis*. Photodichroism and circular dichroism of reaction centers at 100 K. Biochim Biophys Acta 545:296–308

Shuvalov VA, Duysens LN (1986) Primary electron transfer reactions in modified reaction centers from *Rhodopseudomonas sphaeroides*. PNAS 83:1690–1694

Shuvalov VA, Amesz J, Duysen LNM (1986) Picosecond charge separation upon selective excitation of the primary electron donor in reaction centers of *Phodopseudomonas viridis*. Biochim Biophys Acta 851:327–330

Skulachev VP (1972) The driving forces and mechanisms of ion transport through coupling membranes. FEBS Simposia 28:371–385

Skulachev VP (1988) Membrane bioenergetics. Springer, Berlin

Stroebel D, Choquet Y, Popot JL, Picot D (2003) An atypical haem in the cytochrome b_6f complex. Nature 426:413–418

Wakao N, Yokoi N, Isoyama N, Hiraishi A, Shimada K, Kobayashi M, Kise H, Iwaki M, Itoh S, Takaichi S, Sakurai Y (1996) Discovery of natural photosynthesis using Zn-containing bacteriochlorophyll in an aerobic bacterium *Acidiphilium rubrum*. Plant Cell Physiol 37:889–893

Widger WR, Cramer WA, Herrmann RG, Trebst A (1984) Sequence homology and structural similarity between cytochrome *b* of mitochondrial complex III and the chloroplast b_6f complex: position of the cytochrome *b* hemes in the membrane. PNAS 81:674–678

Witt HT (1979) Energy conversion in the functional membrane of photosynthesis. The central role of the electric field. Biochim Biophys Acta 505:335–427

Chapter 3
Organotrophic Energetics

3.1 Substrates of Organotrophic Energetics

Organic substances are the energy sources (substrates) for the majority of heterotrophic organisms. A substantial amount of energy is released in reactions of aerobic catabolism when substrates are oxidized by oxygen. This energy can be transformed into a form convenient for the cell (e.g. into ATP). The substrates used in these processes are carbohydrates, lipids, and proteins.

Polysaccharides, lipids, and especially proteins are extremely diverse, so an organism has to deal with many thousands of different compounds. During the first stage of metabolism of these substrates they are depolymerized, i.e., polymers and complex compounds are cleaved to monomers. Carbohydrates are decomposed into monosaccharides (there are fewer than ten basic monosaccharides); lipids are cleaved to glycerol and fatty acids (there are also about ten of these); 20 different amino acids result from the proteolysis of proteins. Altogether, a rather limited number of monomers (a few dozens) are obtained from thousands of food biopolymers (Fig. 3.1). Hence, we can see that *the first metabolic stage* results in a substantial unification of the substrates used as fuel.

3.2 Short Review of Carbohydrate Metabolism

The second metabolic stage results in a further unification of compounds so that they might be later oxidized by molecular oxygen. Let us consider this process using the example of glucose—the most common sugar. Oxidation of glucose by oxygen (Eq. 3.1) releases a very large amount of energy (2850 kJ/mol), which has to be transferred to a convertible form (ATP):

$$C_6H_{12}O_6 + 6\ O_2 \rightarrow 6\ CO_2 + 6\ H_2O \tag{3.1}$$

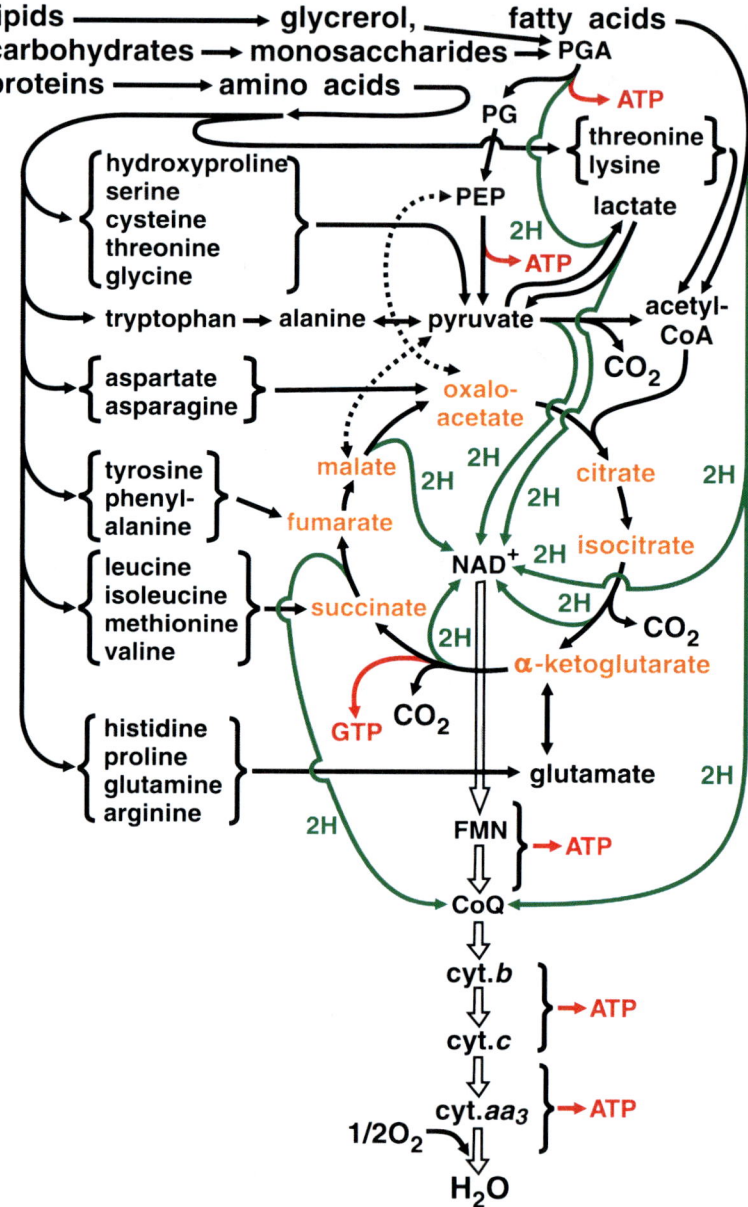

Fig. 3.1 Scheme for unification of "fuel" (reducing equivalents) used in organotrophic energetics. Pathways of hydrogen atom transfer are shown in *green*, reactions of ATP (GTP) synthesis *red*, metabolic pathways of "fuel" unification *black*, and electron transfer along the respiratory chain is marked with *broad arrows*

According to Eq. 3.1, full oxidation of one hexose molecule means transfer of 24 electrons from this molecule to six O_2 molecules. It is obviously not possible to perform this within one stage, and that is why this oxidation takes a few consecutive steps. The transfer of an electron to oxygen is coupled to a substantial release of energy. The conservation of this energy in the form of ATP is a rather complicated process that requires the functioning of respiratory chain enzyme complexes. For this to take place, electrons need to enter the respiratory chain. So, they are first taken from the monomers and transferred to specific cofactor molecules—nicotinamide adenine dinucleotide (NAD^+) or (less often) ubiquinone (CoQ). The NADH and $CoQH_2$ formed during these reactions are oxidized by the respiratory chain complexes and later by molecular oxygen. So, from the bioenergetic perspective, oxidation of glucose and other compounds and the resulting unification of reducing equivalents into just two compounds (NADH and $CoQH_2$) are the main results of the first stages of metabolism of carbohydrates.

Oxidation of glucose usually involves consecutive functioning of two systems: (i) glycolytic oxidation of glucose to pyruvate and (ii) oxidation of pyruvate to CO_2 in the Krebs cycle (the scheme of these metabolic pathways and photo of Hans Krebs are presented in Figs. 3.2 and 3.3, respectively). Glucose is first phosphorylated at the expense of ATP, thus forming glucose-6-phosphate. As mentioned earlier, cell *energy supply* is the main function of carbohydrate metabolism. But already at the first stage the cell needs to *spend* energy that is used to form an ester bond between the sugar and a phosphoric acid residue. The greater part of the energy of ATP cleavage is dissipated as heat, which makes this process irreversible. This fact is supposed to provide regulation of glycolysis. Control over any kind of metabolic process is most often realized via regulation of its *first* reaction. It seems obvious that such regulation will be simplified if the regulated process is *irreversible*. However, there is no ATP consumption when glycogen, rather than glucose, is used as a substrate. In this case it is inorganic phosphate that is the phosphoryl source for the reaction of glucose phosphate formation (the glycogen phosphorolysis reaction is catalyzed by glycogen phosphorylase, where the attachment of phosphate to sugar occurs at the expense of energy of glucose residue bonds in glycogen).

Glucose-6-phosphate is later isomerized to fructose-6-phosphate. Fructose-6-phosphate is again phosphorylated at the expense of ATP, thus forming fructose-1,6-bisphosphate, i.e., one more glycolytic reaction is accompanied by energy *consumption*. Again this is most likely needed for regulation of this metabolic pathway, especially due to the fact that glucose-6-phosphate besides glycolysis can also participate in the pentose phosphate pathway or be used for synthesis of glycogen and other carbohydrates. So, it is no wonder that the main control over glycolysis takes place at the stages that are catalyzed by hexokinase and phosphofructokinase.

In the next stage, fructose-1,6-bisphosphate is cleaved to 3-phosphoglyceric aldehyde and dihydroxyacetone phosphate. The latter is then converted to phosphoglyceric aldehyde (PGA). The aldehyde is in turn oxidized to 3-phosphoglyceric acid (PG). The first oxidative stage of the glucose metabolism pathway occurs just here. As a result, two electrons are transferred from each of the two parts of the

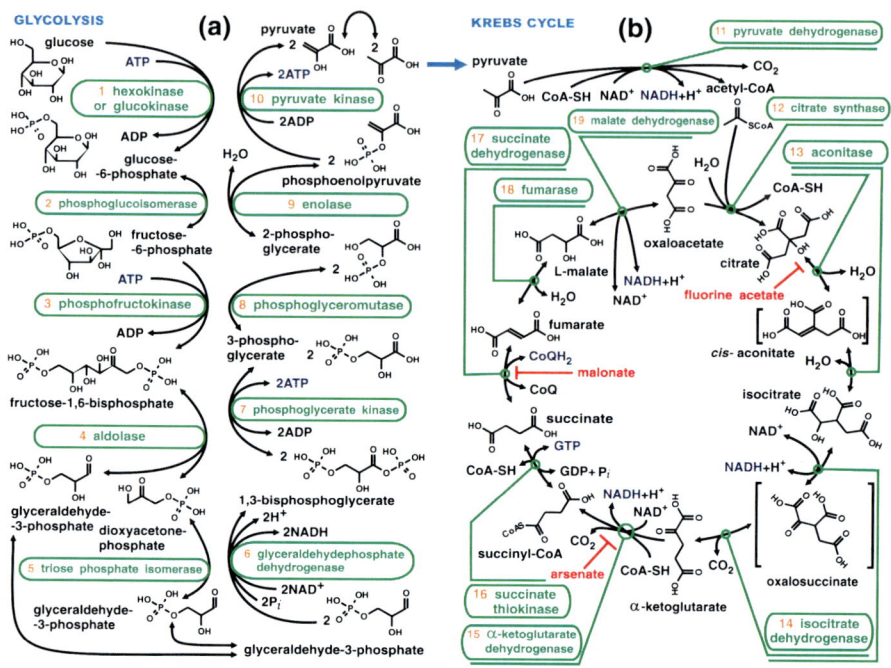

Fig. 3.2 Main pathway of aerobic glucose catabolism: **a** oxidation of glucose to pyruvate; **b** pyruvate oxidation in the Krebs cycle. Under anaerobic conditions, pyruvate is reduced by NADH to lactate (glycolysis; see Fig. 3.6)

Fig. 3.3 Hans Krebs

original sugar molecule to two NAD^+ molecules. Later phosphoglyceric acid is transformed into phosphoenolpyruvate (PEP) and then into pyruvate.

Pyruvate undergoes oxidative decarboxylation, thus forming acetyl-CoA. Two electrons from each pyruvate molecule are transferred to NAD^+ during this process. The acetyl-CoA enters the Krebs cycle by being condensed with oxaloacetic acid to form citric acid, which is then isomerized to isocitric acid. The latter undergoes oxidative decarboxylation (which is accompanied by electron transfer to NAD^+), thus forming α-ketoglutaric acid. The α-ketoglutarate undergoes oxidative decarboxylation once again, this process being accompanied by electron transfer to NAD^+, and it turns into succinic acid (this process takes two consecutive steps). Succinic acid is later oxidized to fumaric acid. But the redox potential of the succinate/fumarate couple (+30 mV) is much more positive than the potential of the $NADH/NAD^+$ couple (−320 mV), and in this case electrons from the organic substrate are transferred not to NAD^+, but to an electron carrier with a more positive redox potential—ubiquinone (CoQ, +60 mV).

Fumaric acid is hydrated to form malic acid, which is in turn oxidized (with electrons being transferred to NAD^+) to oxaloacetic acid, which is ready for the condensation reaction with the next acetyl-CoA molecule.

So, we can see that during glycolysis and the Krebs cycle full oxidation of a glucose molecule to carbon dioxide occurs, and electrons taken from the substrate are used for formation of NADH and ubiquinol ($CoQH_2$):

$$C_6H_{12}O_6 + 10\ NAD^+ + 2\ CoQ + 6\ H_2O$$
$$\rightarrow 6\ CO_2 + 2\ CoQH_2 + 10\ NADH + 10\ H^+ \qquad (3.2)$$

The NADH and $CoQH_2$ can be later oxidized by molecular oxygen via the respiratory chain:

$$10 NADH + 10\ H^+ + 2\ CoQH_2 + 6\ O_2$$
$$\rightarrow 10\ NAD^+ + 2\ CoQ + 12\ H_2O \qquad (3.3)$$

It is in the oxidation of NADH and ubiquinol by oxygen (i.e., in reaction 3.3) that the greater part of the energy is released (we will discuss this later). But reaction 3.2 is also exergonic. Although the amount of energy released is not so substantial in this case, it can also be partly stored in the form of ATP. Let us consider the mechanism of this energy transformation, *substrate phosphorylation*, using oxidation of 3-phosphoglyceric aldehyde (PGA) to 3-phosphoglycerate (PG) oxidation in the glycolysis reaction chain as an example.

3.3 Mechanism of Substrate Phosphorylation

The redox potential of the PGA/PG couple (−550 mV) is much more negative than the potential of the $NADH/NAD^+$ couple (−320 mV), so electron transfer from PGA to NAD^+ is accompanied by the release of a substantial amount of energy (43 kJ/mol) sufficient to carry out synthesis of one ATP molecule from ADP and P_i. Such a synthesis presumes that the exergonic oxidative reaction is

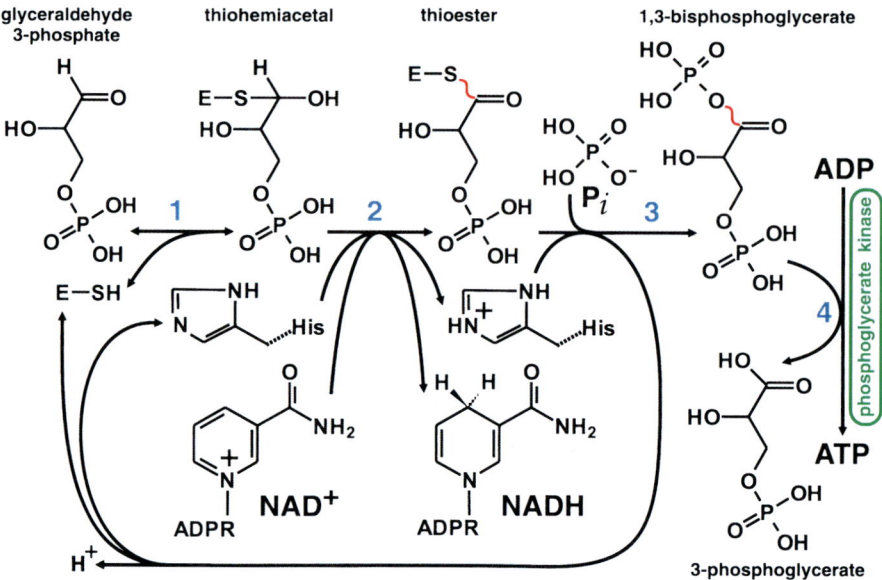

Fig. 3.4 Mechanism of ATP synthesis coupled to oxidation of glyceraldehyde 3-phosphate. High energy bonds are shown in red. His—a histidine residue of glyceraldehyde phosphate dehydrogenase. The role of H^+ ions produced during this process is considered in Sect. 3.5

coupled to the endergonic reaction of phosphorylation of ADP by inorganic phosphate. But from the chemical perspective, such a coupling of two reactions (exergonic and endergonic, i.e., one that releases energy and another one that consumes it) is a rather complicated task. Let us look at the mechanisms of this coupling.

Oxidation of PGA to PG coupled to synthesis of ATP is catalyzed by two enzymes, i.e., glyceraldehyde phosphate dehydrogenase and phosphoglycerate kinase. The sequence of events is shown in Fig. 3.4, and it can be divided into three steps.

(1) Glyceraldehyde phosphate dehydrogenase first binds PGA by forming a covalent bond between the SH-group of its active-site cysteine residue and the aldehyde group of PGA. As a result, a so-called thiohemiacetal is produced. It is important to note that this reaction is reversible, i.e., no energy needs to be spent for this reaction. *Thus, in the first step the substrate molecule is connected to a reaction group that proceeds without any energy consumption.*

(2) An energetically profitable aldehyde group oxidation to a carboxyl group due to transfer of two electrons and one proton to an NAD^+ molecule occurs in the second step. But due to the fact that this is a thiohemiacetal derivative of an aldehyde group that participates in the reaction, it results in formation of a thioester of the carboxyl group, and not the carboxyl group *per se*. Free energy

of oxidation of PGA is not dissipated as heat, but is stored in the form of a strained thioester bond. *In fact, the second step results in appearance of a high energy bond between the substrate and the enzyme at the expense of energy from the oxidation of the substrate (aldehyde) to the product (carboxylic acid).*

(3) Phosphorolysis of the thioester bond leading to the formation of the phosphoanhydride of the substrate carboxyl group occurs in the third step. This process regenerates the original form of the enzyme and is accompanied by replacement of the high-energy thioester bond by the high-energy phosphoanhydride bond. Then the phosphate residue is transferred to ADP, thus forming ATP reaction 4 on Fig. 3.4 catalyzed by phosphoglycerate kinase). This type of mechanism of energy conservation is called *phosphorylation on the substrate level* or *substrate phosphorylation.*

Besides the reaction described above, the substrate phosphorylation mechanism is also used at some other stages of glycolysis and the Krebs cycle, namely when 3-phosphoglycerate is converted into pyruvate, and also during the oxidative decarboxylation of pyruvate and α-ketoglutarate.

Thiamine pyrophosphate serves as a cofactor in stage (1) of oxidative decarboxylation of pyruvate and α-ketoglutarate (Fig. 3.5). At the stage (2), substrate decarboxylation occurs, resulting in formation of a semialdehyde. Then the semialdehyde is oxidized by lipoic acid to an acyl. As a result, reduced lipoic acid is produced (stage 3). Reduced lipoic acid accepts the acyl residue so that a thioester bond is formed (stage 4). An acetyl (or succinyl) group is then transferred from the lipoic acid to the SH-group of CoA (stage 5). This is followed by thioester bond phosphorolysis, which forms acetyl (or succinyl) phosphate (stage 6). The high-energy phosphate group is transferred to imidazole of a histidine residue of the enzyme (stage 7) and then to a GDP molecule (stage 8). The formed GTP can be used to produce ATP (stage 9). It should be noted that in the majority of organisms (with the exception of some bacteria) energy of the thioester bond acetyl $\sim$ S-CoA is not used for ATP synthesis; it is rather used to launch the first reaction of the Krebs cycle or to initiate heme synthesis and some other synthetic processes. The final stage 10 is the regeneration of oxidized lipoic acid and reduction of NAD^+.

The mechanism of energy conservation on the glycolytic transformation of 2-phosphoglycerate to pyruvate seems to be the simplest. There is no need to attach any kind of reaction group to the substrate in this case, for 2-phosphoglycerate already has a phosphoric acid residue. So, one needs only to destabilize in some way the phosphoester bond in 2-phosphoglycerate, i.e., to increase the potential of this phosphate group transfer in order for ATP synthesis to proceed. It is enolase that fulfills this function. This enzyme catalyzes removal of a water molecule from 2-phosphoglycerate, thus forming phosphoenolpyruvate. This reaction increases substantially the energy of the phosphoester bond breakage so that it is more than sufficient for transfer of a phosphate residue to an ADP molecule. The latter reaction is catalyzed by pyruvate kinase. (This and preceding processes are shown as reactions 9 and 10 in Fig. 3.2a).

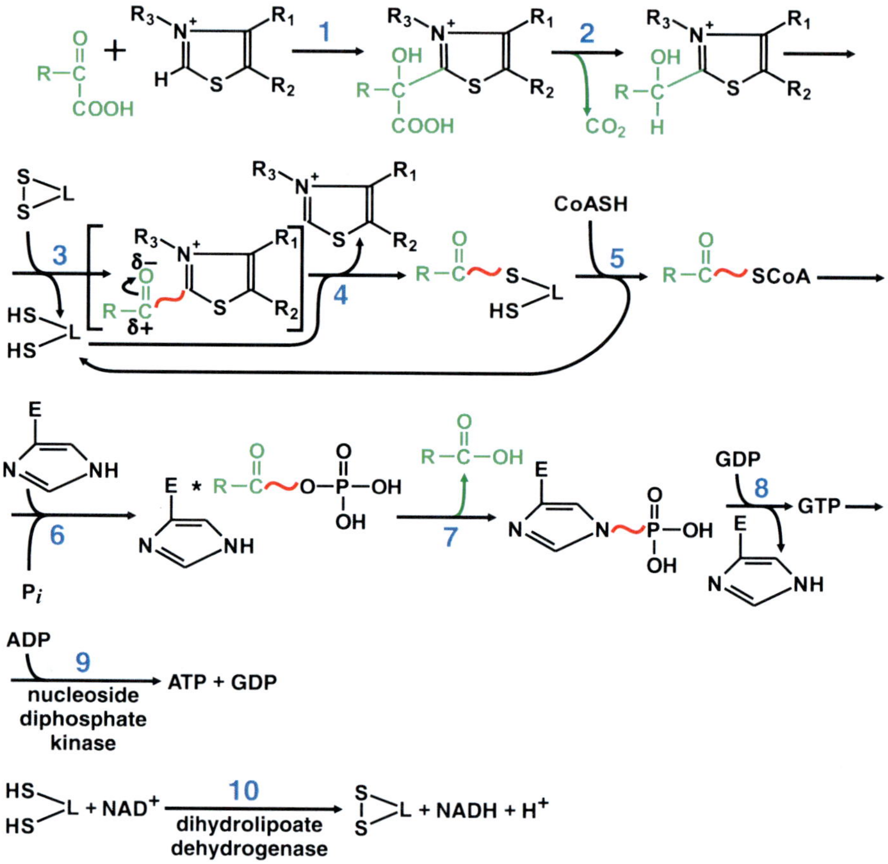

Fig. 3.5 Oxidative decarboxylation of pyruvate and α-ketoglutarate. These ketoacids are shown in *green*; high-energy bond is in *red*

The substrate phosphorylation mechanism in this case seems to be substantially different when compared to oxidation of 3-phosphoglyceroaldehyde or oxidative decarboxylation of α-ketoacids since in this case H_2O instead of 2H is removed from the substrate. However, phosphoenolpyruvate formation can also be considered as a redox reaction. The only difference is that in this case electrons are redistributed not between the substrate molecule and redox-active cofactor, but intramolecularly, i.e., between different atoms of one and the same substrate molecule (in this case reduction of the third carbon atom of 2-phosphoglycerate proceeds due to the oxidation of the second carbon atom). Such an energetically profitable disproportionation of different carbon atoms is the energy source for ATP synthesis during the transformation of 2-phosphoglycerate to pyruvate.

So, we have discussed the main metabolic reactions that are accompanied by ATP synthesis by means of the substrate phosphorylation mechanism. Relative simplicity of this process seems to be the main advantages of this type of energy conservation. We will see later that the mechanism of oxidative phosphorylation (i.e., ATP synthesis at the expense of energy of oxidation of respiratory substrates by molecular oxygen) is much more complex. Then why is not substrate phosphorylation more common? Or, in other words, what are the drawbacks of the substrate phosphorylation mechanism?

First of all, energy from oxidation of the majority of compounds cannot be stored in the form of ATP through this mechanism. For strictly chemical reasons, it seems to be possible only with a rather limited number of compounds (particularly aldehydes and ketones). Another drawback seems to be determined by the strict stoichiometry of this process of energy conservation: exactly one ATP molecule can be synthesized per molecule of oxidized substrate. This fixed ratio imposes certain restrictions on the use of this mechanism and decreases its effectiveness. On one hand, if energy of substrate oxidation exceeds substantially the energy of ATP hydrolysis, this entire excess will be lost as heat. On the other hand, if the energy of substrate oxidation is less than the energy of ATP hydrolysis, then the energy of this reaction cannot be stored through the mechanism of substrate phosphorylation. As we will see later, the mechanism of oxidative phosphorylation in respiratory and photosynthetic electron transfer chains is free of these drawbacks; that is why it plays a more important role in the supply of energy in living cells.

3.4 Energetic Efficiency of Fermentation

In spite of the fact that there are only two ATP molecules synthesized per glucose molecule consumed during glycolysis, this metabolic pathway is the only cell energy source for some organisms. These include various bacteria and some eukaryotes living under anaerobic conditions. These organisms cannot use oxidative phosphorylation due to the lack of oxygen and other electron acceptors, so they have no choice but to be content with the relatively small amount of energy that is stored as a result of substrate phosphorylation. This type of metabolism is called *fermentation*.

The main reactions of catabolism of sugars have already been discussed. It was mentioned that two NADH molecules are formed per glucose oxidized to pyruvate (see Fig. 3.2a). In the case of aerobic organisms, NADH is later oxidized by oxygen, this process leading to regeneration of NAD^+. However, this is impossible under anaerobic conditions. Here, it is necessary to find some electron acceptors for NADH other than O_2. Doing so, it is necessary to maintain the strict redox reaction balance. In other words, the number of NADH molecules produced as a result of fermentation needs to be exactly the same as the number of NAD^+ molecules formed during the reduction of the final electron acceptor that oxidizes NADH.

Let us consider the well-known type of fermentation—*lactic acid fermentation (glycolysis)*—as an example of this phenomenon. This type of catabolism is typical for different lactic acid bacteria, but it is also used in animal skeletal muscle during

Fig. 3.6 Lactic acid fermentation

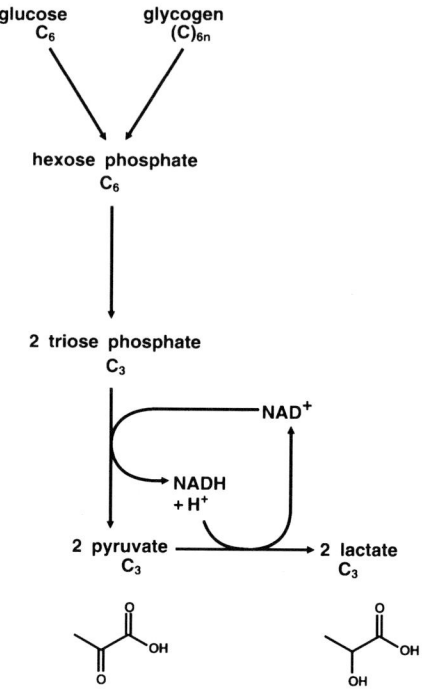

short but intensive exercise. In glycolysis, glucose is converted into two molecules of pyruvic acid. Two ATP molecules (they are the only source of energy for lactate-producing bacteria) and two NADH molecules are also formed. The latter must be oxidized for further glycolytic reactions to proceed. In the case of lactic acid fermentation, oxidation of NADH is achieved by reduction of pyruvate to lactic acid. The general scheme of lactic acid fermentation is presented in Fig. 3.6. This scheme also shows the importance of maintaining the complete redox balance of fermentation. In the case of a compound more reduced compared to glucose (e.g. sorbitol) or more oxidized (e.g. gluconic acid), it is no longer possible to keep that balance. That is why these compounds that seem so similar to glucose are not fermentable substrates for lactic acid-producing bacteria.

Let us try to answer the question of where the energy necessary for ATP synthesis in the course of lactic acid fermentation comes from. When considering glycolysis, it is necessary to note that all the carbon atoms in a glucose molecule have an intermediate degree of reduction (all of them are connected to a hydroxyl or belong to a carbonyl group). Reduction of certain carbon atoms of glucose due to oxidation of other atoms of the same sugar molecule is energetically profitable. This happens when glucose is converted into lactate, where the first carbon atom is oxidized due to the reduction of the third carbon atom.

So, two ATP molecules are formed per consumed glucose molecule during glycolysis. Similar stoichiometry is characteristic also for the well-known fermentation yielding ethanol. Is there a way to increase ATP/glucose stoichiometry

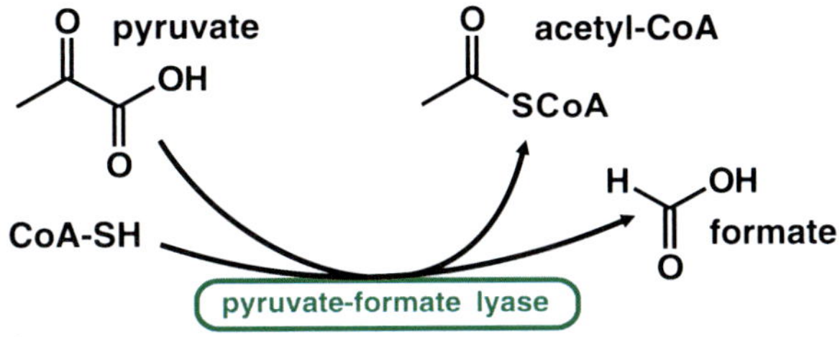

3.4 Energetic Efficiency of Fermentation

Fig. 3.7 Pyruvate-formate lyase reaction

of fermentation? In fact, it is possible. One only needs to increase electron disproportionation between different carbon atoms of the fermentation products. Let us consider this phenomenon using mixed fermentation of enterobacteria as an example. It has already been described that there are two pyruvate molecules, two NADH molecules, and two ATP molecules formed per metabolized glucose molecule decomposed using the canonical glycolytic pathway. It would seem very tempting not to reduce pyruvate to lactate but to obtain acetyl~CoA and later acetate and ATP instead (the way it has already been described for the mechanism of oxidative decarboxylation of pyruvate). Then we would obtain four ATP molecules per glucose consumed. But in this case it would not be possible to oxidize two NADH molecules that have been formed in the course of glycolysis. Even more so, two additional NADH molecules are formed during the functioning of the pyruvate dehydrogenase complex, thus making it impossible to maintain the redox balance of fermentation. That is why enterobacteria chose a different path. They use pyruvate-formate lyase instead of pyruvate dehydrogenase (Fig. 3.7) (Knappe and Sawers 1990). This enzyme forms acetyl~CoA and formate from pyruvate (and not acetyl~CoA, CO_2, and NADH as in the case of the pyruvate dehydrogenase reaction). The further fates of the two acetyl~CoA molecules formed from one glucose are different. One undergoes phosphorolysis, thus forming ATP and acetate. The second acetyl~CoA molecule is reduced twice by NADH (complete oxidation of NADH obtained during glycolysis occurs), thus forming ethanol and free CoA. The described process leads to the synthesis of *three*, rather than two, ATP molecules.

And what would the ideal (from the perspective of energetic output) fermentation look like? It would most likely mean the maximum possible electron disproportionation between the different carbon atoms in the fermentation products, i.e., the following process:

$$C_6H_{12}O_6 \rightarrow 3\ CO_2 + 3\ CH_4 \qquad (3.4)$$

Certain microbial communities are capable of implementing this reaction (methanogenesis). It will be described later in the chapters on bacterial respiratory chains and also sodium energetics (Sects. 5.3 and 12.4, respectively).

3.5 Carnosine

The very low energetic outcome of fermentation seems to be the main drawback of this type of metabolism. It means that a large amount of fermentable substrate needs to be consumed by the cell to be supplied with a sufficient amount of energy, and this will lead to a substantial accumulation of fermentation products, some of which can be toxic. In many cases of anaerobic fermentation, such products are lactic, acetic, succinic, propionic, or butyric acids which completely dissociate to corresponding anions and H^+ ions at neutral pH.

This fact is of particular importance in the case of intense work of skeletal muscles. The muscles exhaust oxygen under these conditions, and a substantial amount of lactic acid accumulates. This acidifies the cytoplasm, and muscle enzymes can be inactivated by excess H^+ ions. To consider how an organism solves this problem we will discuss muscle dipeptides. Peptide history is an interesting chapter in biochemistry. Peptides regulate different processes in our body, and they are very important components of our body. Now, very few people remember that the first peptide was discovered in Russia at the turn of the nineteenth and twentieth centuries—in 1900—by Vladimir Gulevitch (Fig. 3.8). This compound, β-alanyl-histidine (Fig. 3.9), was given the name *carnosine* (Gulevitch and Amiradzhibi 1900). Later it was found that carnosine has an analog where one of the hydrogen atoms in the histidine heterocycle is substituted by a methyl group (β-alanyl-methylhistidine, or *anserine*). These two compounds are present in many muscles in very large amounts. Their total concentration in cytoplasm sometimes reaches 50 mM.

At the very beginning of the twentieth century, 30 years before the discovery of ATP, many metabolites were unknown, and new compounds of biological origin were discovered every year. With the passing of time, the role of the majority of new compounds became clear, but the functions of carnosine and anserine remained unknown. At the end of his life path, Gulevitch bequeathed the study of these compounds to his beloved student, Sergey Severin (Fig. 3.10). Severin remained true to his professor's wish—he dedicated his whole life to the discovery of the physiological role of carnosine and anserine. Many years after Gulevitch's death, in 1953, the function of muscle dipeptides was clarified by his former student. But this function proved to be so simple and apparently so primitive that Severin in fact could not believe that he solved the problem formulated by his mentor. Later, up to the end of his long life (Severin lived for more than 90 years), he continued to look for an answer to the question that had been already answered—what is the physiological role of carnosine and anserine. Let us look at the experiment that became known in physiology as the "Severin phenomenon". An isolated frog muscle was stimulated by an electrode and began to contract. After some time, the muscle became tired and contraction ceased. However, when carnosine was added to the solution, the muscle was able to work for hours without any traces of fatigue (Severin et al. 1953). This was a truly spectacular effect. Severin's coworkers invested much effort in attempts to understand the mechanism

3.5 Carnosine

Fig. 3.8 Vladimir Gulevitch

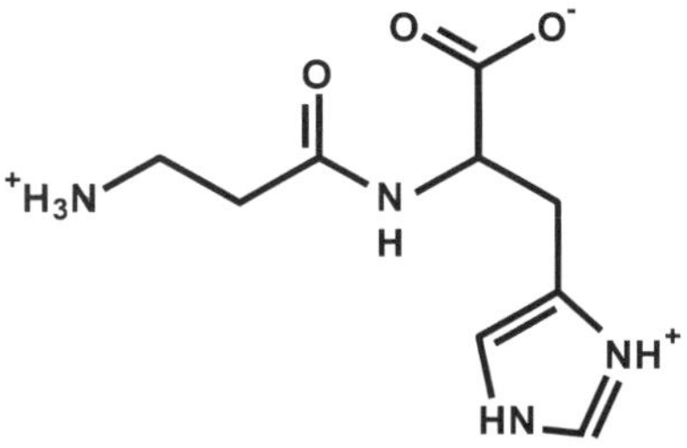

Fig. 3.9 Carnosine (protonated form)

Fig. 3.10 Sergey Severin

of this phenomenon; they measured many different parameters, and it was shown that nothing was changed by carnosine besides one thing—the muscle was able to accumulate very high amounts of lactate if carnosine was added. A simple explanation was that carnosine binds H^+ ions produced by glycolysis. In 1938, Edgar Charles Bate Smith determined the pK_a values of the imidazole nitrogen of a histidine residue in muscle dipeptides. The pK_a of carnosine is 6.95, and the pK_a of anserine is 7.05 (Bate Smith 1938; Deutsch and Eggleton 1938) This means that these compounds have pH-buffering properties at physiological pH values. So, lactate and protonated carnosine are the final glycolysis products in an anaerobic muscle. Protonated carnosine is a harmless compound, and that is why a muscle can work till all the carnosine and anserine molecules are mostly protonated. Bate Smith predicted that the only function of these compounds might be based on pH buffering in the neutral range (Bate Smith 1938). Severin did not agree with such a point of view, as he considered this buffering function to be too insignificant when compared to functions of ATP, NADH, coenzyme A, etc.

And this is what happened next. Severin's observations were published in Russian, they did not get a wide response, and the author himself did not attach a high value to the effect he discovered. He really liked this experiment, told students about it during his lectures, but all the time he was trying to find some other alternative function of carnosine. Carnosine might influence neuromuscular transmission, or it might get phosphorylated, and phosphorylated carnosine might be used in muscles as a macroerg, or it might influence mitochondria in some way... Carnosine was found to participate to some extent in almost all these processes, but Severin could never observe such a spectacular effect as the one described in 1953. Only some minor positive effects of carnosine were discovered. This stimulated the scientist's interest, and step-by-step he studied the influence of carnosine on practically all the "metabolic highways". By the end of his life, Severin had become one of the most versatile of educated biochemists of the world. The majority of Russian biochemists were his direct or indirect students. And 30 years after the original discovery of Severin's phenomenon, his experiment was repeated abroad with Tris buffer instead of carnosine (Effron et al. 1978). Addition of this compound led to a huge increase in the ability of muscle to perform mechanical work; the buffering effect was suggested to be responsible for this phenomenon, and the authors made no reference to Severin's work.

Carnosine is metabolically inert; this property is very important for its role as a specialized pH buffer. Carnosine is easily protonated and deprotonated, and this event *per se* has no effect on other metabolic processes. The fact that carnosine is not just a simple dipeptide seems also to be important. The unusual β-amino acid is part of it, and this fact seems to be one more way to make this substance more inert, which also means that it can no longer be affected by the usual peptidases.

There is another very important property of carnosine—it is a *mobile* pH buffer, which makes pH equilibration all through the giant muscle cell much easier. Histidine residues of proteins, which also participate in pH buffering in the neutral range, are essentially fixed in space (belonging mainly to actin and myosin, the main insoluble muscle proteins). Calculations showed that immobilized pH buffers enhance the danger of local acidification of parts of the cell (Junge and McLaughlin 1987). Also, protonation of proteins practically always influences their biological function (recall the fact that the great majority of enzymes have a pH optimum for activity).

Ultimately, the pH-buffering function of carnosine turned out to be not its only function (Boldyrev 2000). Carnosine is also a good chelator of copper and iron ions, this fact being of substantial importance for cells, as these cations catalyze the transformation of hydrogen peroxide to OH• radical, one of the strongest metabolic poisons (see Chap. 15). Carnosine also protects proteins and DNA from the destructive effects of aldehydes. High concentrations of carnosine probably protect us from intoxication caused by formaldehyde and other aldehydes. Aldehydes are sure to damage carnosine, but it is rather "cheap"—it is just a dipeptide. Carnosine also protects cells from accumulation of amyloid fragments, from undesirable protein–protein interactions, etc. But all these functions are secondary, as they do not explain why it was dipeptides with their ability to buffer the H^+ level near pH 7 that were chosen, and why it is only in muscles that they are present in very high concentrations.

References

Bate Smith EC (1938) The buffering of muscle in rigor; protein, phosphate and carnosine. J Physiol 92:336–343
Boldyrev AA (2000) Problems and perspectives in studying the biological role of carnosine. Biochemistry (Mosc) 65:751–756
Deutsch A, Eggleton P (1938) The titration constants of anserine, carnosine and some related compounds. Biochem J 32:209–211
Effron MB, Guarnieri T, Frederiksen JW, Greene HL, Weisfeldt ML (1978) Effect of tris(hydroxymethyl)aminomethane on ischemic myocardium. Am J Physiol 235:H167–H174
Gulevitch VS, Amiradzhibi S (1900) Ueber das Carnosin, Eine Neue Organische Base des Fleischextractes. Ber Deutsch Chem Ges 33:1902–1903
Junge W, McLaughlin S (1987) The role of fixed and mobile buffers in the kinetics of proton movement. Biochim Biophys Acta 890:1–5
Knappe J, Sawers G (1990) A radical-chemical route to acetyl-CoA: the anaerobically induced pyruvate formate-lyase system of *Escherichia coli*. FEMS Microbiol Rev 6:383–398
Severin SE, Kirzon MV, Kaftanova TM (1953) Effect of carnosine and anserine on action of isolated frog muscles. Dokl Akad Nauk SSSR 91:691–701 (In Russian)

Chapter 4
The Respiratory Chain

The term "respiratory chain" applies to a reaction sequence responsible for the transfer of hydrogen atoms or electrons from respiratory substrates to molecular oxygen.

There are *two types of respiratory chains—coupled to energy transduction and noncoupled*. The biological significance of the first type was recognized in the early 1930s when Vladimir Engelhardt (Fig. 4.1) discovered respiratory phosphorylation (Engelhardt 1930). The role of the second type became clear almost 30 years later. In 1960, one of this book's authors (V.P.S.) suggested that uncoupled respiration can be biologically useful, e.g., under conditions when heat, rather than ATP, becomes the most important for organisms (Skulachev and Maslov 1960).

This chapter describes the energy transduction-coupled respiratory chain in mitochondria and in certain bacteria. In the case of eukaryotic cells, respiration coupled to energy transformation is localized in the inner mitochondrial membrane. In the case of respiring bacteria, the process can be found in the cytoplasmic membrane, mesosomes, or thylakoids.

4.1 Principle of Functioning

As already discussed in Chap. 2 devoted to chlorophyll-mediated photosynthesis, the energy of sunlight is the main energy source for life on Earth. Such autotrophic oxygenic organisms as cyanobacteria and plants use the energy of light quanta for two simultaneous processes, i.e., for energy storage in the form of $\Delta \bar{\mu}_{H^+}$ (which is later transformed into ATP) and for the oxidation of water and transfer of the electrons obtained for $NADP^+$.

ATP and NADPH obtained as a result of photosynthetic electron transfer are used for fixation of carbon dioxide. Carbohydrates, and later lipids, amino acids, and nucleotides are formed at the expense of the energy of light.

Fig. 4.1 Vladimir Engelhardt

Heterotrophic organisms are usually not capable of direct use of light energy. In fact, they utilize the energy that has been accumulated by photosynthesizing plants. Roughly speaking, heterotrophic organisms reverse the reactions of photosynthesis. They first oxidize carbon atoms of carbohydrates (as well as of fatty acids and amino acids) to CO_2 (as described in Chap. 3) and use the electrons obtained for reduction of NAD^+. The NADH formed is then oxidized by molecular oxygen via the respiratory chain so that water is formed. It has already been mentioned that a very large amount of free energy (about 1.1 electron volt per electron transferred from NADH to oxygen) is released when electrons are transferred from NADH to O_2. This energy can be stored by the respiratory chain in the form of $\Delta\bar{\mu}_{H^+}$. The proton-motive force on biological membranes is usually about 200 mV (if the electric potential reaches higher values, it can lead to disruption of the lipid bilayer and the loss of the stored energy). Calculation shows that transfer of four electrons from two NADH molecules to O_2 needs to be coupled to translocation of at least 20 protons across the membrane for the energy to be stored in the form of $\Delta\bar{\mu}_{H^+}$. It is the necessity of such a high H^+/O_2 stoichiometry that was most likely the reason for Nature to have chosen a complicated mechanism of energy conservation when NADH oxidation was split into three discrete energy-releasing steps. Electrons are first transferred from NADH to ubiquinone by NADH: CoQ-oxidoreductase (complex I), the ubiquinol obtained is then oxidized by cytochrome c via $CoQH_2$:cyrochrome c-oxidoreductase (also called bc_1 complex or complex III), and finally the reduced cytochrome c is oxidized by molecular oxygen via cytochrome c oxidase (complex IV). The activity of all of these enzyme complexes is coupled to generation of proton potential that can be utilized to form ATP. In this context, it should be indicated

4.1 Principle of Functioning

Fig. 4.2 Albert Lehninger (*left*) and David E. *Green*

that Albert L. Lehninger and David E. Green (Fig. 4.2) made a crucial contribution to discovery of energy coupling sites in the respiratory chain and purification of the respiratory chain complexes responsible for energy coupling.

The redox potential difference between the substrates and products of the reaction is ~ 380 mV for complex I, ~ 190 mV for complex III, and ~ 570 mV for complex IV (see Fig. 4.3).[1] If we assume the value of $\Delta\bar{\mu}_{H^+}$ equal to ~ 200 mV, we can calculate that complexes I, III, and IV can transfer 2, 1, and 2–3 protons, respectively, per electron transported. Experiments have shown the $H^+/\bar{e}$ ratios equal to 2, 1, and 2 for complexes I, III, and IV, respectively (Wikström 1977, 1984; Galkin et al. 1999).

Comparison of potentials and $H^+/\bar{e}$ stoichiometry makes it obvious that complexes I and III store almost all the energy in the form of $\Delta\bar{\mu}_{H^+}$ released during electron transfer. The energy released at the last (cytochrome oxidase) step of the respiratory chain is substantially higher than at the previous spans of the chain. Nevertheless, $H^+/\bar{e}$ stoichiometry for cytochrome oxidase is the same as for complex I (two H^+ per one $\bar{e}$). So, a substantial part of the energy is dissipated in the form of heat at the last stage of respiration, and in contrast to the reactions of the previous stages, this stage is irreversible. In fact, huge energy drop

---

[1] It is important to note that it is the standard redox potentials (E_0') that are given both in the text and in Fig. 4.3. The real values of redox potentials (E) in a cell can differ substantially from those E_0' values because of the difference in concentrations of corresponding oxidized and reduced forms of the redox compounds involved.

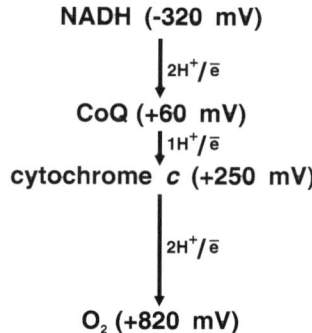

Fig. 4.3 Scheme of electron transfer and $\Delta\bar{\mu}_{H^+}$ generation in the mitochondrial respiratory chain of higher animals

accompanying the cytochrome oxidase reaction makes irreversible the whole metabolism of the aerobic cell.

As shown in Fig. 4.4a, the respiratory chain starts with oxidation of NADH. Several redox centers of the *NADH:CoQ-oxidoreductase* complex (complex I) are involved in this process, namely FMN and a group of FeS clusters. The NADH:CoQ-oxidoreductase reduces CoQ. The $CoQH_2$ produced is then oxidized by the Q-cycle system. The latter includes the FeS cluster of complex III (FeS_{III}; another name, Rieske FeS protein), cytochrome c_1, and two cytochrome b hemes—the high potential b_H and the low potential b_L. The FeS_{III} cluster serves as a direct electron acceptor for $CoQH_2$. From FeS_{III} the electrons are transferred to cytochrome c_1 and then to cytochrome c. The latter is used as a reductant of the last respiratory chain enzyme, cytochrome oxidase, which includes two hemes (a and a_3) and three copper atoms. Cytochrome oxidase reduces oxygen to water.

The oxidation of NADH by oxygen is coupled to the translocation of 10 protons from the mitochondrial matrix (or, in bacteria, from the cytoplasm) to the intermembrane space of mitochondria (or, to the bacterial periplasm in case of gram-negative bacteria and to the outer environment in case of gram-positive bacteria). Thus formed $\Delta\bar{\mu}_{H^+}$ is later used for ATP synthesis via the so-called complex V (proton-motive ATP-synthase, see Sect. 7.1) or for performing some other types of useful work.

The above-described multicomponent arrangement of the respiratory chain has a number of advantages besides the high H^+/O_2 stoichiometry. First of all, it provides the respiratory chain with some flexibility in utilizing substrates with different redox potentials. Reducing equivalents can enter the respiratory chain at different levels depending on the redox potential of the respiratory substrate. If this potential is lower, equal, or slightly higher than that of the $NADH/NAD^+$ couple (-320 mV), the entire respiratory chain sequence can be involved in the oxidation of such a substrate. This is the case for the majority of respiratory substrates. If the redox potential of the substrate is much more negative than that of the $NADH/NAD^+$ couple, a special energy accumulation mechanism of the so-called substrate level phosphorylation is included before the respiratory chain (see Chap. 3).

If the redox potential of the substrate is significantly more positive than that of the $NADH/NAD^+$ couple, the reducing equivalents are transferred to the middle or

4.1 Principle of Functioning

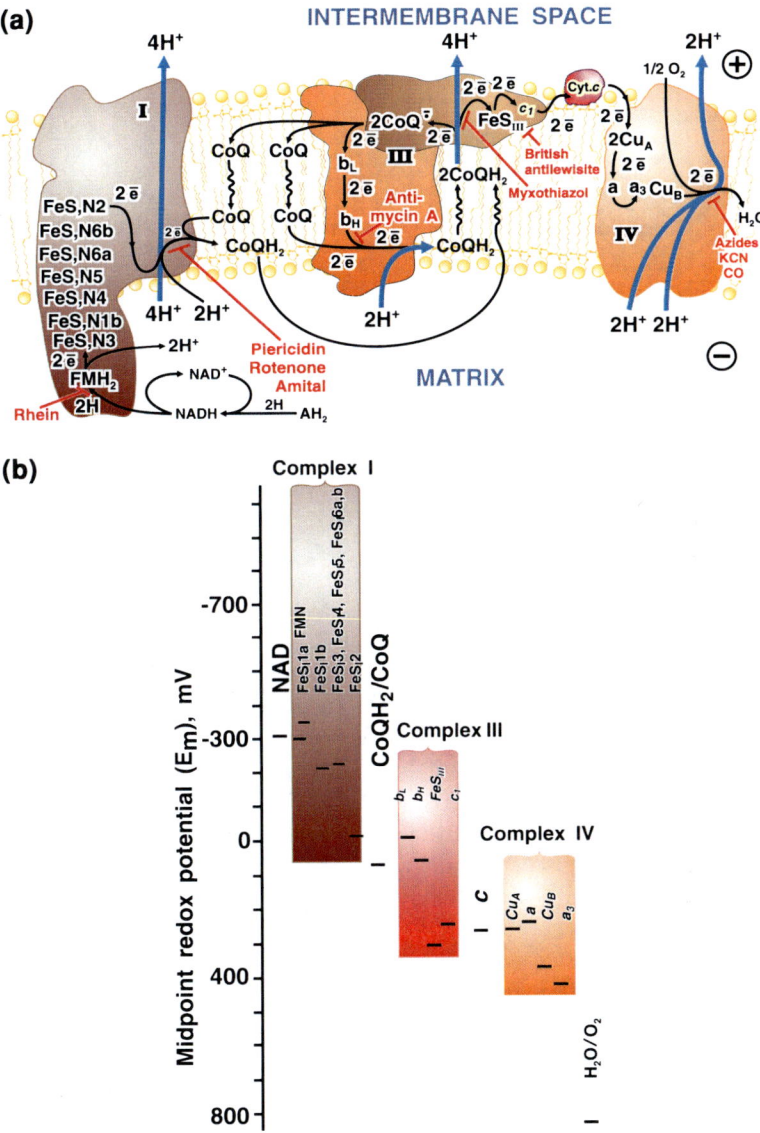

Fig. 4.4 a Mitochondrial respiratory chain. AH$_2$—an NAD$^+$-reducing respiratory substrate. Electron transfer is indicated by *black arrows*, and transmembrane proton transfer by *blue arrows*. Inhibitors are shown in *red*. **b** midpoint potential values of different electron carriers of the mitochondrial respiratory chain

the terminal segment of the respiratory chain. This is the case for a Krebs cycle substrate, succinate (redox potential, +30 mV) as well as for fatty acyl-CoA (redox potential, about −10 mV), the substrate of the first oxidoreduction in the fatty acid β-oxidation system (Sato et al. 1999). Succinate dehydrogenase and acyl-CoA

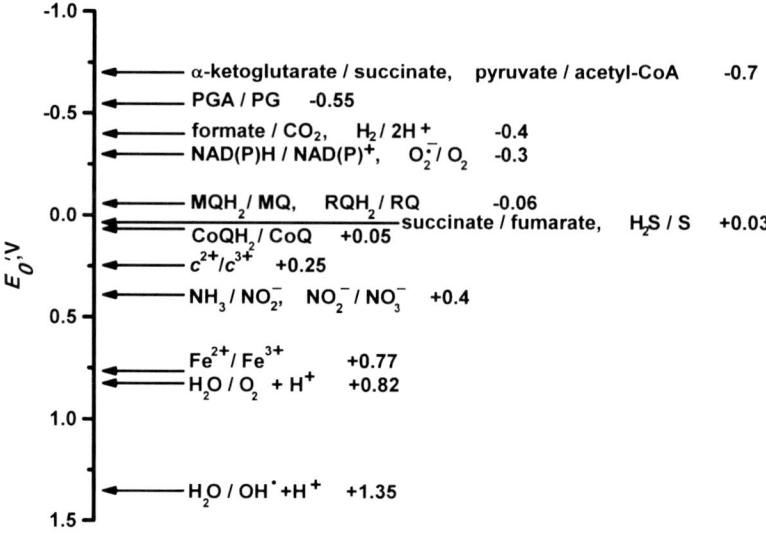

Fig. 4.5 Redox potential scale of different compounds related to respiratory chain

dehydrogenase feed the respiratory chain with electrons transferred directly to ubiquinone, bypassing complex I.

In very rare cases, when the redox potential of the oxidation substrate is more positive than that of CoQ, reducing equivalents enter the respiratory chain at the level of cytochrome c so that only the cytochrome oxidase $\Delta\bar{\mu}_{H^+}$ generator is involved in energy transduction. In animals, ascorbic acid and superoxide anion are oxidized in such a way. As for bacteria, oxidation of methanol, Fe^{2+}, or nitrite are examples of this phenomenon. Figures 4.4b and 4.5 show the redox potentials of some respiratory chain substrates and intermediates [standard potentials (E_0') are given for pH 7.0].

4.2 NADH:CoQ-Oxidoreductase (Complex I)

As was already mentioned, NADH:CoQ-oxidoreductase (complex I) catalyzes electron transfer from NADH to quinone, and this reaction is coupled to translocation of four protons across the membrane (Galkin et al. 1999). Complex I represents the first span of the respiratory chain. Unfortunately, it is also its least studied component. A very complex arrangement of complex I (a huge number of subunits and redox cofactors) and its extreme lability make studies of this enzyme difficult.

4.2.1 Protein Composition of Complex I

Complex I is the most complex component of the respiratory chain. The mammalian enzyme consists of 45 different polypeptides, each of them being present in just one copy. The total mass of the complex is about 980 kDa (Hirst et al. 2003).

Complex I from mitochondria of lower eukaryotes probably has a simpler structure. But even in this case the enzyme consists of at least 35 subunits (e.g., 40 subunits have been identified so far in complex I from the yeast *Yarrowia lipolytica* (Morgner et al. 2008)).

However, many bacterial respiratory chains contain complex I (it is also called NDH-1 in this case) that consists of only 14 subunits.[2] Such a simplified enzyme with mass of just 530 kDa is still capable of performing its main function—it transfers two electrons from NADH to quinone, in a manner coupled to translocation of four protons across the membrane. That is why this bacterial enzyme of 14 subunits is considered to be the minimal form of complex I (Leif et al. 1993). Bacterial complex I subunits are very conservative, their analogs being found in all the H^+-translocating NADH:quinone-oxidoreductases that have been studied so far. Later in this section, we will discuss mainly the structure and functions of this more simply organized bacterial complex I. One question remains unclear though—if analogs of 14 main (bacterial) subunits are sufficient for functioning of complex I, then what is the role of the other 31 subunits of the mitochondrial enzyme? As for now, they are considered to participate in some way in the assembly of complex I and regulation of its activity. It is perhaps worth noting that other enzymes of the mitochondrial respiratory chain also have many additional subunits when compared to bacterial analogs.

Fourteen genes that code the subunits of complex I are united into one operon in the genomes of the majority of prokaryotes. These genes are most often called *nuo* (from the first letters of NADH:ubiquinone-oxidoreductase) in alphabetical order according to their sequence in the operon (this gene order in *nuo*-operon is conservative for various bacteria). Seven of the complex I subunits (NuoA, H, J, K, L, M, and N) are extremely hydrophobic. They are integral membrane polypeptides and form 55 transmembrane α-helices. The seven other subunits are relatively hydrophilic and are attached to the membrane mainly due to protein–protein interactions with the hydrophobic subunits (Leif et al. 1993; Efremov and Sazanov 2011). In the case of eukaryotic complex I, analogs of seven hydrophobic bacterial subunits are coded in the mitochondrial genome and are synthesized by mitochondrial ribosomes. The other 37 subunits are coded in the nucleus, synthesized

---

[2] In the case of enterobacteria, NDH-1 consists of 13 subunits—not because of the lack of one of the main polypeptides, but rather due to fusion of the genes of two subunits (*nuo*C and *nuo*D) into one elongated gene (*nuo*CD).

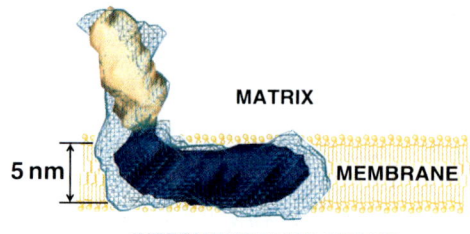

Fig. 4.6 Spatial arrangement of complex I from the bacterium *Escherichia coli* (peripheral domain is shown *yellow*, and membrane domain *blue*) and complex I from the mitochondria of the fungus *Neurospora crassa* (*light-blue* shading) (Guenebaut et al. 1998)

in the cytoplasm, and enter mitochondria after being transported across the two mitochondrial membranes.[3]

Complex I was shown to be L-shaped being composed of two domains with a right angle between them (Fig. 4.6). One of the domains is located inside the membrane, and the other peripheral domain protrudes by about 150 Å from the membrane into the mitochondrial matrix (or cytoplasm in the case of bacteria). The majority of the redox centers of the enzyme are located in this peripheral domain.

4.2.2 Cofactor Composition of Complex I

Already in early studies of complex I this enzyme was shown to contain noncovalently bound FMN and a number of FeS clusters as redox prosthetic groups. Electron paramagnetic resonance (EPR) spectroscopy showed that complex I has at least five FeS clusters (Ohnishi 1998). Two of these (N1a and N1b) are of the [2Fe–2S] type, and the rest (N2, N3, and N4) of the [4Fe–4S] type. However, amino acid sequence analysis of complex I subunits showed that theoretically this enzyme might have up to eight FeS clusters (Table 4.1). X-ray analysis of the hydrophilic part of complex I was carried out recently, and it directly showed the presence of eight FeS clusters in a bacterial complex I (Sazanov and Hinchliffe 2006). Thus, it is supposed that the EPR signals from certain clusters (N5, N6a, and N6b) are much broadened because of strong mutual dipole–dipole interactions, or these EPR-silent clusters have too negative redox potential to be in the

---

[3] We note that this principle (i.e. the fact that the mitochondrial genome possesses only the genes of the most hydrophobic subunits) is characteristic also for other respiratory chain components. Transport of such hydrophobic polypeptides through the cytoplasm and across the outer mitochondrial membrane would be very difficult. That is probably why the corresponding genes were left in the mitochondrial genome instead of being transferred together with the vast majority of other mitochondrial protein genes into the nucleus. This seems to be a likely explanation of why this unique organelle has kept its own genome.

4.2 NADH:CoQ-Oxidoreductase (Complex I)

Table 4.1 Properties of the 14 subunits of bacterial complex I

Subunit	Subcomplex	Molecular mass, kDa	Prosthetic group bonding
NuoA	HP	19	[4Fe–4S] (N2)
NuoB	IP	28	
NuoC	IP	25	
NuoD	IP	65	
NuoE	FP	18	[2Fe–2S] (N1a)
NuoF	FP	45	NADH, FMN, [4Fe–4S] (N3)
NuoG	HP	90	[4Fe–4S] (N4), [4Fe–4S] (N5), [2Fe–2S] (N1b)
NuoH	HP	32	
NuoI	IP	23	[4Fe–4S] (N6a), [4Fe–4S] (N6b)
NuoJ	IP	23	
NuoK	HP	10	
NuoL	HP	45	
NuoM	HP	40	
NuoN	HP	38	

paramagnetic form and thus they cannot be identified by EPR spectroscopy. Moreover, complex I also has a tightly bound ubiquinone that also might function as a redox prosthetic group.

4.2.3 Subfragments of Complex I

When treated with chaotropic agents, complex I can be separated into three parts (subfragments). One of them, a flavoprotein (FP), contains two subunits and includes at least two FeS clusters (N1a and N3), FMN, and also an NADH-binding site. FP can catalyze oxidation of NADH by artificial hydrophilic electron acceptors (for instance, ferricyanide), but not by natural hydrophobic ubiquinones.

The second complex I subfragment, iron-protein (IP), contains five subunits and includes several FeS clusters. This fragment lacks enzymatic activity. Together, FP and IP constitute the peripheral domain of complex I (Galante and Hatefi 1978; Friedrich and Scheide 2000).

The third subfragment (hydrophobic subcomplex, or HP) includes seven subunits of complex I, and similar to IP has no enzymatic activity. It is important to note that HP contains neither flavin nor FeS clusters, hence all known complex I redox prosthetic groups are located in the peripheral domain of the protein.

Separation of complex I into subfragments serves two purposes—the topology of the enzyme can be studied and understanding of the evolutionary origin of such a complex protein can be gained. Such a huge and intricate enzyme complex was probably constructed in the course of evolution due to combining and uniting into one entity parts of different preexisting enzymes (Friedrich and Scheide 2000). Comparison of the primary sequences of complex I subunits and other known proteins has shown that the main part of the IP subfragment originates from water-soluble [Ni–Fe] hydrogenase which catalyzes electron transfer to H^+ ions, thus forming molecular hydrogen. The membrane part of complex I (HP) has certain similarities to the multisubunit complex Mrp of alkalophilic bacilli, which is supposed to function as a $Na^+(K^+)/H^+$-antiporter (Efremov and Sazanov 2011). Unification of these two enzymatic modules probably made it possible to create a protein capable of coupling electron transfer from FeS clusters to quinone and transmembrane proton transport. One more part of complex I, subcomplex FP, functions as an adaptor and provides substrate specificity of this enzyme to its electron donor. It is interesting to note that a number of prokaryotes have enzymes homologous to complex I that contain some other subunits nonhomologous to NuoE and NuoF instead of subcomplex FP. This replacement led to a change in substrate specificity. For instance, in some archaea a similar enzymatic complex oxidizes cofactor $F_{420}H_2$ instead of NADH, in cyanobacteria and chloroplasts it oxidizes NADPH, and in the bacterium *Helicobacter pylori* (which causes gastric ulcer in humans) it uses ferredoxin as the reductant.

4.2.4 Inhibitors of Complex I

Many complex I inhibitors have been described. This wide spectrum of compounds can be divided into two groups according to the site of their action—basically all the known complex I inhibitors prevent the interaction of the enzyme with either NADH or quinone.

Rhein and ADP-ribose are inhibitors of the NADH-dehydrogenase part of the enzyme (Zharova and Vinogradov 1997), while rotenone, piericidin A, amytal, capsaicin (for their formulas, see Appendix 3), and many others hydrophobic compounds are the inhibitors of quinone-reductase step.

The binding sites of the inhibitors of the quinone-reductase part of complex I are located between FeS cluster N2 and CoQ (Tocilescu et al. 2010). These are first of all structural analogs of CoQ, and also many hydrophobic compounds which occupy the binding site of the CoQ isoprenoid "tail" in complex I. Some hydrophobic inhibitors imitate the quinone "head" of CoQ. Amytal of the barbiturate group was one of the first inhibitors to be discovered. The scale of its effective concentrations was soon found to be in the millimolar range, and its inhibitory effect was shown to be caused by nonspecific hydrophobic interactions with the enzyme. Water-soluble vitamin K_3 actuating an electron transfer bypassing barbiturate-sensitive step of complex I and Q-cycle was found to be an antidote for amytal. The use of vitamin K_3 decreases the efficiency of complexes I and III but makes it possible to avoid the fatal consequences of barbiturate overdose.

Rotenone can be considered as the most popular complex I inhibitor. It is also known as a "fish poison" for this compound was used in ponds when there was a need to get rid of any unwanted fish before introducing commercial fish species. It

4.2 NADH:CoQ-Oxidoreductase (Complex I)

Fig. 4.7 Scheme of transmembrane proton potential generation according to scheme of the Mitchell loop (**a**) or proton pump (**b**)

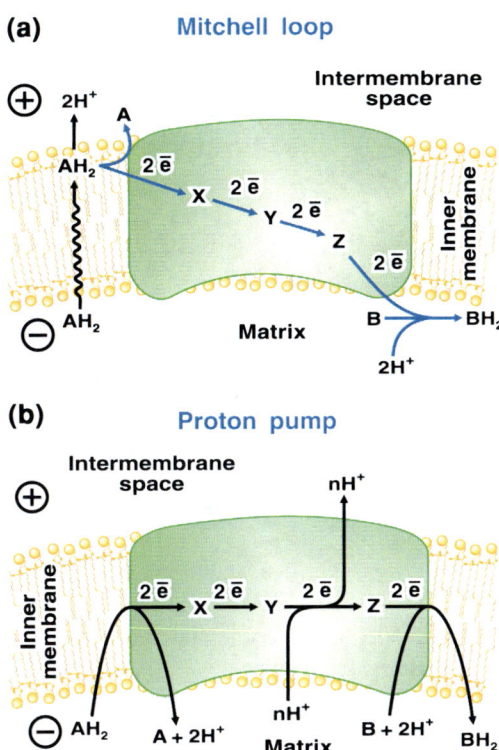

must be mentioned that rotenone is as much a poison for other animals as it is for fish. Capsaicin is another interesting example of complex I inhibitors. This compound is responsible for the burning taste of red pepper. Capsaicin inhibits complex I only at relatively high concentrations, but all the known representatives of this group of enzymes are sensitive to this compound (Yagi 1990). That is why capsaicin is often used for identification of complex I in respiratory chains of different organisms. It is important to note that many detergents (including those often used in everyday life) are strong complex I inhibitors, and thus they are quite dangerous. For instance, dodecyl sulfate and Triton X-100 act in this manner (Grivennikova et al. 2003).

4.2.5 Possible Mechanisms of $\Delta\bar{\mu}_{H^+}$ Generation by Complex I

Before we start describing possible mechanisms of proton potential generation by complex I, let us discuss how redox reactions can be coupled to proton translocation across the membrane.

Two fundamentally different ways to generate $\Delta\bar{\mu}_{H^+}$ are possible in electron-transporting chains. One is based on transmembrane transfer of electrons. In this

Fig. 4.8 Sergio Papa (*left*) and Leonid Sazanov

case, it is such redox cofactors and prosthetic groups as chlorophylls, cytochromes, FeS clusters, copper ions, vitamin K_1, etc., that serve as reducing equivalent carriers. This mechanism was suggested by Peter Mitchell (Mitchell 1966) and was given the name "Mitchell loop". Functioning of such a loop includes uptake of two protons from the inner side of the membrane (in the case of mitochondria—from the matrix-facing side), release of two protons on the outer side of the membrane, and transfer of two electrons across the membrane toward the mitochondrial matrix (Fig. 4.7a). This process is equivalent to translocation of two H^+ ions from the matrix to the external medium, even though there is no direct transmembrane proton movement in this case. This mechanism inevitably implies $H^+/\bar{e}$ stoichiometry of 1 during generation of $\Delta\bar{\mu}_{H^+}$. Generation of proton potential in photosynthetic electron transporting chains as described in Chap. 2 occurs essentially in line with this scheme.

The alternative possible mechanism of $\Delta\bar{\mu}_{H^+}$ generation was proposed by Sergio Papa (Fig. 4.8) and independently by one of this book's authors (V.P.S.) (Papa et al. 1973; Papa and Brunori 2011; Skulachev 1973, 1974). This hypothetical mechanism was given the name "proton pump". The energy of redox reactions in this case needs to be coupled to protein conformational changes and changes in pK values of different proton-accepting groups. The mechanism in question postulates direct transmembrane H^+ ions transfer driven in some way by electron transfer along the membrane (Fig. 4.7b). Such a process should not necessarily have rigid stoichiometry of $H^+/\bar{e} = 1$. Even more so, H^+ ions might be replaced in this case by some other ion, for instance by Na^+. A proton pump mechanism was directly proved when H^+-translocating NADPH-NAD^+ transhydrogenase was studied (see below, Sect. 7.4).

4.2 NADH:CoQ-Oxidoreductase (Complex I)

Let us now discuss the mechanism of proton potential generation by complex I. Many researchers used to think that this enzyme could function according to the Mitchell loop mechanism. According to some theories, FMN or CoQ bound to complex I could transfer hydrogen atoms across the membrane, and many FeS clusters might form a transmembrane path of electron transport. However, there are very serious difficulties in realization of such mechanisms. First of all, let us remember that $H^+/\bar{e}$ stoichiometry for complex I is 2, i.e., it would be necessary to ensure the functioning of two Mitchell loops simultaneously. Even more so, we have already mentioned that practically all the complex I cofactors (besides CoQ) are located in the peripheral domain of the enzyme, i.e., relatively far away from the membrane. That is why it seems rather unlikely that they would be able to drive transmembrane transfer of hydrogen atoms or electrons.

So, complex I probably generates proton potential according to the proton pump mechanism. But it remains unclear how protons are transported across the membrane and in what way their transport is coupled to the redox reactions catalyzed by complex I.

Quite recently, the three-dimensional structure of the peripheral domain of complex I from the thermophilic bacterium *Thermus thermophilus* was established in the group of Leonid Sazanov (Fig. 4.8) by X-ray analysis (Figs. 4.9a, b; Sazanov and Hinchliffe 2006). The number of FeS clusters and their relative position in the protein were clearly demonstrated. The eight main FeS clusters[4] form an electron-conducting chain in the peripheral domain, and this chain is stretched out along the entire domain. The FMN binding site is located between clusters N1a and N3. Taking these data into account, one can suggest the following sequence of electron transfers in this protein. NADH first transfers two electrons to FMN. This is followed by transfers of electrons from $FMNH_2$ along the chain of isopotential FeS clusters (N3, N1b, N4, N5, N6a, and N6b). It is important to note the role of flavin in oxidation of NADH. NADH is known to be a carrier of hydride ion (H^-), i.e., it can donate two (and only two!) electrons and one proton. However, an FeS cluster can usually accept only one electron. On the other hand, flavin can act as either two- and one-electron carrier (the process forming flavosemiquinone in the latter case). So, in the case of complex I as well as many other proteins, flavin acts as an adaptor between two- and one-electron redox cofactors. The role of the N1a cluster still remains unclear. This cofactor is probably not part of the electron transporting chain, for it is located "higher" than the flavin binding site. In the mitochondrial complex I, the N1a cluster has a more negative redox potential than flavin. The N1a cluster is probably needed for temporary oxidation of FMN semiquinone when this cofactor for some reason cannot transfer the second electron to the electron-conducting pathway right after the transfer of the first electron. This phenomenon should substantially decrease the lifetime of

---

[4] Complex I from *Thermus thermophilus*, when compared to other homologous proteins, contains an extra [4Fe–4S] cluster (N7). But this cofactor is not conservative. It seems not to participate in electron transfer from NADH to quinone.

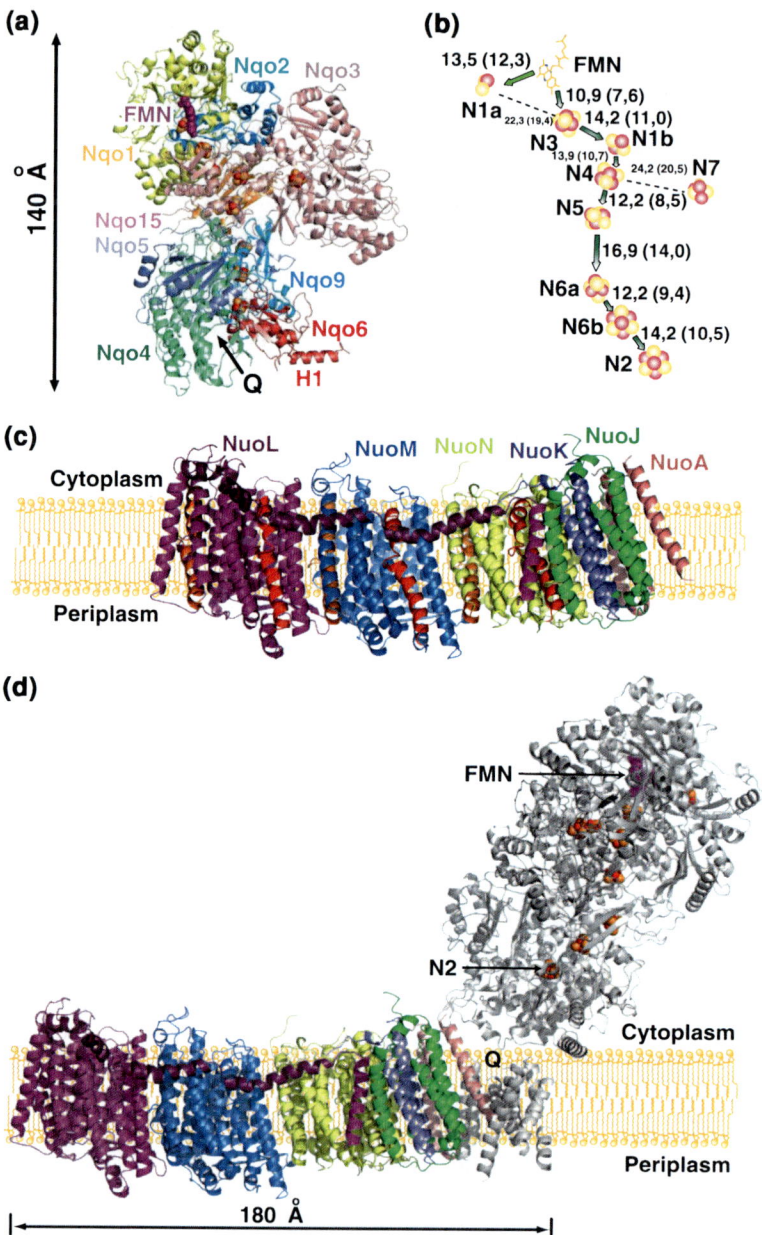

Fig. 4.9 a Three-dimensional structure of the peripheral domain of complex I from *T. thermophilus*. Arrangement of FeS clusters is shown by *red–orange* color. A pocket in the protein structure that is the likely site of ubiquinone binding is indicated by the *arrow* (Sazanov and Hincliffe 2006). **b** Arrangement of FeS clusters in complex I from *T. thermophilus*. Distances between the cluster centers and also between their edges (in brackets) are shown in Å (Sazanov and Hincliffe 2006). **c** Structure of membrane domain of complex I from *E. coli*. Broken helices are shown in *red* and *orange* (Efremov and Sazanov 2011). **d** Membrane domain of complex I from *E. coli* aligned with the membrane domain of the entire complex I from *T. thermophilus*. FMN is shown as magenta spheres and FeS clusters as *red-orange* spheres. The probable quinone-binding site (Q) is indicated (Efremov and Sazanov 2011)

flavosemiquinone and hence the rate generation of reactive oxygen species by complex I since (i) flavosemiquinone easily reduces O_2 to $O_2^{\bullet-}$ and (ii) FMN, which stands at NADH-binding site of complex I, must be accessible for such a small molecule as O_2 (Efremov et al. 2010; Efremov and Sazanov, 2011).

It would be important to determine what stage(s) of electron transfer in complex I might be coupled to protons translocation. The $E_m{'}$ values for FMN and the majority of FeS clusters are in the range from -370 to -240 mV, i.e., they do not differ too much from the potential of the $NAD^+/NADH$ couple. It is only the midpoint potential of iron–sulfur cluster N2 that is more positive (from -20 to -150 mV depending on the studied organism) (Ohnishi 1998). Due to the fact that there is no substantial difference in redox potentials of NADH and the isopotential group of FeS clusters, electron transfer from NADH to N6b is considered not to be accompanied by generation of $\Delta\bar{\mu}_{H^+}$. However, electron transfer from N6b to N2 and from N2 to quinone releases energy sufficient for transmembrane proton transfer. So these reactions are assumed to be coupled to proton potential generation.

The three-dimensional structures of the membrane domain of complex I from *E. coli* (Efremov et al. 2010) (Fig. 4.9c) as well as the entire complexes I from *T. thermophilus* (Efremov et al. 2010) (Fig. 4.9d) and *Y. lipolytica* (Hunte et al. 2010) have been determined quite recently. A pattern of 14 transmembrane helices belonging to homologous subunits NuoL, M, and N was found to repeat three times in the structure of the hydrophobic domain. Each of these subunits contains, in the same positions, two characteristic "broken" helices. Such discontinuous transmembrane helices are thought to be important for the function of many transporters (see Chap. 9). To conserve free energy of NADH:quinone-oxidoreduction in complex I, this energy must be transferred from the peripheral domain of the enzyme to subunits NuoL, M, and N, i.e. over a very long distance (up to ~ 100Å). The resolution of the structure of the membrane part of complex I from *E. coli* and *T. thermophilus* provided an intriguing possibility how this task might be accomplished. It was found that the most distal NuoL subunit in the membrane domain has a unique feature—the large fragment between the 15th and the last, 16th, transmembrane helices is organized as an amphipathic α-helix that runs along almost the entire length of the membrane domain on the N-side, and it is anchored by the 16th transmembrane α-helix in the proximal part of the domain

(Fig. 4.9c, d) (Efremov et al. 2010; Efremov and Sazanov 2011). A similar but somewhat shorter "horizontal" α-helix was observed in the three-dimensional structure of mitochondrial complex I from *Y. lipolytica* (Hunte et al. 2010). A "piston model" was proposed to explain how the energy might be transferred over a long distance in the membrane domain (Efremov et al. 2010). Delivery of electrons from NADH to quinone was postulated to cause conformational changes in the proximal membrane subunits leading to movements of the amphipathic α-helix, which is in contact with transmembrane helices. Such movements occurring along the membrane might result in formation of proton-conducting channels in the NuoL, M, and N subunits and, as a consequence, in transmembrane proton translocation. In this hypothesis, the amphipathic α-helix plays the role of a mechanical piston (analogs to the piston of a steam engine) that propagates long-distance conformational changes and results in proton pumping (Efremov and Sazanov 2011). Further studies are needed to test this hypothesis.

4.3 CoQH$_2$:Cytochrome *c*-Oxidoreductase (Complex III)

NADH:CoQ-oxidoreductase reduces CoQ, which in turn transfers electrons to the next segment of the respiratory chain, *CoQH$_2$:cytochrome c-oxidoreductase* (other names, complex *bc$_1$* and complex III). Complex III transfers electrons from quinol to cytochrome *c*, this reaction being coupled to $\Delta\bar{\mu}_{H^+}$ generation with stoichiometry of $H^+/\bar{e} = 1$.

4.3.1 Structural Aspects of Complex III

CoQH$_2$:cytochrome *c*-oxidoreductase crystals for the enzymes from bovine, hen, and yeast mitochondria have been prepared, and three-dimensional structures for these proteins with resolution of up to 2.3 Å have been determined (Xia et al. 1997; Zhang et al. 1998; Hunte et al. 2000). In contrast to the above-described complex I, the main aspects of *bc$_1$* complex functioning have already been established.

The polypeptide composition of complex III from bovine heart mitochondria is shown in Table 4.2. Complex III of the respiratory chain is a homodimer multi-subunit enzyme. Each of the *bc$_1$* complex monomers contains three catalytic subunits with four redox centers each: these are the low- and high-potential hemes b_L and b_H bound to cytochrome *b*; a [2Fe–2S]-type cluster that combines with a corresponding apoprotein [FeS$_{III}$ or Rieske protein; originally described by E. C. Slater (Fig. 4.10) and called "Slater factor"] and the C heme bound to the apoprotein of cytochrome *c$_1$*.

Besides reducing equivalent carriers, mitochondrial CoQH$_2$:cytochrome *c*-oxidoreductase includes eight more polypeptides that have no prosthetic groups. One (subunit VI) is responsible for CoQ binding. It consists of 110 amino acid

4.3 CoQH$_2$:Cytochrome c-Oxidoreductase (Complex III)

Table 4.2 Complex bc_1 subunits from beef heart mitochondria

Subunit	Molecular mass	Name	Function
I	49.5	Core protein I	Unknown
II	47.0	Core protein II	Unknown
III	44.0	cytochrome b	Binds hemes b_L and b_H
IV	28.0	cytochrome c_1	Binds heme c_1
V	21.5	Rieske protein	Binds [2Fe–2S] cluster
VI	13.5	Q$_{III}$ protein	Binds CoQ
VII	9.5		Unknown
VIII	9.0		Unknown
IX	8.0		Unknown
X	7.0		Binds cytochrome c
XI	6.5		Binds Rieske protein

Fig. 4.10 Britton Chance (*left*) and Edward Charles Slater

residues and contains a substantial number (about 50 %) of α-helix. There are some indications that subunit X is connected to cytochrome c_1 and facilitates its interaction with cytochrome c. In this subunit, 27 % of all the amino acid residues are glutamate and glutamine. An acidic cluster of eight consecutive glutamic acid residues is located near the N-terminus. One can speculate that this negatively charged sequence interacts with cytochrome c, which contains many positively charged amino acid residues. Subunits I and II are the largest; they were given the name "core proteins". This unfortunate name was invented at the time when practically nothing was known of the three-dimensional structure of complex III. A misconception of these proteins as the centers of structural arrangement of bc_1

complex seemed to be supported by their size (they are larger than other complex subunits) and by their lack of redox groups. Later these proteins were found to be peripheral. The smallest polypeptide of complex III (subunit XI) participates in the binding of Rieske protein. Functions of the other polypeptides with no redox centers remain unclear.

Only one of 11 polypeptides that constitute the bc_1 complex of eukaryotes, namely cytochrome b, is coded in the mitochondrial genome. The genes of the other proteins are in the cell nucleus.

It is noteworthy that the respiratory chain of many bacteria also contains an enzyme homologous to the bc_1 complex of eukaryotes but like the bacterial H^+-translocating NADH:quinone-oxidoreductases, prokaryotic bc_1 complexes have far simpler structure than their mitochondrial analogs. Bacterial enzymes consist of just three polypeptides homologous to three catalytic subunits of mitochondrial complex III (cytochrome b, cytochrome $c1$, and Rieske protein).

It has been already discussed (see Sect. 2.3.3) that a very close analog of bc_1 complex, namely b_6f complex, functions in plant chloroplasts. It participates in operation of the photosynthetic redox chain, where it transfers electrons from photosystem II to photosystem I. Considering its mechanism of functioning, b_6f complex is practically a full analog of mitochondrial complex III. It is interesting to note though that b_6f complex contains a number of additional prosthetic groups. The fact that it has three instead of two hemes is the most interesting difference. This additional heme, which was coined c_i, is located next to the b_H heme (Zhang et al. 2004). Its function is still unclear.

4.3.2 X-Ray Analysis of Complex III

Results of X-ray analysis of complex III are presented in Fig. 4.11. The narrower part of this enzyme penetrates the membrane while its wider parts protrude into the mitochondrial matrix and intermembrane space. Eight transmembrane α-helices of cytochrome b form the central hydrophobic domain of bc_1 complex. The second and the fifth α-helical segments were found to contain a pair of histidine residues that are used as axial ligands for the b_L and b_H hemes. The distance between the iron atoms of two hemes is 21 Å, which constitutes about half of the membrane thickness. There is also an additional way to anchor each heme to the protein—this occurs when carboxyl groups of the propionate residues of the heme bind to protonated amino groups of lysine residues of the cytochrome b apoprotein.

Cytochrome c_1 and Rieske protein are connected to the membrane via single transmembrane helices, and their catalytic domains face the intermembrane space of mitochondria. The "core proteins" are parts of a fragment of the enzyme protruding into the mitochondrial matrix. The catalytic domain of the FeS protein was found to be rather flexible. Depending on the reduction level of the bc_1 complex, it can occupy two different positions—next to b_L heme or next to c_1 heme. So, the FeS cluster of Rieske protein might function as a mobile electron carrier between the quinol and cytochrome c_1 (Zhang et al. 1998).

4.3 CoQH$_2$:Cytochrome c-Oxidoreductase (Complex III)

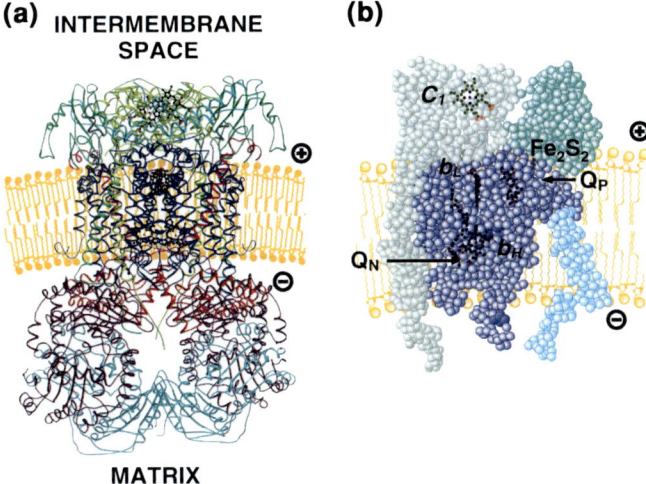

Fig. 4.11 Three-dimensional structure of yeast complex III. **a** Homodimer complex that is composed of catalytic subunits of cytochrome b (*blue*), Rieske protein (*green*), and cytochrome c_1 (*yellow*) with the corresponding prosthetic groups (*black*) and also of eight additional subunits, six of which are indicated by separate colors: core protein I (*purple*), core protein II (*blue-green*), subunit VI (*blue*), VII (*red*), VIII (*pink*), and IX (*gray*). **b** Catalytic subunits of one monomer. Q$_P$ and Q$_N$—quinone binding sites in complex III (Hunte et al. 2003)

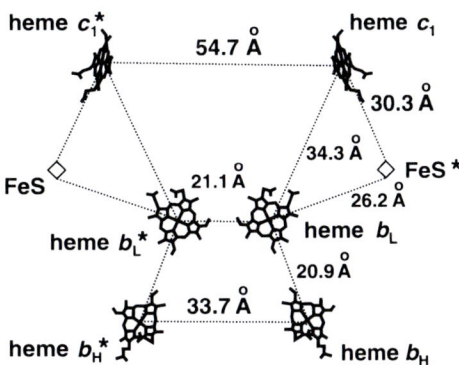

Fig. 4.12 Arrangement of prosthetic groups in the complex III homodimer form. The groups of the left monomer are marked (*). It is noteworthy that the *right* FeS cluster belongs to the *left* monomer due to complicated configuration of the Rieske apoprotein

The X-ray analysis also helped to reveal the two CoQ-binding sites in the bc_1 complex. They were given the names P and N centers (P) and (N) are for positively and negatively charged compartments of the mitochondrion, respectively. Other names are o (out) and i (in) centers.

Figure 4.12 presents the distances between the electron transport centers of complex III. One can see that the arrangement makes it possible to transfer electrons not just within a certain monomer, but also between monomers of the homodimer of this complex (Hunte et al. 2000).

4.3.3 Functional Model of Complex III

The scheme of functioning of the mitochondrial $CoQH_2$:cytochrome c-oxidoreductase has already been presented (see Fig. 4.4a). It is noteworthy that the mechanism of proton potential generation by this enzyme was predicted by Peter Mitchell in 1976 (Mitchell 1976). By now this elegant scheme has been completely confirmed both by various structural data and many biophysical experiments. Let us remember the key postulate of the Q-cycle scheme: paths of each of the two electrons leaving $CoQH_2$ are different. This statement has been directly proved. Quinol oxidation was shown to occur in the P center with two protons being released into the mitochondrial intermembrane space. The first electron is transferred to the FeS cluster of the Rieske protein, while the second electron is transferred to the low potential b_L heme of cytochrome b. The further destinies of two electrons are different. The electron from the FeS cluster is transferred to cytochrome c_1, and then to cytochrome c and cytochrome c oxidase. The electron from the b_L heme is transferred to the high potential heme b_H, and it is later used for the reduction of the quinone molecule in the N center, thus forming semiquinone anion. Oxidation of the second quinol molecule in the P center, results in reduction of one more cytochrome c and one more heme b_L. The latter reduces heme b_H which in turn add electron to ubisemiquinone in the N center, reducing it to quinol. This quinol is later detached from the enzyme and diffuses to the other side of the membrane, so as to be oxidized again in the P center.

Electron transfer from the N center to cytochrome c proceeds along the membrane, and therefore it does not result in energy conservation. However, electron transfer from heme b_L to heme b_H is transmembrane, thus leading to generation of electric potential. So, proton potential generation by complex III follows the mechanism of a Mitchell loop.

The redox potentials of the electron carriers of the $CoQH_2$:cytochrome c-oxidoreductase are compatible with the Q-cycle scheme, i.e. -40 mV for heme b_L, $+40$ mV for heme b_H, $+280$ mV for the [2Fe–2S] cluster, and $+220$ and $+250$ mV for cytochromes c_1 and c, respectively. It is important that the b_H heme redox potential is substantially more positive than that of the b_L heme. This is why electron transfer from b_L to b_H is accompanied with an energy release. This energy is stored in the form of $\Delta\Psi$, for $b_H \rightarrow b_L$ oxidoreduction is directed across the membrane. Electron transfer from the FeS cluster of the Rieske protein to cytochrome c_1 and then to cytochrome c proceeds without any major redox potential changes.

The properties of the mitochondrial $CoQH_2$:cytochrome c- oxidoreductase are in many ways similar to those of homologous enzyme complexes of the photosynthetic redox chains of purple bacteria chromatophores and thylakoids of cyanobacteria or chloroplasts. The bc_1 complex from mammalian mitochondria was reconstructed into proteoliposomes with the photosynthetic reaction center complex from chromatophores of the purple bacterium *Rhodobacter sphaeroides*. The resulting "chimera" was able to catalyze light-driven cyclic electron transfer

Fig. 4.13 Spatial structures of cytochromes c from tuna muscle mitochondria (**a**) and the bacterium *P. denitrificans* (**c**) and also cytochrome c_2 from the bacterium *R. rubrum* (**b**). The heme is shown in *red*

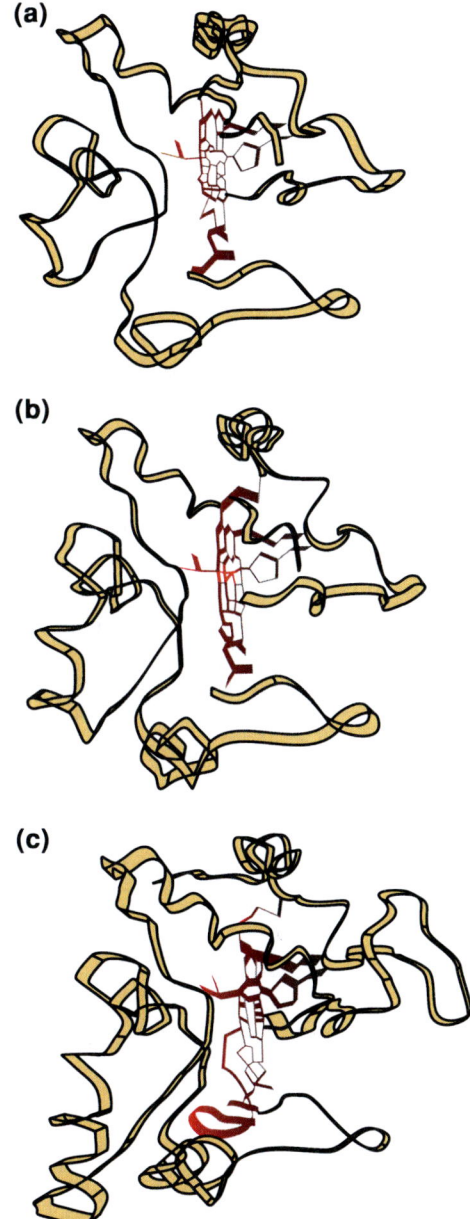

coupled to H^+ translocation across the proteoliposome membrane (Rich and Heathcote 1983).

A unique natural system has been described with one PQH_2:cytochrome c-oxidoreductase functioning in a respiratory chain in the dark and in a photosynthetic chain in the light. This is the thylakoid membrane of cyanobacteria.

When in the light, plastoquinone, b_6f complex, cytochrome c, and plastocyanin provide the connection between photosystems 2 and 1. In the dark, b_6f complex and cytochrome c participate in the respiratory chain between NAD(P)H:plastoquinone-oxidoreductase and cytochrome oxidase.

4.3.4 Inhibitors of Complex III

Several inhibitors of bc_1 complex with various mechanisms of functioning have been found. The toxic antibiotic antimycin A prevents quinone from binding in the N center. Myxothiazol and mucidin inhibit quinol oxidation in the P center. The FeS Rieske cluster seems to be the target for Zn^{2+} ions and for British anti-lewisite (β-mercaptopropanol). Stigmatellin blocks the electron pathway from the FeS Rieske cluster to cytochrome c_1. Structures of these inhibitors are shown in Appendix 3.

A very interesting experiment was conducted by Britton Chance (Fig. 4.10) and his colleagues who studied muscle mitochondria of a girl suffering from severe muscle fatigue. The child was confined to a wheelchair. Chance first analyzed the impact of exercise stress on the creatine phosphate level in her muscles. Muscle activity caused a rapid decrease in creatine phosphate concentration, while in the state of rest it was restored to the initial level extremely slowly. From a bioenergetic perspective, it is the decrease in ATP synthesis rate that might be the simplest explanation for this effect. Muscle biopsy and analysis of activity of the respiratory complexes in isolated mitochondria were carried out so that to test this hypothesis. As a result, deficiency of some complex III subunits causing by strong inhibition of electron transfer complex III was found. It was also shown that vitamin K_3 abolishes such an inhibitor. Injection of vitamin K_3 and ascorbic acid caused a miraculous cure for the patient—within an hour the girl felt strong enough to rise from her wheelchair. Although the maintenance of normal muscle activity required repeated injections of vitamin K_3 every 6 hours, the girl could leave her wheelchair, go to college, successfully graduate, get married, and live a normal life (Eleff et al. 1984; Argov et al. 1986).

4.4 Cytochrome c Oxidase (Complex IV)

Cytochrome c oxidase (complex IV) catalyzes oxidation of cytochrome c by molecular oxygen. This reaction is coupled to generation of $\Delta\bar{\mu}_{H^+}$ with the stoichiometry of 2 $H^+/\bar{e}$.

4.4.1 Cytochrome c

Cytochrome c localized in the mitochondrial intermembrane space functions as a link in the process of reducing equivalent transfer from complex III to complex IV. This water-soluble protein with molecular mass of 12 kDa is composed of 104 amino acid residues. The spatial structure of cytochrome c was established on the basis of X-ray analysis with the resolution of 1.5 Å (Fig. 4.13). This electron carrier seemed to have been studied quite thoroughly. The more amazing then was the discovery made in the last decade of the twentieth century. Cytochrome c was shown to be one of the "death proteins" that participate in a cell suicide (*apoptosis*). We will discuss this phenomenon in one of the last chapters of this book (Chap. 15), and here we will concentrate mainly on the vital function of cytochrome c as the respiratory chain electron carrier.

The His-18 and Met-80 residues play an important role in the functioning of cytochrome c because they are the axial ligands of the iron atom in the heme. Similar to all the analogous cytochromes of C type, mesoheme C is covalently bound to sulfhydryl groups of cysteine residues of the protein (Cys-14 and Cys-17). This is the difference between the c-type cytochromes and those of other types (a, b, d, o, and other), where hemes are connected in a noncovalent manner. The covalent binding of heme C to apocytochromes was suggested to be the result of the fact that majority of the c-type cytochromes are located on the outer side of the mitochondrial or bacterial membrane (Wood 1983). A noncovalently bound heme might be dissociated and diluted in the cytosol (in case of mitochondria) or periplasm and then the outer medium (in the case of bacteria).

The three-dimensional structure of cytochrome c is reminiscent of a mitten that keeps the heme in such a position that the heme edge reach to the surface of the protein (Takano and Dickerson 1980). A corona of positively charged amino acid residues (Lys-13, Lys-72, Lys-79, and Lys-86) is located close to the heme edge. Lysine residues are needed to tack cytochrome c to its partners—cytochrome c_1 and cytochrome oxidase. For instance, cytochrome oxidase has three negatively charged amino acids, which (similar to a magnetic lock) provide the electrostatic attraction of cytochrome c and its fixation in the right position. Oxidation and reduction of cytochrome c are not accompanied by substantial changes in its conformation.

The amino acid sequence of cytochrome c was established long before the first attempts for its X-ray analysis were made. Extraordinary stability, small size, and relative simplicity of cytochrome c purification were reasons for the rapid accumulation of information on the structure of this protein in different organisms. When comparing amino acid sequences of cytochromes c extracted from different sources, genosystematics began to develop their first evolutionary trees. It was found that humans and chimpanzees have identical primary structure of cytochrome c. When compared to human cytochrome c, rhesus macaques have one amino acid change, horses 12, chickens 13, tuna 21, wheat 35, and the fungus *Neurospora* 44 changes.

4.4.2 Cytochrome c Oxidase: General Characteristics

Cytochrome c oxidase (complex IV) has a molecular mass of about 200 kDa. Eukaryotic complex IV consists of 13 polypeptides. The first three are hydrophobic; they are coded by the mitochondrial genome and are considered to be the main catalytic subunits of the enzyme. The accepted numbering of the subunits was done on the basis of SDS electrophoresis data: I (57 kDa), II (26 kDa), III (30 kDa). It is worth noting that bacterial cytochrome oxidases usually consist of just three subunits that are homologous to the three major subunits of eukaryotic complex IV. The other subunits of eukaryotic cytochrome oxidase are rather small, their molecular masses varying from 5 to 17 kDa.

Mitochondrial cytochrome oxidase contains four redox cofactors—two hemes of A-type (a and a_3) and two redox centers containing copper atoms (Cu_A and Cu_B). In addition, complex IV includes a few redox-inert metal ions—Mg^2 (Mn^{2+}), Zn^{2+}, and in some cases also Ca^{2+} or Na^+. Functions of these ions in the oxidase remain unclear. All the cytochrome oxidase redox cofactors are located on subunits I and II (Wikström et al. 1981). The Cu_A factor is located on subunit II, and it is also on this subunit that the cytochrome c binding site is located (the above-mentioned group of negatively charged amino acids, i.e. Asp-136, Asp-187, and Glu-227). Subunit I contains the remaining redox centers of the enzyme—heme a, heme a_3, and Cu_B. Heme a_3 and Cu_B are located very close to each other (the distance between the metal atoms is 4.5 Å), thus forming the so-called heme–copper binuclear center. This center serves as a binding site for the O_2 molecule. The third catalytic subunit carries no redox cofactors and its function remains unclear. In case of the bacterium *Paracoccus denitrificans* the loss of subunit III during the extraction of cytochrome oxidase does not influence the functional properties of this protein.

Let us consider now the redox cofactors of cytochrome oxidase in more detail. The Cu_B center consists of a copper ion bound to the protein by coordination bonds with three histidine residues (via nitrogen atoms of imidazole rings). This ion can be in one of two valence states, namely +1 and +2. So, Cu_B functions as a one-electron carrier. The Cu_A center consists of two copper atoms located very close to each other. The Cu_A center, similar to Cu_B, functions as a one-electron carrier ($Cu^{2+} Cu^{1+} \leftrightarrow Cu^{2+}Cu^{2+}$).

Besides the copper atoms, mitochondrial cytochrome oxidase contains also two hemes—a and a_3. Let us first describe some general properties of heme before studying these cofactors. Iron ions can form six coordinating bonds with ligands in complex compounds (hemes belong to this group). In the case of cytochromes, four coordinating bonds with iron are provided by four nitrogen atoms of the porphyrin ring (see Appendix 2), and these bonds are located in the plane of the ring. The two remaining bonds are usually provided by amino acid residues of the protein (most often by histidine, rather rarely by methionine). Such ligands are called axial—their bonds to iron are located on the opposite sides of the porphyrin ring and are perpendicular to it. The hemes with all the six coordinating bonds of

4.4 Cytochrome c oxidase (Complex IV)

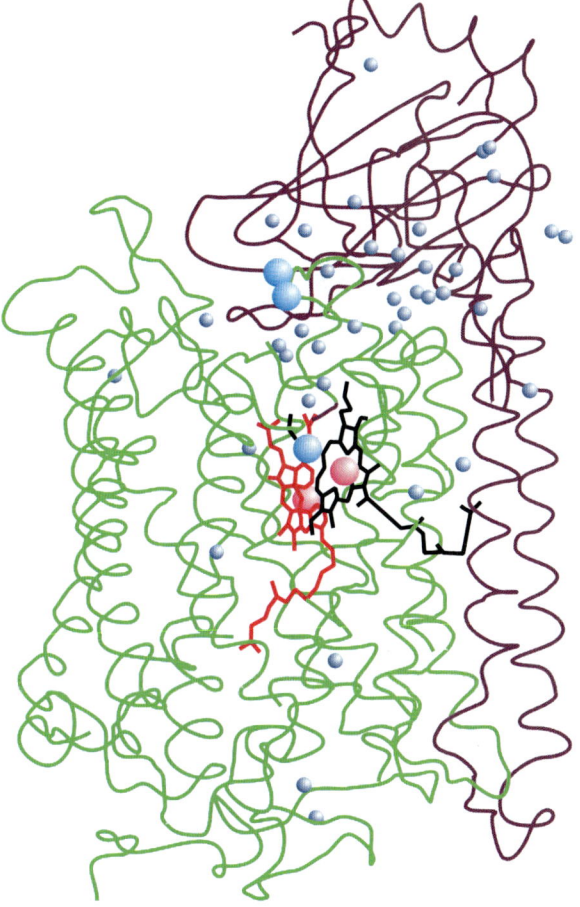

Fig. 4.14 Structure of the two main subunits of cytochrome c oxidase from *P. denitrificans*. Subunit I is colored *green*, subunit II *purple*, heme *a red*, heme a_3 *black*; copper atoms are represented as big *blue* spheres, iron atoms big *pink* spheres, and water molecules small *dark-blue* spheres (Ostermeier et al. 1997)

iron are called *low-spin* (for they have an equilibrium electron configuration that corresponds to the lowest energy level with the minimal spin s = 1/2). Such hemes usually function as electron carriers; all the cytochromes described in the previous chapters as well as heme *a* of cytochrome oxidase belong to this group. Nevertheless, certain hemes (e.g. heme a_3) lack one of the axial ligands. The spin of such hemes is 5/2, so they are called *high-spin*. When small nucleophilic compounds are present in the medium (e.g. O_2, NO, CO, H_2S, HCN, azide, etc.) they can bind to a high-spin heme, thus operating as its second axial ligand. In the case of oxygen, such bonding can be accompanied by electron transfer from the heme to O_2, and hence oxygen-reductase activity will occur. So, high-spin hemes can function not only as electron carriers, but also as the binding sites for small ligands, (like oxygen in myoglobin and hemoglobin, or NO in guanylate cyclase), or as sites of their reduction (e.g. oxygen reduction in cytochrome oxidase or NO reduction in NO-reductase).

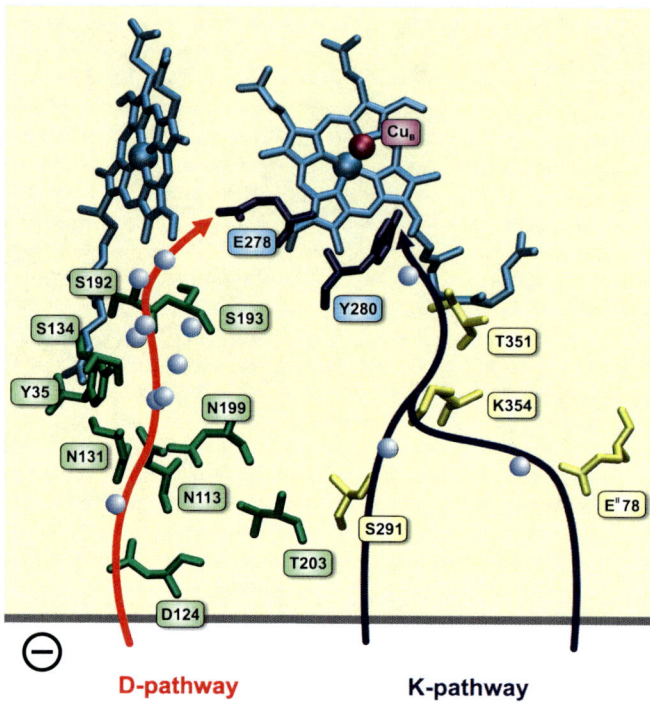

Fig. 4.15 Structure of K- and D-channels in bovine cytochrome oxidase. Small *light-blue* spheres represent water molecules

4.4.3 X-Ray Analysis of Complex IV

The X-ray analysis of cytochrome oxidase crystals was conducted simultaneously by two groups of researchers—one in Japan (laboratory of Shinya Yoshikawa who studied complex IV from bovine heart), and the other in Germany (under the guidance of Nobel Prize winner Hartmut Michel, the enzyme was extracted from the bacterium *P. denitrificans*) (Tsukihara et al. 1996) and (Iwata et al. 1995), respectively. The high-resolution (2.3 Å) three-dimensional structure of cytochrome oxidase was described. It is noteworthy that the structures of the main catalytic subunits were practically identical in the eukaryotic and prokaryotic enzymes. Cytochrome oxidase was shown to be almost completely immersed in the membrane, and its cross-section is reminiscent of a trapezoid (Fig. 4.14). The copper-binding domain of subunit II is the only part of the protein protruding from the membrane depth into the intermembrane space. This domain contains the Cu_A center and is anchored in the membrane by a harpoon-like structure that consists of two hydrophobic transmembrane α-helices. Subunit I forms 12 transmembrane helices, while subunit III forms seven. It has already been mentioned that subunit I carries three redox-active prosthetic groups—hemes *a* and a_3 and copper Cu_B.

4.4 Cytochrome c oxidase (Complex IV)

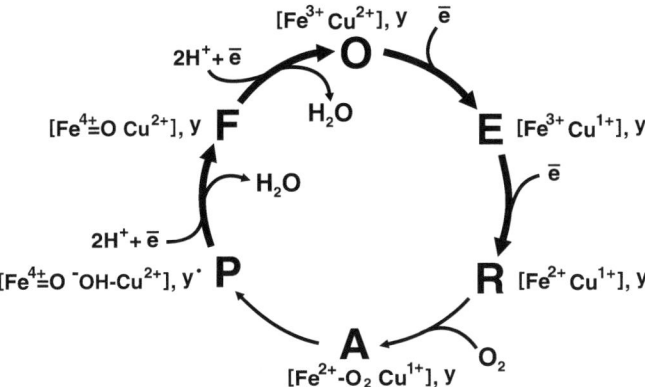

Fig. 4.16 A six step scheme of the process of reduction of molecular oxygen in the binuclear center of cytochrome c oxidase. Y—presumably tyrosine-244 of subunit I

Figures 4.14 and 4.15 show these centers to be located approximately along one line parallel to the membrane plane (*the hemes themselves are located perpendicular to the membrane*); they are closer to its external surface (about 15 Å away from this surface). The X-ray analysis studies showed an interesting feature of the arrangement of the binuclear cytochrome oxidase center. One of the histidine residues (which serves as copper Cu_B ligands), His-240, forms a covalent bond with the phenol ring of a nearby tyrosine residue, Tyr-244. This covalent bond was later found to be formed autocatalytically during the turnover of the enzyme. Such type of posttranslational modification is extremely rare and has been found only in oxidases so far.

Analysis of the three-dimensional structure of complex IV also suggests the existence of two channels for proton transport from the mitochondrial matrix to the binuclear center (see Fig. 4.15). These pathways consist of conservative residues of polar amino acids and firmly bound water molecules. They were coined as K- and D-channels according to the abbreviations of amino acid residues located at the entrance to these pathways, Lys-354 and Asp-124, respectively.

Analysis of the three-dimensional structure of cytochrome oxidase fails to suggest a pathway(s) of proton transfer from the binuclear center to the intermembrane space. However, this part of the protein is not so hydrophobic; the distance between the binuclear center and the outer surface of the membrane is only 15 Å, so it seems possible that proton transfer does not require any special channel in this case.

4.4.4 Electron Transfer Pathway in Complex IV

Equation 4.1 presents the sequence of electron transfer steps in cytochrome oxidase. The electron is first transferred from cytochrome c to the Cu_A center. Then it is transferred from Cu_A to the a heme and further to the binuclear heme–copper [heme a_3–Cu_B] center, the terminal respiratory chain component that reduces O_2:

$$\text{Cytochrome } c \xrightarrow{4\bar{e}} Cu_A \xrightarrow{4\bar{e}} \text{heme } a \xrightarrow{4\bar{e}} [\text{heme } a_3 \; Cu_B] \xrightarrow{4\bar{e}+O_2+4H^+} 2\,H_2O \quad (4.1)$$

However, this equation does not seem to be sufficient for the description of electron transfer in cytochrome oxidase. Let us remember that four electrons are needed to reduce an O_2 molecule to water, while cytochrome c functions as a carrier for just one electron. This means that a special mechanism is needed to accumulate electrons on the oxidase and to reduce O_2. Let us look at this mechanism in more detail (Fig. 4.16):

I The oxidation of the first cytochrome c molecule leads to an electron transfer from cytochrome c to Cu_A and further to a, a_3, and Cu_B. The enzyme changes from an oxidized form (O) to a one-electron reduced form (E). This form (E) is not yet capable of binding oxygen. The exact localization of the first electron on the enzyme in state E has not been established; we will assume the binuclear enzyme center to be in the state ($Fe^{3+}Cu^{1+}$) so as to simplify the discussion.

II Oxidation of the second cytochrome c molecule actuates the following chain of events:

(a) both electrons (the first and the second) can move to the binuclear center and reduce it, thus generating the $Fe^{2+}Cu^{1+}$ form of the enzyme, which was given the name R;

(b) due to the transfer of the two electrons to the binuclear center, it becomes able to bind oxygen; so, the R form changes into the A form (the state of the binuclear center is $Fe^{2+}-O_2 \; Cu^{1+}$);

(c) oxygen binding is followed by electron transfer from cytochrome oxidase to this oxygen.

It was first suggested that this stage involved two-electron reduction of O_2 leading to peroxide bridge formation between the atoms of iron and copper of the binuclear center ($Fe^{3+}-O-O-Cu^{2+}$). This form was given the name P, or peroxide. However, it was later found that it was already at this stage that four-electron reduction of oxygen was taking place (but this form was still left with the name P for a historical reason). There surely appears the question how can this four-electron reduction of oxygen be possible by the 2ē-reduced enzyme? One additional electron for reduction of oxygen was found to be taken from the heme a_3 iron, thus transforming it into the very unusual ferryl state Fe^{4+}. The other electron (and most likely a proton as well) was found to be transferred from some aromatic amino acid residue located nearby, this process leading to formation of the radical form of the residue ($Y^{\bullet}$).[5] Also, one H^+ ion

---

[5] The radical form $Y^{\bullet}$ is considered to be formed from the Tyr-244 residue. Generation of such an unusual and extremely reactive radical is probably the reason for formation of the covalent bond between Tyr-244 and a nearby His-240. This covalent bond probably stabilizes the radical form, which helps to avoid destruction of the protein.

Fig. 4.17 Mårten Wikström (*left*) and Michael Verkhovsky

is taken up, and copper is bound to a hydroxyl anion. So, it now seems that the P form of cytochrome oxidase looks as follows: $Y^\bullet\ Fe^{4+} = O^{2-}\ {}^-HO\text{-}Cu^{2+}$. It should be noted that the redox potentials of the $Y^\bullet/Y$ and Fe^{4+}/Fe^{3+} couples are extremely high ($\geq +\ 0.7$ V).

So, it is already at this stage that the reduction of oxygen to water occurs. Later the enzyme only needs to be returned to its initial state. This happens during the oxidation of the next two cytochrome c molecules.

III During the oxidation of the third cytochrome c molecule, an electron is transferred to the $Y^\bullet$ and another proton is taken from the water phase. This leads to the formation of the ferryl (F) form of the enzyme ($Fe^{4+} = O^{2-}\ Cu^{2+}$) and the first H_2O molecule.

IV The oxidation of the fourth cytochrome c molecule leads to the formation of the second H_2O molecule and regeneration of the original oxidized (O) form of the enzyme ($Fe^{3+}\ Cu^{2+}$), hence the catalytic cycle of cytochrome oxidase is completed.

4.4.5 Mechanism of $\Delta\bar{\mu}_{H^+}$ Generation by Cytochrome c Oxidase

It has been already mentioned that cytochrome oxidase generates $\Delta\bar{\mu}_{H^+}$ with stoichiometry of 2 $H^+/\bar{e}$. When describing the previous components of the respiratory chain, we postulated the existence of two fundamentally different mechanisms of proton potential generation—a Mitchell loop and a proton pump. Both of these mechanisms occur during the functioning of cytochrome oxidase (Wikström 2004).

1. *Mitchell Loop.* Electron transport from cytochrome c to the binuclear center is coupled to electron transfer from the intermembrane space into the membrane by about one-third of its span (see Fig. 4.14). However, protons are also needed for the reduction of oxygen to water. So, transport of protons via K- or D-channels occurs from matrix to binuclear center. As a result of these two processes, generation of $\Delta\bar{\mu}_{H^+}$ with stoichiometry of about 1 $H^+/\bar{e}$ takes place.
2. *Proton Pump.* In 1977, Mårten Wikström (Fig. 4.17) showed that oxidation of cytochrome c by complex IV was accompanied by acidification of the mitochondrial intermembrane space (Wikström 1977). This should not have happened if only a Mitchell loop was operative. So, cytochrome oxidase was shown to generate $\Delta\bar{\mu}_{H^+}$ not only according to the Mitchell loop mechanism, but also working as a proton pump with overall efficiency of about 2 $H^+/\bar{e}$.

Although substantial progress in studies of this phenomenon has been achieved, the precise mechanism of proton pumping by cytochrome oxidase remains unknown. As for today, there are a few interesting theoretical models of this process (Belevich and Verkhovsky 2008). However, their discussion goes beyond the scope of this book. We will only briefly describe the way proton pumping is connected to the functioning of the cytochrome oxidase catalytic cycle.

Reactions O → E and E → R of the cytochrome oxidase catalytic cycle were shown to be characterized by only minor free energy changes. The energy is barely enough for $\Delta\Psi$ generation in the process of electron transfer from Cu_A to a_3. However, the P → F and F → O transitions (i.e. reduction of such high-potential compounds as $Y^\bullet$ and Fe^{4+}) are accompanied by substantial release of energy. That is why the O → E and E → R transitions were first thought not to be coupled to proton pumping, while the P → F and F → O reactions were suggested to be coupled to transmembrane H^+ transport with doubled stoichiometry. However, later it was shown by Michael Verkhovsky (Fig. 17) and coauthors that each of the four electrons transferred from cytochrome c to cytochrome oxidase is coupled to pumping of one proton (Bloch et al. 2004). Understanding of the mechanism of these reactions requires further study.

4.4.6 Inhibitors of Cytochrome Oxidase

It has already been mentioned that high-spin hemes can bind small nucleophilic compounds, such as CO, NO, CN^-, H_2S, and N_3^-. The binding of these ligands leads to the change of the high-spin state of heme into the low-spin state. It also prevents the interaction of heme with such a relatively weak ligand as O_2. That is why all the above-mentioned compounds are strong inhibitors of cytochrome oxidase. It is important to note that many of these inhibitors act as strong poisons.

References

Argov Z, Bank WJ, Maris J, Eleff S, Kennaway NG, Olson RE, Chance B (1986) Treatment of mitochondrial myopathy due to complex III deficiency with vitamins K_3 and C: A ^{31}P-NMR follow-up study. Annal Neurology 19:598–602

Belevich I, Verkhovsky MI (2008) Molecular mechanism of proton translocation by cytochrome c oxidase. Antioxid Redox Signal 10:1–29

Bloch D, Belevich I, Jasaitis A, Ribacka C, Puustinen A, Verkhovsky MI, Wikström M (2004) The catalytic cycle of cytochrome c oxidase is not the sum of its two halves. Proc Natl Acad Sci USA 101:529–533

Efremov RG, Sazanov LA (2011) Structure of the membrane domain of respiratory complex I. Nature 476:414–420

Efremov RG, Baradaran R, Sazanov LA (2010) The architecture of respiratory complex I. Nature 465:441–445

Eleff S, Kennaway NG, Buist NR, Darley-Usmar VM, Capaldi RA, Bank WJ, Chance B (1984) ^{31}P NMR study of improvement in oxidative phosphorylation by vitamins K_3 and C in a patient with a defect in electron transport at complex III in skeletal muscle. Proc Natl Acad Sci USA 81:3529–3533

Engelhardt WA (1930) Ortho- und pyrophosphate im Aeroben und Anaeroben Stoffwechsel der Blutzellen. Biochem Z 251:16–21

Friedrich T, Scheide D (2000) The respiratory complex I of bacteria, archaea and eukarya and its module common with membrane-bound multisubunit hydrogenases. FEBS Lett 479:1–5

Galante YM, Hatefi Y (1978) Resolution of complex I and isolation of NADH dehydrogenase and an iron-sulfur protein. Methods Enzymol 53:15–21

Galkin AS, Grivennikova VG, Vinogradov AD (1999) $H^+/2e^-$ stoichiometry in NADH-quinone reductase reactions catalyzed by bovine heart submitochondrial particles. FEBS Lett 451:157–161

Grivennikova VG, Ushakova AV, Cecchini G, Vinogradov AD (2003) Unidirectional effect of lauryl sulfate on the reversible NADH:ubiquinone oxidoreductase (Complex I). FEBS Lett 549:39–42

Guénebaut V, Schlitt A, Weiss H, Leonard K, Friedrich T (1998) Consistent structure between bacterial and mitochondrial NADH:ubiquinone oxidoreductase (complex I). J Mol Biol 276:105–112

Hirst J, Carroll J, Fearnley IM, Shannon RJ, Walker JE (2003) The nuclear encoded subunits of complex I from bovine heart mitochondria. Biochim Biophys Acta 1604:135–150

Hunte C, Koepke J, Lange C, Rossmanith T, Michel H (2000) Structure at 2.3 Å resolution of the cytochrome bc_1 complex from the yeast *Saccharomyces cerevisiae* co-crystallized with an antibody Fv fragment. Structure 8:669–684

Hunte C, Palsdottir H, Trumpower BL (2003) Protonmotive pathways and mechanisms in the cytochrome bc_1 complex. FEBS Lett 545:39–46

Hunte C, Zickermann V, Brandt U (2010) Functional modules and structural basis of conformational coupling in mitochondrial complex I. Science 329:448–451

Iwata S, Ostermeier C, Ludwig B, Michel H (1995) Structure at 2.8 Å resolution of cytochrome c oxidase from *Paracoccus denitrificans*. Nature 376:660–669

Leif H, Weidner U, Berger A, Spehr V, Braun M, van Heek P, Friedrich T, Ohnishi T, Weiss H (1993) *Escherichia coli* NADH dehydrogenase I, a minimal form of the mitochondrial complex I. Biochem Soc Trans 21:998–1001

Mitchell P (1966) Chemiosmotic coupling in oxidative and photosynthetic phosphorylation. Glynn Research, Bodmin

Mitchell P (1976) Possible molecular mechanisms of the protonmotive function of cytochrome systems. J Theor Biol 62:327–367

Morgner N, Zickermann V, Kerscher S, Wittig I, Abdrakhmanova A, Barth HD, Brutschy B, Brandt U (2008) Subunit mass fingerprinting of mitochondrial complex I. Biochim Biophys Acta 1777:1384–1391

Ohnishi T (1998) Iron-sulfur clusters/semiquinones in complex I. Biochim Biophys Acta 1364:186–206

Ostermeier C, Harrenga A, Ermler U, Michel H (1997) Structure at 2.7 Å resolution of the *Paracoccus denitrificans* two-subunit cytochrome c oxidase complexed with an antibody FV fragment. Proc Natl Acad Sci U S A 94:10547–10553

Papa S, Brunori M (2011) Allosteric cooperativity in respiratory proteins. Biochim Biophys Acta 1807:1251–1252

Papa S, Guerrieri F, Lorusso M, Simone S (1973) Proton translocation and energy transduction in mitochondria. Biochemie 55:703–716

Rich PR, Heathcote P (1983) Light-activated proton-motive force generation in lipid vesicles containing cytochrome bc_1 complex and bacterial reaction centres. Biochim Biophys Acta 725:332–340

Sato K, Nishina Y, Setoyama C, Miura R, Shiga K (1999) Unusually high standard redox potential of acrylyl-CoA/propionyl-CoA couple among enoyl-CoA/acyl-CoA couples: a reason for the distinct metabolic pathway of propionyl-CoA from longer acyl-CoAs. J Biochem 126:668–675

Sazanov LA, Hinchliffe P (2006) Structure of the hydrophilic domain of respiratory complex I from *Thermus thermophilus*. Science 311:1430–1436

Skulachev VP (1973) Redox and hydrolytic generators of electric potential ($\Delta\psi$) in coupling membranes. Inter Congr Biochem Abstr 9:208

Skulachev VP (1974) Enzymic generators of membrane potential in mitochondria. Ann NY Acad Sci 227:188–202

Skulachev VP, Maslov SP (1960) The role of the noncoupled oxidation in thermoregulation. Biokhimiia 25:1058–1064 (in Russian)

Takano T, Dickerson RE (1980) Redox conformation changes in refined tuna cytochrome c. Proc Natl Acad Sci U S A 77:6371–6375

Tocilescu MA, Zickermann V, Zwicker K, Brandt U (2010) Quinone binding and reduction by respiratory complex I. Biochim Biophys Acta 1797:1883–1890

Tsukihara T, Aoyama H, Yamashita E, Tomizaki T, Yamaguchi H, Shinzawa-Itoh K, Nakashima R, Yaono R, Yoshikawa S (1996) The whole structure of the 13-subunit oxidized cytochrome c oxidase at 2.8 Å. Science 272:1136–1144

Wikström M (1977) Proton pump coupled to cytochrome c oxidase in mitochondria. Nature 266:271–273

Wikström M (1984) Two protons are pumped from the mitochondrial matrix per electron transferred between NADH and ubiquinone. FEBS Lett 169:300–304

Wikström M (2004) Cytochrome c oxidase: 25 years of the elusive proton pump. Biochim Biophys Acta 1655:241–247

Wikström M, Krab K, Saraste M (1981) Cytochrome oxidase, a synthesis. Academic Press, London

Wood PM (1983) Why do c-type cytochromes exist? FEBS Lett 164:223–226

Xia D, Yu CA, Kim H, Xia JZ, Kachurin AM, Zhang L, Yu L, Deisenhofer J (1997) Crystal structure of the cytochrome bc_1 complex from bovine heart mitochondria. Science 277:60–66

Yagi T (1990) Inhibition by capsaicin of NADH-quinone oxidoreductases is correlated with the presence of energy-coupling site 1 in various organisms. Arch Biochem Biophys 281:305–311

Zhang Z, Huang L, Chi Y-I, Kim KK, Crofts AR, Berry EA, Kim S-H (1998) Electron transfer by domain movement in cytochrome bc_1. Nature 392:677–684

Zhang H, Primak A, Cape J, Bowman MK, Kramer DM, Cramer WA (2004) Characterization of the high-spin heme x in the cytochrome b_6f complex of oxygenic photosynthesis. Biochemistry 43:16329–16336

Zharova TV, Vinogradov AD (1997) A competitive inhibition of the mitochondrial NADH-ubiquinone oxidoreductase (complex I) by ADP-ribose. Biochim Biophys Acta 1320:256–264

Chapter 5
Structure of Respiratory Chains of Prokaryotes and Mitochondria of Protozoa, Plants, and Fungi

In the previous chapter, we discussed the structure of the respiratory chain from the mitochondria of higher animals. Relatively high efficiency of energy conservation (high value of H^+/O_2 coefficient) is characteristic for this type of respiratory chain, and the electron transfer pathway from NADH to oxygen is linear in this case. The structure of bacterial and archaeal respiratory chains and also that of protozoan, plant, and fungal mitochondria have a number of crucial differences from the electron transport chain of animal mitochondria.

(1) Each of three "canonical" respiratory chain energy coupling sites (complexes I, III, and IV) can be bypassed via alternative noncoupled or partially coupled electron transfer pathways.
(2) Certain enzymes of the bacterial respiratory chains use sodium ions instead of protons as the primary coupling ions (we will discuss this phenomenon in more detail in Chap. 12).
(3) Bacteria are capable of utilizing a far broader list of compounds both as respiratory chain electron donors and acceptors. For instance, many microorganisms, besides oxygen, can reduce such compounds as NO_3^-, NO_2^-, NO, N_2O, different organic S-oxides and N-oxides, fumarate, etc. Along with NADH, succinate, fatty acids, and glycerophosphate, bacterial respiratory chains can directly oxidize molecular hydrogen, formate, malate, lactate, glucose, certain amino acids, Fe^{2+}, NO_2^-, $S_2O_3^{2-}$, etc.
(4) Respiratory chain composition can vary depending on the growth conditions, and in the case of multicellular organisms it can also be tissue-specific.

These differences are probably connected to different conditions of functioning of respiratory chains of animal mitochondria as compared to bacterial respiratory chains and mitochondria of protozoa, fungi, and plants. Animal mitochondria function under constant conditions of their external medium, i.e., cytosol of these cells. In contrast, conditions of the functioning of mitochondria of protozoa, plants, and unicellular fungi, and especially conditions of bacterial growth can undergo considerable changes within relatively short periods of time. That is why respiratory chain structure of these organisms is far more complex and can vary greatly depending on external conditions. This type of electron transport chains is usually characterized by a lower efficiency of energy conservation (lower H^+/O_2 ratio) when compared to the respiratory chains of animal mitochondria. It is their reliability and ability to adapt to changing unfavorable external conditions, and not the efficiency of energy conservation (as in animal mitochondria) that most likely became the main selective factor in evolution of these electron transport chains.

5.1 Mitochondrial Respiratory Chain of Protozoa, Plants, and Fungi

Mitochondrial electron transport chains of these organisms contain the same enzyme complexes that function in animal mitochondria, i.e. NADH:CoQ-oxidoreductase, $CoQH_2$:cytochrome *c*-oxidoreductase, and cytochrome *c* oxidase. Complexes I, III, and IV from mitochondria of protozoa, plants, and fungi are very similar to their homologs from animal mitochondria; they possess only a slightly different set of additional subunits. However, the mitochondria of these organisms contain a number of enzymes that animal mitochondria lack (Rasmusson et al. 2008). A scheme of the composition of the mitochondrial respiratory chain of protozoa, plants, and fungi is presented in Fig. 5.1. NADH from the matrix of these mitochondria can be oxidized not only by complex I, but also by an additional alternative enzyme, inner NADH dehydrogenase (ND_{in}). This peripheral membrane protein has a far simpler structure than complex I. It consists of just one polypeptide (with molecular mass of 40–50 kDa) and contains just one prosthetic group (FAD). Enzymatic activity of ND_{in} (in contrast to complex I) is not coupled to proton translocation across the membrane, i.e., all the energy of the NADH:ubiquinone-oxidoreductase reaction is lost as heat.

Animal mitochondria are able to oxidize cytoplasmic NADH. For this process to occur, they use different shuttle schemes (they will be described in more detail later in Sect. 11.1). However, mitochondria of protozoa, plants, and fungi oxidize external NADH (and sometimes also NADPH) in the presence of Ca^{2+} ions *directly* with relatively high rates. This process is catalyzed by another (external) NADH dehydrogenase (ND_{ex}). This enzyme is similar to ND_{in} in its primary sequence and its main properties. However, its amino acid sequence contains a different address, which directs this protein not to the matrix, but to the mitochondrial intermembrane space. So, ND_{ex} is attached to the external surface of the inner mitochondrial membrane; that is why it oxidizes not the intramitochondrial, but the cytoplasmic NADH. Like ND_{in}, the activity of ND_{ex} is not coupled to $\Delta\bar{\mu}_{H^+}$ generation. In contrast to ND_{in}, ND_{ex} can bind Ca^{2+} ions. So, in the case of the mitochondrial respiratory chain of protozoa, plants, and fungi the first coupling site can be shunted by the operation of two energy-noncoupled enzymes—ND_{in} and ND_{ex}.

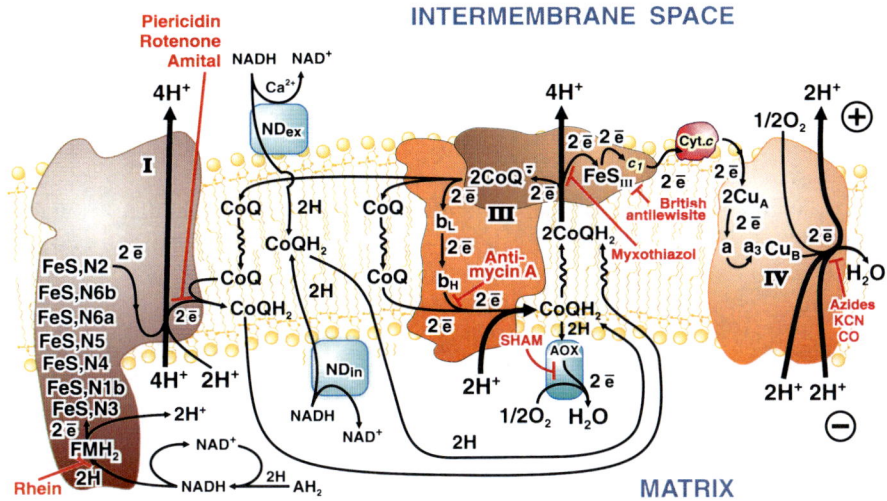

Fig. 5.1 Respiratory chain of protozoa, plants, and fungi. For explanation, see the text

Already during early studies it was shown that oxygen consumption in plant mitochondria (in contrast to animal mitochondria) is cyanide-resistant. This activity was later shown to be connected to the presence of a special enzyme—alternative cyanide-resistant terminal oxidase (AOX) inherent in the mitochondria of protozoa, plants, and fungi.[1] This protein consists of just one polypeptide of molecular mass of about 40 kDa. The alternative oxidase is a peripheral membrane protein (it has no transmembrane α-helices) and, similarly to ND_{in}, is attached to the inner side of the inner mitochondrial membrane (Affourtit et al. 2002). The genes coding all the alternative enzymes of the mitochondrial respiratory chain (AOX, ND_{in}, and ND_{ex}) are located in the nuclear genome.

The primary sequence of AOX has certain similarities to the sequences of different monooxygenases and, like these enzymes, the alternative oxidase contains two atoms of nonheme iron as redox cofactors. These atoms form a binuclear center (similar to the heme–copper binuclear center of complex IV), which is probably the site of O_2 reduction. AOX is different from complex IV in a few other important properties. For instance, it is directly ubiquinol, and not cytochrome c, that serves as an electron donor for this enzyme (i.e., AOX is not a cytochrome c oxidase, but a quinol oxidase). Also, alternative oxidase activity is not coupled to transmembrane proton translocation. So, all the energy of the reaction catalyzed by AOX is dissipated as heat (as occurs with ND_{in} and ND_{ex}). Hence, AOX shunts the second and third energy coupling sites of the respiratory chain.

It is important to note that the alternative oxidase activity is controlled by a large number of cell metabolism parameters. First of all, high activity of plant

---

[1] AOX probably also functions in mitochondria of certain representatives of the animal kingdom.

AOX can be observed only in the presence of α-ketoacids such as pyruvate or α-ketoglutarate. AOX activity also depends strongly on the level of CoQ reduction. In contrast to the bc_1 complex, the alternative oxidase activity can be observed only when the value of the $CoQH_2/CoQ$ ratio is very high. So, it is only the excess of ubiquinol (that for some reasons cannot be oxidized by "canonical" respiratory chain activity) that is utilized by AOX.

So, the mitochondria of protozoa, plants, and fungi, besides a regular "coupled" respiratory chain (similar to the respiratory chain of animal mitochondria), contain an alternative electron transport chain that is not coupled, and it intersects the "canonical" chain only at the level of ubiquinone. What is the physiological sense of the alternative respiratory chain if its activity does not lead to energy storage, but only to its dissipation? One of the possibilities is that such an uncoupled chain may be used for heat production (as occurs, for instance, in the inflorescence of thermogenic plants; we will discuss this phenomenon later when describing thermogenesis mechanisms, see Sect. 10.3). However, the alternative respiratory chain was found to be active not only in thermogenic, but also in other plants and many protozoa and fungi. It is important to note that the absence of transmembrane proton translocation allows noncoupled enzymes to avoid limiting of their electron transfer rate by proton potential generation, i.e., to avoid the respiratory control effect. That is why the noncoupled alternative branch of the respiratory chain can increase the respiration rate, which in turn can be used not only for heat production, but also for decrease in intracellular oxygen concentration (Bertsova et al. 2001), regulation of metabolic flows (Shi et al. 2002), decrease in reactive oxygen species production (Popov et al. 1997), or some other, yet unknown functions. It should be stressed that combination of coupled and noncoupled respiratory chains may be responsible for resistance of phosphorylating oxidation reliable to xenobiotics toxic for initial or, alternatively, to terminal spans of the "canonical" chain. For example, a respiratory chain composed of noncoupled ND_{in} (and/or ND_{ex}) and complex IV will produce ATP under conditions when complex I is poisoned. Similarly, combination of complex I and AOX can avoid inhibition of respiratory phosphorylation with poisons killing complex IV (Skulachev 1988).

5.2 Structure of Prokaryotic Respiratory Chains

Bacterial respiratory chains are extremely diverse, and it seems quite difficult to find an electron transport chain of one microorganism that would serve as a standard example. So we will describe respiratory enzymes of the few most thoroughly studied bacteria and will discuss the main principles of their functioning.

5.2 Structure of Prokaryotic Respiratory Chains

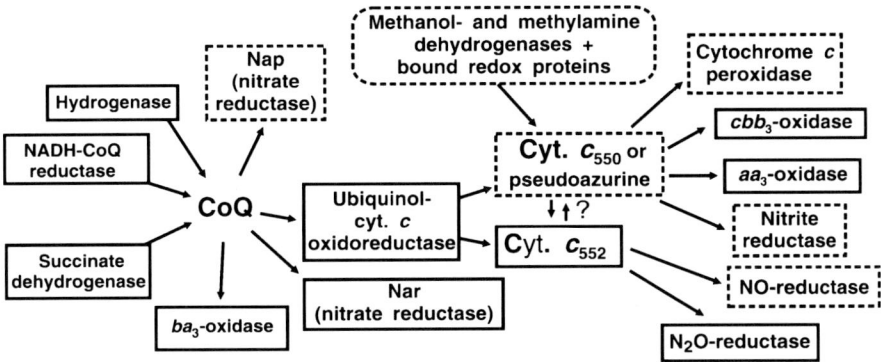

Fig. 5.2 Respiratory chain of the soil bacterium *P. denitrificans*. See explanation in the text

5.2.1 Respiratory Chain of **Paracoccus denitrificans**

Paracoccus denitrificans belongs to the group of α-proteobacteria and seems to be the closest "relative" of mitochondria among the free-living microorganisms. Its respiratory chain structure is in many ways similar to mitochondria, this fact being the reason for this bacterium becoming the favorite model for oxidative phosphorylation studies (van Verseveld and Bosma 1987). *P. denitrificans* contains proteins which are homologous to mitochondrial complexes I, III, and IV, i.e. H^+-translocating NADH:CoQ-oxidoreductase, bc_1 complex, and cytochrome c oxidase of aa_3 type. The respiratory complexes from *P. denitrificans* have only one main difference from their mitochondrial analogs—a substantially lesser number of subunits. However, in contrast to mitochondria, oxygen reduction in *P. denitrificans* can be catalyzed not only by a "canonical" aa_3-type oxidase, but also by two additional enzymes—a cbb_3-type cytochrome c oxidase and a ba_3-type quinol oxidase (Fig. 5.2). Both of these enzymes are homologous to aa_3 oxidase. They contain a heme–copper binuclear center, and they are proton pumps ($H^+/\bar{e} > 1$). However, these enzymes also have a number of properties different from those of aa_3-type oxidases. For instance, the ba_3-type enzyme oxidizes not cytochrome c but directly ubiquinol. This means that the second coupling site of the respiratory chain is shunted by ba_3. The structure of this oxidase is similar to that of the aa_3-type enzyme. The only exception is that ba_3 lacks the Cu_A redox cofactor, and its role is most likely performed by quinol.

The cbb_3-type cytochrome c oxidase catalyzes the same reaction as the aa_3-type enzyme (i.e., it is a H^+-translocating cytochrome c oxidase), but its primary sequence has only minimal similarity to the sequences of the "canonical" oxidases (Pitcher and Watmough 2004). However, the cbb_3-type oxidase is homologous to NO-reductases, i.e., one of the enzymes of anaerobic electron transfer. This fact suggests the possible evolutionary connections between enzymes of anaerobic and aerobic metabolism.

Paracoccus denitrificans is a soil bacterium, and the denitrification in soil (i.e., reduction of oxidized forms of nitrogen to N_2) seems to be the main ecological niche of this microorganism. This process is an important stage of nitrogen circulation in nature, and unfortunately it leads to the depletion of soil sources of bound nitrogen. Denitrification can be described as follows:

$$NO_3^- \rightarrow NO_2^- \rightarrow NO \rightarrow N_2O \rightarrow N_2 \qquad (5.1)$$

The anaerobic redox chain of *P. denitrificans* contains enzymes that catalyze all the reactions of this process (Ferguson, 1994). The first stage of denitrification (nitrate reduction to nitrite) is catalyzed by two different nitrate-reductases (Nap and Nar, see Fig. 5.2). The Nar protein is a membrane enzyme and includes an unusual cofactor that contains a molybdenum atom. Quinol serves as an electron donor for Nar. This enzyme is a $\Delta \bar{\mu}_{H^+}$ generator that functions as a Mitchell loop with the stoichiometry of 1 $H^+/\bar{e}$. The second nitrate-reductase (Nap) is located in the periplasm, and its activity is not coupled to proton potential generation. The next three stages of denitrification are catalyzed by nitrite-reductase, NO-reductase, and N_2O-reductase, respectively. Cytochrome *c* serves as an electron donor for all of these enzymes. Activities of nitrite-reductase, NO-reductase, and N_2O-reductase are not coupled to $\Delta \bar{\mu}_{H^+}$ generation. However, due to cytochrome *c* oxidation, these enzymes create a possibility for proton potential generation under anaerobic conditions in the first two coupling sites of the respiratory chain (i.e., by complexes I and III). So, functioning of these enzymes leads (though indirectly) to the increased efficiency of energy use under the conditions of anaerobic growth of *P. denitrificans*. The overall denitrification process serves as an example of an anaerobic electron transfer, i.e., a bacterial respiratory chain using other than O_2 compounds as the terminal electron acceptors.

5.2.2 Respiratory Chain of Escherichia coli

The aerobic respiratory chain of *E. coli* (Fig. 5.3) is energetically less efficient than the mitochondrial electron transport chain (Ingledew and Poole 1984). It lacks bc_1 complex and any cytochromes of *c* type, i.e., there are only two coupling sites, i.e. NADH:quinone-oxidoreductase and quinol oxidase. Oxidation of NADH in the *E. coli* respiratory chain is catalyzed by two different enzymes. One of them (NDH-1) is homologous to mitochondrial complex I.[2] Enzymatic activity of NDH-1 is coupled to $\Delta \bar{\mu}_{H^+}$ generation (Bogachev et al. 1996). There is also an alternative NADH:quinone-oxidoreductase (NDH-2) in the respiratory chain of *E. coli*. This protein is similar to ND_{in} of plant mitochondria both in its primary sequence and

---

[2] Its properties were thoroughly discussed in the previous chapter.

5.2 Structure of Prokaryotic Respiratory Chains

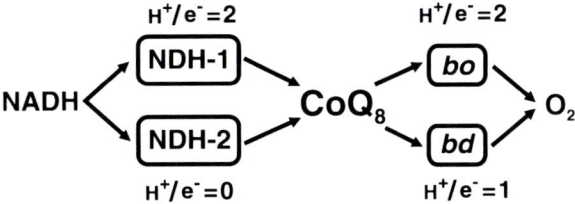

Fig. 5.3 Respiratory chain of the enterobacterium *E. coli*

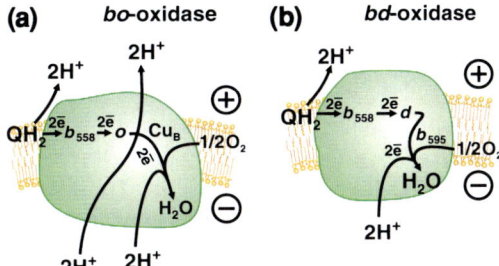

Fig. 5.4 Scheme of functioning of terminal oxidases of *bo*-type (**a**) and *bd*-type (**b**) in *E. coli*

main properties. NDH-2 consists of one polypeptide, contains one redox cofactor (FAD), and its enzymatic activity is not coupled to transmembrane proton translocation.

Quinol, being reduced by NDH-1 or NDH-2, is then oxidized by one of two alternative terminal quinol oxidases—a *bo*- or *bd*-type. The *bd*-type enzyme becomes the main oxidase of the *E. coli* respiratory chain when the cells grow under microaerobic conditions (i.e., under low O_2 concentration), while the *bo*-type oxidase prevails under high O_2 concentration. The *bo*-type quinol oxidase consists of three subunits and contains three redox cofactors: copper Cu_B and two hemes—a low-spin heme B and a high-spin heme O. The heme O and Cu_B form a binuclear center which serves (similar to other heme–copper oxidases) as a site for oxygen binding and reduction. Although the *bo*-type oxidase is a quinol oxidase, its primary sequence and especially its three-dimensional structure are very similar to aa_3-type cytochrome *c* oxidase. So, it is not surprising that the *bo*-type oxidase functions as a proton pump and generates $\Delta\bar{\mu}_{H^+}$ with stoichiometry of 2 $H^+/\bar{e}$ (Puustinen et al. 1991) (Fig. 5.4a).

The second terminal oxidase of *E. coli* (cytochrome *bd*) is a rather unusual enzyme. This protein was found to be homologous neither to oxidases with heme–copper binuclear center (mitochondrial cytochrome *c* oxidases and similar bacterial enzymes), nor cyanide-resistant oxidases with iron–iron binuclear center (alternative oxidases of plant mitochondria, protozoa, fungi, bacteria, and chloroplasts). So, the *bd*-type quinol oxidase is a representative of a third type of terminal oxidases. This *E. coli* enzyme consists of two subunits and includes three prosthetic groups: a low-spin heme b_{558} and two high-spin hemes—heme b_{595} and a chlorin heme *d*. High-spin hemes b_{595} and *d* are located very close to each other in the protein, and they probably form a heme–heme binuclear center similar to the heme–copper or iron–iron binuclear centers of other oxidases[3] (Borisov et al. 2011).

Prosthetic groups of *bd* oxidase are supposed to be located in the hydrophobic part of the protein parallel to the membrane plane and closer to the periplasmic surface of this membrane (Fig. 5.4b). Electrons from quinol are first transferred to the heme b_{558} and then to the binuclear center, where oxygen is reduced. The *bd*-type oxidase does not function as a proton pump, and this enzyme generates $\Delta\bar{\mu}_{H^+}$ only according to the Mitchell loop mechanism. Protons that are produced during quinol oxidation are released to the periplasm, while protons necessary for water formation are taken from the cytoplasm. This process generates proton potential with the stoichiometry of $H^+/\bar{e} = 1$ (Puustinen et al. 1991). It is interesting to note that electric membrane potential produced in this case is not due to transmembrane electron transfer (as, for example, in photosynthetic reaction centers), but almost completely due to the transport of protons, this fact being the consequence of the location of the electron transport pathway parallel to the membrane surface.

So, there are two types of enzymes present in the respiratory chain of *E. coli*—enzymes homologous to the respiratory chain proteins of animal mitochondria (NDH-1 and *bo*-type quinol oxidase) and enzymes with either decreased efficiency of energy conservation (*bd*-type oxidase) or completely noncoupled to proton translocation (NDH-2). Once again, we face the question of the physiological sense of alternative respiratory chain enzyme functioning. As for now, we do not have a definite answer to this question, but there is a possibility that in the case of *E. coli* the low respiration efficiency is compensated by its high resistance to poisonous compounds. Complexes III and especially I are extremely sensitive to the inhibitory effects of hydrophobic xenobiotics. But this type of substances is not toxic in the case of *E. coli* because complex I can be shunted by NDH-2, and complex III is altogether missing. This example illustrates a general biological principle according to which biological system efficiency is sacrificed to its higher stability. We will illustrate this principle later with one more example—energetics of the nitrogen-fixing bacterium *Azotobacter vinelandii*.

Escherichia coli cells can grow under either aerobic or anaerobic conditions. In the absence of oxygen, additional redox enzymes are induced that can use electron acceptors other than O_2. For instance, *E. coli* cells can reduce NO_3^- to NO_2^-. This activity is catalyzed by nitrate-reductase—an enzyme similar to the Nar nitrate-reductase from *P. denitrificans* described above. It generates $\Delta\bar{\mu}_{H^+}$ with stoichiometry $H^+/\bar{e} = 1$. Besides nitrates, *E. coli* can reduce various organic N-oxides (i.e., trimethylamine *N*-oxide) to amines or various organic S-oxides (i.e., dimethyl sulfoxide) to sulfides. *Escherichia coli* cells can also reduce fumarate to succinate due to quinol oxidation under anaerobic conditions.[4] None of these reactions are

---

[3] Similar arrangement of oxygen-reductase centers is an interesting example of convergence of different types of terminal oxidases of independent evolutionary origin.

[4] This process is catalyzed by a quinol-fumarate reductase complex similar (but not identical) to succinate-CoQ reductase complex (other names: succinate dehydrogenase and complex II).

coupled to transmembrane proton transfer, but due to quinol oxidation these enzymes allow NDH-1 to generate proton potential during quinone reduction under anaerobic conditions.

The reduction of fumarate to succinate faces certain difficulties when ubiquinone is used as an electron carrier from NDH-1 to fumarate-reductase. The E_0' of the ubiquinone/ubiquinol couple (+60 mV) is more positive than that of the fumarate/succinate couple (+30 mV), which creates a difficulty for the fumarate-reductase reaction. That is why anaerobic growth of *E. coli* induces replacement of ubiquinone with menaquinone (vitamin K) (Jones and Garland 1982; Shestopalov et al. 1997). The E_0' of the menaquinone/menaquinol couple is -80 mV, which facilitates the reduction of fumarate. This fact also makes our symbiont *E. coli* a potential source of vitamin K for the human organism.

5.2.3 Redox Chain of Ascaris Mitochondria: Adaptation to Anaerobiosis

Many bacteria use fumarate as a terminal electron acceptor. But there is an example of a similar mechanism being used also in mitochondria of an invertebrate. This is the case of the imago of *Ascaris*—a helminth that parasitizes the anaerobic parts of mammalian intestines. Biologists were surprised by the fact that *Ascaris* cells contained a great number of mitochondria: these worms live in the absence of oxygen, so mitochondria do not seem to be necessary. It has shown that the activity of the respiratory chain terminal segments is really very low in the *Ascaris* mitochondria, but its initial segment was found to be very active. The Krebs cycle enzymes responsible for tricarboxylic acid interconversions are absent, while the enzymes responsible for metabolism of dicarboxylates are present in substantial number. Energy conservation by *Ascaris* mitochondria was shown to be due to reduction of fumarate (Seidman and Entner 1961). Both fumarate itself and hydrogen atoms necessary for its reduction are produced as a result of dismutation of two molecules of malate (Fig. 5.5). It is the glycolysis reaction sequence that serves as the malate source. Glycolysis leads to pyruvate formation from glucose; malate can be obtained from pyruvate by means of its carboxylation. Transfer of reducing equivalents from NADH to quinone is catalyzed by complex I. Energy is stored at this stage in the form of $\Delta\bar{\mu}_{H^+}$. The quinol produced is then oxidized by fumarate-reductase (Kita and Takamiya 2002). It should be noted that ubiquinone is replaced by its analog with more negative redox potential in the mitochondria of *Ascaris* imago, like it occurs in the anaerobic respiratory chain of *E. coli*. The only difference is that it is not menaquinone, but another quinone derivative, rhodoquinone, that is used in this case (see Appendix 2).[5]

---

[5] It is noteworthy that the scheme employed by *Ascaris* imago mitochondria is hardly an invention of this invertebrate. Some bacteria convert glucose to two molecules of lactate. One of

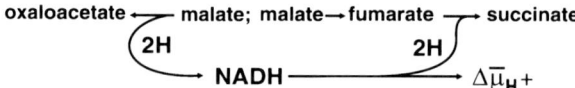

Fig. 5.5 Malate dismutation in *Ascaris* mitochondria

Ascaris larvae that live under aerobic conditions (in blood) have mitochondria with a typical respiratory chain and all the enzymes of the Krebs cycle.

5.2.4 Respiratory Chain of Azotobacter vinelandii

Azotobacter vinelandii belongs to the group of soil nitrogen-fixing bacteria (i.e., it can reduce molecular nitrogen to ammonium ions). The majority of nitrogen-fixing bacteria cannot assimilate N_2 in the presence of high oxygen concentration because nitrogenase complex, responsible for nitrogen reduction, is irreversibly inactivated when interacting with O_2. *A. vinelandii* is an exception to this rule: this microorganism can effectively fix molecular nitrogen even in the presence of high O_2 concentration in the growth medium. However, the nitrogenase complex of *A. vinelandii* is as sensitive to oxygen as similar enzymes of other bacteria. So, there must be some special mechanism that allows *A. vinelandii* to protect its nitrogenase from oxygen-induced inactivation.

In 1969, Howard Dalton and John Postgate suggested a "respiratory protection" hypothesis that explained the ability of *Azotobacter* to fix N_2 even in the presence of high ambient oxygen concentration (Dalton and Postgate 1969). According to this hypothesis, the respiration rate in *A. vinelandii* cells is increased to such a degree that it leads to a substantial drop in O_2 concentration in the bacterial cytoplasm so that the remaining oxygen already cannot inactivate nitrogenase. Further research proved that *A. vinelandii* cells possess very high respiration rates under conditions of high oxygen concentration; in fact, these rates are the highest among all studied bacteria. The respiratory chain of *A. vinelandii* consists mainly of components that are already well known to us—two NADH:quinone-oxidoreductases (NDH-1 and NDH-2), bc_1 complex, and at least two terminal oxidases (a cbb_3-type cytochrome *c* oxidase and a *bd*-type quinol oxidase; Fig. 5.6). When growing under conditions of low oxygen concentration and excess of fixed nitrogen, the *A. vinelandii* respiratory chain consists mainly of the enzymes having high coupling efficiency: NDH-1, bc_1 complex, and cbb_3-type cytochrome *c* oxidase. The overall efficiency of proton potential generation under these conditions

---

(Footnote 5 continued)
them is oxidized to acetate while the other is converted to fumarate, which is further reduced to succinate. The reducing equivalents for succinate production are passed from the former lactate molecule to a quinone via NADH and complex I.

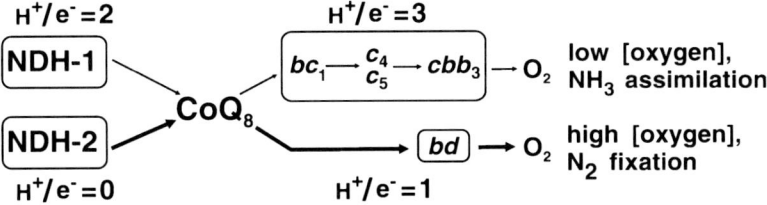

Fig. 5.6 Respiratory chain structure in the soil nitrogen-fixing bacterium A. vinelandii. See explanation in the text

is equal to the value observed in the mitochondrial respiratory chain, namely 5 H$^+$/ē. However, increased aeration and especially deficiency of fixed nitrogen induce the respiratory chain enzymes that are noncoupled or have decreased efficiency of energy conservation: NDH-2 and *bd*-type quinol oxidase (Poole and Hill 1997; Bertsova et al. 2001). The overall coupling efficiency of this chain is only 1 H$^+$/ē. Hence, induction of these enzymes causes a fivefold lowering of the efficiency of proton potential generation by the entire respiratory chain.

It is due to such low efficiency that the activity of the electron transport respiratory chain of A. vinelandii is not limited by the generated proton potential when it becomes necessary to activate respiratory protection of the nitrogenase complex. Besides that, NDH-2 and the *bd*-type quinol oxidase have very high turnover rates. For instance, in the case of NDH-2 from *E. coli*, the specific activity of the enzyme is at least 2–3 orders of magnitude higher than the activity of NDH-1. Very fast turnovers of NDH-2 and *bd*-type quinol oxidase are most likely connected to the lack of proton pumps activity of these enzymes. There is a price that needs to be paid for the kinetic advantages of the respiratory chain consisting of NDH-2 and *bd*-type quinol oxidase. It is not surprising that the growth yield (biomass growth normalized by the amount of substrate consumed) is very low under conditions of high aeration.

5.2.5 Oxidation of Substrates with Positive Redox Potentials by Bacterial Respiratory Chains

The bacterium *Acidithiobacillus* (*Thiobacillus*) *ferrooxidans* can serve as an example of this type of energetics; it can use ferrous ion (Fe^{2+}) as the only energy source, this ion being oxidized to ferric ion (Fe^{3+}) by the oxygen in air (Ferguson and Ingledew 2008). The redox potential of the Fe^{2+}/Fe^{3+} couple is +770 mV, and that of the H$_2$O/O$_2$ couple is +820 mV at neutral pH. So, the potential difference between the oxidant and the reductant is only about 50 mV, a value far too low to generate the $\Delta\bar{\mu}_{H^+}$ necessary for ATP formation by H$^+$-ATP synthase (about 150 mV).

Fig. 5.7 Fe^{2+} oxidation by the respiratory chain of *Acidithiobacillus ferrooxidans*

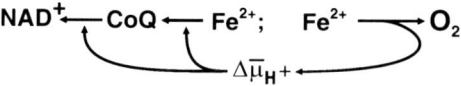

The solution to this problem was found when the pH-dependence of the redox potential of the H_2O/O_2 couple was taken into account. Because hydrogen ions participate in the formation of water from O_2 ($O_2 + 4\bar{e} + 4\ H^+ \rightarrow 2\ H_2O$), the redox potential becomes more positive when the medium is acidified, which moves the balance of the reaction of oxygen reduction to the right. This change can be expressed quantitatively as 60 mV per pH unit. For instance, at pH = 3 the redox potential of the H_2O/O_2 couple would be 820 mV + (60 mV × 4) = 1060 mV. This value is 290 mV more positive than the redox potential of the Fe^{2+}/Fe^{3+} couple, which does not depend on pH because H^+ ions do not participate in the oxidation of ferrous ions. Such a potential difference is sufficient for ATP synthesis. That is why iron bacteria live only at acidic pH values (the pH optimum for *A. ferrooxidans* is between 3 and 1).

However, ATP synthesis is the only one problem that needs to be solved by *A. ferrooxidans*. Another deals with the necessity to use an electron donor with positive redox potential to provide biosyntheses with reducing equivalents. *Acidithiobacillus ferrooxidans* belongs to the group of chemolithotrophic bacteria. It is capable of synthesizing all of its components from CO_2, H_2O, NH_3, and mineral salts. For these processes to be realized, a mechanism for reduction of $NAD(P)^+$ by Fe^{2+} is needed. This type of reduction consumes energy because the redox potential for the $NAD^+/NADH$ couple (−320 mV) is far more negative than the potential of the Fe^{2+}/Fe^{3+} couple.

It is reverse electron transfer in the middle and initial segments of the respiratory chain that is used by *A. ferrooxidans* to overcome this difficulty. This thermodynamically unfavorable electron transport from high to low redox potential is catalyzed with the expense of energy of $\Delta\bar{\mu}_{H^+}$ generated in the terminal segment of the respiratory chain.

According to the scheme of energy supply of iron bacteria, the electrons taken from Fe^{2+} ions enter the cytochrome oxidase system, which transfers electrons to oxygen, this process being coupled to $\Delta\bar{\mu}_{H^+}$ generation. The generated proton potential is later used not only for ATP synthesis, but also for all the other types of work inherent in bacterial membrane. The $\Delta\bar{\mu}_{H^+}$ is also utilized for reversing the $CoQH_2$:cytochrome *c*-oxidoreductase and NADH:CoQ-oxidoreductase reactions, which thus become $\Delta\bar{\mu}_{H^+}$ consumers (instead of being $\Delta\bar{\mu}_{H^+}$ generators, Fig. 5.7).

Experiments carried out with the two species of iron bacteria by Aleem (1966) and in our group by Tikhonova et al. (1967) showed that Fe^{2+} ions reduce NAD^+ due to reverse electron transfer that is supported by the energy produced in the terminal segment of the respiratory chain. *Acidithiobacillus ferrooxidans* was also shown to have huge amounts of cytochromes *c* and cytochrome oxidase, while the components of the middle and the initial segments of the respiratory chain were found to be present in far lower quantities. Such hypertrophy of the cytochrome oxidase system provides the explanation for the extremely high rate of oxygen consumption per mg protein, which turned out to be seven times higher than in the case of bovine heart mitochondria. This phenomenon becomes quite understandable if one takes into account the fact that iron bacteria have only one $\Delta \bar{\mu}_{H^+}$ generator that provides energy for all the functions of their cells.

Ferric ion is not the only electron donor with a positive redox potential that is utilized by bacteria of the *Acidithiobacillus* genus. They are also capable of oxidizing thiosulfate to sulfate via cytochrome c and cytochrome oxidase.

$$S_2O_3^{2-} + 2\ O_2 + H_2O \rightarrow 2\ SO_4^{2-} + 2\ H^+ \qquad (5.2)$$

This process is coupled to ATP synthesis, and protonophorous uncouplers block this synthesis. A situation similar to the one in *Acidithiobacillus* was also observed in nitrite-oxidizing *Nitrobacter*. Nitrite turns into nitrate (redox potential +420 mV), and the electrons taken from nitrite are transferred to cytochrome oxidase and then to O_2. This process provides energy for ATP synthesis and electron transfer from NO_2^- to NAD^+.

5.2.6 Respiratory Chain of Cyanobacteria

The respiratory chain of cyanobacteria has a unique structure. Simultaneous functioning of two electron transporting chains in the thylakoid membranes of cyanobacteria—photosynthetic and respiratory ones—is a main feature of these microorganisms. Even more so, these two chains have several common components, namely $b_6 f$ complex, plastoquinone, and cytochrome c_6 (or plastocyanin) (Hart et al. 2005). The respiratory chain of cyanobacteria consists of the proteins that have already become well known to us: several NAD(P)H:quinone-oxidoreductases of NDH-1[6] and NDH-2 type, $b_6 f$ complex, and cytochrome c oxidase of aa_3 type. Together with the "canonical" complex IV, certain cyanobacteria also contain a bd-type quinol oxidase or an alternative cyanide-resistant oxidase. It is important to note that plastoquinone was found to be the only quinone of the respiratory chain of cyanobacteria.

The locations of photosynthetic and respiratory chains in cyanobacterial cells are slightly different. The components of the respiratory chain are located both in the cytoplasmic and thylakoid membranes, while the photosynthetic chain was found to function only in the thylakoid membrane.

---

[6] It has been already mentioned in the preceding chapter that NDH-1 of cyanobacteria contains an unusual FP-fragment, and this protein is likely to function as a NADPH:plastoquinone-oxidoreductase.

5.2.7 Respiratory Chain of Chloroplasts

Oxygen consumption by chloroplasts (chlororespiration) was first described over 40 years ago, but the molecular mechanisms of this phenomenon were discovered only very recently. Chloroplasts were shown to contain certain components of the respiratory chain similar to those of cyanobacteria (Peltier and Cournac 2002). Chloroplast membranes of some (but not of all) plants contain NDH-1-type NAD(P)H:plastoquinone-oxidoreductase. NDH-2-type NAD(P)H:plastoquinone-oxidoreductases are probably also present in these membranes. Furthermore, these organelles were found to possess an oxidase homologous to cyanide-resistant oxidase of cyanobacteria.[7] So, a respiratory chain comprised of a plastoquinone-reductase and plastoquinol oxidase probably functions in chloroplasts.

The activity of the chloroplast respiratory chain is rather low (electron transfer along this chain constitutes less than 1 % of the electron transfer along the photosynthetic chain). It participate in the regulation of the photosynthetic chain activity rather than chloroplast energy supply that is considered to be the main role of this respiratory chain. For instance, respiratory chain activity in chloroplasts can influence the level of plastoquinone reduction, and simultaneous functioning of photosystem I and NDH-1 (NDH-2) can result in cyclic electron transfer around this photosystem.[8]

5.3 Electron Transport Chain of Methanogenic Archaea

It was mentioned in Chap. 3 that fermentation provides the energy supply for bacteria under anaerobic conditions. Glucose can be fermented to acetate, CO_2 and H_2 (Eq. 5.3), and less often to lactate, ethanol, succinate, propionate, formate, etc.

$$\text{Glucose} + 2\ H_2O \rightarrow 2\ CH_3COOH + 2\ CO_2 + 4\ H_2, \qquad (5.3)$$

For this reaction, the standard Gibbs free energy at pH = 7 ($\Delta G^{0'}$, see Appendix 1) is equal to -216 kJ/mol.

Reaction 5.3 is accompanied by energy release, which can be stored in the form of ATP during fermentation, usually through the mechanism of substrate phosphorylation. But the products of glucose fermentation still contain a rather large

---

[7] Cyanide-resistant oxidase of chloroplasts manifests more similarities to AOX of cyanobacteria than to mitochondrial AOX of the same plants.

[8] The respiratory chain seems to operate in plastids other than chloroplasts (e.g., etioplasts). NDH-1 seems to be the only energy-transducing enzyme of these organelles.

5.3 Electron Transport Chain of Methanogenic Archaea

amount of usable energy. The following process would be energetically most favorable in this case.

$$\text{Glucose} \rightarrow 3\ CO_2 + 3\ CH_4, \tag{5.4}$$

$$\Delta G^{0'} = -418\ \text{kJ/mol}$$

So far, no single microorganism capable of using reaction 5.4 has been discovered. However, this process is carried out in many anaerobic niches of the terrestrial biosphere. It was found to be conducted by bacterial communities, each participant of which is specialized in catalyzing particular stages of cleavage of glucose to carbon dioxide and methane (Schink 1997). Glucose is first fermented by bacteria and some unicellular eukaryotes to such compounds as acetate, ethanol, formate, CO_2, H_2, etc. These compounds are later transformed to methane by methanogenic archaea.

We should note that methanogenesis plays an extremely important role in carbon cycling in the terrestrial biosphere. Over 10^9 tons of methane are thought to be produced by methanogens per year. The greater part of this gas is later transferred to the aerobic layers of the Earth's atmosphere, where methane can be oxidized by molecular oxygen in cells of different methylotrophic bacteria (this process is also very important for the biosphere because CH_4 contributes to the greenhouse effect). A small part of the produced methane is accumulated in underground stores. It is this part that is intensely used by humans as one of our main fuel sources.

The enzymes of methanogens can catalyze the following main types of reactions (Thauer 1998).

$$CO_2 + 4\ H_2 \rightarrow CH_4 + 2\ H_2O \tag{5.5}$$

$$\Delta G^{0'} = -131\ \text{kJ/mol}$$

$$CH_3OH + H_2 \rightarrow CH_4 + H_2O \tag{5.6}$$

$$\Delta G^{0'} = -113\ \text{kJ/mol}$$

In other words, methanogens reduce one-carbon compounds using molecular hydrogen, thus forming methane and water.

Methanogens can also metabolize certain fermentation products without using H_2.

$$CH_3COOH \rightarrow CH_4 + CO_2 \tag{5.7}$$

$$\Delta G^{0'} = -36\ \text{kJ/mol}$$

$$4\ HCOOH \rightarrow CH_4 + 3\ CO_2 + 2\ H_2O \tag{5.8}$$

$$\Delta G^{0'} = -145\ \text{kJ/mol}$$

$$4\ CH_3OH \rightarrow 3\ CH_4 + CO_2 + 2\ H_2O \tag{5.9}$$

$$\Delta G^{0\prime} = -107\ kJ/mol$$

Energetically favorable disproportionation of different carbon atoms in C_2-compounds or different molecules of C_1-compounds occurs in the course of those reactions. For instance, in reaction 5.7 one C-atom of acetate can be oxidized to CO_2 at the expense of reduction of the second atom of the same molecule to CH_4; or in reaction 5.9, one methanol molecule can be oxidized to CO_2 at the expense of reduction of three other methanol molecules to methane.

Methanogens are highly specialized organisms; they use only a strictly limited number of compounds as growth substrates: acetate, H_2, CO_2, and some C_1-compounds (formate, methanol, methylamine, and methylthiols). All the currently known methanogens belong to the kingdom of archaea, and they all have a unique set of enzymes and cofactors that are basically absent from bacteria and eukaryotes discussed above.

Methanogenic archaea can store the energy released in reactions 5.5–5.9 in the form of ATP. However, in contrast to fermentation, in this case ATP is synthesized at the expense of $\Delta \bar{\mu}_{H^+}$ or $\Delta \bar{\mu}_{Na^+}$ as intermediates, and not due to substrate phosphorylation reactions (Deppenmeier 2004; Schäfer et al. 1999). This means that (similar to respiratory phosphorylation) transmembrane H^+ (or Na^+) ion transfer takes place at the expense of energy of certain methanogenesis reactions. The formed proton (sodium) potential is later used for ATP synthesis due to the activity of A_0A_1-ATP synthase—the archaea-specific analog of V_0V_1-ATP synthase of bacteria and eukaryotes (see Sect. 7.1).

Let us consider the possible mechanism of storage of energy of methanogenesis reactions in the form of $\Delta \bar{\mu}_{H^+}$ or $\Delta \bar{\mu}_{Na^+}$.

5.3.1 Oxidative Phase of Methanogenesis

Methanogenesis reactions can be divided into two stages—"oxidative" and "reducing". The former stage involves electron transfer from a unique cofactor of methanogens, HS-CoM (formulas of this and other cofactors of methanogenesis are shown in Appendix 2), to a carbon atom, leading to the formation of oxidized CoM and a methane molecule. Coenzyme M (CoM, 2-mercaptoethanesulfonate) contains a reactive sulfhydryl group, and this cofactor is oxidized to methyl-S-CoM at the initial stages of the oxidative phase (Fig. 5.8). If methanol, methylamine, or methylthiols are used as growth substrates, a direct methyl group transfer to CoM occurs. However, if acetate or CO_2 are used, formation of CH_3-CoM takes several stages. A methyl group is first accepted by the special cofactor H_4MPT (tetrahydromethanopterin). Then the methyl group is transferred from CH_3-H_4MPT to CoM.

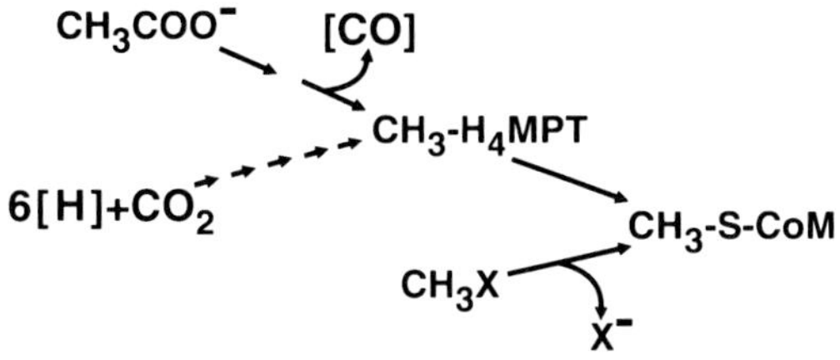

Fig. 5.8 Pathways of CH_3–S–CoM formation from acetate, CO_2, and some reduced C_1 compounds (CH_3–X), such as methanol, methylthiol, and methylamine in methanogenic Archaea

$$CH_3 - H_4MPT + HS - CoM \rightarrow H_4MPT + CH_3 - S - CoM \quad (5.10)$$

This exergonic reaction is coupled to transmembrane sodium ion translocation, i.e., the released energy can be stored in the form of $\Delta\bar{\mu}_{Na^+}$ (this reaction will be described in more detail in Sect. 12.3). So there is a reaction coupled to energy storage in the oxidative part of methanogenesis when acetate or CO_2 (but not methanol, methylamine, or methylthiols) are used.[9]

In the next stage, CH_3–S–CoM interacts with another methanogen-specific cofactor that contains an SH-group,—coenzyme B (7-mercaptoheptanoylthreonine phosphate).

$$CH_3 - S - CoM + HS - CoB \rightarrow CH_4 + CoM - S - S - CoB \quad (5.11)$$

Release of methane and formation of heterodisulfide CoM–S–S–CoB occur as a result of this reaction. Reaction 5.11 is accompanied by a substantial release of energy. However, this reaction was shown to be coupled neither to $\Delta\bar{\mu}_{H^+}$ ($\Delta\bar{\mu}_{Na^+}$) formation nor to ATP synthesis. Apparently, the energy released at this stage can serve as a factor changing the balance of further endergonic stages of the reducing part of methanogenesis in the direction of product formation.

5.3.2 Reducing Phase of Methanogenesis

The reducing phase of methanogenesis comprises oxidation of different substrates coupled to reduction of heterodisulfide to CoM–SH + CoB–SH. These reactions are catalyzed by several membrane enzymes organized in an energy-storing electron transport chain that uses heterodisulfide as the terminal electron acceptor (Deppenmeier 2004; Schäfer et al. 1999). Molecular hydrogen (in the case of methanogens catalyzing reactions 5.5–5.6) or a reduced form of a specific archaean cofactor F_{420}[10] (in the case of reactions 5.8–5.9) serve as primary electron donors for this chain.

---

[9] It should be noted that the ATP hydrolysis energy and the energy of transmembrane sodium potential are used at the initial stages of methanogenesis for the activation of CH_3COOH and CO_2 molecules, respectively. So, the oxidative part of methanogenesis on the whole does not lead to a net energy storage.

[10] F_{420} is a deazaflavin derivative (see Appendix 2, Fig. A2.11).

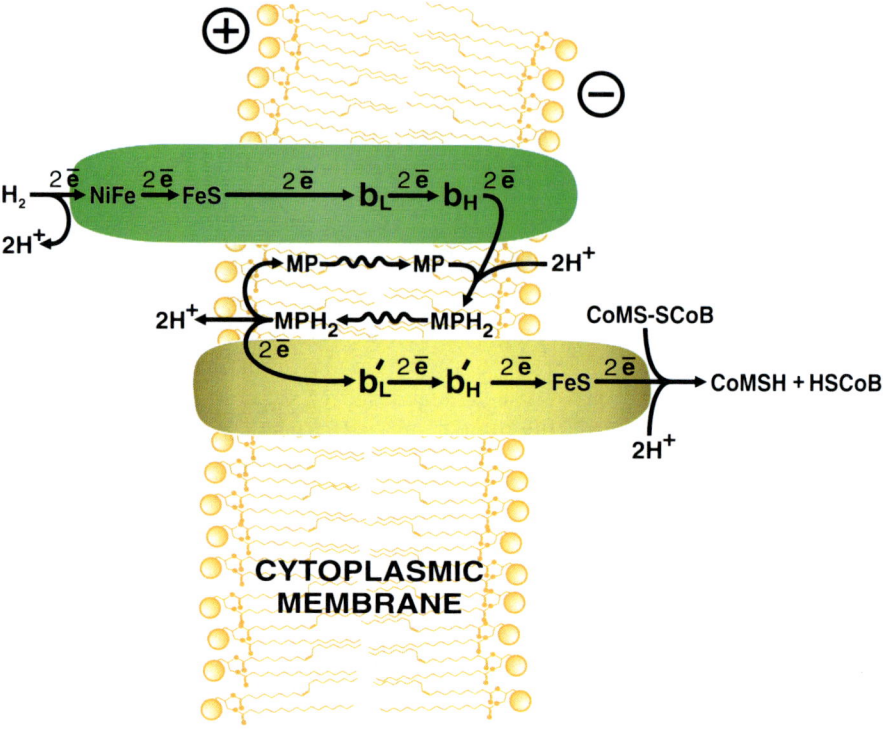

Fig. 5.9 An example of the electron transport chain of methanogenic archaea

Let us discuss the possible mechanisms of energy storage during the reduction of heterodisulfide by molecular hydrogen. This reaction takes two stages. At first, a special enzyme, F_{420}-independent hydrogenase, oxidizes H_2 and transfers electrons to the archaean lipophilic cofactor methanophenazine (MP), which is a functional (but not structural) analog of the quinones of the bacterial and eukaryotic respiratory chains (Abken et al. 1998). This hydrogenase consists of three subunits containing two B hemes, a FeS cluster, and a special prosthetic group composed of a nonheme iron and a nickel atom. According to the scheme presented in Fig. 5.9, H_2 oxidation occurs at the NiFe center, then electrons are transferred via the FeS cluster to cytochrome b where methanophenazine is reduced. The protons formed during the oxidation of H_2 are released to the outer medium; at the same time, protons are taken from the cytoplasm during the formation of MPH_2 from MP. So the hydrogenase generates proton potential according to the Mitchell loop mechanism with stoichiometry of $H^+/\bar{e} = 1$.

Thus formed MPH_2 can be later oxidized by the next enzyme of the methanogen electron transport chain—heterodisulfide-reductase. This enzyme consists of two subunits and contains two FeS clusters and two B hemes (in the case of *Methanosarcina barkeri* the redox potentials of the hemes are −180 and −23 mV). Oxidation of MPH_2 occurs at the cytochrome subunit of the heterodisulfide-reductase. Protons are released to the outer medium, and electrons are transferred across the membrane via B hemes to FeS clusters and further to heterodisulfide, thus forming CoM–SH and CoB–SH. Two protons are taken up from the cytoplasm during the latter reaction. So heterodisulfide-reductase (similar to hydrogenase) generates proton potential according to the Mitchell loop mechanism with stoichiometry of $H^+/\bar{e} = 1$. The energy of the H_2:heterodisulfide-oxidoreductase reaction is stored in the form of $\Delta\bar{\mu}_{H^+}$ in the course of these two stages.

In the case of methanogens growing in the presence of various C_1-compounds in the absence of H_2 (reactions 5.8–5.9), it is the reduced form of cofactor F_{420} ($E_0' = -360$ mV) that becomes the main electron donor for the electron transport chain. Cofactor F_{420} serves as a hydrophilic hydride-ion carrier in archaea. This cofactor is reduced by C_1-compounds[11] and is oxidized by an electron transport chain. Thus, F_{420} is a functional analog of NAD^+ in methanogens.

Oxidation of reduced F_{420} is catalyzed by $F_{420}H_2$:methanophenazine-oxidoreductase. It has been already discussed in the preceding chapter that this enzyme is homologous to the mitochondrial respiratory chain complex I, and its activity is coupled to transmembrane H^+ ion translocation. Thus, formed MPH_2 is then oxidized by heterodisulfide-reductase (as occurs when H_2 is used).

So the reactions of the reducing phase of methanogenesis are coupled to $\Delta\bar{\mu}_{H^+}$ generation. Thus formed proton potential can be later used for ATP synthesis by A_0A_1-ATP synthase. These reactions are the main energy source for methanogenic archaea.

References

Abken HJ, Tietze M, Brodersen J, Bäumer S, Beifuss U, Deppenmeier U (1998) Isolation and characterization of methanophenazine and function of phenazines in membrane-bound electron transport of *Methanosarcina mazei* Göl. J Bacteriol 180:2027–2032

Affourtit C, Albury MS, Crichton PG, Moore AL (2002) Exploring the molecular nature of alternative oxidase regulation and catalysis. FEBS Lett 510:121–126

Aleem MI (1966) Generation of reducing power in chemosynthesis. 3. Energy-linked reduction of pyridine nucleotides in *Thiobacillus novellus*. J Bacteriol 91:729–736

Bertsova YV, Bogachev AV, Skulachev VP (2001) Noncoupled NADH: ubiquinone oxidoreductase of *Azotobacter vinelandii* is required for diazotrophic growth at high oxygen concentrations. J Bacteriol 183:6869–6874

Bogachev AV, Murtasina RA, Skulachev VP (1996) H^+/e^- stoichiometry for the NADH dehydrogenase I and dimethyl sulfoxide reductase in anaerobically grown *Escherichia coli* cells. J Bacteriol 178:6233–6237

Borisov VB, Gennis RB, Hemp J, Verkhovsky MI (2011) The cytochrome *bd* respiratory oxygen reductases. Biochim Biophys Acta 1807:1398–1413

---

[11] $F_{420}H_2$ can be formed also during the oxidation of H_2 due to the activity of F_{420}-dependant hydrogenase.

Dalton H, Postgate JR (1969) Growth and physiology of *Azotobacter chroococcum* in continuous culture. J Gen Microbiol 56:307–319

Deppenmeier U (2004) The membrane-bound electron transport system of *Methanosarcina* species. J Bioenerg Biomembr 36:55–64

Ferguson SJ (1994) Denitrification and its control. Antonie Van Leeuwenhoek 66:89–110

Ferguson SJ, Ingledew WJ (2008) Energetic problems faced by micro-organisms growing or surviving on parsimonious energy sources and at acidic pH: I. *Acidithiobacillus ferrooxidans* as a paradigm. Biochim Biophys Acta 1777:1471–1479

Jones RW, Garland PB (1982) Function of quinones in energy conserving systems. In: Trumpover BC (ed) Academic Press, New York, pp 465–476

Hart SE, Schlarb-Ridley BG, Bendall DS, Howe CJ (2005) Terminal oxidases of cyanobacteria. Biochem Soc Trans 33:832–835

Ingledew WJ, Poole RK (1984) The respiratory chains of *Escherichia coli*. Microbiol Rev 48:222–271

Kita K, Takamiya S (2002) Electron-transfer complexes in *Ascaris* mitochondria. Adv Parasitol 51:95–131

Peltier G, Cournac L (2002) Chlororespiration. Annu Rev Plant Biol 53:523–550

Pitcher RS, Watmough NJ (2004) The bacterial cytochrome cbb_3 oxidases. Biochim Biophys Acta 1655:388–399

Poole RK, Hill S (1997) Respiratory protection of nitrogenase activity in *Azotobacter vinelandii*—roles of the terminal oxidases. Biosci Rep 17:303–317

Popov VN, Simonian RA, Skulachev VP, Starkov AA (1997) Inhibition of the alternative oxidase stimulates H_2O_2 production in plant mitochondria. FEBS Lett 415:87–90

Puustinen A, Finel M, Haltia T, Gennis RB, Wikström M (1991) Properties of the two terminal oxidases of *Escherichia coli*. Biochemistry 30:3936–3942

Rasmusson AG, Geisler DA, Møller IM (2008) The multiplicity of dehydrogenases in the electron transport chain of plant mitochondria. Mitochondrion 8:47–60

Schäfer G, Engelhard M, Müller V (1999) Bioenergetics of the archaea. Microbiol Mol Biol Rev 63:570–620

Schink B (1997) Energetics of syntrophic cooperation in methanogenic degradation. Microbiol Mol Biol Rev 61:262–280

Seidman I, Entner N (1961) Oxidative enzymes and their role in phosphorylation in sarcosomes of adult *Ascaris lumbricoides*. J Biol Chem 236:915–919

Shestopalov AI, Bogachev AV, Murtasina RA, Viryasov MB, Skulachev VP (1997) Aeration-dependent changes in composition of the quinone pool in *Escherichia coli*. Evidence of post-transcriptional regulation of the quinone biosynthesis. FEBS Lett 404:272–274

Skulachev VP (1988) Membrane bioenergetics. Springer, Berlin

Shi NQ, Cruz J, Sherman F, Jeffries TW (2002) SHAM-sensitive alternative respiration in the xylose-metabolizing yeast *Pichia stipitis*. Yeast 19:1203–1220

Thauer RK (1998) Biochemistry of methanogenesis: a tribute to Marjory Stephenson. Microbiology 144:2377–2406

Tikhonova GV, Lisenkova LL, Doman NG, Skulachev VP (1967) Electron transfer pathways in iron-oxidizing bacteria *Thiobacillus ferrooxidans*. Biokhimiia 32:725–734 (in Russian)

van Verseveld HW, Bosma G (1987) The respiratory chain and energy conservation in the mitochondrion-like bacterium *Paracoccus denitrificans*. Microbiol Sci 4:329–333

Chapter 6
Bacteriorhodopsin

6.1 Principle of Functioning

In 1971, Dieter Oesterhelt and Walther Stoeckenius (Fig. 6.1) reported the discovery of the new retinal-containing protein *bacteriorhodopsin* (Oesterhelt and Stoeckenius 1971) in the membranes of the extreme halophilic archaean *Halobacterium salinarium* (which was first erroneously classified as *H. halobium*). Further studies showed that this protein transforms the energy of light into $\Delta\bar{\mu}_{H^+}$, i.e. it is a *light-dependent H^+-pump*. Antanas Jasaitis (Fig. 6.2) and coworkers in our laboratory showed that the $\Delta\bar{\mu}_{H^+}$ generated by bacteriorhodopsin in the *H. salinarium* membrane is able to support ATP synthesis (Belyakova et al. 1975). Substantial progress in understanding the mechanism of the light-induced transmembrane proton transfer by bacteriorhodopsin has been achieved over the years.

Bacteriorhodopsin has a number of properties that facilitate its studies. These are: convenient spectral properties of the protein in the visible range; the possibility to launch transmembrane protons transfer by a laser flash; the ability of bacteriorhodopsin to form both 2-dimensional crystals in purple membrane and 3-dimensional crystals in artificial lipidic cubic phase; the possibility to use the method of site-specific mutagenesis; relatively small size and exceptional stability of the protein over an extremely wide range of organic solvent concentrations, pH values, temperatures, pressures, and many other conditions. Today, bacteriorhodopsin has become the most thoroughly studied proton pump, and a full understanding of the mechanism of the functioning of this protein facilitates the studies of other proton pumps.

Bacteriorhodopsin consists of a single 26 kDa polypeptide (the smallest among the known $\Delta\bar{\mu}_{H^+}$ generators). Bacteriorhodopsin molecules usually form trimers in vivo (though when in monomer form, the protein can also work as a proton pump). Bacteriorhodopsin is localized in special areas of the *H. salinarium* cytoplasmic membrane, so-called *purple membrane* that can be up to 0.5 μm in diameter. The purple membrane is practically saturated with bacteriorhodopsin; the protein constitutes about 75 % of this membrane substance (the rest comprises phospholipids, sulfolipids, and carotenoids). Bacteriorhodopsin is characterized by regular

Fig. 6.1 Dieter Oesterhelt (*left*) and Walther Stoeckenius

Fig. 6.2 Antanas Jasaitis

(crystal) packing in the purple membrane. There are no other proteins in this membrane, which in fact represents 2-D crystals of bacteriorhodopsin.

Such processes as intermolecular transfer of reducing equivalents or hydrolysis of covalent high-energy bonds, always involved in reactions catalyzed by other $\Delta\bar{\mu}_{H^+}$ generators, are absent from bacteriorhodopsin.

According to the current point of view, *the mechanism of bacteriorhodopsin functioning* can be described in the following way. A photon absorbed by retinal residue (a chromophore covalently bound with the aldimine bond to the ε-amino group of a lysine residue of the bacteriorhodopsin polypeptide) induces "all-trans → 13-cis" isomerization of retinal (Fig. 6.3). This process is accompanied by a large acidic pK_a shift of the aldimine Schiff base nitrogen protonated in the dark. As a result, the Schiff base loses H^+, which appears to be released to the aqueous phase *outside* the bacterial cell. Then, the pK_a value of the Schiff base returns to its initial level, i.e. the affinity of the base to protons is once again increased. As a result,

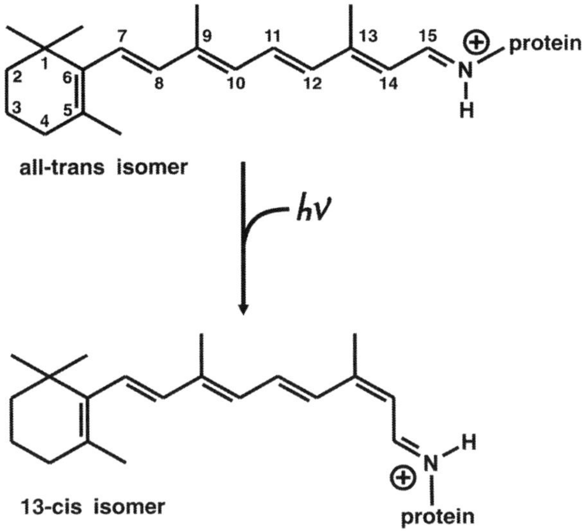

6.1 Principle of Functioning

Fig. 6.3 Light-dependent "all-trans → 13-cis" isomerization of retinal residue in bacteriorhodopsin

the Schiff base is reprotonated by an H⁺ ion coming from the *inside* of the cell (from the cytoplasm). Then retinal returns to its initial conformation (a "13-cis → all-trans" transition). Thus, the light-induced turnover of bacteriorhodopsin is coupled to the translocation of H⁺ across the cytoplasmic membrane of *H. salinarium*.

6.2 Structure of Bacteriorhodopsin

Bacteriorhodopsin was the first $\Delta\bar{\mu}_{H^+}$ generator whose primary structure was described. This was done by Yuri Ovchinnikov (Fig. 6.4), Nazhmutin Abdulaev and colleagues in 1978 (Ovchinnikov et al. 1978) and confirmed by Har Gobind Khorana (Fig. 6.4) and colleagues in 1979 (Gerber et al. 1979). The protein was found to be composed of *248 amino acid residues forming seven hydrophobic sequences* of about 24–28 residues each, interrupted by six short hydrophilic sequences. Moreover, there is a hydrophilic "tail" including about 20 amino acids at the C-end of the polypeptide.

A spatial model (~7 Å resolution) of bacteriorhodopsin was published by Henderson (Fig. 6.5) and Unwin in 1975 (Henderson and Unwin 1975); the resolution of their method was later increased to ~3 Å (Grigorieff et al. 1996). In this work, the authors made use of a unique feature of bacteriorhodopsin, which forms in vivo 2-D crystals in purple membrane. With the aid of electron microscopy and the electron diffraction technique, they obtained bacteriorhodopsin images as shown in Fig. 6.6a, b, c. The bacteriorhodopsin polypeptide chain was shown to cross the membrane seven times, forming seven α-helical columns (marked by alphabetic letters from A to G), each about 3.5 nm in height and built mainly of hydrophobic amino acid residues. Hydrophilic links connecting the α-helical

Fig. 6.4 Yuri Ovchinnikov (*left*) and Har Gobind Khorana

Fig. 6.5 Richard Henderson

hydrophobic segments are absent from the model shown in Fig. 6.6b, since the resolution of the method was too low. The N- and C-ends of the bacteriorhodopsin polypeptide chain are located on the opposite sides of the cytoplasmic membrane: the N-end is outward directed and C-end inward.

The 3-D structure of bacteriorhodopsin was later determined also by the X-ray analysis method ($\sim$1.4 Å resolution) (Pebay-Peyroula et al. 1997; Schobert et al. 2002). This more detailed model of the protein structure made it possible to identify some important elements that indicate the possible mechanism of its functioning (Lanyi 2004). The retinal moiety was found to be connected to *Lys*-216 located in the *G* α-helical column of the bacteriorhodopsin molecule. The Schiff base is located in the middle of the core hydrophobic part of the protein (Fig. 6.7). Two aspartic acid residues—*Asp*-85 and *Asp*-212—are located next to the Schiff base. The *Asp*-85 residue is connected to the outer medium via a proton-transferring pathway that consists of the *Arg*-82, *Glu*-194, and *Glu*-204 residues,

Fig. 6.6 Spatial arrangement of bacteriorhodopsin.
a Electron scattering density map of the purple membrane (projection onto the plane of the membrane); a single bacteriorhodopsin molecule is outlined with a *dashed line*, a cell of a two-dimensional crystal with a *full line*; **b** side view of the model of a single bacteriorhodopsin molecule based on electron scattering density maps; and **c** a scheme illustrating positions of seven α-helical segments in the bacteriorhodopsin molecule. Electron microscopy studies (Henderson and Unwin 1975)

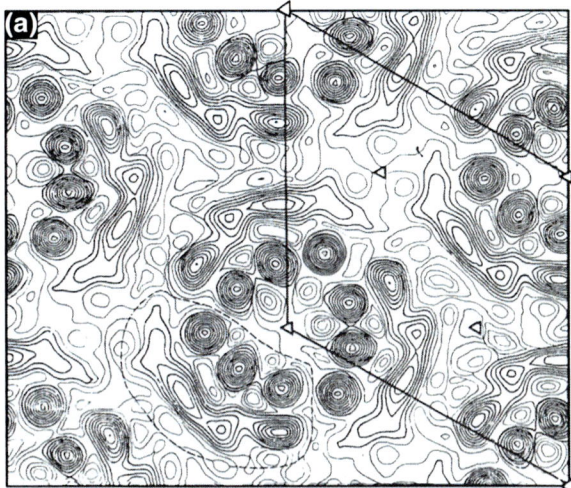

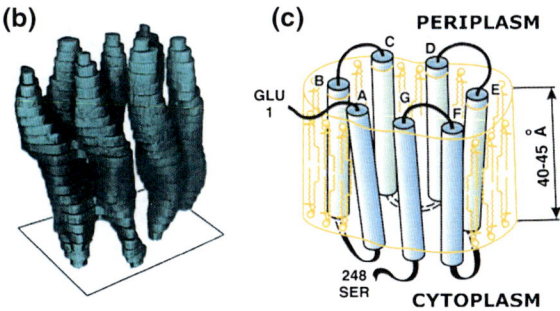

and also of seven or eight firmly bound water molecules. Thus, the Schiff base deprotonation in the bacteriorhodopsin photocycle can lead to export of the released proton to the extracellular medium. In contrast, only one polar amino acid residue (*Asp*-96) is located between the Schiff base and cytoplasmic membrane surface in bacteriorhodopsin. The distances between the cytoplasmic membrane surface and *Asp*-96 and further between *Asp*-96 and the Schiff base (10–12 Å) are too far for the direct proton transfer. Hence substantial conformational changes should occur in order for the Schiff base to be reprotonated by cytoplasmic H^+ ions. These changes really take place. They result in the formation of a very narrow water-filled cleft leading from the cytoplasm to *Asp*-96 and further to the Schiff base. Only about 10 H_2O molecules can find space in this cleft. Certain structural properties of bacteriorhodopsin indicate the possible mechanism of such conformational changes. The *B*, *C*, and *F* columns are bent due to the presence of proline residues in these α-helices. Column *G* also has a mobile structural element. The α-helix in this segment is interrupted by one turn of π-helix ("π-bulge"). This element is formed due to the fact that oxygen carbonyl atoms in the peptide group of *Ala*-215 and *Lys*-216 form hydrogen bonds not with the hydrogen atoms of the

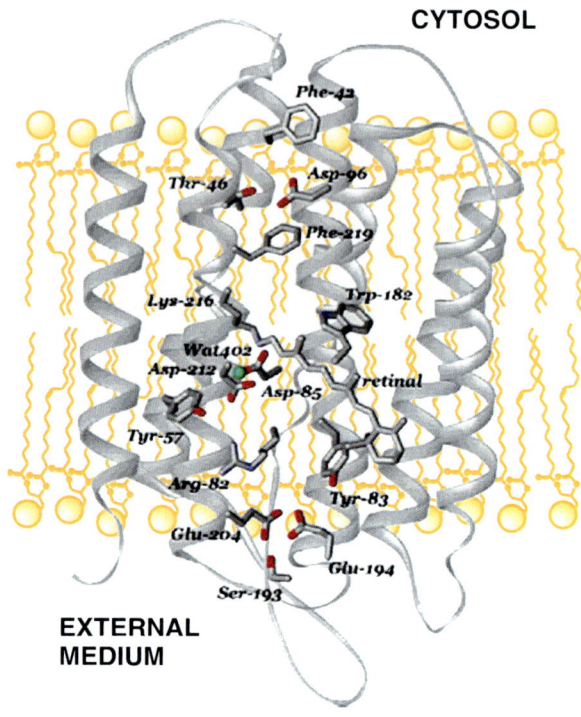

Fig. 6.7 Structure of bacteriorhodopsin showing the main functional amino acid residues, based on X-ray data (Lanyi 2004). Wat$_{402}$ (the *green ball* in the center of the membrane) represents the bound water molecule in the area of Asp-85 and Asp-212

following peptide groups (as in α-helices), but with water molecules. Thus, a π ↔ α transition is possible in this place, a feature facilitating light-induced structural changes in the cytoplasmic part of bacteriorhodopsin, resulting in reprotonation of the Schiff base by cytoplasmic H$^+$ ions (Lanyi 2004).

6.3 Bacteriorhodopsin Photocycle

Absorption of a photon by bacteriorhodopsin causes a *cyclic* event which involves a number of intermediate states of the protein (Fig. 6.8). Each of them can be identified by a characteristic spectral shift. As early as 1971, Oesterhelt and Stoeckenius described a key intermediate of the bacteriorhodopsin photocycle absorbing at a much shorter wavelength (412 nm) than the initial form of bacteriorhodopsin bR (568 nm) (Oesterhelt and Stoeckenius 1971). This spectral shift is similar to that accompanying the photoconversion of animal rhodopsin to the so-called metarhodopsin *II* (*MII*), and the short wavelength bacteriorhodopsin intermediate was named M_{412}. Later, two intermediates between bR_{568} and M_{412} were discovered. They were designated by the letters of the alphabet preceding M. Thus, the following chain of the light-induced events was assumed: $bR_{568} \rightarrow K_{590} \rightarrow L_{550} \rightarrow M_{412}$. The $M_{412} \rightarrow bR_{568}$ transition during the photocycle dark phase was also found to

Fig. 6.8 The bacteriorhodopsin photocycle (modified from Lozier et al. 1975 and Drachev et al. 1986)

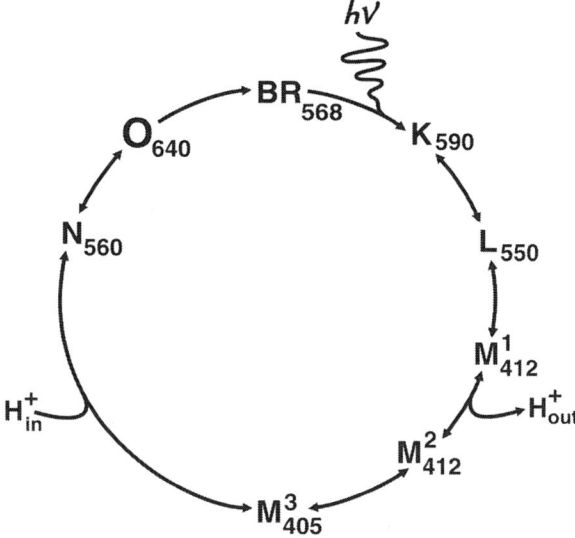

include two intermediates ($M_{412} \rightarrow N_{560} \rightarrow O_{640} \rightarrow bR_{568}$). The majority of intermediates of this photocycle were first described by Richard Lozier and Roberto Bogomolni in Stoeckenius' laboratory (Lozier et al. 1975) and by Drachev and Kaulen in our group (Drachev et al. 1986). Further studies identified several different intermediates showing absorption maxima at 412 nm or slightly shorter (they were given the names M^1_{412}, M^2_{412}, and M^3_{405}), a fact leading to further complication of the scheme of the bacteriorhodopsin photocycle (Fig. 6.8).

Storage of bacteriorhodopsin for several minutes in the dark gives rise to so-called *dark adaptation* when about half of the protein molecules convert into a new component absorbing light at 548 nm and containing 13-*cis* retinal. This dark-adapted bR_{548} is inactive as a proton pump, but illumination induces a fast $bR_{548} \rightarrow bR_{568}$ transition (*light adaptation*) to the regular active form of bacteriorhodopsin (Harbison et al. 1984).

6.4 Light-Dependent Proton Transport by Bacteriorhodopsin

The development and concerted use of fast measuring techniques, i.e. spectroscopy and electrometry, and very short laser flashes for launching the bacteriorhodopsin photocycle, have led to a substantial progress in the understanding of the functional activity of this protein. Spectral measurements at the absorption maxima of the photocycle intermediates allow their formation and decay to be monitored. Sensitive pH indicators can signal H^+ release and uptake by bacteriorhodopsin molecules simultaneously actuated by a short light flash, and vectorial transfer of protons was shown in vesicular systems containing oriented bacteriorhodopsin in

Fig. 6.9 Andrey Kaulen (*left*) and Lel Drachev

their membranes (Lozier et al. 1976). Direct measurements of charge displacement in bacteriorhodopsin incorporated into a phospholipid-impregnated collodion film yield information about the transfer of H^+ or other charged components inside this $\Delta\bar{\mu}_{H^+}$-generator molecule (Drachev et al. 1984; Skulachev 1988).

Drachev and Kaulen (Fig. 6.9) in our laboratory (Drachev et al. 1974, 1984) measured the following three parameters in their experiments: (i) formation and decomposition of the M_{412} intermediate measured as a light absorption increase and decrease at 412 nm, respectively; (ii) pH change in the medium, measured as a light absorption change of the pH indicator *p*-nitrophenol; and (iii) $\Delta\Psi$ generation and discharge caused by charge displacement in bacteriorhodopsin (it was measured by a voltmeter connected to two electrodes separated by a collodion film covered from one side with bacteriorhodopsin-containing proteoliposomes). A 15-ns laser flash was applied to induce a single turnover of bacteriorhodopsin. A suspension of open purple membranes was used in measurements of the pH changes and absorption at 412 nm.

The formation and decomposition kinetics of M_{412} were shown to almost coincide with the release and absorption of H^+ by the purple membranes, respectively. Fortunately, M_{412} formation and H^+ release are much faster ($t_{1/2} = 250$ μs) than M_{412} decay and H^+ uptake ($t_{1/2} = 30$ ms), so that it was easy to correlate these two sets of events with the two main electrogenic processes observed in the purple membranes-collodion film system (the $t_{1/2}$ values are given for 5 °C at which this experiment was carried out). The formation of M_{412} and H^+ release was correlated to the fast electrogenic stage developing in the microsecond timescale, whereas M_{412} decay and H^+ uptake were correlated, respectively, to the second (ms) stage. The ratio of magnitudes of the first and second stages of $\Delta\Psi$ generation was 1:4. The micro- and millisecond stages are in the direction that corresponds to positive charges transfer from the cytoplasmic to the outer membrane surface (Fig. 6.10).

6.4 Light-Dependent Proton Transport by Bacteriorhodopsin

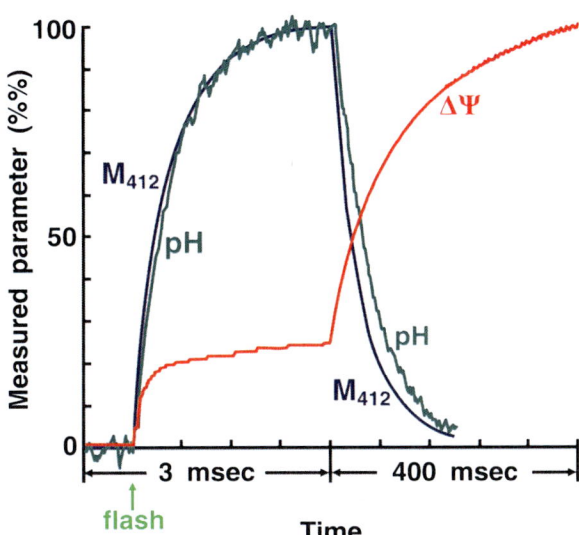

Fig. 6.10 Kinetics of three functional parameters of bacteriorhodopsin: (1) formation and decay of the M_{412} intermediate of the bacteriorhodopsin photocycle, (2) release and uptake of H$^+$ by open bacteriorhodopsin sheets, and (3) generation and dissipation of $\Delta\Psi$ in the purple sheet-collodion film system. *Vertical arrow* shows a 532-nm laser flash ($t_{1/2}$ = 15 ns). Conditions: 100 mM NaCl, 1 mM Mes (pH 6.8), and t = 5 °C (Drachev et al. 1984)

A chain of molecular events leading to the light-dependent transmembrane proton translocation was described later (Kalaidzidis et al. 2001; Lanyi 2004). This was made possible by analysis of different mutant forms of bacteriorhodopsin with individual amino acid residues substituted. The most important data were obtained when the 3-D structures of not only the dark-adapted form of bacteriorhodopsin bR_{548}, but also some of its photocycle intermediates, were determined with high resolution (Lanyi 2001). Below, the most probable sequence of events happening during the transmembrane proton translocation by bacteriorhodopsin will be described.

Three proton donor (and proton acceptor) groups, which are the main participants of this process, were identified in bacteriorhodopsin. These are: the nitrogen of the Schiff base and two aspartate residues, *Asp*-85 and *Asp*-96 (these residues are located on the outer and the cytoplasmic sides from the Schiff base, respectively, see Fig. 6.7). In the initial bR_{568} form, the nitrogen atom of the Schiff base is protonated (the pK$_a$ of this group is >12), *Asp*-85 is deprotonated (pK$_a$ = 2.5), and *Asp*-96 is protonated (pK$_a$ = 10). The retinal moiety is in an *all-trans* conformation.

Light absorption induces the following main events in bacteriorhodopsin.

1. $bR_{568} \rightarrow K_{590}$ *transition*

Absorption of a photon by the *all-trans* retinal moiety results in its isomerization to 13-*cis* retinal. This photoisomerization leads to a substantial change of the configuration of this molecule in a solution. However, the new conformation of retinal cannot be realized during the first stage of the bacteriorhodopsin photocycle because of structural restrictions imposed on this prosthetic group by the retinal-binding center of the protein. This is why the conformation of retinal in K_{590} intermediate was called "twisted". Thus, the energy of the absorbed photon is stored in the form of a strained conformation of the retinal moiety at the first stage of the photocycle.[1]

2. $K_{590} \rightarrow L_{550}$ transition

During the next stage, retinal residue changes from the strained conformation to the relaxed 13-*cis* conformation. In fact, twisted retinal residue "pushes apart" amino acid residues of the protein which acquires a new strained conformation with a changed retinal-binding site (at this stage, energy is transferred from retinal to the protein).

3. $L_{550} \rightarrow M^{I}_{412}$ transition

The Schiff base and *Asp*-85, when in the new protein conformation, change their affinities to H$^+$ so that their pK$_a$ shift to approximately 8 (the reasons for these changes are still unknown). This event initiates proton transfer from the Schiff base to Asp-85.

4. $M^{I}_{412} \rightarrow M^{II}_{412}$ transition

Deprotonation of the *Glu*-204 residue on the outer bacteriorhodopsin surface occurs at this stage. As a result, a proton is released to the outer medium.

5. $M^{II}_{412} \rightarrow M^{III}_{405}$ transition

At this stage, substantial conformational changes occur in the bacteriorhodopsin molecule. The structure of the mobile elements of the α-helices is probably changed (see above) and a cleft is formed between α-columns in the cytoplasmic part of bacteriorhodopsin. It is due to these changes that water penetrates the space between *Asp*-96 and the Schiff base, facilitating proton transfer between these two groups.

6. $M^{III}_{405} \rightarrow N_{560}$ transition

The change in the surroundings of the *Asp*-96 residue shifts its pK$_a$ to ~7. This in turn results in proton transfer from *Asp*-96 to the Schiff base. Thus, reprotonation of the bacteriorhodopsin prosthetic group occurs at this stage.

7. $N_{560} \rightarrow O_{640}$ transition

Reverse isomerization of 13-cis retinal to the energetically more favorable *all-trans* state occurs at this stage. Also, the *Asp*-96 residue is reprotonated due to the proton uptake from the cytoplasm, and the cleft in the cytoplasmic part of the bacteriorhodopsin molecule closes.

---

[1] Two intermediates in the bacteriorhodopsin photocycle before K_{590} were discovered using spectroscopy with sub-picosecond resolution—a "truly initial" intermediate I_{480}, and J_{625} which is formed from I_{480} within 700 fs ($7 \cdot 10^{-13}$ s).

8. $O_{640} \to bR_{568}$ transition

At the last stage, proton transfer from *Asp*-85 to *Glu*-204 occurs, completing the bacteriorhodopsin photocycle.

Thus, absorption of one photon by a bacteriorhodopsin molecule leads to translocation of one proton across the membrane. It has to be stressed that in spite of a substantial progress achieved in our understanding of the mechanism of proton translocation by bacteriorhodopsin, certain key reactions remain unknown. A conformational change that switches the Schiff base availability from the external to the internal (cytoplasmic) proton translocating pathway seems to be the most important one. It also remains unclear why the Schiff base reprotonation is achieved due to the proton transfer from *Asp*-96, and not through the return of the H^+ from *Asp*-85.

6.5 Other Retinal-Containing Proteins

6.5.1 Halorhodopsin

Another light-dependent ion pump, halorhodopsin, was discovered during studies of *H. salinarium* (Mukohata and Kaji 1981). Membranes of this microorganism were found to contain another retinal-containing ion pump besides bacteriorhodopsin. Halorhodopsin (*hR*) is a retinal protein that pumps Cl^- ions from the outer medium to the cytoplasm at the expense of light energy (Schobert and Lanyi 1982). It is a membrane protein of about 27 kDa molecular mass, i.e. similar in size to bacteriorhodopsin. Incorporation of halorhodopsin into liposomes produces vesicles competent in light-driven Cl^- ions pumping. Their association with a planar phospholipid membrane gives rise to a system generating a photoelectric current as a result of the transmembrane electrogenic movement of Cl^-. Halorhodopsin is localized in regions other than the bacteriorhodopsin purple sheets of *H. salinarium*, and its amount in cells is about 100-fold lower than that of bacteriorhodopsin.

The amino acid sequences of these two retinal containing proteins show about 70 % difference; similarity is rather high in α-helical segments, the number of which in halorhodopsin is the same as in bacteriorhodopsin (seven). Similar to bacteriorhodopsin, the retinal residue is bound to halorhodopsin by an aldimine bond with an ε-amino group of a lysine residue.

The 3-D structure of halorhodopsin determined by X-ray analysis is very similar to that of bacteriorhodopsin (Kolbe et al. 2000). The most important difference is that halorhodopsin does not contain charged amino acids residues playing the key role in H^+ transfer by bacteriorhodopsin (*Asp*-85 and *Asp*-96).

The photocycle halorhodopsin is similar to that of bacteriorhodopsin; its first stage includes isomerization of the *all-trans* retinal to the 13-*cis* isomer. However, in contrast to bacteriorhodopsin, the Schiff base remains protonated through the entire halorhodopsin photocycle.

The mechanism of halorhodopsin operation is not yet clear. The 3-D structure of this protein suggests that in the ground state, a Cl^- ion is bound close to the positively charged, protonated Schiff base of halorhodopsin. The light-induced isomerization of the retinal residue probably causes the Schiff base (together with the electrostatically bound Cl^- ion) to move from the inward to the outward Cl^--conducting pathway. It has to be stressed that there are hardly any polar amino acid residues of halorhodopsin that are involved in the Cl^- transfer between the Schiff base and the cytoplasmic membrane surface. Thus, substantial conformational changes need to occur in the cytoplasmic part of halorhodopsin during the photocycle. These changes will lead to the formation of a cleft postulated also for the structure of some intermediates of the bacteriorhodopsin photocycle (see above).

Yet, it remains unclear how it is possible that two proteins of similar structure and properties (bacteriorhodopsin and halorhodopsin) have specificity to such different ions (H^+ and Cl^-) and even more so—transfer them in the *opposite* direction as a result of light absorption. The situation is even more intriguing since point mutation D85T in bacteriorhodopsin results in appearance of the light-dependent Cl^- pumping from the outer medium to the cytosol (in halorhodopsin, Asp-85 is replaced with threonine), whereas addition of small pH buffer to halorhodopsin makes this protein competent in the light-dependent pumping of H^+ from cytosol to the outer medium.

The biological function of halorhodopsin consists, first of all in the return to the cytoplasm of the Cl^- ions lost by the cell due to Cl^- electrophoresis down an electric potential gradient created by bacteriorhodopsin and/or respiratory $\Delta\bar{\mu}_{H^+}$ generators. Some leakage of Cl^- is apparently unavoidable since $\Delta\Psi$ across the halobacterial membrane can reach 200 mV and the cytoplasmic Cl^- concentration level can be as high as 4 M.

Halorhodopsin is responsible for the return of a lesser part of chlorine ions lost by the cell. Their greater part returns via Cl^-, H^+-symporter localized in the same membrane. Both systems can also increase the number of intracellular Cl^- ions during increase in cell volume prior to cell division.

It is interesting that another halophilic archaeon, *Natronomonas pharaonis*, does not contain bacteriorhodopsin, and halorhodopsin is the only light-dependent ion pump in this microorganism. Thus, halorhodopsin is probably used not only for the return of chlorine ions to the cytoplasm, but also for the light energy storage in the form of $\Delta\bar{\mu}_{Cl^-}$. Later, the transmembrane potential of Cl^- ions is probably converted to $\Delta\bar{\mu}_{H^+}$ by means of the mentioned Cl^-, H^+-symporter.

6.5.2 Distribution of Bacteriorhodopsin and its Analogs in Various Microorganisms

For a long time, bacteriorhodopsin could only be found as a relict in the representatives of the oldest organisms that survived in certain extreme niches of the Earth's biosphere—*H. salinarium* and closely related halophilic and thermophilic archaean species. However, recently a gene of a protein similar to bacteriorhodopsin was discovered in certain marine proteobacteria. The expression of this gene in *E. coli* cells growing in the presence of added retinal resulted in the synthesis of proteorhodopsin—a protein capable of light-dependent transmembrane proton translocation (Béjà et al. 2000). Thus, proteins of the bacteriorhodopsin family have proven to be far more common than it was originally thought, and they probably play some role in the Earth's biosphere (Béjà et al. 2001).

However, it is obvious that bacteriorhodopsin photosynthesis is less efficient than its chlorophyll-mediated analog. H^+/photon stoichiometry is 1, 2, and 1.5 for bacteriorhodopsin photosynthesis, cyclic photoredox chain of purple bacteria, and noncyclic photoredox chain of cyanobacteria or chloroplasts, respectively. Moreover, noncyclic chain is also used to reduce $NADP^+$ by electrons transformed from H_2O. It is not surprising that bacteriorhodopsin functioning *per se* cannot support the autotrophic type of metabolism. For instance, *H. salinarium* cells are not capable of CO_2 fixation—they need organic substrates for growth.[2]

6.5.3 Sensory Rhodopsin and Phoborhodopsin

Halophilic archaea perceive red light as an attractant signal, while blue is a repellent signal.

Attractant and repellent effects of low intensity light are provided by two retinal-containing pigments—*sensory rhodopsin (sR)* and *phoborhodopsin (pR)*. These protein photoreceptors demonstrate certain similarity to bacteriorhodopsin. For instance, *sR* and *pR* contain seven transmembrane segments. Their retinal residue is connected to an ε-amino group of a lysine residue of the *G* segment.

Sensory rhodopsin has a light absorption maximum at 590 nm which is shifted to 370 nm under illumination (Bogomolni and Spudich 1982). The light causes *all-trans* → *13-cis* isomerization of the retinal residue. Illumination of sR_{370} rapidly converts it back to the sR_{590} form. In the dark, this conversion also occurs but slowly (this process then takes about 1 s, i.e. it is 100-fold slower than in the case

---

[2] Chlorophyll-containing photosystems also possess better light-absorbing properties, especially when one takes into account light-harvesting complexes (antennae) which seemed to be absent in case of bacteriorhodopsin (However, it was recently found that a bacteriorhodopsin homolog from the bacterium *Salinibacter ruber* contains a carotenoid molecule, which functions as a primitive light-harvesting antenna (Balashov and Lanyi 2007)).

of bacteriorhodopsin and halorhodopsin). Apparently, the formation of a rather long-lived transient is necessary for initiating a chain of events resulting in multiplication of the attractant signal (Sasaki and Spudich 2008).

As shown by experiments, red light of low intensity sensitizes H. salinarium cells to the repellent action by blue light. This effect was interpreted within the framework of a concept considering sR_{370} (which is obtained from sR_{590}) to be a sensor of repellent (blue) light.

Further studies showed that halophilic archaea do possess a fourth retinal protein in addition to bR, hR, and sR, and this chromoprotein mediates (together with the sR_{370} form of sensory rhodopsin) the repellent signal of blue light (Takahashi et al. 1985). Its spectral maximum is at 480 nm. This pigment called phoborhodopsin (pR) is converted into intermediates absorbing at 450 nm and 360 nm when illuminated. The latter, long-lived transient ($t_{1/2} = 0.25$ s) apparently sends a repellent signal to the taxis system. It is remarkable that all four retinal-containing proteins of H. salinarium respond to photon absorption by *all-trans* → *13-cis* isomerization of the retinal residue.

The concentration and localization of sR and probably pR in the membrane are similar to those of halorhodopsin: they are minor components localized in areas of the H. salinarium cytoplasmic membrane other than bacteriorhodopsin sheets. Sensory rhodopsins must be connected to an efficient system of signal amplification; 20 photons are sufficient to induce the attractant effect, and 1–2 photons induce the repellent effect.

Interestingly, the attractant effect of the red light is provided not only by the sensory rhodopsin sR, but also by a regular bacteriorhodopsin in H. salinarium. Kim Lewis from the laboratory of one of this book's authors (V.P.S) showed that cyanide (inhibitor of respiration) and dicyclohexylcarbodiimide (DCCD) inhibiting H^+-ATPase substantially increase the attractant response of halophilic archaea to red light of high intensity (Baryshev et al. 1981). Sergei I. Bibikov from the same laboratory showed this effect to be absent from a strain lacking bacteriorhodopsin, but to be present in the bacteriorhodopsin-containing strains lacking sR and pR (Bibikov et al. 1993). These relationships were explained assuming that halobacteria possess a proton potential monitoring sensor ("protometer") operating in such a way that an increase and decrease in $\Delta\bar{\mu}_{H^+}$ are perceived as attractant and repellent signals, respectively. In the light, bacteriorhodopsin generates $\Delta\bar{\mu}_{H^+}$, and the cell registers the increase in $\Delta\bar{\mu}_{H^+}$ level by means of "protometer", sending the attractant signal to the bacterial flagella in case of $\Delta\bar{\mu}_{H^+}$ increase. Thus bacteriorhodopsin, while not being a specialized photosensor, still participates in perception of light as an attractant stimulus when the light intensity is high enough to actuate a substantial number of bacteriorhodopsin molecules, and hence to influence the $\Delta\bar{\mu}_{H^+}$ level. It is essential that under conditions of high light intensity, sR-mediated light measurement turns out to be impossible because of too high sensitivity of sR. In other words, in the case of prokaryotes, bacteriorhodopsin and sensory rhodopsin play the same roles as, respectively, cone cells (vision in high light) and rod cells (twilight vision) in our eyes. It seems quite understandable that switching off light-independent $\Delta\bar{\mu}_{H^+}$ generators (respiratory chain and H^+-

ATPase) by corresponding inhibitors is favorable for "protometer"-mediated light sensing by bacteriorhodopsin (Bibikov et al. 1993).

6.5.4 Animal Rhodopsin

Animal rhodopsin, discovered about a century before bacteriorhodopsin, is a photosensory pigment of photoreceptor cells in eyes. It is responsible for photon absorption and initiation of light-stimulus processing.

Bovine rhodopsin is composed of 348 amino acid residues, i.e., it contains 100 more amino acids than bacteriorhodopsin. The amino acid sequence of animal rhodopsin, obtained by Ovchinnikov and coworkers, turned out to be quite different from that of bacteriorhodopsin (Ovchinnikov 1982); however, it perfectly fits the model of seven α-helical columns penetrating the membrane. Seven hydrophobic sequences of the same length as in bacteriorhodopsin are connected with hydrophilic sequences of various lengths. One hundred additional amino acids found in the animal protein are localized in longer (than in bacteriorhodopsin) terminal sequences and in water-exposed loops between the helices. The loop connecting the fifth and sixth segments is especially large.

Besides the seven-column arrangement of the membranous part of the proteins, some other structural similarities were revealed when archaean and animal rhodopsins were studied. In particular, in both cases retinal is connected to the protein via an ε-amino group of a lysine residue (in case of bacteriorhodopsin and animal rhodopsin it is *Lys*-216 and 296, respectively) located in the closest to the C-end seventh (G) α-helix. The Schiff base is protonated in the dark in both proteins. Animal rhodopsin (similar to bacteriorhodopsin and halorhodopsin) was shown to generate $\Delta\Psi$ at the expense of a single light flash. However, in contrast to the bacterial retinal proteins, continuous illumination fails to generate steady-state electric current.

The first steps of the photocycle are also very similar in the two types of rhodopsins. The following consecutive spectral shifts were found in both types of rhodopsins: (1) a very fast (ps), long wavelength shift, (2) its reversal in the microsecond time scale, and (3) a slower, short wavelength shift. In both bacterial and animal rhodopsins, the light induces isomerization of the retinal residue, but instead of *all-trans* → *13-cis* isomerization described in halophilic archaea, the 11-*cis* → *all-trans* transition occurs in the animal pigment.

The functional differences between the bacteriorhodopsin and animal rhodopsin are mainly related to the fate of short wavelength intermediates (M_{412} or *MII*, respectively). In both cases, formation of these intermediates is accompanied by Schiff base deprotonation and H^+ release into the aqueous phase.

However, in the case of animal rhodopsin, *MII* lives minutes instead of milliseconds for the M_{412} intermediate of the bacteriorhodopsin photocycle.

MII cannot convert spontaneously back to the initial form of rhodopsin. Instead, it produces free *all-trans* retinal and opsin, which requires addition of 11-*cis* retinal to regenerate the rhodopsin.

Animal rhodopsin converted to *MII* acquires the ability for specific interaction with another protein called *transducin*. This interaction induces replacement of GDP by GTP in transducin. Such an event initiates the following processes. A complex of transducin and GTP activates a cGMP-specific phosphodiesterase, causing, as a result, a decrease in the cGMP concentration in the photoreceptor cell. The latter event entails closing Na^+ channels in the outer membrane of these cells. The channels in question were found to require cGMP to be open. Closing of the channels lowers the membrane conductance and therefore increases $\Delta\Psi$ across the plasma membrane (this $\Delta\Psi$ is continuously generated by the Na^+/K^+-ATPase of the plasma membrane of the photoreceptor cells). A $\Delta\Psi$ increase means excitation of the photoreceptor cell, which is recognized by special neurons transmitting the light signal to a corresponding center in the brain.

All these processes are assumed to participate in amplification of the light signal. For instance, *one* molecule of the long-lived intermediate *MII* facilitates GDP–GTP exchange in *hundreds* of transducin molecules. Moreover, *one* phosphodiesterase activated by the GTP–transducin complex hydrolyzes *hundreds* of molecules of cGMP before the active phosphodiesterase–transducin–GTP complex decomposes due to the hydrolysis of the bound GTP to GDP and P_i. Taking these facts into account, one can understand how the rod cell can be excited by such a small amount of light energy as a single photon.

Amplification, however, does not appear to be necessary and is even undesirable at high light intensities when, for example, a million photons, rather than a single photon should be sensed. In this case, a linear rather than exponential relationship between the light intensity and its effect seems to be required. It has to be noted that the electric potential difference generated by rhodopsin in the cone cells plasma membrane is of the same direction as the one generated by Na^+/K^+-ATPase (the cell interior is negative). This means that the ability of rhodopsin for light-induced $\Delta\Psi$ generation may be used for perception of a strong and sudden increase in illumination, and this response will be much faster than in the case of a complicated system of light signal amplification used in rod cells in low light intensity. Another interesting feature is that the response to intense light will spontaneously disappear over time, as rhodopsin is a photoelectric generator of a single action: photon absorption leads to the splitting of the retinal moiety from rhodopsin, and in order for the initial protein form to be regenerated, 11-*cis* retinal needs to be once again attached to aporhodopsin. It is noteworthy that in cones and rods, rhodopsin is localized in the outer cell membrane and photoreceptor disks, respectively. The latter are not connected to the outer cell membrane; the rhodopsin-generated $\Delta\Psi$ increase can directly excite cones but not rods, the fact consistent with well-known functional difference between cones responding to bright light, and rods responding to dim light. It should be stressed that cones, besides direct rhodopsin-mediated excitation in response to strong and sudden increase in a light intensity, are equipped, like rods, by a mechanism of signal amplification which may be used for perception of steady changes in low light intensities (Skulachev 1988).

References

Balashov SP, Lanyi JK (2007) Xanthorhodopsin: proton pump with a carotenoid antenna. Cell Mol Life Sci 64:2323–2328

Baryshev VA, Glagolev AN, Skulachev VP (1981) Sensing of $\Delta\bar{\mu}_{H^+}$ in phototaxis of *Halobacterium halobium*. Nature 292:338–340

Béjà O, Aravind L, Koonin EV, Suzuki MT, Hadd A, Nguyen LP, Jovanovich SB, Gates CM, Feldman RA, Spudich JL, Spudich EN, DeLong EF (2000) Bacterial rhodopsin: evidence for a new type of phototrophy in the sea. Science 289:1902–1906

Béjà O, Spudich EN, Spudich JL, Leclerc M, DeLong EF (2001) Proteorhodopsin phototrophy in the ocean. Nature 411:786–789

Belyakova TN, Kadzyaukas YuP, Skulachev VP, Smirnova IA, Chekulaeva LN, Jasaitis AA (1975) Generation of electrochemical potential of H^+ ions and photophosphorylation in the cells of *Halobacterium halobium*. Dokl Akad Nauk SSSR 223:483–486

Bibikov SI, Grishanin RN, Kaulen AD, Marwan W, Oesterhelt D, Skulachev VP (1993) Bacteriorhodopsin is involved in halobacterial photoreception. Proc Natl Acad Sci U S A 90:9446–9450

Bogomolni RA, Spudich JL (1982) Identification of a third rhodopsin-like pigment in phototactic Halobacterium halobium. Proc Natl Acad Sci U S A 79:6250–6254

Drachev LA, Jasaitis AA, Kaulen AD, Kondrashin AA, Liberman EA, Nemecek IB, Ostroumov SA, Semenov AYu, Skulachev VP (1974) Direct measurement of electric current generation by cytochrome oxidase, H^+-ATPase and bacteriorhodopsin. Nature 249:321–324

Drachev LA, Kaulen AD, Skulachev VP (1984) Correlation of photochemical cycle, H^+ release and uptake, and electric events in bacteriorhodopsin. FEBS Lett 178:331–335

Drachev LA, Kaulen AD, Skulachev VP, Zorina VV (1986) Protonation of a novel intermediate P is involved in the M-bR step of the bacteriorhodopsin photocycle. FEBS Lett 209:316–320

Gerber GE, Anderegg RJ, Herlihy WC, Gray CP, Biemann K, Khorana HG (1979) Partial primary structure of bacteriorhodopsin: sequencing methods for membrane proteins. Proc Natl Acad Sci U S A 76:227–231

Grigorieff N, Ceska TA, Downing KH, Baldwin JM, Henderson R (1996) Electron-crystallographic refinement of the structure of bacteriorhodopsin. J Mol Biol 259:393–421

Harbison GS, Smith SO, Pardoen JA, Winkel C, Lugtenburg J, Herzfeld J, Mathies R, Griffin RG (1984) Dark-adapted bacteriorhodopsin contains 13-cis, 15-syn and all-trans, 15-anti retinal Schiff bases. Proc Natl Acad Sci U S A 81:1706–1709

Henderson R, Unwin PN (1975) Three-dimensional model of purple membrane obtained by electron microscopy. Nature 257:28–32

Kalaidzidis IV, Kaulen AD, Radionov AN, Khitrina LV (2001) Photoelectrochemical cycle of bacteriorhodopsin. Biochemistry (Mosc) 66:1220–1233

Kolbe M, Besir H, Essen LO, Oesterhelt D (2000) Structure of the light-driven chloride pump halorhodopsin at 1.8 Å resolution. Science 288:1390–1396

Lanyi JK (2001) X-ray crystallography of bacteriorhodopsin and its photointermediates: insights into the mechanism of proton transport. Biochemistry (Mosc) 66:1192–1196

Lanyi JK (2004) Bacteriorhodopsin. Annu Rev Physiol 66:665–688

Lozier RH, Bogomolni RA, Stoeckenius W (1975) Bacteriorhodopsin: a light-driven proton pump in *Halobacterium halobium*. Biophyl J 15:955–962

Lozier RH, Niederberger W, Bogomolni RA, Hwang S, Stoeckenius W (1976) Kinetics and stoichiometry of light-induced proton release and uptake from purple membrane fragments, *Halobacterium halobium* cell envelopes, and phospholipid vesicles containing oriented purple membrane. Biochim Biophys Acta 440:545–556

Mukohata Y, Kaji Y (1981) Light-induced membrane-potential increase, ATP synthesis, and proton uptake in Halobacterium halobium, R1mR catalyzed by halorhodopsin: Effects of N, N'-dicyclohexylcarbodiimide, triphenyltin chloride, and 3,5-di-tert-butyl-4-hydroxybenzylid-enemalononitrile (SF6847). Arch Biochem Biophys 206:72–76

Oesterhelt D, Stoeckenius W (1971) Rhodopsin-like protein from the purple membrane of *Halobacterium halobium*. Nat New Biol 233:149–152

Ovchinnikov YuA (1982) Rhodopsin and bacteriorhodopsin: structure-function relationships. FEBS Lett 148:179–191

Ovchinnikov YA, Abdulaey NG, Feigina MY, Lobanov NA, Kiselev AV, Nasimov IA (1978) Primary structure of bacteriorhodopsin. Bioorg Chemistry 4:1573–1574 (In Russian)

Pebay-Peyroula E, Rummel G, Rosenbusch JP, Landau EM (1997) X-ray structure of bacteriorhodopsin at 2.5 angstroms from microcrystals grown in lipidic cubic phases. Science 277:1676–1681

Sasaki J, Spudich JL (2008) Signal transfer in haloarchaeal sensory rhodopsin-transducer complexes. Photochem Photobiol 84:863–868

Schobert B, Lanyi JK (1982) Halorhodopsin is a light-driven chloride pump. J Biol Chem 257:10306–10313

Schobert B, Cupp-Vickery J, Hornak V, Smith S, Lanyi J (2002) Crystallographic structure of the K intermediate of bacteriorhodopsin: conservation of free energy after photoisomerization of the retinal. J Mol Biol 321:715–726

Skulachev VP (1988) Membrane bioenergetics. Springer, Berlin

Takahashi T, Tomioka H, Kamo N, Kobatake Y (1985) A photosystem other than PS370 also mediates the negative phototaxis of *Halobacterium halobium*. FEMS Microbiol Lett 28:161–164

Part III
$\varDelta\overline{\mu}_{H^+}$ Consumers

Chapter 7
$\Delta\bar{\mu}_{H^+}$-Driven Chemical Work

Five enzymatic systems have been described that can be regarded as $\Delta\bar{\mu}_{H^+}$ consumers performing chemical work: H^+-ATP synthase; H^+-pyrophosphate synthase; H^+-transhydrogenase; reverse NADH:CoQ-oxidoreductase (complex I); reverse CoQH$_2$:cytochrome c-oxidoreductase (complex III).

In the first two cases, energy is utilized to form a high energy phosphoanhydride bond (ATP synthesis from ADP and P$_i$ or PP$_i$ synthesis from P$_i$); in the other three cases energy is used to drive uphill transfer of reducing equivalents. Among these five systems, H^+-ATP synthase is certainly the most important, being responsible for the interconversion of two biological energy currencies, i.e. the membrane-linked ($\Delta\bar{\mu}_{H^+}$) and the water-soluble (ATP). This is why H^+-ATP synthase is one of the most intensively studied enzymes.

7.1 H^+-ATP Synthase

H^+-ATP synthase is an enzyme catalyzing $\Delta\bar{\mu}_{H^+}$-driven phosphorylation of ADP by inorganic phosphate in mitochondria, chloroplasts, and respiring or photosynthetic bacteria.

In many anaerobic bacteria, such as *Lactococcus casei* or *Enterococcus hirae*, a very similar enzyme catalyzes the reverse process (ATP → ADP + P$_i$ + $\Delta\bar{\mu}_{H^+}$). This function, being the only one for this enzyme in *L. casei* and *E. hirae*, plays a minor role in aerobic and photosynthetic cells and subcellular organelles. (This phenomenon will be discussed in more detail at the end of this chapter, see Sect. 7.2.1).

7.1.1 Subunit Composition of H^+-ATP Synthase

The H^+-ATP synthase complexes from various sources have very similar structure in spite of the substantial variety of primary $\Delta\bar{\mu}_{H^+}$ generators that are found in

Table 7.1 Subunit composition of the E. coli H$^+$-ATP synthase

Complex E. coli	Subunit E. coli	Mitochondrial analog of subunit	Molecular mass, kDa	Number of amino acid residues	Number of subunits per F_oF_1
F_1	α	α	55.0	513	3
	β	β	50.0	459	3
	γ	γ	31.5	287	1
	δ	OSCP	19.5	177	1
	ε	δ and ε	15.0	138	1
F_o	a	a	30.0	271	1
	b	b	17.0	156	2
	c	c	8.0	79	10

living organisms (see previous chapters). H$^+$-ATP synthase can be easily dissociated into *two subcomplexes, one responsible for transmembrane proton transport* (F_o) *and the other for ATP synthesis or hydrolysis* (F_1).[1]

For the *E. coli* H$^+$-ATP synthase, the subunit composition, amino acid sequence, and operon structure have been established (Bragg 1984). It was found that the *E. coli* factors F_1 and F_o are composed of five (named alphabetically from α to ε) and three (*a, b,* and *c*) types of subunits, respectively (see Table 7.1). The stoichiometry of subunits was shown to be $3α:3β:γ:δ:ε:a:2b:10c$.[2] The molecular masses of F_1, F_o and the total H$^+$-ATP synthase complex from *E. coli* are 381, ~145, and ~525 kDa, respectively.

The F_oF_1 complex of *E. coli* is encoded by a single operon called *unc* or *atp*. It is about seven thousand base pairs in size. The order of the genes corresponds to the following order of transcription of mRNA coding, i.e., *I, a, b, c, δ, α, γ, β,* and ε subunits where *I* is a hypothetical protein with unknown functions that is absent from purified F_oF_1.

Mitochondrial H$^+$-ATP synthase is also composed of subcomplex F_1 (six types of main subunits), subcomplex F_o (three types of main subunits), and several additional subunits (Walker et al. 1985). The F_1 factor contains 3α, 3β, γ, δ, and ε subunits and the oligomycin sensitivity conferring protein (OSCP) protein. The major subunits (α and β) resemble very much their *E. coli* analogs. For instance, β subunits of F_1 factor from bovine and yeast mitochondria, from *E. coli*, and from chloroplasts have about 70 % conservative sequences. This is also true for the α subunit.

A comparison of minor F_1 subunits from *E. coli* and animal mitochondria showed that the δ and ε subunits are different, while the γ subunits are similar (Walker et al. 1982) (see also Sect. 7.1.2). In the first approximation, we can say that mitochondrial subunits δ and ε correspond to the ε subunit of *E. coli*. Mitochondrial factor F_1 was shown to contain an additional polypeptide, the so-called

---

[1] Besides ATP synthases of F_oF_1-type, ATP synthases (ATPases) of V_oV_1- and E_1E_2-types may be also found in living organisms. The last sections of this chapter deal with these enzymes.

[2] The number of *c*-type subunits of F_o factor will be discussed in more detail later.

7.1 H⁺-ATP Synthase

OSCP. This protein has many homologies with the *E. coli* δ subunit, but some differences between these two proteins were also found. OSCP is slightly larger (21 kDa instead of 19.3 kDa of the *E. coli* δ subunit); it has a hydrophobic segment close to the N-terminus, and there are several pieces in the sequence homologous not to the δ, but to the *b* subunit of the *E. coli* enzyme.

The structure of the F_o factor from animal mitochondria is similar to that from *E. coli.* Polypeptides *a, b,* and *c* are its main subunits. Subunit *b* from animal mitochondria was shown to be slightly different from the bacterial analog, and animal F_o factor probably contains only one *b*-type polypeptide (instead of two in bacteria). The following stoichiometry of the main subunits of the animal ATP synthase is now generally accepted: $3\alpha:3\beta:\gamma:\delta:\varepsilon:OSCP:a:b:8c$.

Several additional polypeptides in *mitochondrial H⁺-ATP synthase* are absent from the bacterial enzyme. These are: *A6L* subunit of F_o factor; factor F_6 (9 kDa); 9.5 kDa protein inhibitor of F_1 factor; 18.5 kDa *d* subunit (which, together with F_6 and OSCP is required for correct binding of F_1 and F_o); and several other subunits.

In animals and yeasts, all the F_1 subunits are encoded in the nuclear genome and synthesized in the cytoplasm. Among F_o components of the yeast enzyme, subunits *a, c,* and *A6L* are encoded by the mitochondrial genome; all the others are encoded by the nuclear genome. In DNA of animal mitochondria, there are two overlapping genes encoding subunits *a* and *A6L* of F_o; all the other subunits (including *c*) are encoded in the nucleus.

The subunit composition of the *chloroplast H⁺-ATP synthase* (it is usually referred to as CF_oCF_1) is, in principle, similar to that of bacteria and mitochondria (Richter et al. 2000). Again, factor F_1 has the $3\alpha:3\beta:\gamma:\delta:\varepsilon$ structure. The ε subunit is homologous to the bacterial one. However, the CF_o factor contains not three, but four types of main subunits—*a, b, b',* and *c* (subunits *b* and *b'* are homologous) with the suggested stoichiometry of $a:b:b':14c$. In other words, CF_o contains not a homodimer of *b* subunits (as in bacteria), but a heterodimer of two different (though rather similar) subunits *b* and *b'*.

7.1.2 Three-Dimensional Structure and Arrangement in the Membrane

The H⁺-ATP synthase complex is so large that it protrudes from the membrane surface for a rather long distance. The protruding part (factor F_1) faces the cytoplasm in bacteria, the matrix in mitochondria, and the stroma in chloroplasts.

Negative staining of inside-out submitochondrial vesicles reveals *mushroom-like knobs about 9 nm in diameter* covering the entire outer membrane surface. Racker (Fig. 7.1) showed that the knobs vanished after mechanical agitation of a suspension of vesicles in solution containing ethylenediamine tetraacetate (EDTA), a chelator for Mg^{2+} ions (Racker 1965). This process was accompanied by the disappearance of ATPase and ATP synthase activity of the vesicles and the appearance in the solution of 9 nm spherical particles that possessed ATPase activity. Reconstitution of these

Fig. 7.1 Efraim Racker

spherical particles with the membrane vesicles was shown to restore ATP-synthase and ATPase activity in the latter. From these observations, Racker concluded that the knobs represent the catalytic part of the H$^+$-ATP synthase, i.e., *factor F_1*. A part of H$^+$-ATP synthase responsible for H$^+$-ion transfer (*factor F_o*), was shown to remain in the membrane after the detachment of factor F_1. This fact led to a substantial increase in proton conductivity of the membrane. The results of a computer analysis of electron micrographs of factor F_1 are summarized in Fig. 7.2. The diameter of the frontal projection of factor F_1 is about 10 nm.

The 3D structure of the main part of mitochondrial factor F_1 has been established with the high resolution of 2.4 Å by X-ray analysis (Abrahams et al. 1994).

Fig. 7.2 Structure of the catalytic part of H$^+$-ATP synthase: isolated factor F_1 from beef heart mitochondria (negative contrast); electron micrographs of 379 projections of the five main types were averaged using an image analysis system (Boekema et al. 1986)

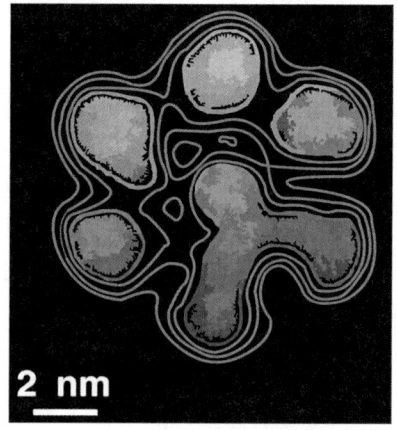

7.1 H⁺-ATP Synthase

Fig. 7.3 Nobel Laureate John Walker

This work was recognized by the Nobel Prize in Chemistry in 1997 to John Walker (Fig. 7.3). As shown in Fig. 7.4, the greater part of F_1 is composed of subunits α, β, and γ with stoichiometry of 3:3:1. The α and β subunits form a spherical globule (Stock et al. 2000). The central part of this globule consists of very long α-helical

Fig. 7.4 Structure of the main part of factor F_1 from mitochondrial H⁺-ATP synthase. Different subunits are marked with the following colors: α—*red*, β—*yellow*, γ—*blue*, δ—*green*, ε—*purple* (Stock et al. 2000)

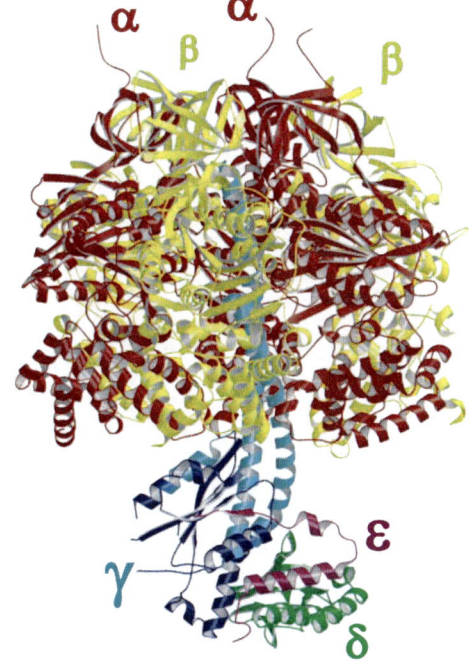

parts of the γ subunit; subunits α and β are located alternately around the γ subunit (this structure is often compared to that of an orange). It seems important to note that the γ subunit occupies an asymmetrical position in the center of the globule. Subunit γ protrudes significantly beyond the $3\alpha:3\beta$-globule on the one of its sides. This protruding part of subunit γ together with subunits δ and ε form the *"central stalk"* of factor F_1 that connects it to factor F_o.

Subunits α and β contain one nucleotide-binding center each. A nucleotide-binding domain has also been identified in γ subunit. However, under physiological conditions this domain is most likely not able to bind nucleotides.

Only indirect information is available concerning the 3D structure of the rest of ATP synthase. These data were obtained using such methods as electron diffraction, NMR, and X-ray diffraction analysis of isolated subunits, and different possibilities of their cross-linking in F_oF_1 complex. A scheme of the F_oF_1 complex structure is presented at Fig. 7.5.

The 3D structure of the c subunit from the *E. coli* enzyme in a hydrophobic solvent was determined by NMR. This subunit was shown to consist of two transmembrane α-helical columns that are connected to each other on the cytoplasmic membrane side by a loop of polar amino acid residues (Fig. 7.6). The C-terminal α-helical column contains a conservative aspartate residue (Asp-61) that plays a key role in transmembrane proton transport. This residue seems to be located approximately in the middle of the hydrophobic layer of the membrane. In factor F_o the c-type subunits form an oligomer consisting of 8–14 c-type subunits. Recently, the structure of this c oligomer of Na^+-ATP synthase from the bacterium *Ilyobacter tartaricus* has been determined by X-ray diffraction (Meier et al. 2005). The α-helical columns of c subunits form in the membrane a kind of repeating

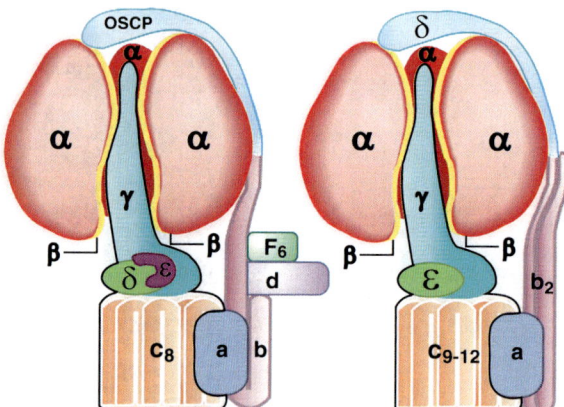

Fig. 7.5 Scheme illustrating current understanding of the structure of F_oF_1-ATP synthase from eukaryotic mitochondria (*left*) and bacteria (*right*) (Stock et al. 2000). The two β subunits (*yellow*) are basically covered by α subunits; the third β subunit is not shown. The A6L subunit as well as other small subunits of the mitochondrial factor F_o absent from the bacterial subcomplex are also not shown

7.1 H⁺-ATP Synthase

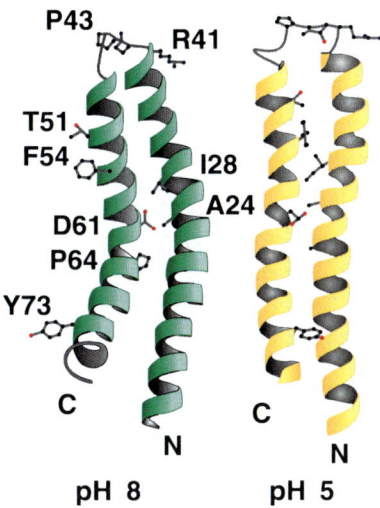

Fig. 7.6 Structure of c subunit of F_oF_1-ATP synthase from *E. coli* in a hydrophobic solvent, determined by NMR for deprotonated (*green*, pH 8) and protonated (*yellow*, pH 5) forms (Rastogi and Girvin 1999)

system consisting of two rings, one being inside the other (Fig. 7.7). The external and internal rings consist of C-terminal and N-terminal α-helical columns, respectively. The number of c subunits is supposed to vary in ATP synthases of different organisms. For instance, ATP synthase from animal mitochondria is considered to contain 8 c subunits; the yeast mitochondria enzyme 10 c subunits; the chloroplast ATP synthase 14 c subunits; and the sodium ATP synthase from *I. tartaricus* 11 c subunits (Stock et al. 2000). We will discuss later that the variable stoichiometry of c subunits can play an important physiological role for it leads to a variable ratio of H⁺/ATP.

The oligomer of c subunits is connected to factor F_1 through the central stalk of the F_oF_1 complex (Fig. 7.8). It is important to note that the central stalk is located

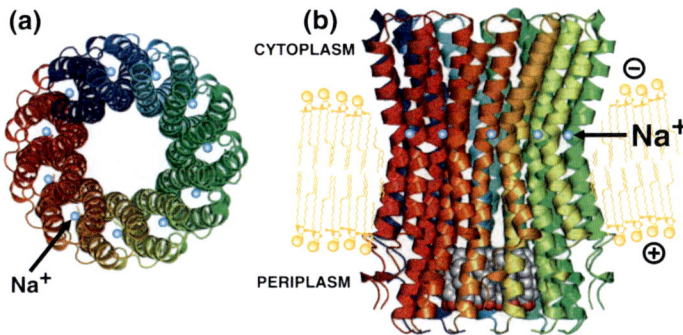

Fig. 7.7 Structure of the c ring of Na⁺-ATP synthase from *I. tartaricus*: **a** view perpendicular to the membrane plane, **b** view parallel to the membrane plane. Individual c subunits are marked with different colors. *Blue spheres* represent bound sodium ions (Meier et al. 2005)

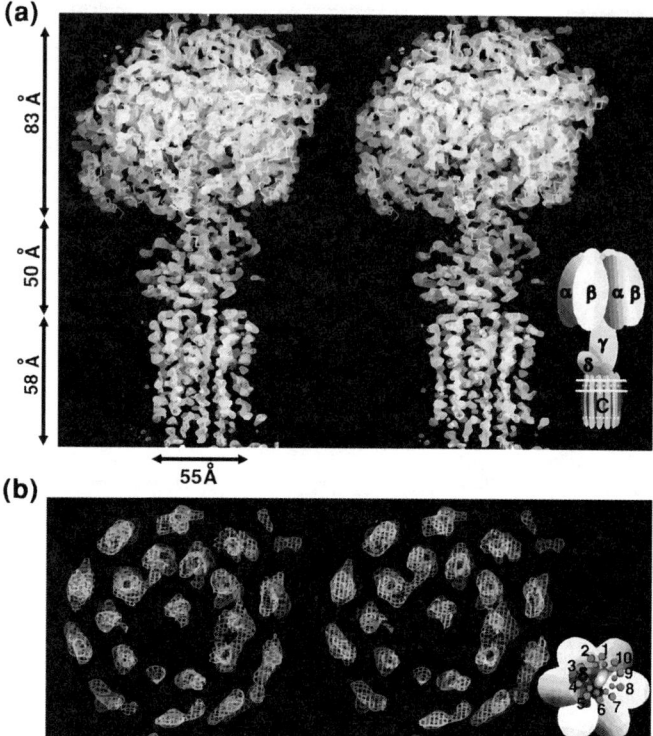

Fig. 7.8 Stereo image of the F_1-c complex from *S. cerevisiae* mitochondria. Side view (**a**) and bottom view (**b**). Schemes showing the arrangement of subunits are presented in the *lower right* corners. Peripheral stalk is not seen (Stock et al. 2000)

asymmetrically in relation to the c oligomer; as a result, it interacts with the polar loops of only six subunits of the oligomer.

Analysis of primary amino acid sequences suggests that the a subunit is hydrophobic and forms five α-helical transmembrane columns that contain several conservative residues of charged amino acids (these are Arg-210, His-245, Glu-196, and Glu-219 in the case of *E. coli*). These residues might participate in the formation of a transmembrane proton transport pathway that is created due to the interaction of subunits a and c.

Subunit b is more hydrophilic, and in the case of *E. coli* it probably crosses the membrane only once. Formation of the so-called *peripheral stalk*, which also binds factors F_o and F_1, seems to be the main function of this subunit (see Fig. 7.5). The peripheral stalk of the *E. coli* enzyme apparently contains two b subunits and one δ subunit. The peripheral stalk of mitochondrial ATP synthase is probably formed of the subunits b, OSCP, d, and F_6 with stoichiometry 1:1:1:1.

7.1.3 ATP hydrolysis by Isolated Factor F_1

Factor F_1, when detached from the membrane sector of H^+-ATP synthase, *retains the ability to hydrolyze ATP to ADP and phosphate*. The reaction proceeds vigorously but looses sensitivity to oligomycin, diethylstilbestrol, and low concentrations of dicyclohexylcarbodiimide (DCCD), differing in this respect from ATP hydrolysis by the native F_oF_1 complex.

ATP hydrolysis by factor F_1 was shown to start with a water molecule attacking directly the anhydride bond of ATP. This proceeds *without any covalent intermediate formed*. Both the data of experiments on isotope exchange and the fact that ATP hydrolysis is resistant to vanadate, hydroxylamine, and other acyl-phosphate reagents support this statement.

As we have already discussed in the previous section, six potential nucleotide-binding sites were shown to be present in major subunits of factor F_1, i.e., a site per each α and β subunit.[3] Biochemical experiments showed F_1 to be able to bind six adenine nucleotide molecules. However, it is only the sites on the β subunits that exchange bound and free nucleotides with rates comparable to the turnover time of the enzyme. Apparently, these three sites fulfill the catalytic functions, while the nucleotide sites of α subunits might have a structural and/or a regulatory role (e.g., they might participate in modulation of ATP synthase activity).

It seems quite remarkable that three catalytic sites of factor F_1 have different affinity for nucleotides (Boyer 2002). One of the sites ("T-site", from *tight*) manifests extremely high affinity for ATP or ADP + P_i with binding constant for them of about 10^{-12} to 10^{-8} M. The two other sites ("L-"and "O-" sites, from *loose* and *open*, respectively) bind nucleotides only when their concentration is relatively high ($K_s = 10^{-6}$ to 10^{-3} M); they also probably have different selectivity to ATP and ADP (Fig. 7.9). This diversity of the catalytic sites of factor F_1 is caused by the different conformational states of different α and β subunits of this protein, which seems to be the result of asymmetrical location of the γ subunit inside the $3\alpha:3\beta$ globule.

When the concentration of labeled ATP is low, i.e., when only one nucleotide-binding site (the T-site) is saturated with ATP, factor F_1 hydrolyzes ATP very slowly (so-called *unisite catalysis*) (Grubmeyer and Penefsky 1981). When excess of unlabeled ATP is added, "cold" ATP is bound in the L-site. This causes a manifold increase in the ATP^{32} hydrolysis reaction rate as a result of a significant acceleration of the release of the ATPase reaction products (ADP and P_i^{32}) from the T-site. Thus, binding of the second ATP molecule in the L-site changes the affinity of the T-site for nucleotides; the T-site transfers to the O-state (in other words, different sites of factor F_1 manifest negative cooperation in nucleotide

---

[3] It should be stressed that in reality all the nucleotide-binding sites are located in the area of α and β subunit contacts. But in order to simplify the description, we will attribute these sites to those subunits which possess more amino acid residues participating in the formation of the corresponding sites.

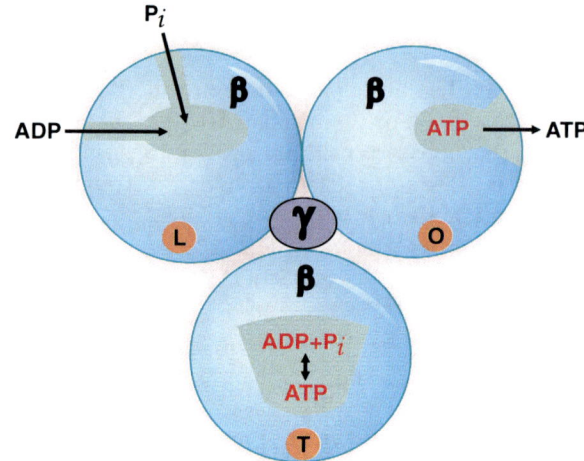

Fig. 7.9 Scheme of catalytic sites of F_oF_1-ATP synthase. The L-site binds the substrates of the ATP-synthase reaction (ADP + P_i). The T-site has very high affinity to ATP or ADP + P_i and catalyzes their energy-independent interconversion. The $\Delta\bar{\mu}_{H^+}$-driven release of synthesized ATP (or ATP binding in case of its hydrolysis) occurs in the O-site

binding and positive cooperation in the catalysis reaction). ATP hydrolysis is probably accompanied by consecutive changes in conformation of the nucleotide-binding sites: O → T → L... For instance, all the three sites undergo identical conformational changes in the course of catalysis, but they are all in different conformations at any given time. Such a mechanism of ATP hydrolysis by factor F_1 was suggested to be the result of consecutive changes of γ subunit position in relation to different α and β subunits, e.g. of its *rotation* inside the $3\alpha:3\beta$-globule.

ATP hydrolysis by F_oF_1 complex is regulated by different parameters of the cell energy status. Andrei Vinogradov (Fig. 7.10) and coworkers reported ATP

Fig. 7.10 Andrei Vinogradov

hydrolysis by isolated mitochondrial F_1 to be strongly inhibited by one of the reaction products, ADP. This effect is substantially increased in the presence of azide and is prevented by sulfite. The elimination of ADP from the inhibited factor F_1 (which is part of the membrane-bound F_oF_1 complex) was found to be $\Delta\bar{\mu}_{H^+}$-dependent. This means that ADP inhibition can occur only when $\Delta\bar{\mu}_{H^+}$ is low, i.e. when there is danger of the cell ATP pool being exhausted due to the reversal of the H$^+$-ATP synthase reaction (Vasilyeva et al. 1980). In the case of mitochondria, a special polypeptide, *protein inhibitor*, may fulfill the same function as ADP (Frangione et al. 1981). This protein inhibitor also suppresses ATPase activity in isolated factor F_1 as well as in the membrane complex F_oF_1 when $\Delta\bar{\mu}_{H^+}$ is low.

7.1.4 Synthesis of Bound ATP by Isolated Factor F_1

In 1982, Feldman and Sigman showed that *purified factor CF_1 isolated from chloroplasts could synthesize ATP from tightly bound ADP in response to the addition of inorganic phosphate* (Feldman and Sigman 1982). The synthesized ATP, persistently bound to CF_1, could be released into the aqueous phase only after denaturing treatment of CF_1. Later, a suspension of isolated mitochondrial factor F_1 was shown to catalyze the synthesis of bound ATP from medium containing ADP and P_i.

Theoretical calculations made in the group of one of this book's authors (V.P.S.) indicate that the equilibrium constant of ATP hydrolysis in the T-site of factor F_1 is no more than 100 and is probably close to about one (Skulachev and Kozlov 1982). This means that the synthesis of *bound* ATP in the T-site requires a much smaller amount of energy than that of *free* ATP in solution (or in other words, the T-site affinity for ATP is much higher than for ADP + P_i). The point is that the electronic structure of the anhydride bond in bound ATP is substantially different from the one in free ATP because of interactions with numerous ligands belonging to the protein (Fig. 7.11). One can speculate that during ATP synthesis by H$^+$-ATP synthase the energy of $\Delta\bar{\mu}_{H^+}$ is expended not at the stage of phosphoanhydride bond synthesis, but at the stage of release of tightly bound ATP from the catalytic site. This possibility was first suggested by Boyer (Fig. 7.12) (Boyer et al. 1973). This also means that during ATP hydrolysis the energy is released on the binding of ATP to the T-site, and not on the phosphoanhydride bond cleavage.

7.1.5 F_o-Mediated H$^+$ Conductance

Removal of F_1 from thylakoids, chromatophores, or inside-out subbacterial or submitochondrial vesicles always results in *H$^+$ leakage through F_o*. Proteoliposomes containing F_o are also leaky for protons. The leakage is abolished by a specific

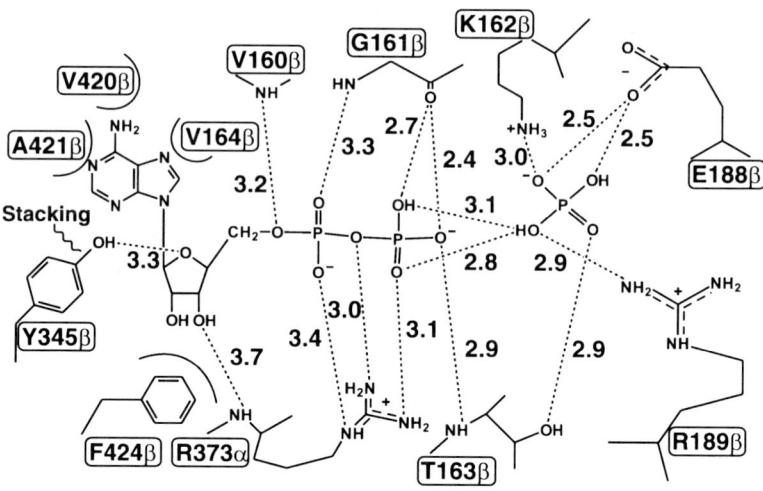

Fig. 7.11 Binding of ADP and P_i in the active site of factor F_1. The numbers near the *dotted lines* represent the distance in angstroms

Fig. 7.12 Nobel Laureate Paul D. Boyer

F_o modifier, DCCD or by a specific F_oF_1ATPase inhibitor, oligomycin. Reconstitution of the F_oF_1 complex suppresses the leakage as well (Friedl et al. 1983).

In bacteria, it has been shown that H^+ conductance does not occur in F_1-stripped membranes from mutant strains in which any of the three F_o subunits has been lost. It seems very probable that it is the a and c subunits that play the leading role in proton transport, and the transmembrane proton transport pathway is formed at the site of contact of these subunits. There is an aspartate or glutamate residue in the c subunits

7.1 H⁺-ATP Synthase

of F_o from different organisms that is absolutely necessary for protons transfer (in *E. coli* it is *Asp*-61). It is located about 1.8 nm from the membrane surface, i.e., in the middle of the hydrophobic core of the membrane. Its replacement with *Asn* or *Gln* completely abolishes both the H⁺ transport and $\Delta\bar{\mu}_{H^+}$-powered ATP synthesis (Hoppe et al. 1982). Transmembrane α-helical columns of the *a* subunit also contain several conservative charged amino acid residues (Arg-210, His-245, Glu-196, and Glu-219). These residues probably participate in the formation of the proton transport pathway.

Sensitivity of F_o to DCCD is a characteristic feature of this protein. DCCD is a covalent modifier of protonated carboxylic groups. At low concentrations, it binds selectively to *Asp*-61 of the *c* subunits of ATP synthase (Negrin et al. 1980). This modification of *Asp*-61 completely abolishes the H⁺-conducting function of F_o. It seems worth noting that modification of only *one* *c* subunit in F_o seems to be sufficient for complete inactivation of the enzyme.

Oligomycin was shown to be another effective inhibitor of mitochondrial factor F_o (besides the already mentioned DCCD). Yeast mutants were obtained with H⁺-conductance through factor F_o that are resistant to this antibiotic. Two loci of the *a* subunit were shown to be responsible for this resistance. Both were shown to be conservative in humans and mice. However, only one of the conservative loci was found in the *a* subunit of *E. coli*, a fact that is in accordance with data on the weak effect of oligomycin on the ATP synthase of this bacterium.

Factor F_o is probably not a simple pore because the rate of H⁺ translocation through F_o obeys saturation kinetics with optimum at pH 7–9. These data indicate the presence of several H⁺ binding sites. Factor F_o probably has two H⁺-binding sites located on the two sides of the membrane. The highest values of H⁺ conductance were obtained for factor F_o from chloroplasts and photosynthesizing bacteria: $\sim 6{,}000$ H⁺ per second per F_o at $\Delta\Psi = 100$ mV (Junge 1987). However, even this value seems to be too small for a channel-type H⁺-conducting mechanism.

A functional model of factor F_o was suggested (Junge et al. 2001) that explains all the mentioned structural features of F_o, kinetic characteristics of proton transport by this factor, and also its DCCD-induced inactivation (Fig. 7.13). This model is based on the assumption of two proton half-channels (located on the opposite membrane sides) being present in factor F_o. Proton transfer between these

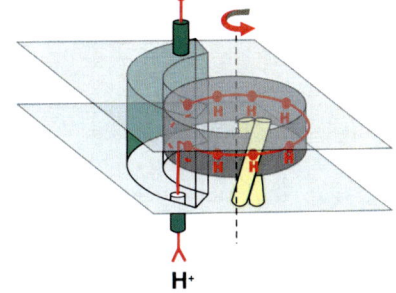

Fig. 7.13 Tentative scheme of the mechanism of proton transport by factor F_o. Subunits *a* and *c* are colored *green* and *yellow*, respectively (Junge et al. 2001)

two half-channels proceeds parallel to the membrane surface due to the transfer of the protonated aspartate (glutamate) residue of a c subunit. Such movement of a carboxylic group can be achieved due to rotation of the c ring relative to the a subunit. During this process, the given carboxylic group has to be deprotonated when interacting with the a subunit, and it has to be protonated when leaving this contact. Hence, proton transport via factor F_o has to be coupled to c ring rotation.

7.1.6 Possible Mechanism of Energy Transduction by F_oF_1-ATP Synthase

The main principles of the functioning of ATP synthase have been already described in the preceding sections of this chapter. First of all, *ADP and phosphate bound to factor F_1 can produce bound ATP in an energy-independent fashion; energy is required to expel the bound ATP from the F_1 catalytic site to the aqueous medium.*

The release of this F_1-bound ATP is driven by the energy of $\Delta\bar{\mu}_{H^+}$. This statement was proved in experiments on the inhibitory effects of a nonhydrolyzable ATP analog—5-adenylyl imidodiphosphate (AMPPNP)—on the ATPase activity of submitochondrial vesicles. Inhibition of ATPase activity could be completely abolished by a short preincubation of the AMPPNP-treated vesicles with succinate under aerobic conditions. If an uncoupler was added *before* succinate, inhibition remained the same as without succinate (Skulachev and Kozlov 1982). This effect was accounted for assuming that $\Delta\bar{\mu}_{H^+}$ generated by succinate oxidation removes AMPPNP from the catalytic site.

How can the energy of $\Delta\bar{\mu}_{H^+}$ be utilized to change the affinity of the ATP synthase catalytic sites to nucleotides? Let us remember that proton transport takes place in factor F_o, while nucleotide-binding sites are located in factor F_1. A very unusual mechanism resembling an electric motor was invented by nature to carry out this function. The proton potential energy is first transformed to mechanical energy (rotation of the H^+-ATP synthase rotor), which is then spent on performance the chemical work (change in affinity of ATP synthase catalytic sites to nucleotides) (Stock et al. 2000).

It has been already stated that γ subunit occupies an asymmetric position in the center of the $3\alpha:3\beta$ globule, which leads to differences in its interactions with different α and β subunits. This seems to be the reason for differences in the properties of the three (T, O, and L) nucleotide-binding sites of ATP synthase. Because ATP hydrolysis is accompanied by consecutive conformational changes of nucleotide-binding sites, it seems logical to suggest that ATP hydrolysis can be coupled to the *rotation of the γ subunit* inside the $3\alpha:3\beta$ globule. This hypothesis was first suggested by Paul Boyer (he received the Nobel Prize in Chemistry in 1997 for this idea) and was later confirmed by a very elegant direct experiment carried out by the Japanese researcher Masasuke Yoshida and his colleagues.

7.1 H$^+$-ATP Synthase

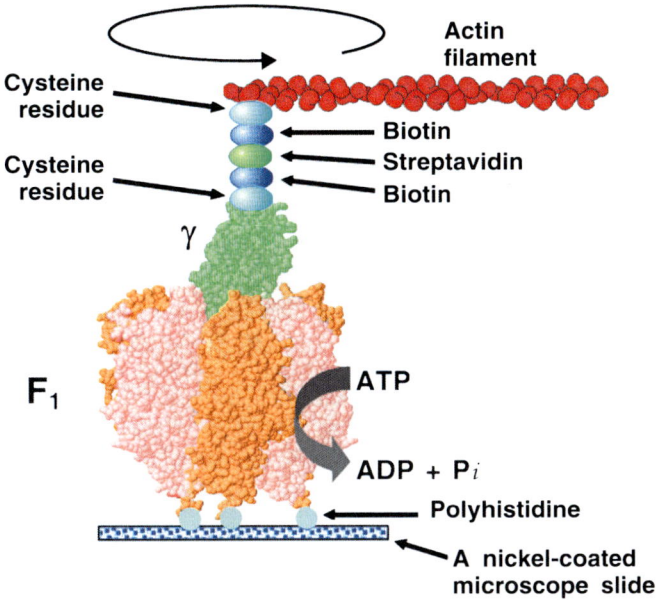

Fig. 7.14 Scheme of experiment visualizing rotation of factor F_1 γ subunit during hydrolysis of ATP. Factor F_1 subunits are colored as follows: α—*pink*; β—*orange*; γ—*green* (Noji et al. 1997)

A simplified version of the bacterial factor F_1 comprised of just three types of subunits, i.e., α, β, and γ with stoichiometry of $3\alpha{:}3\beta{:}\gamma$, was employed. Using molecular genetic methods, Yoshida removed all the cysteine residues from these subunits. Moreover, one cysteine residue was introduced at the edge of the γ subunit. Also, a sequence of six histidine residues capable of high-affinity binding of nickel ions was attached to the N-terminus of the β subunit. Then the protein was fixed by this polyhistidine tag on a glass surface covered with nickel ions, and a very long actin filament fluorescently labeled with fluorescein molecules was attached to the single cysteine residue of the γ subunit using the biotin-binding protein streptavidin (Fig. 7.14). Using a light microscope, Yoshida directly observed the counterclockwise rotation of the fluorescent actin filament after the addition of ATP. The full 360° turn was shown to be composed of three consecutive steps, each being a 120° turn. The ATP-induced rotation could be observed only in the presence of Mg^{2+} and was shown to be sensitive to inhibitors of the ATPase reaction (Noji et al. 1997). So direct evidence of the energy of ATP hydrolysis being transformed into mechanical energy of γ subunit rotation during the functioning of factor F_1 was obtained.

Some data suggest proton transport across factor F_o to be coupled to the rotation of the c oligomer, which means also the rotation of the central stalk connected to it. So, it is through the rotation mechanism that proton transport can be coupled to ATP synthesis (or hydrolysis) by this enzyme. The ring of c subunits of factor F_o together with the subunits γ and ε[4] (i.e., with the central stalk of factor F_1) form the rotor of this motor, while all the other subunits of H$^+$-ATP synthase serve as the stator (see above, Fig. 7.5). The role of the *peripheral stalk* that also connects factors F_o and F_1 becomes clear within this hypothesis. This additional connection of the two factors is probably needed for the attachment of the $3\alpha:3\beta$ globule to the membrane part of the enzyme stator, preventing the rotation of the $3\alpha:3\beta$ globule together with the central stalk (Stock et al. 2000).

It should be stressed that many questions concerning the functioning of the F_oF_1-ATP synthase remain unclear. First of all, we do not understand the mechanism of the coupling of proton transport in factor F_1 to the rotation of the c ring. Electrostatic interaction of the deprotonated anion *Asp-61* on the c subunit (rotor) and the cation of one of the positively charged amino acid residues of the a subunit (stator) is suggested to play the key role in this process.

7.1.7 H$^+$/ATP Stoichiometry

It has been firmly established that more than one proton needs to be transported across the membrane for one ATP molecule to be synthesized by H$^+$-ATP synthase. This conclusion follows from measurements of $\Delta\bar{\mu}_{H^+}$ and intracellular concentrations of ATP, ADP, and phosphate. There is a general consensus that at physiological concentrations of substrates and products of the H$^+$-ATP synthase, the energy price for ATP formation is about 44 kJ/mol (10.5 kcal/mol). In terms of $\Delta\bar{\mu}_{H^+}$, this is equivalent to 455 mV if the H$^+$/ATP ratio is 1. This value is at least twofold higher than the maximal $\Delta\bar{\mu}_{H^+}$ observed in experiments. So, the *minimal* H$^+$/ATP ratio should be assumed to be two.

Experimental determination of H$^+$/ATP stoichiometry is mainly based on the measurement of the so-called *P/O ratio* in mitochondria oxidizing different respiratory substrates. This parameter represents the ratio of synthesized ATP (or esterified phosphate) to the oxygen consumed. The P/O criterion was first suggested by Belitser (Fig. 7.15) in 1939 (Belitser and Tsibakova 1939). In the case of animal mitochondria oxidizing NAD$^+$-dependent substrates or succinate, the P/O values obtained in experiments was usually ≤ 3 or ≤ 2, respectively (Hinkle 2005). As NADH and succinate oxidation by the respiratory chain enzymes is coupled to 10 and 6 protons translocated across the membrane, respectively (see Chap. 4), we can calculate the number of protons used by H$^+$-ATP synthase for the formation of one ATP molecule from ADP and phosphate, which should be transported to mitochondrial matrix to form intramitochondrial ATP. The latter is then exported to extramitochondrial solution. This calculation results in a value of about four. However, one should take into account that in experiments on isolated

---

[4] This is true for the bacterial type enzyme; in the case of the mitochondrial enzyme, the central stalk is composed of subunits γ, δ, and ε.

Fig. 7.15 Vladimir Belitser

mitochondria, an *extramitochondrial* ADP is phosphorylated by an *extramitochondrial* phosphate. This means that $\Delta\bar{\mu}_{H^+}$ is consumed not only by H^+-ATP synthase, but also by ATP^{4-}/ADP^{3-}-antiporter and $H_2PO_4^-,H^+$-symporter. As a result, the exchange of external ADP and phosphate for internal ATP is accompanied by one H^+-ion import to the mitochondrion. So, the H^+/ATP stoichiometry for the F_oF_1 complex per se (i.e., for ATP synthesis in mitochondrial matrix) should be $4 - 1 = 3$.

If we consider the hypothesis of the rotary mechanism of functioning of the enzyme to be true, then the 360° rotation of the F_oF_1 rotor should be coupled to the synthesis of three ATP molecules (because there are *three* catalytic nucleotide-binding sites in this enzyme). On the other hand, the 360° rotation of the mitochondrial F_oF_1 rotor should lead to transmembrane transport of eight protons (for the c oligomer is comprised of 8 c subunits in case of animal mitochondria). So, the structural data support the assumption that the H^+/ATP stoichiometry for the ATP synthase from animal mitochondria should be $8/3 \approx 2.7$, which corresponds well enough to the experimental data (≤ 3). This example also illustrates the importance of such parameter as the number of subunits that comprise the c oligomer. The c-subunit stoichiometry should have an extremely important physiological meaning since it determines such essential cell parameter as H^+/ATP ratio. It was already mentioned that chloroplast ATP synthase is probably comprised of 14 c subunits[5] (Seelert et al. 2000), and Na^+-ATP synthase from

---

[5] These data are in accordance with the experimentally measured H^+/ATP stoichiometry for chloroplast H^+-ATP synthase. In the case of plants, ADP is photophosphorylated on the external side of the thylakoid membrane. The ATP obtained is mainly utilized in the chloroplast stroma

I. tartaricus of 11 *c* subunits (Meier et al. 2005). So, such an extremely important parameter as the "exchange rate" of the biological convertible energy currencies ($\Delta \bar{\mu}_{H^+}, \Delta \bar{\mu}_{Na^+}$ and ATP) can vary in different organisms.

It seems interesting to note that if we take a mechanical approach to the question of H$^+$/ATP stoichiometry (i.e., assume it to be equal to a whole number), we expect the number of *c* subunits in the enzyme to be a multiple of three. However, in majority of the studied cases, this rule is not fulfilled, which suggests a fractional H$^+$/ATP ratio.

7.2 H$^+$-ATPases as Secondary $\Delta \bar{\mu}_{H^+}$ Generators

In this section, we will consider the functioning of various ATPases as proton potential generators. The most widespread $\Delta \bar{\mu}_{H^+}$ generators use the energy of light or respiration for the generation of proton potential. ATP does not participate in this process. But sometimes the path from the energy source to $\Delta \bar{\mu}_{H^+}$ appears to be more complex—the energy is first converted to ATP and only then utilized for $\Delta \bar{\mu}_{H^+}$ generation by H$^+$-motive ATPases that can be defined as "secondary $\Delta \bar{\mu}_{H^+}$ generators".

As a rule, H$^+$-ATPases appear to be necessary if neither light nor respiratory energy are available for a membrane that performs some form of $\Delta \bar{\mu}_{H^+}$-driven work. Anaerobic bacteria, obtaining energy only by *means of substrate phosphorylation (glycolysis)*, exemplify this situation. Here, the ATP formed by glycolytic enzymes can be utilized by H$^+$-ATPases localized in the cytoplasmic membrane of the cell. The generated $\Delta \bar{\mu}_{H^+}$ supports *osmotic work*, namely, the accumulation of metabolites inside the cell via H$^+$,metabolite-symporters (Fig. 7.16), and *mechanical work* (flagellum rotation). Moreover, the H$^+$-ATPases can fulfill one more important function, namely *regulation of intracellular pH*. In fact, they prevent acidification of cytoplasm of glycolyzing cells by pumping of glycolysis-produced H$^+$ ions from cytoplasm to the outer medium.

From a formal standpoint, any enzyme competent in $\Delta \bar{\mu}_{H^+}$ generation at the expense of the energy of ATP can be regarded as H$^+$-ATPase. An H$^+$-ATP synthase that is usually employed to form ATP in a $\Delta \bar{\mu}_{H^+}$-driven fashion can under certain conditions hydrolyze ATP and generate $\Delta \bar{\mu}_{H^+}$. For this reason, this enzyme is often called H$^+$-ATPase. In our opinion, however, it would be more appropriate to use such a name when dealing with enzymes whose *main* biological function is *ATP hydrolysis rather than synthesis*. By introducing this limitation, we can refer to several enzymes differing in structures (those of classes F_oF_1, V_0V_1, and E_1E_2) and reaction mechanisms but performing one and the same function, to the

---

(Footnote 5 continued)
during the synthesis of glucose. No porters participate in this process, and the H$^+$/ATP ratio was shown to be about four protons per ATP molecule.

7.2 H^+-ATPases as Secondary $\Delta\bar{\mu}_{H^+}$ Generators

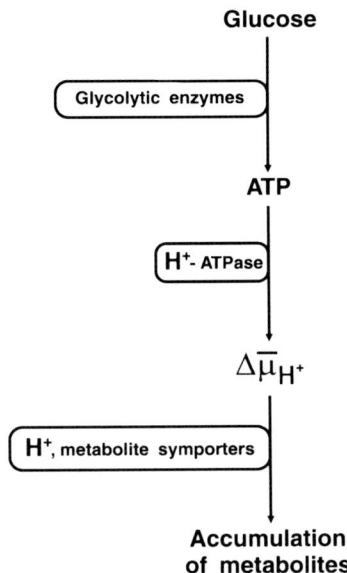

Fig. 7.16 Scheme of one of the pathways of energy transformation in anaerobic bacteria synthesizing ATP solely by means of substrate phosphorylation

category of H^+-ATPases (see Table 7.2). Let us consider these types of enzymes in more detail.

7.2.1 F_oF_1-Type H^+-ATPases

These ATPases function in the cytoplasmic membrane of the majority of anaerobic bacteria. When taking into account the structural and functional properties of these enzymes as well as their subunit composition, they seem to be a version of the above-described F_oF_1-type H^+-ATP synthases. Even more so, membranes containing H^+-ATP synthase are capable of generating $\Delta\bar{\mu}_{H^+}$ at the expense of ATP hydrolysis by the reversal of the reaction catalyzed by this enzyme. This effect can be observed, for instance, in certain bacteria utilizing both the energy of light or respiration and that of glycolysis.

H^+-ATPases can be easily divided into two subcomplexes: the proton-conducting factor F_o and catalytic factor F_1. ATP hydrolysis by enzymes of this type is sensitive to azide, DCCD, aurovertin, and several other inhibitors of H^+-ATP synthases. The reaction catalyzed by H^+-ATPases is reversible. For instance, these enzymes can synthesize ATP at the expense of energy of $\Delta\Psi$ that was artificially created through K^+ ion outflow from bacteria in the presence of valinomycin.

H^+-ATPases from such anaerobic bacteria as *Lactobacillus casei* and *Enterococcus hirae* are examples of the most thoroughly studied representatives of enzymes of this type.

Table 7.2 ATPases as secondary $\Delta\bar{\mu}_{H^+}$ generators

Enzyme	Location	Mechanism	Functions, supported by formed $\Delta\bar{\mu}_{H^+}$
H^+-ATPase	Cytoplasmic membrane of obligate anaerobic bacteria	F_oF_1 and V_oV_1	Uphill transport of solutes, pH regulation in cell
H^+-ATPase	Outer cell membrane of plants and fungi	E_1E_2	Uphill transport of solutes, pH regulation in cell
H^+-ATPase	Tonoplast of plant and fungal cells	V_oV_1	Uphill transport of solutes, pH regulation in vacuole
H^+-ATPase	Outer cell membrane of certain animal tissues	E_1E_2 and V_oV_1	Secretion of H^+; uphill transport of solutes
H^+-ATPase	Membranes of secretory vesicles, endosomes, lysosomes, Golgi apparatus, and endoplasmic reticulum	V_oV_1	Uphill transport of solutes, pH regulation in intracellular vesicles
H^+-ATP synthase (reverse)	Cytoplasmic membrane of respiring or photosynthetic bacteria	F_oF_1	Uphill transport of solutes, pH regulation in a cell, rotation of flagellum in absence of O_2 and light
H^+-ATP synthase (reverse)	Inner membrane of mitochondria	F_oF_1	Uphill transport of solutes to anaerobic mitochondria
H^+/K^+-ATPase	Outer membrane of stomach epithelial cell	E_1E_2	Secretion of H^+

7.2 H⁺-ATPases as Secondary $\Delta\bar{\mu}_{H^+}$ Generators

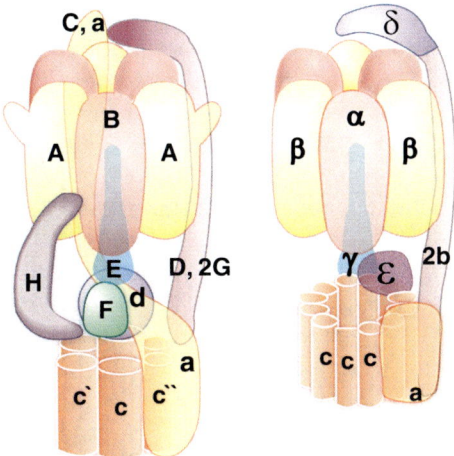

Fig. 7.17 Comparative view of the structure of ATPases of eukaryotic V_0V_1-type (*left*) and prokaryotic F_oF_1-type (*right*) (Lolkema et al. 2003)

7.2.2 V_0V_1-Type H⁺-ATPases

H⁺-ATPases of V_0V_1 type are in many ways similar to the above-described F_oF_1-type H⁺-ATP synthases (H⁺-ATPases). However, they have some specific structural and functional properties. H⁺-ATPase from tonoplast (a membrane that separates the vacuolar contents from the cytoplasm in plant and fungal cells) is an example of V_0V_1-type enzyme (Wagner et al. 2004). Similar enzymes were discovered in such eukaryotic membrane organelles as secretory vesicles, endosomes, lysosomes, and Golgi apparatus. The V_0V_1-type H⁺-ATPases can also be found in plasmalemma of certain eukaryotic cells, and also in the cytoplasmic membrane of a limited number of obligate anaerobic prokaryotes (e.g., in some clostridia) (Lolkema et al. 2003).

The V_0V_1-type ATPases consist of two subcomplexes, one of which is responsible for transmembrane proton transport (V_0), and the other for ATP hydrolysis (V_1). In contrast to F_oF_1-type ATP synthases, subcomplexes V_0 and V_1 lose their functional properties once the V_0V_1-type ATPase complex is dissociated.

Subcomplex V_1 from vacuolar V_0V_1 ATPase is composed of eight subunits (from A to H). Subunits A and B contain nucleotide-binding sites[6] and form a 3A:3B globule similar to 3α:3β globule of F_oF_1-type ATP synthases. Subunits C, E, F, and H (with stoichiometry of 1:1:1:1) probably form the central stalk of the V_0V_1-type enzyme, while subunits D and G (with stoichiometry of 1D:2G) form its peripheral stalk (Fig. 7.17).[7]

The V_0 factor consists of five subunits: *a, d, c, c′,* and *c″*. Subunits *c, c′,* and *c″* were found to be homologous. They show substantial similarity to the *c* subunit of

---

[6] Catalytic sites are located in subunits A.
[7] Two peripheral stalks are found in the vacuolar V_0V_1-type ATPase.

F_oF_1-type ATP synthases. However, subunits c and c' of factor V_0 are twice as large as subunit c of factor F_o; in fact, they are a duplicated version of subunit c of factor F_o containing four (rather than two) transmembrane α-helices. Subunit c'' is about 2.5 times larger than subunit c of factor F_o; it contains five transmembrane columns. Only one conservative glutamate (aspartate) residue is preserved in all the c subunits of factor V_0. The c oligomer of factor V_0 was previously assumed to contain a number of transmembrane α helices similar to that of the c oligomer of factor F_o. If that were the case, then V_0V_1 ATPase would have had a two times lower stoichiometry of H^+/ATP when compared to the F_oF_1-type enzyme (because of a reduced number of active carboxyl groups). It was suggested that V_0V_1 ATPase, due to duplication and fusion of c subunits (which led to the loss of one carboxylate and a reduced H^+/ATP stoichiometry) could function in vivo only as H^+-ATPase. However, a recent X-ray study has shown the Na^+-V_0V_1 ATPase from *Enterococcus hirae* to contain an oligomer composed of 10 c subunits (Murata et al. 2005), i.e., the Na^+/ATP ratio for this enzyme has to be the same as the H^+/ATP ratio for H^+-ATP-synthase from *E. coli*.

H^+-transport and ATPase activity of the V_0V_1-type enzymes are inhibited by a unique set of compounds that includes nitrate, N-ethylmaleimide, trialkyltin, and high DCCD concentrations. Oligomycin and DCCD at low concentrations (inhibitors F_oF_1-type of H^+-ATP synthase) show no inhibitory effect on the tonoplast H^+-ATPase.

7.2.3 E_1E_2-Type H^+-ATPases

H^+-ATPases of E_1E_2 type (another name, P type) consist of a very long polypeptide (about 100 kDa) that can be accompanied by a smaller subunit fulfilling regulatory rather than catalytic function (Chow and Forte 1995). The reaction starts with the phosphorylation of one of the aspartate residues of the large subunit, which causes its conformational shift called the $E_1 \rightarrow E_2$ transition. This transition is accompanied by a decrease in free energy of the hydrolysis of the phosphorylaspartate. The dephosphorylation of the enzyme regenerates the original E_1 form. This mechanism is in many ways similar to that of Na^+/K^+-ATPase and Ca^{2+}-ATPase (see Sect. 12.5.2). E_1E_2-ATPases have been described for the outer cell membranes of plants and fungi. A similar structure has been found in the H^+/K^+-ATPase of the stomach epithelium. E_1E_2-ATPase activity was found to be sensitive to vanadate, hydroxylamine, and diethylstilbestrol but resistant to the main inhibitors of F_oF_1- and V_0V_1-type ATPases (oligomycin, azide, nitrate, etc.).

7.2.3.1 H^+-ATPase of the Outer Cell Membrane of Plants and Fungi

The outer cell membrane (plasmalemma) of plants and fungi contains an enzyme exporting H^+ ions from the cytoplasm at the expense of the energy of ATP (Palmgren 2001). The H^+ transport is electrogenic. The enzyme belongs to the

7.2 H⁺-ATPases as Secondary $\Delta\bar{\mu}_{H^+}$ Generators

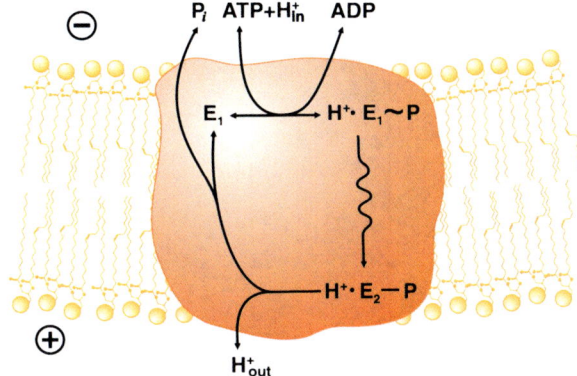

Fig. 7.18 Mechanism of E_1E_2-type H⁺-ATPase. H^+_{in} and H^+_{out} correspond to intracellular and extracellular H⁺ ions, respectively

E_1E_2-type of ATPases forming a vanadate- and hydroxylamine-sensitive aspartyl-phosphate intermediate when ATP is hydrolyzed. Diethylstilbestrol and DCCD are also inhibitory. This H⁺-ATPase is a single polypeptide with a molecular mass slightly above 100 kDa. Its content in the plasmalemma reaches 15 % of the total protein. The mechanism of ATP hydrolysis is shown in Fig. 7.18. H⁺-ATPases from plant and fungi outer membranes have been reconstituted into proteoliposomes, and the reversibility of ATP hydrolysis was demonstrated in this system ($\Delta\Psi$ artificially created by valinomycin-mediated K⁺ diffusion was used for ATP synthesis in these experiments).

There must be some physiological significance to the fact that the pH optimum of plasmalemma H⁺-ATPase is between 6.0 and 7.0. At pH above 7.0, the activity of the enzyme decreases drastically. The impression is that the enzyme is used to prevent acidification of the cytoplasm by removing excess H⁺ ions when the intracellular pH drops below neutrality. Apparently, *intracellular pH regulation* is a function of the plasmalemma H⁺-ATPase. However, it is obvious that this ATPase also performs another function, namely, formation of $\Delta\bar{\mu}_{H^+}$, which is then utilized for the *uphill import of metabolites into the cell* that are transported via H⁺,metabolite-symporters (see Chap. 9).

7.2.3.2 H⁺/K⁺-ATPase of Gastric Mucosa

A unique $\Delta\bar{\mu}_{H^+}$-generating ATPase was discovered in the *gastric mucosa outer cell membrane* (Ganser and Forte 1973). It catalyzes an electroneutral exchange of intracellular H⁺ for extracellular K⁺, in this way acidifying the gastric juice. The enzyme is called H⁺/K⁺-ATPase. It has been purified and reconstituted into proteoliposomes.

H⁺/K⁺-ATPase was shown to belong to E_1E_2 type; the enzyme contains three polypeptides of molecular mass of about 100 kDa (Shin et al. 1997). A phosphorylated intermediate is formed in the course of the hydrolysis reaction. The enzyme activity is inhibited by vanadate and hydroxylamine. A peculiar feature of

Fig. 7.19 Mechanism of H⁺/K⁺-ATPase of the stomach epithelium

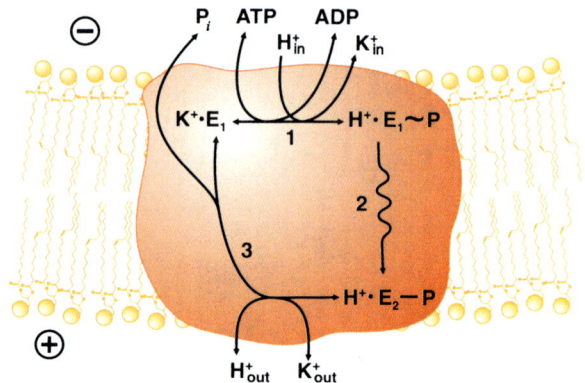

H⁺/K⁺-ATPase is that it seems to be composed of three similar but nonidentical 100 kDa subunits.

Some evidence has suggested the following *mechanism of H⁺/K⁺-ATPase* functioning. A complex of K⁺ and the enzyme in its E_1 conformation binds ATP and H⁺ on the cytoplasmic surface of plasmalemma surrounding the gastric mucosa cell. As a result, K⁺ is released into the cytoplasm. In a triple complex ($E_1 \bullet$ATP$\bullet$H⁺), a reversible transfer of a phosphoryl from ATP to a carboxylic group occurs so that an intermediate H⁺$\bullet E_1 \sim$ P complex is formed. Then a conformational change (the $E_1 \rightarrow E_2$ transition) occurs, resulting in a translocation of the bound H⁺ across the barrier of the membrane. The H⁺$\bullet E_2$-P complex cannot phosphorylate ADP. Substitution of extracellular K⁺ for the bound H⁺ leads to H⁺ release into the extracellular medium. The binding of K⁺ induces hydrolysis of E_2-P. The subsequent $E_2 \rightarrow E_1$ transition completes the cycle (Fig. 7.19). A similar enzyme has been described in the epithelium of the colon.

7.2.4 Interrelations of Various Functions of H⁺-ATPases

H⁺-ATPases usually perform two main functions: membrane energization and maintenance of necessary pH in one of the compartments divided by a membrane.

In anaerobic bacteria utilizing a glycolytic substrate as the only energy source, glycolytic *ATP is hydrolyzed by H⁺-ATPase so as to form* $\Delta\bar{\mu}_{H^+}$, which is then used for the uphill import of metabolites into the cell.

However, even here pH regulation may also be of some significance: H⁺-ATPase pumps H⁺ ions out of the cell, thus preventing acidification of the cytoplasm that may result from anaerobic bacteria catalyzing neutral carbohydrate fermentation to acidic final products. The same is true for plasmalemmal H⁺-ATPase of plants and fungi, which is activated when cytoplasmic pH falls below 7.0. Under such conditions, *removal of H⁺ ions from the cytoplasm* rather than $\Delta\bar{\mu}_{H^+}$ generation for osmotic work *becomes the master function of this H⁺-ATPase*.

The function of some other H$^+$-ATPases of animal tissues is the *acidification of the vesicle interior*. This is inherent in lysosomal, endosomal, and perhaps also the Golgi apparatus and endoplasmic reticulum H$^+$-ATPases. In these systems, *$\Delta\Psi$ formed by the H$^+$-ATPases is converted to ΔpH due to electrophoretic influx of an anion*, e.g. Cl$^-$. In lysosomes, acidification of the inner solution is required to create optimal conditions for the operation of the lytic enzymes having pH optima in the acidic range. In fact, lysosomal enzymes are inactive at neutral pH typical of cytosol. As a result, the cytosolic components cannot be digested when these enzymes are accidently released from lysosomes. Acidic pH also seems necessary for the functioning of endosomes and some other cytoplasmic inclusions possessing H$^+$-ATPases.

A different role is played by H$^+$-ATPases of the plasma membrane of some types of renal cells and bladder epithelium: they are involved in *trans-cellular transport of acidic equivalents*. Here again, $\Delta\Psi \rightarrow \Delta$pH transition seems to be necessary. In the case of the gastric mucosa H$^+$/K$^+$-ATPase, likewise specialized in the transport of acidic equivalents, the problem of $\Delta\bar{\mu}_{H^+} \rightarrow \Delta$pH transduction does not arise since this enzyme exchanges H$^+$ ions for an equal number of K$^+$ ions. This is why $\Delta\Psi$ is not formed and ΔpH becomes the primary form of $\Delta\bar{\mu}_{H^+}$.

Apparently, this ATPase transfers only one H$^+$ per ATP. As a result, a huge pH gradients can be formed between the cytoplasm and the outer medium.

In chromaffin and other secretory granules, the $\Delta\bar{\mu}_{H^+}$-*generating function of the H$^+$-ATPase is predominant because accumulation of the secreted compounds inside the granule is powered by $\Delta\bar{\mu}_{H^+}$*.

Finally, we would like to note that *H$^+$-ATPases (H$^+$-ATP synthases)* of all the three above-described types simultaneously function in the majority of eukaryotic cells. For instance, F_oF_1-H$^+$-ATP synthases are present in mitochondria and chloroplasts, V_0V_1-H$^+$-ATPases in other membrane organelles, and E_1E_2-type ATPases in plasmalemma.

7.3 H$^+$-Pyrophosphate Synthase (H$^+$-Pyrophosphatase)

It has been already discussed in Sect. 1.4.1 that ATP is hydrolyzed to AMP and pyrophosphate (PP$_i$) when there is a necessity to spend more energy than can be obtained via ATP hydrolysis to ADP and P$_i$ (this is often needed for such metabolic processes as DNA, RNA, or polypeptides synthesis, fatty acid, or glucose activation, etc.).

$$\text{DNA synthesis:} \quad (\text{dNMP})_n + \text{dNTP} \rightarrow (\text{dNMP})_{n+1} + \text{PP}_i \quad (7.1)$$

$$\text{RNA synthesis:} \quad (\text{NMP})_n + \text{NTP} \rightarrow (\text{NMP})_{n+1} + \text{PP}_i \quad (7.2)$$

$$\begin{aligned}&\textit{protein synthesis:}\\&\quad \text{amino acid} + \text{tRNA} + \text{ATP} \rightarrow \text{aminoacyl-tRNA} + \text{AMP} + \text{PP}_i\end{aligned} \quad (7.3)$$

glycogen and starch synthesis:

$$\text{glucose-1-phosphate} + \text{ATP} \rightarrow \text{ADP-glucose} + \text{PP}_i \quad (7.4)$$

fatty acid activation:

$$R - COOH + HS - CoA + ATP \rightarrow R - CO \sim S - CoA + AMP + PP_i \quad (7.5)$$

Pyrophosphate formed in these reactions is irreversibly hydrolyzed by a *soluble pyrophosphatase* in the majority of cells:

$$PP_i + H_2O \rightarrow 2\ P_i \quad (7.6)$$

This reaction is needed so as to keep the pyrophosphate concentration at a very low level, which facilitates ATP-dependent biosyntheses accompanied by PP_i formation.

However, in certain cases pyrophosphate is not immediately cleaved by soluble pyrophosphatase. Soluble pyrophosphatase is either absent from plant cells and certain microorganisms, or its activity is very low. In such cells, the steady-state pyrophosphate concentration can be rather high (e.g., the cytoplasmic concentration of PP_i in higher plants is usually about 0.3 mM), and the energy of PP_i hydrolysis can be utilized for the performance of some kinds of work.

Tonoplasts of plant vacuoles were found to contain a special enzyme, membrane-bound H^+-translocating pyrophosphatase, which catalyzes transmembrane proton transport coupled to pyrophosphate hydrolysis (Maeshima 2000). There are substantial gradients of pH, ions, and metabolites between the vacuole and the cytosol. It was suggested that the V_0V_1-type H^+-ATPase is the main enzyme responsible for the formation of $\Delta\bar{\mu}_{H^+}$, which is the driving force of the uphill transport of different compounds. However, there is another proton pump functioning in plant tonoplasts, H^+-pyrophosphatase. Its activity is substantially higher than that of V_0V_1-type ATPase in certain plant tissues.

A very similar enzyme was also found in some prokaryotes (both bacteria and archaea). In 1966–1969, Baltscheffsky (Fig. 7.20) and her colleagues described the formation of inorganic pyrophosphate by *Rhodospirillum rubrum* chromatophores exposed to light (Baltscheffsky et al. 1966; Baltscheffsky 1969). Later, it was found that in the dark the pyrophosphate can be used, like ATP, by these chromatophores as an energy source. Proton control of pyrophosphatase activity was also demonstrated in chromatophores; this activity was shown to increase eightfold with $\Delta\bar{\mu}_{H^+}$ dissipation. Experiments in the group of one of this book's authors (V.P.S) showed that PP_i hydrolysis can generate $\Delta\Psi$ in *Rh. rubrum* chromatophores and in proteoliposomes reconstituted from purified pyrophosphatase and phospholipids (Isaev et al. 1970). These data warrant the conclusion that *Rh. rubrum* membrane H^+-translocating pyrophosphatase (H^+-pyrophosphate synthase) catalyzes reversible energy interconversion between $\Delta\bar{\mu}_{H^+}$ and inorganic pyrophosphate.

At first glance, the function of H^+-pyrophosphate synthase in *Rh. rubrum* cells may be: (i) to form pyrophosphate at the expense of light- or respiration-produced

7.3 H$^+$-Pyrophosphate Synthase (H$^+$-Pyrophosphatase)

Fig. 7.20 Margareta Baltscheffsky

energy or; (ii) to generate $\Delta\bar{\mu}_{H^+}$ at the expense of pyrophosphate hydrolysis. However, in case (i) the further destiny of synthesized pyrophosphate remains unclear. In contrast to ATP, pyrophosphate cannot operate as a convertible energy currency for cells. Pyrophosphate most probably serves as an energy buffer for $\Delta\bar{\mu}_{H^+}$ and ATP in *Rh. rubrum* cells (this function will be discussed in more detail in Sect. 11.3). In any case, H$^+$-pyrophosphate synthase from *Rh. rubrum* should have an important biological function. Its activity in chromatophores is rather high, and it comprises up to 15 % of the H$^+$-ATP synthase activity.

The primary sequence of H$^+$-pyrophosphatase has been determined for a substantial number of plants as well as for different prokaryotes (Baltscheffsky et al. 1999). In spite of H$^+$-pyrophosphate synthase catalyzing a reaction very similar to the one catalyzed by H$^+$-ATP synthase ($\Delta\bar{\mu}_{H^+}$-dependent formation of phosphoanhydride bond), these two enzymes possess no noticeable homology. Furthermore, H$^+$-pyrophosphatase (in contrast to H$^+$-ATPase) is a relatively simple protein. This enzyme consists of only one polypeptide of about 70 kDa molecular mass in all the studied cases (even though H$^+$-pyrophosphatase in supposed to function in vivo as a homooligomer). This polypeptide is extremely hydrophobic; it contains 14–16 potential α-helical transmembrane columns connected by very short hydrophilic sequences.

The membrane-bound H$^+$-pyrophosphatase has little similarity with the soluble pyrophosphatases. Certain similarities between these two enzymes can be observed only in the area of the hydrophilic loop between transmembrane columns V and VI of H$^+$-pyrophosphatase, which most likely forms the binding site for MgPP$_i$.

H$^+$-pyrophosphatase is active only in the presence of magnesium ions. Aminomethylene diphosphonate selectively inhibits membrane-bound H$^+$-pyrophosphatases (but not the soluble pyrophosphatases) (Baykov et al. 1993). Oligomycin has no inhibitory effect in this case. DCCD decreases H⁺-pyrophosphatase activity in chromatophores and suppresses $\Delta\Psi$ generation by this enzyme reconstituted into proteoliposomes.

The molecular properties of this enzyme and the mechanism of the pyrophosphatase reaction remain unclear. The protein seems to be the least studied proton pump. No phosphorylated intermediates of the H⁺-pyrophosphatase reaction have been described. The H⁺/PP$_i$ stoichiometry for this enzyme is supposed to be only 1 (even though values 0.5 and 2 are also discussed). Such a low (in comparison to H⁺-ATP synthase) stoichiometry might be due to the fact that intracellular [P$_i$] is substantially higher than [PP$_i$], while intracellular [ADP] is much lower than [ATP]. That means that under in vivo conditions (e.g., in a bacterial cell) pyrophosphate synthesis requires far less energy than ATP synthesis. Protonophore-sensitive ATP synthesis utilizing the energy of pyrophosphate hydrolysis, as well as PPi synthesis at the expense of the energy of ATP hydrolysis, were demonstrated in membranes of the anaerobic bacterium *Syntrophus gentianae*. The synthesis of one ATP molecule required hydrolysis of approximately three pyrophosphate molecules, while hydrolysis of only one ATP molecule was sufficient to support the synthesis of three pyrophosphate molecules (Schöcke and Schink 1998). Such a low H⁺/PP$_i$ stoichiometry for H⁺-pyrophosphate synthase is probably the reason for the relatively simple structure of this enzyme when compared to such a complex protein complex as F_oF_1-ATP synthase with the H⁺/ATP stoichiometry of 8/3 (see above).

7.4 H⁺-Transhydrogenase

NAD(H) and NADP(H) are the main reducing equivalent carriers in the cell cytoplasm of the majority of living organisms. The only structural difference between these two cofactors is one phosphate residue. However, this minor difference appears to be sufficient to metabolically isolate NAD(H) and NADP(H) from each other. NAD⁺ is practically always used for substrate oxidation in catabolic reactions, while NADPH is mainly utilized in reductive processes involved in anabolic reactions. That is why NADP(H) is usually reduced in a cell ([NADP⁺] ≪ [NADPH]), while NAD(H) is mainly present in its oxidized form ([NAD⁺] ≥ [NADH]).

Such reactions as glucose oxidation in the pentose phosphate pathway or isocitrate oxidation by NADP⁺-dependant isocitrate dehydrogenase in the Krebs cycle are important mechanisms of NADP⁺ reduction to NADPH. Besides these reactions, there is also a possibility of direct hydride ion (H⁻) transfer from NADH to NADP⁺ (Eq. 7.1), which is catalyzed by enzymes called *transhydrogenases*:

$$NADH + NADP^+ \leftrightarrow NAD^+ + NADPH \tag{7.7}$$

7.4 H⁺-Transhydrogenase

Fig. 7.21 Lars Ernster

This activity was first described in 1952 by Colowick et al. (1952). Because NADP(H) is mainly present in its reduced state while NAD(H) is oxidized, $NADP^+$ reduction at the expense of NADH oxidation is possible only when coupled to utilization of an additional energy source. In 1963, Danielson and Ernster (Fig. 7.21) found that in inside-out submitochondrial particles *energization strongly shifts the equilibrium* of reaction 7.1 to the right, i.e., in the direction of *NADPH formation* (Danielson and Ernster 1963). As a result, the [NADPH] × [NAD⁺]/[NADP⁺] × [NADH] ratio reached 500, whereas in nonenergized vesicles it was close to 1, consistent with the almost equal standard redox potentials (E_0') of the NADH/NAD⁺ and NADPH/NADP⁺ couples.

In 1966, Mitchell proposed that *transhydrogenase is an additional energy-coupling site in the respiratory chain*. It was suggested that oxidation of NADPH by NAD⁺ (the forward reaction) generated $\Delta\bar{\mu}_{H^+}$ of the same direction as respiration, whereas the reverse process consumed respiration-produced $\Delta\bar{\mu}_{H^+}$ (Mitchell 1966).

Experiments conducted by Efim Liberman et al. and the group of one of the authors (V.P.S.) confirmed this hypothesis (Dontsov et al. 1972). Inside-out submitochondrial vesicles, *Rh. rubrum* chromatophores, and proteoliposomes containing purified transhydrogenase were shown to generate $\Delta\Psi$ ("+" inside the vesicle) when oxidizing NADPH by NAD⁺. NADH oxidation by NADP⁺ also caused $\Delta\Psi$ formation but of the opposite direction ("−" inside the vesicle). Both the first and the second types of electrogenesis were shown to be dissipated when the concentrations of the substrates and products of this reaction were equalized. In both cases, the generated $\Delta\Psi$ was a linear function of the ratio ln([NADPH] × [NAD⁺]/[NADP⁺] × [NADH]).

So the transhydrogenase, when in inside-out submitochondrial vesicles and chromatophores, catalyzes the following process:

$$NADP^+ + NADH + n\,H_{in}^+ \leftrightarrow NADPH + NAD^+ + n\,H_{out}^+. \tag{7.8}$$

When respiratory or photosynthetic generators produce $\Delta\bar{\mu}_{H^+}$ favorable for exit of H^+_{in}, the reaction shifts to the right. One transhydrogenase turnover is coupled with transmembrane transfer of one proton, i.e., the stoichiometry $H^-/H^+ = 1$.

A shift in the equilibrium of the transhydrogenase reaction under energized conditions is a direct consequence of the energetics of transhydrogenase and other $\Delta\bar{\mu}_{H^+}$ generators of the respiratory chain. Free energy change (ΔG) in the process of conversion of substrate (S) to product (P) can be calculated as:

$$\Delta G = \Delta G_0 + RT \cdot \ln\frac{[S]}{[P]} \tag{7.9}$$

where ΔG_0 is the energy change when $[S] = [P]$. In all the $\Delta\bar{\mu}_{H^+}$ generators other than transhydrogenase, ΔG_0 is rather large (usually about 30–45 kJ·mol^{-1}). In the transhydrogenase reaction, however, ΔG_0 is close to zero. Therefore, this enzyme can effectively compete with respiration or ATPase in $\Delta\bar{\mu}_{H^+}$ formation only if [NADPH] × [NAD$^+$] are several orders of magnitude higher than [NADP$^+$] × [NADH], a most unlikely situation for the living cell.

Membrane-bound H$^+$-transhydrogenase has been found in the inner mitochondrial membrane of many eukaryotes (first of all in higher animals), and also in cytoplasmic membrane of the majority of bacteria.[8] The mitochondrial enzyme consists of just one polypeptide of ∼120 kDa molecular mass, but in vivo the transhydrogenase functions as a homodimer (Jackson 2003). There are three domains in the protein—dI, dII, and dIII. The dI fragment is a typical nucleotide-binding domain (of the "Rossmann fold" type) that binds specifically NAD$^+$ and NADH. This specificity is reached due to hydrogen bond formation between the conservative aspartate residue of the dI domain and the 2′-OH group of the NAD(H) ribose residue. The dIII fragment is also a nucleotide-binding domain of Rossmann type, but this fragment is capable of binding only NADP$^+$ and NADPH. In this case, specificity is reached due to the interaction of conservative lysine, arginine, and serine residues of the dIII domain with the phosphate residue in the NADP(H) molecule. The dII fragment is a very hydrophobic polypeptide. It is practically completely immersed in the membrane. It is composed of 12–14 (depending on the object) transmembrane α-helical columns connected to each other by very short hydrophilic loops. The dI and dIII domains are supposed to catalyze hydride ion transfer between NADH and NADP$^+$, while the dII fragment

---

[8] Besides the membrane-bound transhydrogenase, the cells may also have a soluble transhydrogenase that catalyzes hydride ion transfer from NADPH to NAD$^+$ in an energy-independent fashion. We will not discuss this enzyme in this book.

is probably responsible for transmembrane proton transport coupled to the former reaction.

Bacterial H$^+$-transhydrogenases have a very similar structure (when compared to mitochondrial enzyme), though these enzymes consist not of one long polypeptide, but of its fragments, two (*E. coli*) or three (*Rh. rubrum*) shorter proteins that constitute the above-described dI, dII, and dIII domains (Bizouarn et al. 2000).

The dI and dIII domains can be extracted and purified. These fragments were shown to spontaneously form a stable complex (the 3D structure has been established for this complex). This dI:dIII complex can bind nucleotides and catalyze rapid hydride ion transfer between them. The *direct* stereospecific transfer of H$^-$ between the A-position of the NADH nicotinamide ring and the B-position of the NADP$^+$ nicotinamide ring occurs in the course of this reaction. The crucial point in this transhydrogenase mechanism is that there is no isotope exchange between the transferred hydride ion and the protons of the medium. So, *the protons transported across the membrane by the enzyme are not the same as the protons transferred between the nicotinamide nucleotides in the form of hydride-ion* during the transhydrogenase reaction.

The shift of the transhydrogenase reaction in the dI:dIII complex in the direction of NADPH formation is probably achieved due to the fact that the affinity of dIII to NADPH is substantially higher than its affinity to NADP$^+$ (and the affinity of dI to NADH is much lower than its affinity to NAD$^+$). However, under steady-state conditions the transhydrogenase activity of the dI:dIII complex is extremely low because of a very slow release of the reaction products from the nucleotide-binding sites. Together, these data support the suggestion that in H$^+$-transhydrogenase the proton potential energy is spent not on the reaction of hydride ion transfer per se, but on the change in the enzyme affinity to the reaction products.

So, we observe an interesting analogy between the work of H$^+$-transhydrogenase and the above-described H$^+$-ATP synthase. Both enzymes couple two processes (a chemical reaction and a transmembrane proton transport), which spatially take place rather far from each other. Hence, for these reactions to be coupled there should be some mechanism of energy transfer between different parts of the protein due to conformational changes, and in both enzymes energy of $\Delta\bar{\mu}_{H^+}$ is utilized not for the corresponding chemical reactions, but for the affinity changes of the proteins to the reaction products.

The mechanism of proton transport in H$^+$-transhydrogenase as well as the way this reaction is coupled to the change of the enzyme affinity to nucleotides remain obscure. It is clear, however, that the mechanism cannot be described by the Mitchell loop scheme and, hence, should represent a version of a redox-linked proton pump that carries out transmembrane transfer H$^+$ ions other than protons involved in the oxidoreduction (Skulachev 1988).

The master function of H$^+$-transhydrogenase undoubtedly consists in the maintenance of the high [NADPH]/[NADP$^+$] ratio *essential for supporting reductive syntheses*, which usually include NADPH-oxidizing step(s). In fact, $\Delta\bar{\mu}_{H^+}$ produced by a respiratory or photosynthetic redox chain and consumed by

Fig. 7.22 Role of H⁺-transhydrogenase in cell energetics

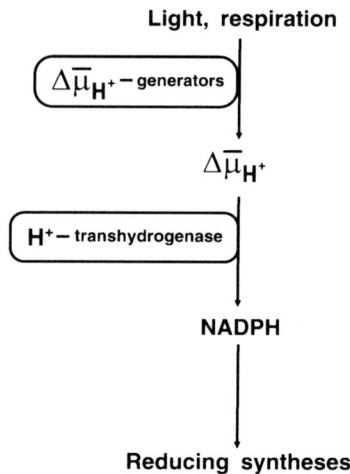

NADH → NADPH transhydrogenation is ultimately utilized as an additional driving force for reducing synthesis (Fig. 7.22).

This mechanism is inherent in mitochondria and the vast majority of those bacteria that do not have a noncyclic photosynthetic chain. On the other hand, in chloroplasts and cyanobacteria the problem of achieving high [NADPH]/[NADP⁺] ratio is solved by reduction of NADP⁺ as the final electron acceptor of the photosynthetic redox chain. Here, H⁺-transhydrogenase is absent.

Mitochondrial H⁺-transhydrogenase probably fulfills an additional function. Two types of enzymes are known to catalyze isocitrate oxidation in mitochondria from animal muscle tissues: NAD⁺- and NADP⁺-dependent isocitrate dehydrogenases. Simultaneous functioning of these two dehydrogenases can lead to a futile cycle dissipating the energy of $\Delta\bar{\mu}_{H^+}$. On the other hand, the existence of such a cycle provides a possibility for Krebs cycle regulation, the membrane potential value being the main regulatory factor (Sazanov and Jackson 1994).

7.5 Other Systems of Reverse Transfer of Reducing Equivalents

In some cases *H⁺-transhydrogenase is employed jointly with other systems of reverse transfer of reducing equivalents along the respiratory chain.* This is apparently the case in *Rh. rubrum* chromatophores oxidizing substrates of about zero redox potential (sulfur, S^{2-}, etc.). In the light, $\Delta\bar{\mu}_{H^+}$ (produced by the cyclic photosynthetic redox chain) can be utilized (i) to reverse NADH:CoQ-reductase and, hence, to reduce NAD⁺ by sulfur or sulfide, and (ii) to oxidize the formed NADH by NADP⁺ via the H⁺-transhydrogenase reaction.

7.5 Other Systems of Reverse Transfer of Reducing Equivalents

There are examples when a reverse transfer of reducing equivalents appears to be one of the main consumers of $\Delta\bar{\mu}_{H^+}$ of the cell. This is the case in bacteria utilizing substrates of positive redox potential as the only energy source. It has already been discussed (see Sect. 5.2.5) that *Acidithiobacillus ferrooxidans* oxidizes Fe^{2+} to Fe^{3+} ($E_0 = +0.77$ V) by oxygen and builds all the cell components of CO_2, H_2O, and NH_3. To do this, it is apparently necessary to reverse not only the NADH:CoQ-oxidoreductase but also the $CoQH_2$-cytochrome c-oxidoreductase.

Reverse electron transfer seems to be unnecessary for chloroplasts and is of limited (most likely regulatory) significance for mitochondria.

References

Abrahams JP, Leslie AGW, Lutter R, Walker JE (1994) Structure at 2.8 Å resolution of F_1-ATPase from bovine heart mitochondria. Nature 370:621–628
Baltscheffsky H, Von Stedingk LV, Heldt HW, Klingenberg M (1966) Inorganic pyrophosphate: formation in bacterial photophosphorylation. Science 153:1120–1122
Baltscheffsky M (1969) Energy conversion-linked changes of carotenoid absorbance in *Rhodospirillum rubrum* chromatophores. Arch Biochem Biophys 130:646–652
Baltscheffsky M, Schultz A, Baltscheffsky H (1999) H^+-PPases: a tightly membrane-bound family. FEBS Lett 457:527–533
Baykov AA, Dubnova EB, Bakuleva NP, Evtushenko OA, Zhen RG, Rea PA (1993) Differential sensitivity of membrane-associated pyrophosphatases to inhibition by diphosphonates and fluoride delineates two classes of enzyme. FEBS Lett 327:199–202
Belitser VA, Tsibakova ET (1939) The mechanism of phosphorylation associated with respiration. Biokimiia 4:516–526 (in Russian)
Bizouarn T, Fjellström O, Meuller J, Axelsson M, Bergkvist A, Johansson C, Göran Karlsson B, Rydström J (2000) Proton translocating nicotinamide nucleotide transhydrogenase from *E. coli*. Mechanism of action deduced from its structural and catalytic properties. Biochim Biophys Acta 1457:211–228
Boekema EJ, Berden JA, van Heel MG (1986) Structure of mitochondrial F_1-ATPase studied by electron microscopy and image processing. Biochim Biophys Acta 851:353–360
Boyer PD (2002) Catalytic site occupancy during ATP synthase catalysis. FEBS Lett 512:29–32
Boyer PD, Cross RL, Momsen W (1973) A new concept for energy coupling in oxidative phosphorylation based on a molecular explanation of the oxygen exchange reactions. Proc Natl Acad Sci U S A 70:2837–2839
Bragg PD (1984) The ATPase complex of *Escherichia coli*. Can J Biochem Cell Biol 62: 1190–1197
Chow DC, Forte JG (1995) Functional significance of the beta-subunit for heterodimeric P-type ATPases. J Exp Biol 198:1–17
Colowick SP, Kaplan NO, Neufeld EF, Ciotti MM (1952) Pyridine nucleotide transhydrogenase. I. Indirect evidence for the reaction and purification of the enzyme. J Biol Chem 195:95–105
Danielson L, Ernster L (1963) Energy-dependent reduction of triphosphopyridine nucleotide by reduced diphosphopyridine nucleotide, coupled to the energy-transfer sytem of the respiratory chain. Biochem Z 338:188–205
Dontsov AE, Grinius LL, Jasaitis AA, Severina II, Skulachev VP (1972) A study on the mechanism of energy coupling in the redox chain. I. Transhydrogenase: the fourth site of the redox chain energy coupling. J Bioenerg 3:277–303
Feldman RI, Sigman DS (1982) The synthesis of enzyme-bound ATP by soluble chloroplast coupling factor 1. J Biol Chem 257:1676–1683

Frangione B, Rosenwasser E, Penefsky HS, Pullman ME (1981) Amino acid sequence of the protein inhibitor of mitochondrial adenosine triphosphatase. Proc Natl Acad Sci U S A 78:7403–7407

Friedl P, Hoppe J, Gunsalus RP, Michelsen O, von Meyenburg K, Schairer HU (1983) Membrane integration and function of the three F_o subunits of the ATP synthase of *Escherichia coli* K12. EMBO J 2:99–103

Ganser AL, Forte JG (1973) K^+-stimulated ATPase in purified microsomes of bullfrog oxyntic cells. Biochim Biophys Acta 307:169–180

Grubmeyer C, Penefsky HS (1981) Cooperatively between catalytic sites in the mechanism of action of beef heart mitochondrial adenosine triphosphatase. J Biol Chem 256:3728–3734

Hinkle PC (2005) P/O ratios of mitochondrial oxidative phosphorylation. Biochim Biophys Acta 1706:1–11

Hoppe J, Schairer HU, Friedl P, Sebald W (1982) An Asp-Asn substitution in the proteolipid subunit of the ATP-synthase from *Escherichia coli* leads to a non-functional proton channel. FEBS Lett 145:21–29

Isaev PI, Liberman EA, Samuilov VD, Skulachev VP, Tsofina LM (1970) Conversion of biomembrane-produced energy into electric form. 3. Chromatophores of *Rhodospirillum rubrum*. Biochim Biophys Acta 216:22–29

Jackson JB (2003) Proton translocation by transhydrogenase. FEBS Lett 545:18–24

Junge W (1987) Complete tracking of transient proton flow through active chloroplast ATP synthase. Proc Natl Acad Sci U S A 84:7084–7088

Junge W, Pänke O, Cherepanov DA, Gumbiowski K, Müller M, Engelbrecht S (2001) Intersubunit rotation and elastic power transmission in F_oF_1-ATPase. FEBS Lett 504:152–160

Lolkema JS, Chaban Y, Boekema EJ (2003) Subunit composition, structure, and distribution of bacterial V-type ATPases. J Bioenerg Biomembr 35:323–335

Maeshima M (2000) Vacuolar H^+-pyrophosphatase. Biochim Biophys Acta 1465:37–51

Meier T, Polzer P, Diederichs K, Welte W, Dimroth P (2005) Structure of the rotor ring of F-type Na^+-ATPase from *Ilyobacter tartaricus*. Science 308:659–662

Mitchell P (1966) Chemiosmotic coupling in oxidative and photosynthetic phosphorylation. Biol Rev Camb Philos Soc 41:445–502

Murata T, Yamato I, Kakinuma Y, Leslie AG, Walker JE (2005) Structure of the rotor of the V-type Na^+-ATPase from *Enterococcus hirae*. Science 308:654–659

Negrin RS, Foster DL, Fillingame RH (1980) Energy-transducing H^+-ATPase of *Escherichia coli*. Reconstitution of proton translocation activity of the intrinsic membrane sector. J Biol Chem 255:5643–5648

Noji H, Yasuda R, Yoshida M, Kinosita K (1997) Direct observation of the rotation of F_1-ATPase. Nature 386:299–302

Palmgren MG (2001) Plant plasma membrane H^+-ATPases: powerhouses for nutrient uptake. Annu Rev Plant Physiol Plant Mol Biol 52:817–845

Racker E (1965) Mechanisms in bioenergetics. Academic Press, London, p 259

Rastogi VK, Girvin ME (1999) Structural changes linked to proton translocation by subunit c of the ATP synthase. Nature 402:263–268

Richter ML, Hein R, Huchzermeyer B (2000) Important subunit interactions in the chloroplast ATP synthase. Biochim Biophys Acta 1458:326–342

Sazanov LA, Jackson JB (1994) Proton-translocating transhydrogenase and NAD- and NADP-linked isocitrate dehydrogenases operate in a substrate cycle which contributes to fine regulation of the tricarboxylic acid cycle activity in mitochondria. FEBS Lett 344:109–116

Schöcke L, Schink B (1998) Membrane-bound proton-translocating pyrophosphatase of *Syntrophus gentianae*, a syntrophically benzoate-degrading fermenting bacterium. Eur J Biochem 256:589–594

Seelert H, Poetsch A, Dencher NA, Engel A, Stahlberg H, Müller DJ (2000) Proton-powered turbine of a plant motor. Nature 405:418–419

Shin JM, Besancon M, Bamberg K, Sachs G (1997) Structural aspects of the gastric H, K ATPase. Ann N Y Acad Sci 834:65–76

Skulachev VP (1988) Membrane bioenergetics. Springer, Berlin
Skulachev VP, Kozlov IA (1982) H^+-ATPase: a substrate translocation concept. Curr Top Membr Transp 16:285–301
Stock D, Gibbons C, Arechaga I, Leslie AG, Walker JE (2000) The rotary mechanism of ATP synthase. Curr Opin Struct Biol 10:672–679
Vasilyeva EA, Fitin AF, Minkov IB, Vinogradov AD (1980) Kinetics of interaction of adenosine diphosphate and adenosine triphosphate with adenosine triphosphatase of bovine heart submitochondrial particles. Biochem J 188:807–815
Wagner CA, Finberg KE, Breton S, Marshansky V, Brown D, Geibel JP (2004) Renal vacuolar H^+-ATPase. Phys Rev 84:1263–1314
Walker JE, Fearnley IM, Gay NJ, Gibson BW, Northrop FD, Powell SJ, Runswick MJ, Saraste M, Tybulewicz VL (1985) Primary structure and subunit stoichiometry of F_1-ATPase from bovine mitochondria. J Mol Biol 184:677–701
Walker JE, Runswick MJ, Saraste M (1982) Subunit equivalence in *Escherichia coli* and bovine heart mitochondrial F_1F_o ATPases. FEBS Lett 146:393–396

Chapter 8
$\Delta\bar{\mu}_{H^+}$-Driven Mechanical Work: Bacterial Motility

For a long time, the molecular mechanism of mechanical work in biological systems was believed to be associated exclusively with ATPases of the actomyosin type. Actomyosin-like ATPases were shown to participate not only in muscle contraction in animals but also in the motility of spermatozoa and in some plants, fungi, and protozoa, as well as in the movement of some intracellular structures. In particular, an ATP-dependent mechanism was shown to be operative in eukaryotic flagella.

When it was found that flagella are also involved in bacterial motility the simplest suggestion seemed to be that it is also ATP-driven. However, all attempts to detect ATPase activity of prokaryotic flagella failed.[1] Moreover, it was demonstrated that the structures of prokaryotic and eukaryotic flagella are entirely different. Finally, it was found that the bacterial flagellum is a type of biological transducer that utilizes $\Delta\bar{\mu}_{H^+}$ (or in some cases $\Delta\bar{\mu}_{Na^+}$, see Sect. 13.2) to perform mechanical work with no ATP involved.

It is noteworthy that living organisms (in contrast to technical devices of our civilization) practically never use the wheel model for mechanical work. The bacterial flagellum seems to be the only exception to this rule.[2]

The bacterial flagellum was shown to perform rotational motion (and not beating as in the case of eukaryotic flagella) lengthwise to its long axis. Because the filament is usually a left-handed spiral, the rear polar flagellum pushes the cell when rotating counterclockwise. Clockwise rotation pulls the cell in the opposite direction, and it moves the flagellum forward.

There are bacteria possessing many laterally located flagella (e.g., *E. coli*) instead of one polar flagellum. Counterclockwise rotation of *E. coli* flagella causes them to wind into a cord, the rotation of which causes the progressive motion of the cell. Clockwise rotation unwinds the cord, and the flagella start rotating

---

[1] Minor ATPase activity of flagella has been discovered recently, but is was shown to be necessary for the assembly of these organelles, and not for the movement *per se*.
[2] Rotation of certain subunits of H^+-ATP synthase (see preceding chapter) is used as a mechanism of carrying out chemical (rather than mechanical) work.

independently and the cell stops (or to be more precise, it starts performing short chaotic movements in different directions ("tumbling").

8.1 $\Delta\bar{\mu}_{H^+}$ Powers the Flagellar Motor

In 1956, Peter Mitchell noted that it is theoretically possible to drive locomotion of bacteria by an ionic gradient (Mitchell 1956). The bacterial flagellum was suggested to play the role of a giant ion channel. Later, this particular scheme was ruled out, again theoretically, as ineffective. Nevertheless, a more general idea of ion-driven locomotion was later proved experimentally.

In 1974, Julius Adler and colleagues (Larsen et al. 1974a) observed that an *E. coli* mutant deficient in oxidative phosphorylation required respiration for motility, although it could not be coupled to ATP synthesis because of the mutation in the ATP synthase (the ATP level was maintained by glycolysis). Moreover, a strong lowering of the ATP level did not stop the respiration-supported motility of the mutant, while the addition of an uncoupler did.

In 1975, experiments conducted by one of this book's authors (V.P.S.) and his coworker Kim Lewis confirmed the Adler's observation for quite another bacterium, *Rhodospirillum rubrum*. In this case too, bacterial motility did not correlate with the ATP level (Glagolev and Skulachev 1978). But there was a good correlation between the rate of motility and the level of membrane potential produced by the photosynthetic redox chain. On the basis of these observations, we suggested that *the motility of Rh. rubrum is supported by* $\Delta\bar{\mu}_{H^+}$, *and not by ATP*. Later, three different laboratories (including the group of the book's author) found that bacteria paralyzed by exhaustion of endogenous sources for $\Delta\bar{\mu}_{H^+}$ generation become motile for several minutes when artificial $\Delta\Psi$ or ΔpH is imposed (Skulachev 1977; Matsura et al. 1977; Manson et al. 1977).

In *Bacillus subtilis*, motility appeared at about 30 mV $\Delta\bar{\mu}_{H^+}$ and then increases with increasing potential to 60 mV. Further increase in $\Delta\bar{\mu}_{H^+}$ exerted no further effect upon the locomotion rate. A $\Delta\bar{\mu}_{H^+}$ threshold was also found in *E. coli*, but here a 200 mV $\Delta\bar{\mu}_{H^+}$ was necessary for the maximal speed to be obtained. This maximal speed of locomotion varies for different species. The highest value (140 μm/s) was observed for *Bdellovibrio bacteriovorus* (Stolp 1968). The majority of other bacteria usually swim at speeds of 20–40 μm/s. The flagellar rotation rate is usually 10–50 revolutions/s in the case of *E. coli*, reaching 300 revolutions/s when operating without load. In some other bacteria the flagellar rotation rate can be as high as 1,700 revolutions per second.

An important feature of the flagellar motor is that *it can rotate the flagellum both clockwise and counterclockwise in spite of the fact that the* $\Delta\bar{\mu}_{H^+}$ *direction remains constant*. This means that the motor is equipped with a gearshift. The latter is controlled by the taxis system in such a way that repellents induce a switch-over from counterclockwise to clockwise rotation, while attractants inhibit this switch-over (Larsen et al. 1974b).

8.1 $\Delta\bar{\mu}_{H^+}$ Powers the Flagellar Motor

One of the parameters sensed by the taxis machinery is the $\Delta\bar{\mu}_{H^+}$ level. As a result, $\Delta\bar{\mu}_{H^+}$ appears to be not only the driving force, but also a regulator of the direction of the rotor rotation. This complicates mechanistic analysis of bacterial motility. To overcome this difficulty, two approaches have been used.

In studies by Howard Berg and coworkers, some mutants of *Streptococcus* defective in taxis but still motile were investigated (Manson et al. 1980). It was shown that the direction of flagellar rotation depended on the polarity of the artificially imposed $\Delta\bar{\mu}_{H^+}$. This effect was strong evidence in favor of a concept according to which $\Delta\bar{\mu}_{H^+}$ is an immediate energy source for the rotation of the bacterial flagellum. Indeed, if there were any intermediates between $\Delta\bar{\mu}_{H^+}$ and rotation, such as a chemical high-energy compound, the oppositely directed $\Delta\bar{\mu}_{H^+}$ would hardly be effective in supporting the rotation in the opposite direction (Skulachev 1988).

Another approach was used by Michael Eisenbach and Julius Adler. They prepared cell envelopes of *E. coli* and *Salmonella typhimurium* still retaining intact flagella. The envelopes were tethered to a microscope cover using antibodies to flagellin. Thus, the tethered filament of the flagellum and its rotor was immobilized, and a switch-on of the H^+-motor caused the stator and, hence, the envelope, to rotate. Using this system deprived of the cytoplasmic mechanism of taxis, the authors found that both respiration-produced and artificially imposed $\Delta\bar{\mu}_{H^+}$ of natural direction support counterclockwise rotation. Artificially imposed $\Delta\bar{\mu}_{H^+}$ of the opposite direction did not rotate the envelope, apparently because of some irreversible damage to the system (Eisenbach and Adler 1981).

8.2 Structure of the Bacterial Flagellar Motor

The bacterial flagellum is one of the most complex organelles of prokaryotic cells. It consists of more than 20 different proteins with their stoichiometry in the flagellum varying from several to >10,000 copies per complex. This assembly and regulation of the organelle requires the activity of more than 30 additional proteins.

A typical bacterial *flagellum* consists of a *filament* and a *basal body* inlaid in the cell wall (Berg 2000; Bardy et al. 2003). The filament is a rigid, usually left-handed helix several micrometers long (sometimes up to 20 μm) and 12–20 nm in diameter. It is composed of a structural protein, flagellin [3] (its molecular mass varies in different species from 25 to 60 kDa), which exhibits no enzymatic activity. In the filament, flagellin (FliC[4]) molecules are organized in such a way

---

[3] In the case of certain bacteria, several different flagellin isoforms comprise the filament. The physiological function of these isoforms is not yet clear.

[4] Hereafter, the presented nomenclature of flagellum proteins is the one used for the bacterium *S. typhimurium*.

that a hollow tube is formed. The assembly of the filament involves the consecutive attachment of flagellin monomers from its distal end, which is located very far away from the cell. Flagellin molecules are assumed to be delivered to the distal end of the filament through the inner hollow tube of this structure, which avoids substantial losses of the protein due to its diffusion to the outer medium. The filament tube is covered on its far end by a special protein, FliD (another name, HAP2), which prevents the depolymerization of flagellin monomers and their diffusion into the outer medium. ATP hydrolysis by a special ATPase composed of 6 FliI subunits and one FliH subunit (see below) is the driving force of the flagellin-transport process. It is quite interesting that the amino acid sequence of FliI is homologous to the α subunit of F_oF_1-ATP synthase, and six subunits of this protein form a structure resembling the $3\alpha : 3\beta$-globule of factor F_1 of the ATP-synthase complex (Imada et al. 2007). The FliH subunit is also homologous to the b and δ subunits of F_oF_1 ATP synthases (it is probably a structural homolog of the *peripheral stalk* of the F_oF_1 complex), which implies the use of a rotating mechanism for flagellin transport (see the preceding Sect. 7.1).

The filament is usually attached to the basal body by a *hook*, i.e., a special curved structure of diameter somewhat larger than that of the filament. This hook consists of a large number of copies of only one protein—FlgE—attached to the filament by two additional proteins—FlgL and FlgK (HAP1 and HAP3). The *basal body* is a system of four (sometimes two) *rings* perpendicular to the filament and a *rod* (comprising proteins FlgB, FlgC, FlgF, and FlgG) transpiercing the rings and connected with the hook. The diameter of the rod varies from 8 to 12 nm, and that of rings from 18 to 33 nm (Fig. 8.1a–c).

The basal body of lateral flagella of *gram-negative* bacteria consists of four rings. Rings *L*, *P*, *M*, and *S* were originally supposed to be part of the flagellum. However, further research showed that rings *M* and *S* represent two different domains of one protein, FliF, i.e., they form one ring, which was given the name *MS*. Later, more careful extraction of the basal body revealed one more additional ring that was found to be a part of this structure. This *C* ring is located on the cytoplasmic side of the inner membrane (Fig. 8.1b). According to our current understanding, there are four rings in the lateral basal body of flagella of *gram-negative* bacteria—*L*, *P*, *MS*, and *C*. The basal body of *gram-positive* bacteria contains only two rings—*MS* and *C*.

The *L* ring is located inside the outer membrane and is composed of several copies of lipoprotein FlgH. The *P* ring is located under the *L* ring and consists of the FlgI protein. The *MS* ring is immersed in the cytoplasmic membrane and consists of the FliF protein. The functions of all these rings are not yet known. The *C* ring is located in the cytoplasm, and it consists of the proteins FlhA, FlhB, FliH, FliJ, FliI, FliN, FliG, FliM, FliO, FliP, FliQ, and FliR. The proteins FliN, FliG, and FliM are responsible at least for the switching of the rotation direction of the flagellum (see below), while proteins FlhA, FlhB, FliH, FliJ, FliI, FliO, FliP, FliQ, and FliR are required for the transmembrane transport of the flagellar proteins during its assembly (see above).

8.2 Structure of the Bacterial Flagellar Motor

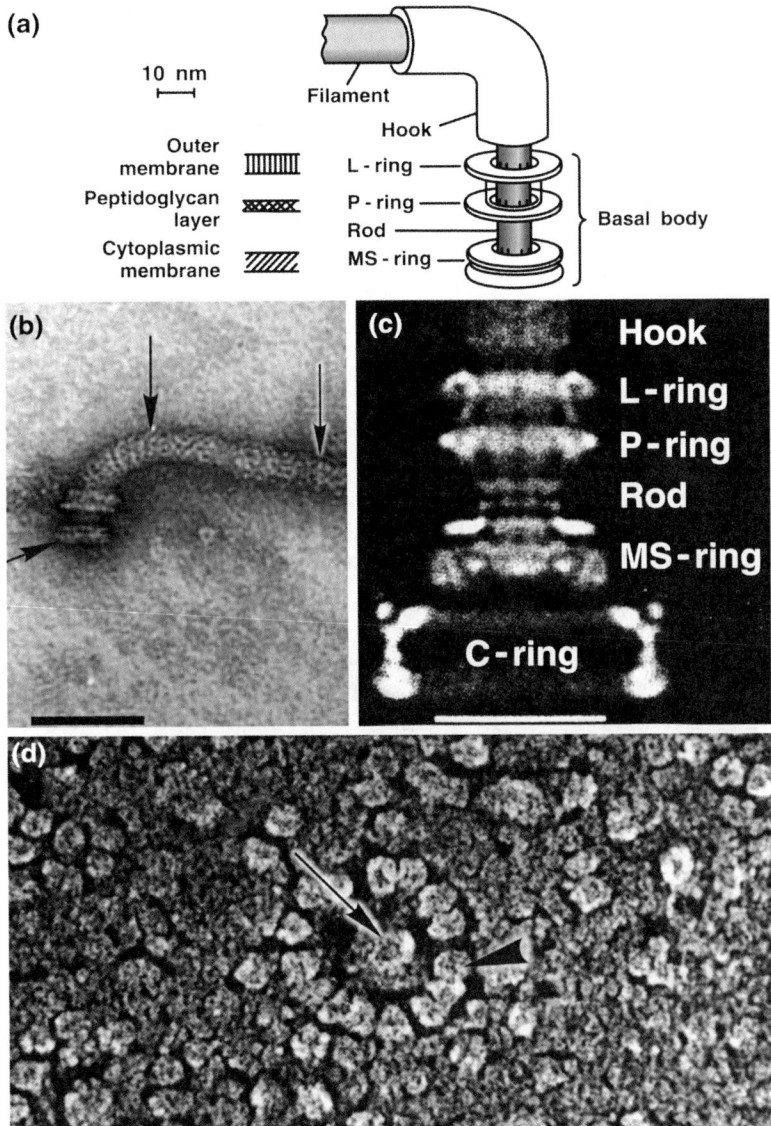

Fig. 8.1 a A scheme of structure of the bacterial flagella basal body (for *E. coli*, C ring is not shown); **b** electron micrograph of *E. coli* flagellum; arrows indicate (*left* to *right*) the basal body, the rod, and the filament; the bar is 50 nm; the C ring is lost during the isolation procedure (Skulachev 1988; Metlina 2001); **c** electron micrograph of basal body of the *E. coli* flagellum (Berg 2000); **d** Mot complexes (*arrow head*) around basal body (*arrow*)

In the bacterial membrane one can see special protein structures surrounding the basal body (Fig. 8.1d). They never accompany the basal body when it is isolated from the cell. These structures are composed of a complex of integral membrane proteins MotA and MotB. The MotB protein forms one transmembrane α-helix in its N-terminus, while its main part is located in the periplasm. A peptidoglycan-binding domain is located on the C-terminus of MotB. This protein is supposed to be tightly bound to the bacterial cell wall. This fact suggests MotB to be an element of the stator of the flagellar motor, for the stator of such a big organelle capable of generating a torque of 4,000 picoNewton nanometers, has to be firmly fixed in the only stiff bacterial structure, its cell wall.

There are four transmembrane α-helixes in the MotA protein, the rest of it (its greater part) being located in the cytoplasm. MotA and MotB were shown to form a stable complex. The MotA–MotB complex is also known to interact with the C-terminal domain of the FliG protein, which is part of the *C* ring. The number of MotA–MotB complexes in one flagellum has not yet been established. Structural studies of various bacterial flagella provide the stoichiometry of 10–16 MotAB complexes per organelle, while functional studies of this motor predict the existence of eight torque generators.

Mutations in the transmembrane parts of MotA and MotB proteins lead to the loss of $\Delta\bar{\mu}_{H^+}$-dependent rotation of the flagellum. MotA and MotB are considered to be responsible for transmembrane proton transport; they are thought to be "stator" components, while the basal body and the filament represent the "rotor" part of the flagellar motor. The latter conclusion is supported by studies of the consequences of genetic cross-linking of various protein components of the flagellum. Functionally active flagella were obtained with fused FliF-FliG and FliM-FliN proteins, this fact being the basis for a conclusion about the role of *MS* and *C* rings as rotor components of the organelle.

8.3 A Possible Mechanism of the H⁺-motor

Intact *Streptococcus* cells tethered to a microscope cover glass by their flagella were studied in Berg's group. An artificial $\Delta\Psi$ was created by a K⁺ diffusion potential, and artificial ΔpH was created by acidifying or alkalinizing the incubation medium. In these experiments the authors obtained one very important result: *torques of flagellar motor energized by artificial ΔΨ or ΔpH showed no dependence on temperature in the studied range of 4 °C–38 °C* (Khan and Berg 1983). This fact suggests that the torque generation does not require conformational changes of protein subunits and other events usually related to enzyme action, which is always strongly inhibited by such a large decrease in the temperature. Another feature distinguishing the motor from enzymes dealing with H⁺ ions is the absence of an isotope effect when D₂O substitutes for H₂O (Khan and Berg 1983).

The mechanism of the work of the flagellum has not yet been determined. In 1978, Kim Lewis (Glagolev) and one of this book's authors (V.P.S.) suggested a mechanism for the bacterial H⁺ motor (Glagolev and Skulachev 1978) illustrated in Fig. 8.2. According to the contemporary interpretation of this scheme, rings *MS*

8.3 A Possible Mechanism of the H⁺-motor

and C (rotor) and MotAB complexes (stator) form the functional part of the bacterial motor.

In the stator there are supposed to be two proton "wells" crossing only part of the hydrophobic membrane core. At the bottom of the inward proton well, there has to be a proton-accepting group X, belonging to the rotor, i.e., to MS or C rings. The appearance of a positively charged X on its protonated form results in its coulombic attraction to a negatively charged proton acceptor group Y^- (e.g. an ionized carboxyl group of a dicarboxylic amino acid residue belonging to the stator). This leads to the rotation of the rotor and a drawing together of X^+H and Y^-. Proton transfer from X^+H to Y^- and further to the cytoplasm via the outward proton channel completes the chain of events. Now a new H^+ ion from the periplasm can protonate the next X group situated at the bottom of the inward H^+ well.

The conservative aspartate residue of the only transmembrane α-helix of the MotB protein (Asp32 in the case of *E. coli*) seems to be the best candidate for the role of the Y group of the stator. Mutations of this amino acid lead to the complete loss of $\Delta\bar{\mu}_{H^+}$-dependant rotation of the flagellum. However, it has not yet been possible to find a good candidate for the role of the X group of the rotor. There are no conservative residues of positively charged amino acids in the FliG, FliM, and FliN proteins, and mutations of the charged amino acid residues in the C terminus of the FliG protein (the domain interacting with the MotAB complex, see above) do not result in the loss of the functional activity of the flagellum.

The above-suggested scheme has a number of other problems that remain to be solved. For instance, this model assumes the existence of a long chain of groups X at the periphery of the MS or C rings so that many protons are necessary to be transferred across the membrane to cause a single turnover of the rotor. According to calculations made by Berg, about 1,200 H^+ ions should be transported across the membrane to the cytoplasm per single rotor turnover (Berg et al. 1982). In the framework of the scheme presented in Fig. 8.2, this means that the number of X groups in the rotor should be about $1200/n$, where n represents the number of MotAB complexes in the flagellum, i.e., the number of X groups should be

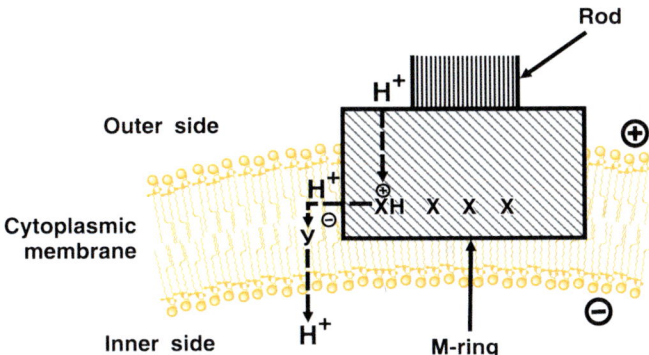

Fig. 8.2 Possible arrangement of the bacterial proton motor (Glagolev and Skulachev 1978). Explanations are provided in the text

relatively high (~ 150), which is rather difficult to imagine if we take into account the data obtained on the structure of the basal body.

In any case, the C ring seems to be better than the MS ring as the candidate for the role of *X* group carrier. In this case there is no direct contact between rotor and stator in the cytoplasmic membrane; it also helps to avoid the breakage of many protein–protein bonds in the hydrophobic membrane layer during rotor functioning. It would be easier to understand this mechanism if we postulate that rotor–stator interactions take place in the cytosol, while only the H^+-ion transfer (causing the change of the spatial location of the part of the H^+-channel located in the cytoplasmic part of the system) occurs in the membrane. In such a case, *the rotor rotation might take place in the membrane phospholipid layer* without any contact with those stator sites that are immersed in the membrane. In any other cases, there appears the problem of the increased passive H^+ leakage through the membrane in the sites of the rotor–stator contact.

As already mentioned, there is a gearshift in the flagellar motor allowing the direction of rotation to be changed in spite of the constant direction of $\Delta\bar{\mu}_{H^+}$. It is the C ring comprising the FliG, FliM, and FliN proteins that fulfills this function in the organelle. The mechanism of this phenomenon should include the ability to change the positions of the proton wells in the stator so that one of them would be favorable for the clockwise and the other for counterclockwise rotation of the rotor driven by proton transfer.

The movement of H^+ ions along proton wells seems to be a physical process since operation of the H^+ motor does not depend on temperature. This may be diffusion of H^+ or H_3O^+ in water-containing pores. Another possibility is migration of H^+ along bound water molecules or along a chain of proteolytic groups without any conformational changes in the protein molecule.

8.4 $\Delta\bar{\mu}_{H^+}$-Driven Movement of Non-Flagellar Motile Prokaryotes and Intracellular Organelles of Eukaryotes

Besides the flagellar motor-driven movements, certain bacteria are capable of other types of motility. The mechanisms of these alternative motility types have not been thoroughly studied yet.

Experiments conducted in the group of one of this book's authors (V.P.S.) showed that *the gliding motility of trichomes of the multicellular filamentous cyanobacteria Phormidium uncinatum on the surface of a substrate is driven by* $\Delta\bar{\mu}_{H^+}$ *rather than ATP* (Glagoleva et al. 1980; Skulachev 1980). This conclusion was supported by inhibitor analysis of the gliding supported by photosynthetic, respiratory, or glycolytic energy sources and by the fact that artificially imposed $\Delta\Psi$ or ΔpH were competent in supporting uncoupler-sensitive motility in the presence of DCCD, which inhibited $\Delta\bar{\mu}_{H^+}$-powered ATP formation.

Phormidium uncinatum does not have flagella. Its motility is assumed to be caused by fibrils located under the outer membrane of the trichome (Halfen and Castenholz 1970, 1971). If this is the case, the fibrils should be connected with the $\Delta\bar{\mu}_{H^+}$-bearing cytoplasmic membrane of the trichome-forming cyanobacterial cells.

However, the gliding of parasites of animal tissues (mycoplasmas) and plant tissues (spiroplasmas) seems to be $\Delta\bar{\mu}_{H^+}$-independent. Not only flagella but also the cell wall are absent from these bacteria, which is separated from the medium by a single (cytoplasmic) membrane. Mycoplasmas and spiroplasmas were shown to be motile in the presence of a protonophore added at high concentration (1 mM dinitrophenol) (Bredt 1973). The gliding of *Spiroplasma* ceased in the presence of inhibitors of glycolytic phosphorylation as well as DCCD. The question of the driving force of this gliding process remains open. It could be ATP, or $\Delta\bar{\mu}_{Na^+}$, or some other ion gradient.

An interesting observation was published by John Waterbury et al. who studied the motility of small unicellular marine cyanobacteria assigned to the genus *Synechococcus* that were discovered in the Atlantic Ocean. They were found to be capable of swimming through liquids at a speed of 25 μm/s. Light microscopy revealed that the motile cells display many features of bacterial flagellar motility. Electron microscopy, however, failed to reveal flagella (Waterbury et al. 1985).

A similar mechanism might be responsible for a *rotary movement of chloroplasts* (see Skulachev 1980 for Ref.). Rotation of chloroplasts was discovered in 1838 in France by Donne, who studied drops of protoplasm squeezed out of *Chara* algal cells. This phenomenon was rediscovered several times in the twentieth century. E. H. Mozenok in the group of one of the authors (V.P.S.) repeated Donne's experiment to reveal the energy source for chloroplast rotation. *Nitella* from Lake Baikal was investigated. It was found that in drops of the protoplasm, chloroplasts can rotate, with the mean rate of rotation being one revolution per 2–3 s. Sometimes the direction of rotation changed. The duration of a period when the direction of rotation was constant could be as long as an hour. Rotation of chloroplasts was also observed in intact algal cells. In some cases, one or several chloroplasts became detached from a motionless chloroplast multilayer located below the cytoplasmic membrane; carried along by the protoplasmic stream, they were found to rotate at the same rate as in drops of protoplasm (Skulachev 1980).

Inhibitor analysis of chloroplast rotation in protoplasm drops was undertaken. For comparison, the effect of the same inhibitors on the ATP-dependent movement of protoplasm (cyclosis) was carried out in the same alga. The resulting data indicated that chloroplast rotation in the light is resistant to DCCD. In the dark, DCCD is inhibitory. Switching on the light immediately actuated chloroplast rotation arrested by DCCD in the dark. Rotation was shown to be inhibited by NH_4^+, specifically discharging ΔpH, the main constituent of $\Delta\bar{\mu}_{H^+}$ in chloroplasts. If NH_4^+ concentration was not too high, its inhibiting action was relieved by an increase in light intensity. Protonophoric uncouplers were also inhibitory. Cytochalasin B, an inhibitor of the ATP-dependent movement of intracellular structures, had no effect upon chloroplast rotation and completely arrested cyclosis. All these relationships could be predicted by assuming that *chloroplast rotation is powered by* $\Delta\bar{\mu}_{H^+}$. Again, as in the above experiments with *Synechococcus*, electron microscopy failed to reveal flagellar structures attached to the chloroplast envelope.

Taking into account the possible origin of chloroplasts from cyanobacteria, one can speculate that chloroplast rotation is supported by the same yet unknown mechanism that moves *Synechococcus*. Since extracellular flagella are absent, we should consider some other mechanisms, e.g., undulation of the chloroplast (*Synechococcus*) outer membrane. Such undulation might cause cytosol movement in narrow gaps between chloroplasts immobilized in multilayers, thereby accelerating intracellular transport of the chloroplast-produced glucose, and this might be the biological function of this mechanism. Cyclosis can perform such a role only for the chloroplast layer facing the free cytosol space.

8.5 Motile Eukaryote: Prokaryote Symbionts

A review on $\Delta\bar{\mu}_{H^+}$-linked motility would be incomplete if we failed to mention rare but impressive examples when a motionless eukaryotic cell employs symbiotic motile bacteria to gain the ability for active movement. *Mixotricha paradoxa*, a large flagellate from the gut of a termite, is a case in point. It was revealed by Cleveland and coworker that *its movement is brought about by coordinated undulations of the membranes of many thousands of spirochetes that cover most of the body surface of M. paradoxa* (Cleveland and Grimstone 1964). It is interesting that *M. paradoxa* has four flagella of its own used not to move the organism but for some other yet unknown purpose, e.g., to steer such a movement, organizing the pacemaker. This might simply be done by the flagellum touching some spirochetes that respond to the touch by a reversal of locomotion. In other words, *M. paradoxa* seems to employ their flagella as switches to hurry on spirochetes, an intriguing speculation waiting for experimental proof.

A similar example of the $\Delta\bar{\mu}_{H^+}$-driven motility of a eukaryote–prokaryote symbiotic system was described by Sidney Tamm, who studied a devescovinid, another flagellate protozoan living in the gut of a termite (Tamm 1982). Locomotion of this protozoan, like that of *Mixotricha*, is powered neither by the cell's own flagella nor by its axostyle but by the *flagella of thousands of ectosymbiotic rod-shaped bacteria* that lie in special pockets formed by the outer membrane of the host. These peritrichous bacteria bear typical prokaryotic flagella on their exposed surface. Locomotion could still be observed when rotation of the devescovinid flagella was inactivated by amphotericin B or nystatin, the channel-forming polyene antibiotics requiring cholesterol to be operative and therefore specifically discharging only eukaryotic membranes. On the other hand, protonophoric uncouplers stopped the motility, probably by lowering $\Delta\bar{\mu}_{H^+}$ across the bacterial membranes.

In both *Mixotricha paradoxa* and the devescovinid, the attached prokaryotes are used to organize the gliding motility of the host cell. The rate of this motility (up to 150 μm/s) is much higher than that of gliding of prokaryotes, e.g., cyanobacteria (up to 10 μm/s). Thus, the locomotive system used by these protozoa appears to be very effective. The length of the devescovinid is about 0.1 mm, that of *Mixotricha paradoxa*, about 0.5 mm. Some motile cyanobacteria can be 1 mm long.

No precedent has been described for $\Delta\bar{\mu}_{H^+}$ to power the locomotion of a large organism. Here, we always find an ATP-dependent mechanism of the actomyosin type.

It is difficult to imagine a device in which a $\leq$20-μm-long bacterial flagellum is employed to perform large-scale mechanical work necessary for locomotion of multicellular animals. However, when studying the actomyosin mechanism, certainly very important for animals and humans, we should keep in mind that diminutive forms of life have chosen quite a different mechanism for motility. Accordingly, they had to invent a rotating ring; so, the wheel is not a human invention. This ring is the heart of a more complex device, the basal body, which is a molecular prototype of our electric motor (not a human invention either!) connected with a rigid filament (the screw of our motorboat).

This or a slightly modified engine is used not only to swim but also to glide by cyanobacterial trichomes and even by some eukaryotic flagellates bearing symbiotic bacteria. In fact, these flagellates tried to "build a steamer from many motor boats". It is not surprising that such an attempt was a dead end of evolution.

References

Bardy SL, Ng SY, Jarrell KF (2003) Prokaryotic motility structures. Microbiology 149:295–304
Berg HC (2000) Constraints on models for the flagellar rotary motor. Philos Trans R Soc Lond B Biol Sci 355:491–501
Berg HC, Manson MD, Conley MP (1982) Dynamics and energetics of flagellar rotation in bacteria. Symp Soc Exp Biol 35:1–31
Bredt W (1973) Motility of mycoplasmas. Ann NY Acad Sci 225:246–250
Cleveland LR, Grimstone AV (1964) The fine structure of the flagellate *Mixotricha paradoxa* and its associated micro-organisms. Proc R Soc Lond B 159:668–686
Eisenbach M, Adler J (1981) Bacterial cell envelopes with functional flagella. J Biol Chem 256:8807–8814
Glagolev AN, Skulachev VP (1978) The proton pump is a molecular engine of motile bacteria. Nature 272:280–282
Glagoleva TN, Glagolev AN, Gusev MV, Nikitina KA (1980) Proton motive force supports gliding in cyanobacteria. FEBS Lett 117:49–53
Halfen LN, Castenholz RW (1970) Gliding in a blue-green algae: a possible mechanism. Nature 225:1163–1165
Halfen LN, Castenholz RW (1971) Gliding motility of blue-green algae: *Oscillatoria princeps*. J Phycol 7:133–145
Imada K, Minamino T, Tahara A, Namba K (2007) Structural similarity between the flagellar type III ATPase FliI and F_1-ATPase subunits. Proc Natl Acad Sci U S A 104:485–490
Khan S, Berg HC (1983) Isotope and thermal effects in chemiosmotic coupling to the flagellar motor of *Streptococcus*. Cell 32:913–919
Larsen SH, Adler J, Gargus JJ, Hogg RW (1974a) Chemomechanical coupling without ATP: the source of energy for motility and chemotaxis in bacteria. Proc Natl Acad Sci U S A 71:1239–1243

Larsen SH, Reader RW, Kort EN, Tso WW, Adler J (1974b) Change in direction of flagellar rotation is the basis of the chemotactic response in *Escherichia coli*. Nature 249:74–77

Manson MD, Tedesco P, Berg HC, Harold FM, Van der Drift C (1977) A protonmotive force drives bacterial flagella. Proc Natl Acad Sci U S A 74:3060–3064

Manson MD, Tedesco PM, Berg HC (1980) Energetics of flagellar rotation in bacteria. J Mol Biol 138:541–561

Matsura S, Shioi J, Imae Y (1977) Motility in *Bacillus subtilis* driven by an artificial protonmotive force. FEBS Lett 82:187–190

Metlina AL (2001) Procariotic flagella as system of biological motility. Usp Biol Chem 41:229–282 (in Russian)

Mitchell P (1956) Hypothetical thermokinetic and electrokinetic mechanisms of locomotion in microorganisms. Proc R Phys Soc Edinb 25:32–34

Skulachev VP (1977) Transmembrane electrochemical H^+-potential as a convertible energy source for the living cell. FEBS Lett 74:1–9

Skulachev VP (1980) Membrane electricity as a convertible energy currency for the cell. Can J Biochem 58:161–175

Skulachev VP (1988) Membrane bioenergetics. Springer, Berlin

Stolp H (1968) *Bdellovibrio bacteriovorus*—a predatory bacterial parasite. Naturwissenschaften 55:57–63 (in German)

Tamm SL (1982) Flagellated ectosymbiotic bacteria propel a eucaryotic cell. J Cell Biol 94:697–709

Waterbury JB, Willey JM, Franks DG, Valois FW, Watson SW (1985) A cyanobacterium capable of swimming motility. Science 230:74–76

Chapter 9
$\Delta\bar{\mu}_{H^+}$-Driven Osmotic Work

9.1 Definition and Classification

The term *"osmotic work"* refers to all the events that occur when a solute is transported across a membrane from lower to higher solute concentration. Such uphill transport is accompanied by the appearance of a transmembrane osmotic imbalance due to changes in the concentration of the solute molecules in at least one of the membrane-separated compartments.

Generally speaking, formation of ion gradients by specific ion pumps such as $\Delta\bar{\mu}_{H^+}$ generators, ion-motive ATPases, etc., can also be regarded as osmotic work. However, these processes are usually excluded from consideration in such a context. Instead, the main attention is given to uphill transport of metabolites that are driven by the energy of $\Delta\bar{\mu}_{H^+}$, $\Delta\bar{\mu}_{Na^+}$, ATP, or phosphoenolpyruvate (PEP). Below, we will discuss $\Delta\bar{\mu}_{H^+}$-driven osmotic work.

In mitochondria, all the known systems of uphill transport are $\Delta\bar{\mu}_{H^+}$-linked. In plasmalemma and tonoplast of plant or fungal cells, the majority of such systems are likewise $\Delta\bar{\mu}_{H^+}$-linked. On the other hand, in the outer membrane of the animal cell, $\Delta\bar{\mu}_{Na^+}$- or ATP-driven transport is much more typical, but in certain, in fact rather special cases, they can be powered by $\Delta\bar{\mu}_{H^+}$. In bacteria, one can find all the above-mentioned driving forces, $\Delta\bar{\mu}_{H^+}$ and ATP being more common if we exclude marine, halophilic, and alkalophilic species usually employing $\Delta\bar{\mu}_{Na^+}$ instead of $\Delta\bar{\mu}_{H^+}$. For instance, in *E. coli*, among 16 systems for uphill transport of amino acids there are 8 utilizing ATP, 5 utilizing $\Delta\bar{\mu}_{H^+}$, 1 utilizing $\Delta\bar{\mu}_{Na^+}$, and 2 supported by $\Delta\bar{\mu}_{H^+}$ and $\Delta\bar{\mu}_{Na^+}$ (Table 9.1). In the same *E. coli*, glucose and lactose transports are mediated by PEP- and $\Delta\bar{\mu}_{H^+}$-dependent systems, respectively.

For solute S, accumulated by a $\Delta\bar{\mu}_{H^+}$-driven system in a negatively charged compartment (e.g. in the mitochondrial matrix or the bacterial cytoplasm) the accumulation ratio ($[S]_{in}/[S]_{out}$) will obey the equation:

$$RT \cdot \ln\frac{[S]_{in}}{[S]_{out}} = -(n + Z) \cdot F \cdot \Delta\Psi - nRT \cdot \ln\frac{[H^+]_{in}}{[H^+]_{out}}, \qquad (9.1)$$

Table 9.1 Driving forces for accumulation of amino acids in *E. coli* cells

System number	Amino acid	Transport driving force
1	Ala, Gly	$\Delta\bar{\mu}_{H^+}$
2	Thr, Ser	$\Delta\bar{\mu}_{H^+}$
3	Leu, Ile, Val	ATP
4	Leu, Ile, Val	ATP
5	Leu, Ile, Val	$\Delta\bar{\mu}_{H^+}$
6	Leu, Ile, Val	$\Delta\bar{\mu}_{H^+}$
7	Phe, Tyr, Trp	$\Delta\bar{\mu}_{H^+}$
8	Pro	$\Delta\bar{\mu}_{H^+}$, $\Delta\bar{\mu}_{Na^+}$
9	Lys, Arg, Orn	ATP
10	Arg	ATP
11	Gln	ATP
12	Gln	ATP
13	Gln	$\Delta\bar{\mu}_{H^+}$, $\Delta\bar{\mu}_{Na^+}$
14	Gln	$\Delta\bar{\mu}_{H^+}$, $\Delta\bar{\mu}_{Na^+}$
15	Cys	ATP
16	Cys	ATP

where n is the number of H$^+$ ions symported with one S molecule, and Z is the number of positive charges on S.

9.2 $\Delta\Psi$ As Driving Force

If a transport system accumulates a cation of a strong base inside a negatively charged compartment, or exports an anion of a strong acid from such a compartment, $\Delta\Psi$ can be the only driving force with no ΔpH involved. In this case, the accumulation ratio of the transported cation will be described by Eq. (9.2):

$$RT \cdot \ln \frac{[C^{Z+}]_{in}}{[C^{Z+}]_{out}} = - Z \cdot F \cdot \Delta\Psi. \qquad (9.2)$$

On the basis of the Nernst equation, it is easy to calculate that a tenfold gradient of a monovalent cation C^+ requires a $\Delta\Psi$ of about 60 mV to be formed (at 25 °C).

Such simple electrophoretic behavior characterizes the system of valinomycin-mediated transport of K$^+$ ions. A similar mechanism is responsible for Ca^{2+} accumulation by mitochondria and bacteria (Fig. 9.1., system 1). The only difference is that in this case coefficient Z is equal to 2, so that 30 mV $\Delta\Psi$ is sufficient to maintain the tenfold gradient.

Both valinomycin and mitochondrial Ca^{2+}-uniporter have a rather simple problem to solve. They should recognize the transported ion and facilitate its diffusion across the hydrophobic membrane core. No special device to transduce $\Delta\Psi$ energy into an ion concentration gradient is necessary, since the overall process consists of the movement of the ion in the electric field (electrophoresis).

Fig. 9.1 Some examples of $\Delta \bar{\mu}_{H^+}$-dependent porters: *1* electrophoretic uniport of a strong base cation (Ca^{2+}) to the negatively charged compartment; *2* symport of protons and a weak acid anion (succinate) to the alkaline compartment down the ΔpH gradient; *3* antiport of a cation of strong base (Ca^{2+}) and protons down the ΔpH gradient; *4* symport of neutral solute (lactose) and protons down $\Delta \Psi$ and ΔpH, contributions of which are equal; *5* antiport of a weak base cation (RNH_3^+) and two protons, contribution of ΔpH being twofold larger than that of $\Delta \Psi$

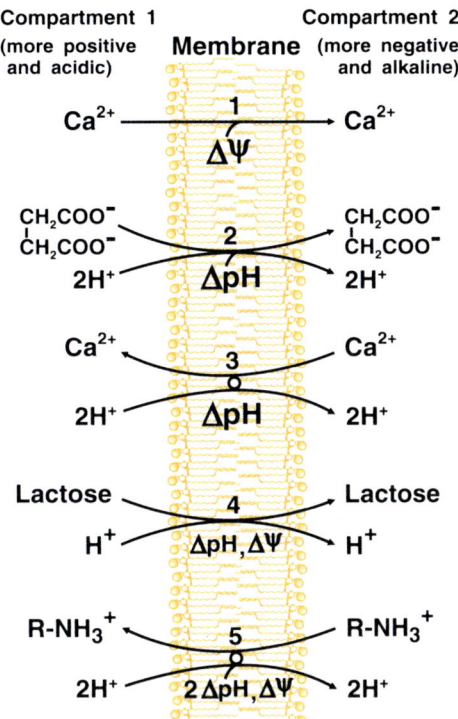

A more complicated mechanism is apparently employed to accumulate K^+ ions in *E. coli*. As revealed by Epstein and Laimins (1980), there are at least two independent mechanisms for K^+ import, namely, a K^+-transporting system (TrkA) and the K^+-ATPase (*Kdp* system). The former operates at a normal $[K^+]_{out}$ level, whereas the latter is induced by a strong decrease in $[K^+]_{out}$. TrkA utilizes $\Delta \bar{\mu}_{H^+}$ and is allosterically activated by ATP. There is an indication that TrkA is a K^+,H^+-symporter (Stewart et al. 1985). There is no doubt that K^+-ATPase utilizes the energy of ATP. However, $\Delta \Psi$ should also be involved, for K^+-ATPase operates as a K^+-uniporter carrying K^+ from the positively to the negatively charged compartment. A similar mechanism seems to be responsible for 50,000-fold accumulation of K^+ in *Methanobacterium thermoautotrophicum* growing in low potassium media (Schönheit et al. 1984).

In *E. coli*, K^+ ions are vital for the maintenance of the osmotic pressure in the cell. It was found that *E. coli* responds to an *increase in osmolarity of the medium* by K^+ influx via the TrkA (Kdp) system. The uptake of K^+ increases $[K^+]_{in}$, which prevents formation of osmotic imbalance between the cytoplasm and the hypertonic medium. It is not surprising that *osmolarity decrease in the medium* causes K^+ efflux. However, K^+ flows out of the cell via a different system, due to the functioning of mechanosensitive channels MscL and MscS that form large pores in the membrane when in a state of hypoosmotic shock. These pores provide passive outflow of ions and small molecules from the cell (Pivetti et al. 2003).

A simple electrophoretic process can apparently explain some anion fluxes across coupling membranes. In thylakoids, Cl^- goes to the interior, which is charged positively by the photosynthetic redox chain. In mitochondria, ATP anion efflux is also electrophoretic. However, here the system is complicated by the fact that $ATP^{4-}_{in}/ADP^{3-}_{out}$-antiport occurs (for more detail, see Sect. 9.7.2).

9.3 ΔpH As Driving Force

The electrophoretic mechanism of solute transport results in cations and anions being accumulated in negatively and positively charged compartments, respectively. However, it is obvious that separation of solutes into cations and anions is hardly the aim of a living system. One way to overcome, at least partially, the limitation of a simple electrophoretic principle is to convert the $\Delta\Psi$ component of the proton potential to ΔpH. $\Delta\Psi$ is known to be the *primary form of* $\Delta\bar{\mu}_{H^+}$, because the electric capacity of the membrane is much lower than the pH buffer capacity of the membrane-bathing solutions. To convert $\Delta\Psi$ to ΔpH, it is necessary to discharge the electric potential difference by an electrophoretic ion flux. If $\Delta\Psi = 0$, Eq. (9.1) can be simplified as follows:

$$\frac{[S]_{in}}{[S]_{out}} = n \cdot \frac{[H^+]_{out}}{[H^+]_{in}}. \tag{9.3}$$

The simplest case is when a weak acid or a weak base is accumulated in the more alkaline or more acidic compartment, respectively. To do this, only two conditions should be fulfilled: the membrane should be *permeable* for the *uncharged* form of the acid (base), and it must be *impermeable* for the *charged* form of this acid (base). Accordingly, the following simple events described by Eqs. (9.4) and (9.5) occur:

$$AH_{out} \rightarrow AH_{in}, \tag{9.4}$$

$$AH_{in} + OH^-_{in} \rightarrow A^-_{in} + H_2O_{in}. \tag{9.5}$$

Consumed OH^-_{in} is regenerated by the $\Delta\bar{\mu}_{H^+}$ generators removing H^+ from, e.g., the mitochondrial matrix.

Exactly in this way, acetate is taken up by energized mitochondria. Since acetic acid in its protonated form, CH_3COOH, is a small uncharged molecule, it enters mitochondria without any carriers, moving down ΔpH, neutralizes intramitochondrial OH^- ions, and accumulates inside in the form of the acetate anion (CH_3COO^-). The latter is charged and thus cannot cross the mitochondrial membrane and escape from the matrix.

A similar mechanism, but operating in the opposite direction, can give rise to NH_3 efflux:

9.3 ΔpH As Driving Force

$$(NH_3)_{in} \rightarrow (NH_3)_{out,} \tag{9.6}$$

$$(NH_3)_{out} + H^+_{out} \rightarrow (NH_4^+)_{out} \tag{9.7}$$

For weak acids or bases that are lipid-insoluble, corresponding carriers are required that are specific to the uncharged form of the solute and incapable of transporting its anion or cation. This mechanism is exemplified by a mitochondrial succinate carrier which binds succinate^{2-} ($^-$OOC-CH$_2$-CH$_2$-COO$^-$) on the outer surface of the inner membrane and symports it with 2 H$^+_{out}$ into the matrix space. This is equivalent to uniport of undissociated succinic acid (HOOC-CH$_2$-CH$_2$-COOH), the concentration of which in neutral solutions is very low (see above, Fig. 9.1, system 2).

Thus, the existence of $\Delta\bar{\mu}_{H^+}$ in its secondary form (ΔpH) allows weak acids and bases to be involved in uphill transport in the direction determined by the pH gradient. Yet just as in the case of the ΔΨ-driven transport of cations and anions, it is hardly possible to obtain a reasonable composition of, e.g., the bacterial cytoplasm by accumulation of anions of weak acids and cations of strong bases, and by export of cations of weak bases and anions of strong acids. It seems obvious that in some cases the direction of the solute fluxes must be opposite.

For cations of strong bases, the problem can be solved by using cation/H$^+$-antiporters. For example, mitochondria (Williams and Fry 1979) and some bacteria (Zimniak and Barnes 1980), *in addition to the electrophoretic system for Ca^{2+} accumulation, also have an electroneutral Ca^{2+}/2H$^+$-antiporter* exporting Ca^{2+} from the mitochondrial matrix or the bacterial cytoplasm down ΔpH (Fig. 9.1, system 3).

Thus, mitochondria and bacteria can not only accumulate Ca^{2+} but also actively export it. It is clear that *these two systems must be alternatively actuated* since their cooperation would result in Ca^{2+} circulation dissipating energy of $\Delta\bar{\mu}_{H^+}$.

Similarly, anions of strong acids can be accumulated inside the matrix in a ΔpH-dependent manner if they are symported with H$^+$. This can be exemplified by phosphate transport systems of mitochondria and bacteria. Electroneutral H$_3$PO$_4$ is practically absent from aqueous solution at physiological pH. Phosphate anions must be exported from the mitochondrial matrix and bacterial cytoplasm down ΔΨ. Nevertheless, phosphate is accumulated, not exported. This results from the operation of the carrier that symports H$_2$PO$_4^-$ with H$^+$.

9.4 Total $\Delta\bar{\mu}_{H^+}$ As Driving Force

Sometimes both ΔΨ and ΔpH are driving forces for uphill transport of a solute. This is the case, for instance, when the transported compound does not contain either ionized atoms or mobile protons. Such a compound can be symported with H$^+$ (or antiported against H$^+$) down total $\Delta\bar{\mu}_{H^+}$. Solute, H$^+$-symport is responsible for the accumulation of lactose in bacteria (Fig. 9.1, system 4; for details, see Sect. 9.7.1). Vacuoles of sugar beet taproot were shown to accumulate sucrose in a $\Delta\bar{\mu}_{H^+}$-driven manner, the process being mediated by sucrose/H$^+$-antiporter (Briskin et al. 1985). The reason for using an antiporter instead of a symporter is that the directions of $\Delta\bar{\mu}_{H^+}$ in the vacuole and in a bacterium are opposite.

For lactose, H$^+$-symport, as well as for sucrose/H$^+$-antiport, $\Delta\Psi$ and ΔpH should be equally effective as the driving forces. In such cases, an increase in the H$^+$/solute stoichiometry simply enhances the gradient of the substance resulting from the transport process. In other cases, an increase in stoichiometry is required to obtain the necessary direction of transport. Accumulation of the positively charged catechol amines (RNH$_3^+$) in chromaffin granules of adrenal medulla is an example. Here, H$^+$-ATPase pumps H$^+$ into the granules, so that their interior is charged positively and acidified (see Sect. 9.7.2). When accumulated in granules, RNH$_3^+$ is antiported against 2 H$^+$ (see Fig. 9.1, system 5). So, import of one RNH$_3^+$ into the granule interior is coupled to the transport of *two* protons and only *one* net positive charge in the opposite directions. These data suggest that at equal $\Delta\Psi$ and ΔpH, the contribution of ΔpH will be twofold larger than that of $\Delta\Psi$ (see above, Eq. (9.1)). On the other hand, in systems like TrkA (see Sect. 9.2) when a cation (K$^+$) is symported with H$^+$, the contribution of $\Delta\Psi$ is twofold larger than ΔpH since the transport of one K$^+$ is accompanied by the movement of *two* net positive charges (K$^+$ and H$^+$) and only *one* proton.

As long as the transport processes are reversible, the *metabolically generated gradient* of a solute exported from the cell by a $\Delta\bar{\mu}_{H^+}$-linked system can, in principle, be an *energy source for* $\Delta\bar{\mu}_{H^+}$ *formation*. For example, in anaerobic *Enterococcus hirae*, glycolytic lactate production results in the formation of a lactate gradient ([lactate]$_{in}$>[lactate]$_{out}$). The efflux of lactate was found to be catalyzed by lactate, nH$^+$-symporter where n is equal to 2 at low [lactate]$_{out}$ and [H$^+$]$_{out}$, and is equal to 1 when the lactate and H$^+$ levels in the medium increase (Simpson et al. 1983). This means that cooperation of glycolysis and the symporter alkalinizes the cytoplasm and charges it negatively when outer concentrations of lactate and protons are low:

$$\text{glucose} \rightarrow 2 \text{ lactate}_{in}^- + 2 \text{ H}_{in}^+; \tag{9.8}$$

$$2 \text{ lactate}_{in}^- + 4 \text{ H}_{in}^+ \rightarrow 2 \text{ lactate}_{out}^- + 4 \text{ H}_{out}^+. \tag{9.9}$$

Accumulation of lactate and H$^+$ in the medium modifies the mechanism in such a way that the amounts of exported lactate and H$^+$ become equal and, hence, $\Delta\bar{\mu}_{H^+}$ generation ceases. Such regulation prevents the cell from being overloaded with lactic acid.

Fig. 9.2 Transport cascade of the inner mitochondrial membrane. The respiratory chain (*1*) pumps H$^+$ ions from the matrix; H$_2$PO$_4^-$,H$^+$-symporter (*2*) and ATP^{4-}/ADP^{3-}-antiporter (*3*) mediate the influx of H$_2$PO$_4^-$ and ADP and the efflux of ATP; HPO$_4^{2-}$/malate^{2-} antiport (*4*) results in malate influx; (malate^{2-}/citrate^{3-}+H$^+$) antiport (*5*) gives rise to citrate accumulation in the matrix

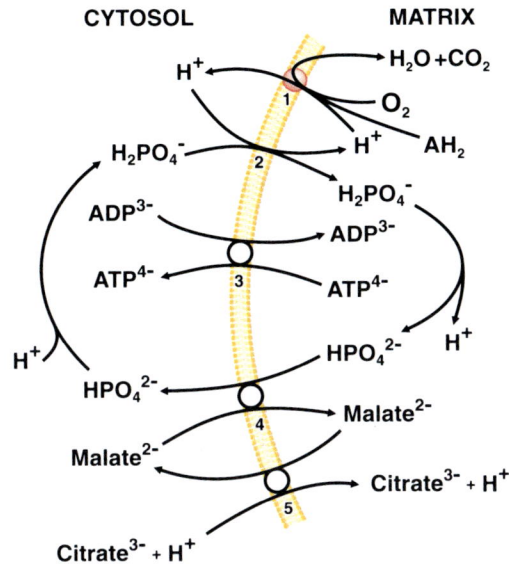

9.5 $\Delta\bar{\mu}_{H^+}$-Driven Transport Cascades

Several systems have been described where $\Delta\bar{\mu}_{H^+}$ *powers the formation of a gradient of a solute* (S_1) *which is then used for uphill transport of another solute* (S_2).

In bacteria, the role of the S_1 intermediate can be performed by Na$^+$. In this case, $\Delta\bar{\mu}_{H^+}$ drives electroneutral Na$^+_{in}$/H$^+_{out}$ or electrogenic Na$^+_{in}$/nH$^+_{out}$-antiport so that the formed ΔpNa can be used to drive accumulation of various metabolites in the cytoplasm by means of Na$^+_{out}$, solute-symporters. This is the main type of osmotic work in alkalophilic, halophilic, and many marine microorganisms.

In mitochondria, a long $\Delta\bar{\mu}_{H^+}$-dependent cascade was described for the transport of phosphate and Krebs cycle intermediates. The initial event is $\Delta\Psi$ formation by the respiratory chain. Then $\Delta\Psi$ is discharged by a flux of ionized penetrants (electrophoretic ATP$^{4-}_{in}$/ATP$^{3-}_{out}$ exchange, Ca^{2+}- or K$^+$-uniport, etc.) and ΔpH is formed. The ΔpH is utilized to form ΔpP$_i$ via H$_2$PO$_4^-$, H$^+$-symport directed from the outside to the inside. The ΔpP$_i$ powers the accumulation of malate by means of an HPO$_4^{2-}$/malate^{2-}-antiporter. (The same carrier can also transport some other dicarboxylates.) The last step of the cascade is import of citrate via malate^{2-}/citrate^{3-},H$^+$-antiporter (Fig. 9.2). Such a complicated system allows a very sophisticated regulation of the mitochondrial metabolic pattern to be organized.

9.6 Carnitine: An Example of a Transmembrane Group Carrier

In certain cases, only a part of a metabolite molecule is transported across the membrane. This applies to the system of fatty acid transport mediated by *carnitine*.

Carnitine studies have a long history, an instructive one for students of metabolite transport processes. Carnitine (Fig. 9.3) was discovered by the Russian biochemists Gulevitch and Krimberg in 1905. At that time, many biochemical compounds were described, and so the discovery of yet another substance of biological origin was not a sensation.

In the 1930s, when particularly impressive progress was made in studying the functions of low molecular mass compounds, the similarity between carnitine and choline inspired a number of studies that, however, failed to identify the role of carnitine in the living cell.

In the 1940s, the possible role of small molecules such as vitamins was extensively investigated and, at last, the first indication that carnitine was really a substance of biological importance was obtained. It was shown to be a growth factor for the mealworm *Tenebrio molitor*, and carnitine was called vitamin B_T (*T* for *Tenebrio*) (Carter et al. 1952). Later it was found that carnitine-deficient larvae died "fat" when they were starved because they were unable to utilize fat stores to survive.

Unfortunately for the students of carnitine, attempts to demonstrate the vitamin function of carnitine in organisms more important for our life than the mealworm were without success. This fact, certainly, did not stimulate any investigations in carnitine since humankind is not interested in the well-being of mealworms. Thus, until the middle of the century, when the great majority of low molecular mass cell substances had already found their place on the metabolic map, carnitine still remained in a blank spot. In 1955, i.e., 50 years after the discovery of carnitine, Stanley Friedman and Gottfried Fraenkel found that carnitine is reversibly acetylated by *acetyl-CoA* (Friedman and Fraenkel 1955). Acetyl-CoA holds a central position in metabolism, and one might expect that acetylcarnitine is an intermediate of acetyl residue transfer to some acetyl acceptor. However, this was not the case. The only reaction of acetylcarnitine utilization found was a transfer of acetyl back to CoA. The same was found to be true for fatty acylcarnitine, the second and the last representative of the carnitine derivative family. So, the impression was that carnitine acylations are dead ends of metabolism.

In the same year, 1955, Irving Fritz pointed out the stimulating effect of carnitine on fatty acid oxidation by liver homogenate (Fritz 1955). This effect of carnitine was rather surprising for a compound that was absent from the metabolic highways. Therefore, that observation awaited an explanation for quite some time.

In the 1960s, when membrane studies began, carnitine was also investigated from this point-of-view. It was found that the above-mentioned stimulating effect of carnitine is localized at the mitochondrial level, and Fritz suggested that carnitine is somehow involved in the transport of fatty acids across the mitochondrial

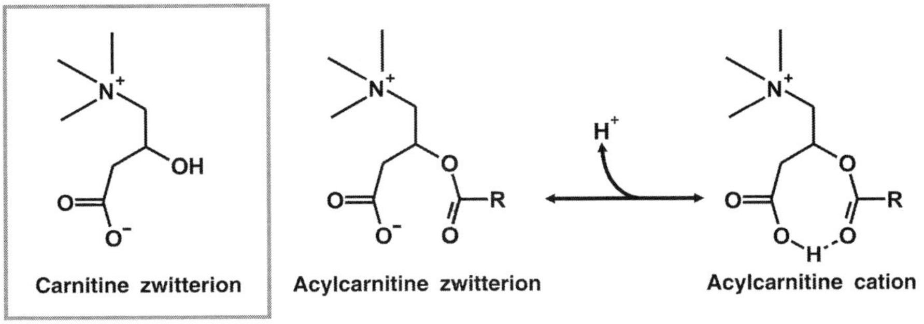

9.6 Carnitine: an Example of a Transmembrane Group Carrier

Fig. 9.3 Carnitine and acylcarnitine in deprotonated (zwitterion) and protonated (cationic) forms. R, a fatty acyl

membrane (Fritz 1963). This idea was supported by the findings that (i) fatty acylcarnitine transferase, catalyzing acyl transfer between carnitine and CoA, is localized on both sides of the hydrophobic barrier of the inner mitochondrial membrane, and (ii) this membrane is permeable for fatty acylcarnitine, but not for fatty acyl-CoA (Bremer 1983). Thus, the function of carnitine (Cn) as the fatty acyl carrier was summarized in the following chain of events (Eqs. (9.10–9.12)):

$$\text{acyl–CoA}_{out} + \text{Cn}_{out} \rightarrow \text{acyl} - \text{Cn}_{out} + \text{CoA}_{out}; \quad (9.10)$$

$$\text{acyl} - \text{Cn}_{out} \rightarrow \text{acyl} - \text{Cn}_{in}; \quad (9.11)$$

$$\text{acyl} - \text{Cn}_{in} + \text{CoA}_{in} \rightarrow \text{Cn}_{in} + \text{acyl} - \text{CoA}_{in}. \quad (9.12)$$

The answers to two questions remained obscure in the framework of this concept. First, why such a hydrophilic molecule as carnitine was chosen in evolution as a fatty carrier for hydrophobic acyls? Second, what happens to Cn_{in}? (It was found that carnitine *per se* is a non-penetrant for mitochondria. This means that Cn_{in} cannot escape the matrix by simply moving down ΔpCn (Bremer 1983)).

Both these problems were solved in the 1970s. For bioenergeticists, this period was highlighted by the struggle over the chemiosmotic hypothesis of Peter Mitchell, which culminated in his triumph. In particular, it was found that any uphill import of solutes into mitochondria is $\Delta\bar{\mu}_{H^+}$-driven. This is why in 1970 one of this book's authors (V.P.S.) and his colleagues suggested that the role of carnitine might be connected in some way with the utilization of $\Delta\bar{\mu}_{H^+}$ (Severin et al. 1970). The fact is that free fatty acids are hydrophobic anions and, hence, might be exported from (rather than imported into) mitochondria in a $\Delta\Psi$-dependent manner. Their protonated forms might return to mitochondria moving down Δp (fatty acid) and ΔpH. Such circulation should discharge $\Delta\bar{\mu}_{H^+}$ (uncoupling). On the other hand, neither carnitine nor acylcarnitine can be an anion. In the deprotonated form, they are zwitterions (the net charge is zero); in the protonated form they are cations, which are more hydrophobic than the zwitterions. This means that in combination with carnitine, fatty acids should lose their uncoupling activity and acquire the ability to be accumulated inside mitochondria where fatty acid

β-oxidation occurs. Such an accumulation, if it did occur, can be described as a symport of fatty acyl and H$^+$ down the overall $\Delta\bar{\mu}_{H^+}$:

$$acyl-Cn_{out} + H^+_{out} \rightarrow acyl-Cn-H^+_{out}; \qquad (9.13)$$

$$acyl-Cn-H^+_{out} \rightarrow acyl-Cn-H^+_{in}; \qquad (9.14)$$

$$acyl-Cn-H^+_{in} \rightarrow acyl-Cn_{in} + H^+_{in}. \qquad (9.15)$$

In agreement with this scheme, Dmitry Levitsky found that a planar phospholipid membrane is permeable for the palmitoylcarnitine cation (Levitsky and Skulachev 1972). As subsequent experiments showed, a discharge of the mitochondrial membrane potential by uncouplers strongly inhibited the oxidation of palmitoylcarnitine added after the uncoupler. On the other hand, the uncoupler stimulated oxidation of palmitoylcarnitine when the latter was added 3 min before the uncoupler. Apparently palmitoyl carnitine accumulated inside energized mitochondria during these 3 min in an amount sufficient to support active respiration during the entire time of the polarographic experiment.

Thus, there came the answer to the question why it is carnitine that plays the role of a fatty acyl carrier. *All three functional groups of carnitine are necessary* for the function of fatty acyl, H$^+$-symporter: *hydroxyl*, to combine with the fatty acyl residue; *carboxyl*, to reversibly bind the H$^+$ ion; and *quaternary nitrogen*, to charge protonated acylcarnitine positively (see Fig. 9.3).

This concept failed to answer one final question: what is the fate of the intramitochondrial carnitine released in the reaction with CoA$_{in}$? This problem was solved in 1976 by Shri Pande and Rehana Parvin and independently by Rona Ramsay and Philip Tubbs, who found stoichiometric carnitine/fatty acylcarnitine antiport across the inner mitochondrial membrane (Pande and Parvin 1976; Ramsay and Tubbs 1976).

A general scheme of events resulting in the transport of extramitochondrial fatty acids into the matrix is given in Fig. 9.4. The process starts in the outer mitochondrial membrane where acyl-CoA synthase is operating. This enzyme esterifies CoA with fatty acids at the expense of the energy of ATP (reaction 1). In the intermembrane space or on the outer surface of the inner membrane, acyl-CoA is attacked by outer fatty acylcarnitine transferase so that acyl is transferred from CoA to carnitine (reaction 2). The resulting acylcarnitine is protonated by H$^+_{out}$. The pool of H$^+_{out}$ is regenerated by respiratory $\Delta\bar{\mu}_{H^+}$ generators that pump protons from the matrix to the intermembrane space (reactions 3 and 4). The protonated acylcarnitine cation moves across the inner membrane down $\Delta\Psi$ and ΔpH (reaction 5) and is deprotonated in the matrix (reaction 6). Acylcarnitine transfers its acyl group to intramitochondrial CoA, this being catalyzed by the inner fatty acylcarnitine transferase (reaction 7). The formed inner carnitine is exchanged for outer acylcarnitine via the carnitine/acylcarnitine-antiporter (reaction 8).

An interesting feature of this system is that the level of the matrix carnitine regulates the distribution of the flow of acylcarnitine between the $\Delta\bar{\mu}_{H^+}$-driven (i.e. energy-expending) pathway 5 and the energy-independent antiporter-mediated

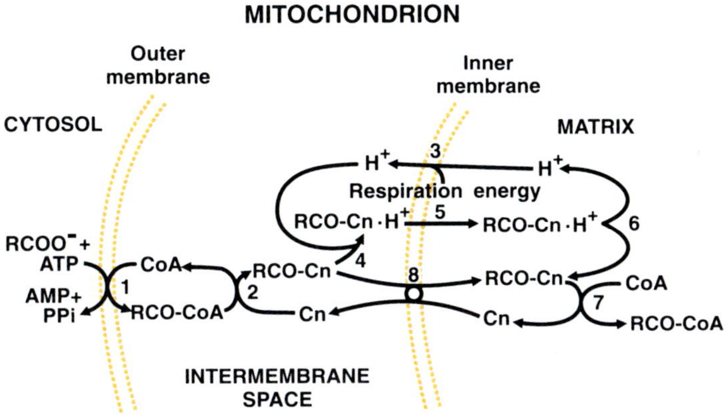

9.6 Carnitine: an Example of a Transmembrane Group Carrier

Fig. 9.4 Carnitine-mediated transport of fatty acyl groups into the mitochondrial matrix: *1* acyl-CoA synthase; *2* outer carnitine acyltransferase; *3* $\Delta\bar{\mu}_{H^+}$ generators of the respiratory chain; *4* protonation of the acylcarnitine carboxyl group; *5* $\Delta\bar{\mu}_{H^+}$-driven influx of protonated acylcarnitine; *6* deprotonation of the acylcarnitine carboxyl group; *7* inner carnitine acyltransferase; *8* carnitine/acylcarnitine-antiporter

pathway 8. Apparently, as the cell begins to oxidize fatty acids, the $\Delta\bar{\mu}_{H^+}$-driven pathway is predominant since intramitochondrial carnitine is below the K_m of the antiporter for carnitine (this K_m is rather high, i.e. 1 mM). At this stage, the energy of $\Delta\bar{\mu}_{H^+}$ is spent to accelerate fatty acid oxidation. Large-scale uphill import of acylcarnitine, followed by utilization of fatty acyls in the system of β-oxidation, leads to the accumulation of carnitine in the matrix to a level sufficient for antiporter saturation. Now, $\Delta\bar{\mu}_{H^+}$ consumption at the stage of transport of fatty acyl residues can be decreased, provided antiport is faster than uniport. It is noteworthy that functioning of the antiporter per se does not require $\Delta\bar{\mu}_{H^+}$.

It should be noted that all these considerations cannot be applied to acetylcarnitine, which is too hydrophilic to be a penetrating cation. It was postulated that *the main function of acetylcarnitine is that of a buffer of high-energy acetyls, just as creatine phosphate buffers ATP level* (Skulachev 1972). It does not exclude though electrophoretic accumulation of protonated acetylcarnitine in mitochondria if we assume the existence of a special protein carrier of this cation. Such a carrier could also be used for the transport of protonated acylcarnitine, for the concentration and permeability of this compound through artificial membranes are very small at neutral pH.

9.7 Some Examples of $\Delta\bar{\mu}_{H^+}$-Driven Carriers

In principle, the formation of a concentration gradient of a solute at the expense of $\Delta\bar{\mu}_{H^+}$ might not need any special energy-transducing mechanisms. As already mentioned, acetate can accumulate inside mitochondria, while ammonia can be exported from mitochondria down ΔpH even without any carriers because mitochondrial membranes are permeable to CH_3COOH and NH_3. The carrier only accelerates these fluxes. On the other hand, the involvement of a carrier is quite necessary when the transported solute or chemical groups per se cannot move across the membrane in the proper direction. *The living cell always tries to avoid spontaneous events, preferring to deal with protein-mediated processes* to achieve a high specificity and a controlled reaction rate that can be regulated within wide limits, from very high down to zero values. This principle can be applied, in particular, to $\Delta\bar{\mu}_{H^+}$-linked osmotic work. *The great majority of transport mechanisms are carrier mediated.*

In rare cases, these carriers are very simple, such as *carnitine*. Such peptide antibiotics as channel-forming *gramicidin* that transports Na^+, K^+, and H^+, as well as the K^+-uniporter *valinomycin*, K^+/H^+-antiporter *nigericin*, or Na^+/H^+-antiporter *monensin* also look like rather simple molecules in comparison with proteins. However, one should remember that both carnitine and the above-mentioned antibiotics perform, in fact, primitive functions.

In the case of carnitine, it is the initiation of a process that then becomes mediated by a protein, namely, the carnitine/acylcarnitine-antiporter. The role of antibiotics is apparently restricted to damaging cell membranes of species other than the producer. When the function of a carrier is more complicated, it is always of protein nature.

The *carrier proteins* are integral membrane polypeptides that may be assisted by peripheral or water-soluble proteins. In the above-mentioned well-studied example of *E. coli* amino acid transport (see Table 9.1), it was shown that porters 1, 2, 5, 7, 8, and 14 do not require any assistants. In the other systems listed in Table 9.1, there are specific transport-assisting proteins in the periplasm. In systems 3, 4, and 6, these proteins act as buffers of the transported amino acids. They reversibly bind these amino acids, in this way sequestering them in the periplasm. As a result, the import of amino acids into the cell does not cause an immediate decrease in the amino acid concentration in the periplasm. This should stabilize the import rate.

Another situation is found in the case of systems 9 and 10. Here, the assisting proteins are necessary for transport per se. They bind corresponding amino acids in the periplasm and then form a complex with an integral protein of the cytoplasmic membrane, which plays, in fact, the role of "apo-carrier". For systems 9 and 10, it is the assisting protein that determines the specificity and affinity of the carrier to the transported solute. Below we will consider some of the $\Delta\bar{\mu}_{H^+}$-driven transport systems that have been studied in detail.

9.7.1 Escherichia coli Lactose, H^+-Symporter

The lactose, H^+-symporter from *E. coli* is the most thoroughly investigated carrier of bacterial origin. It was discovered by Cohen and Rickenberg in 1955. 8 years later, Mitchell suggested that lactose is symported with H^+ down the $\Delta \bar{\mu}_{H^+}$ gradient (Mitchell 1963). In 1970, Ian West demonstrated that lactose intake by *E. coli* cells is really accompanied by H^+ ion intake (West 1970). At the same time, it was found that the *lacY* gene product is a membrane protein. In 1978, Teather et al. cloned the *lacY* gene into a recombinant plasmid, which made it possible to elucidate its nucleotide sequence and the amino acid sequence of the lactose, H^+-symporter, and to carry out the amplification and in vitro synthesis of this protein (Teather et al. 1978). In 1980, the purified symporter was reconstituted in proteoliposomes (Newman and Wilson 1980).

Experiments with intact bacteria, inside-out subbacterial vesicles, and proteoliposomes revealed that passive downhill diffusion of lactose results in the formation of $\Delta \bar{\mu}_{H^+}$ (positive and acidic on the side of lower lactose concentration) that inhibits the lactose flux. Uncouplers abolished this inhibition.

On the other hand, uphill lactose flux across a natural membrane driven by the respiratory chain-produced $\Delta \bar{\mu}_{H^+}$ was found to be inhibited by uncouplers. This effect was first demonstrated on intact *E. coli* cells, and later also on proteoliposomes containing symporter and cytochrome *bo* purified from this bacterium (Kaback 1983).

For many years, the lactose, H^+-symporter has been a classical model for studies of the protein-mediated carrier mechanism. It is a very hydrophobic protein with a molecular mass of 46.5 kDa; it contains 417 amino acid residues. Many mutant forms of this enzyme have been obtained (Kaback and Wu 1997). It seems quite amazing that only six amino acid residues of the lactose, H^+-symporter are absolutely irreplaceable. Recently crystals of one such mutant form (Cys154Gly) were obtained; this made it possible to determine the high-resolution three-dimensional structure of this carrier (Abramson et al. 2003).

The lactose, H^+-symporter contains 12 transmembrane α-helices (Fig. 9.5). The structure of these helices is unusually bent, this fact leading to an original infundibular form of the protein. A very large hydrophilic cavity can be observed in the structure of this symporter; it is at the bottom of this cavity that the lactose- and proton-binding sites are located (approximately at the middle of the lipid bilayer).

As already mentioned, the three-dimensional structure was determined only for the Cys154Gly mutant of the lactose, H^+-symporter. This mutant form is in a way "frozen" in one of the protein conformations; it is due to this fact that its crystallization was successful. It had been established earlier that the lactose, H^+-symporter undergoes substantial conformational changes during lactose transfer, and as a result the accessibility of various amino acid residues on opposite membrane sides to water changes greatly. These observations indicate that the hydrophilic cavity with the bound lactose can be open (depending on the protein conformation) either to the cytoplasm (it is in this conformation that Cys154Gly mutant is fixed) or to the periplasm (Fig. 9.6). Figure 9.7 illustrates a mechanism of transmembrane transfer of lactose that was suggested on the basis of these data.

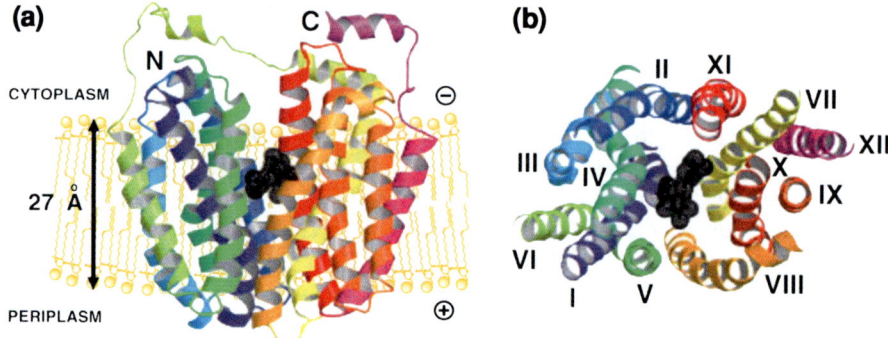

Fig. 9.5 Structure of lactose, H⁺-symporter from *E. coli* in a complex with substrate analog (*black*) (Abramson et al. 2003). **a** *side view* (parallel to the membrane surface); **b** *top view* (from the cytosol side)

Fig. 9.6 Structural differences between the open inward **a** and open outward **b** conformations of lactose, H⁺-symporter. The substrate analog is colored black (Abramson et al. 2003)

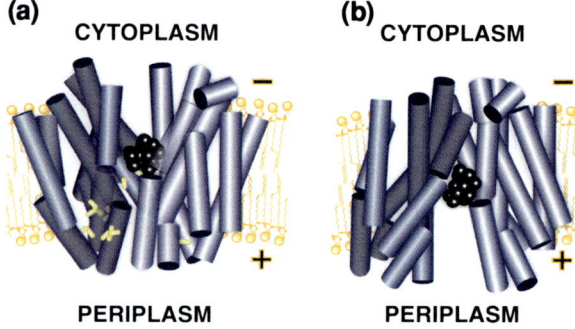

Fig. 9.7 Mechanism of lactose, H⁺-symport: **1–6**, sequence of stages; CH and C⁻ represent protonated and deprotonated forms of a symporter functional group; L stands for lactose

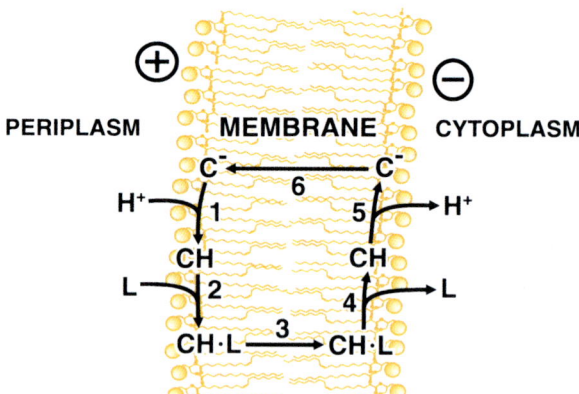

9.7.2 Mitochondrial ATP/ADP-Antiporter

Among mitochondrial transport systems, ATP/ADP-antiporter (other names: adenine nucleotide translocase, adenine nucleotide carrier) has been best studied. This antiporter catalyzes exchange of one ADP molecule for one ATP molecule between the extra- and intramitochondrial spaces. As found by Martin Klingenberg (Fig. 9.8) and Hagai Rottenberg, the antiporter-mediated distribution of ATP and ADP between the matrix and the extramitochondrial spaces appears to be a function of the respiration-produced $\Delta\Psi$ (Klingenberg and Rottenberg 1977). Valinomycin, by discharging $\Delta\Psi$, equalizes ATP/ADP ratios inside and outside the mitochondria and alters the antiport kinetics, whereas nigericin, which collapses ΔpH, is without effect. These facts support the assumption of electrogenic character of ATP/ADP exchange. In contrast to the majority of other nucleotide-binding proteins, ATP/ADP-antiporter binds free forms of ATP and ADP, and not their magnesium complexes. Since in the absence of magnesium at neutral pH ATP bears one negative charge more than ADP, and the direction of the electric field in energized mitochondria is favorable for anion export, it was concluded that $ATP_{in}^{4-}/ADP_{out}^{3-}$-antiport takes place (Klingenberg and Rottenberg 1977). This conclusion was confirmed by experiments with proteoliposomes.

The turnover number of the ATP/ADP-antiporter is rather low (about 25 per second). On the other hand, the amount of this antiporter in the membrane is very large (5–12 % of the inner mitochondrial membrane protein).

The glycosides *atractylate* and *carboxyatractylate* as well as *bongkrekic acid* are specific inhibitors of this antiporter.

The biological role of the ATP/ADP-antiporter is obvious. The extramitochondrial compartments of the cell should receive ATP synthesized by means of oxidative phosphorylation inside the mitochondria. The electrophoretic character of the nucleotide antiport across the energized mitochondrial membrane (i) facilitates ATP release from mitochondria and ADP uptake by them and (ii) increases the energy charge of ATP hydrolysis in cytosol due to an elevation of the ATP/ADP ratio in this cell compartment. The latter effect shifts all the ATP-dependent processes to ATP utilization, and thus some additional useful work can be performed.

When oxidative phosphorylation is not operative (e.g. because of anaerobiosis), the same antiport mechanism prevents glycolytic ATP from being exhausted by reverse H^+-ATP-synthase since, to enter the mitochondrion, ATP must go against an electric field produced by the H^+-ATPase reaction as well as by the $ATP_{out}^{4-}/ADP_{in}^{3-}$ exchange.

It seems interesting to compare substrate specificity of the ATP/ADP-antiporter and the H^+-ATP-synthase. The first interacts only with adenine nucleotides, while the second one also with guanine nucleotides. This means that the GTP formed inside mitochondria as a result of oxidative phosphorylation cannot be immediately pumped out of mitochondria. This can take place only after the transfer of a phosphoryl from GTP to ADP catalyzed by nucleotide diphosphate kinase. The

Fig. 9.8 Martin Klingenberg

same is true for substrate-level phosphorylation coupled to oxidative decarboxylation of α-ketoglutarate, which forms GTP. This means that by inhibiting the matrix nucleoside diphosphate kinase, a mitochondrion can prevent its GTP pool from being exhausted by extramitochondrial energy consumers. This should also facilitate protein synthesis reactions in mitochondria, for they are supported by the energy of GTP.

The ATP/ADP-antiporter is a hydrophobic membrane protein (molecular mass, 32 kDa) comprises 297 amino acid residues (in case of the bovine enzyme). According to the analysis of its amino acid primary sequence, it consists of three repeating sequences, each of them comprises two hydrophobic α-helices connected by a long hydrophilic loop (i.e. there are six transmembrane α-helices in the complete protein). Homologous repeats in the structure of this carrier indicate its most probable origin: it must have appeared in evolution as a result of triplication of some precursor gene. The antiporter operates as a homodimer in vivo.

The three-dimensional structure of the ATP/ADP-antiporter has been recently determined by X-ray analysis (Pebay-Peyroula et al. 2003). Six transmembrane segments of this protein form a rather compact structure in the lipid bilayer (Fig. 9.9). All of these α-helices are substantially bent in relation to the axis perpendicular to the membrane surface. Each second α-helix has an acute distortion approximately in the middle of the bilayer due to the presence of proline residues. Similar to lactose permease, there is a large hydrophilic cone in the protein (Fig. 9.10). It is surfaced on the inside with positively charged amino acid residues (Arg79, Arg104, Arg137, Arg234, Arg235, Arg279; Lys22, Lys32, Lys91, Lys95, and Lys106), which probably provide recognition, release from magnesium, and binding of ATP^{4-} and ADP^{3-}. The binding site of these

Fig. 9.9 Three-dimensional structure of ATP/ADP-antiporter (Pebay-Peyroula et al. 2003)

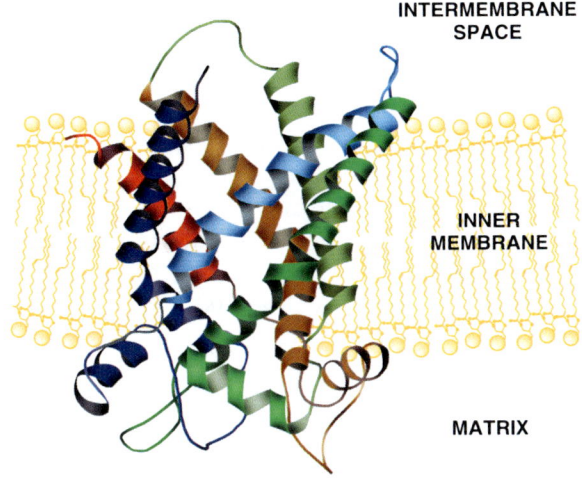

Fig. 9.10 Surface charge of ATP/ADP-antiporter (viewed from the intermembrane space side). Positive and negative charges are colored *blue* and *red*, respectively (Pebay-Peyroula et al. 2003)

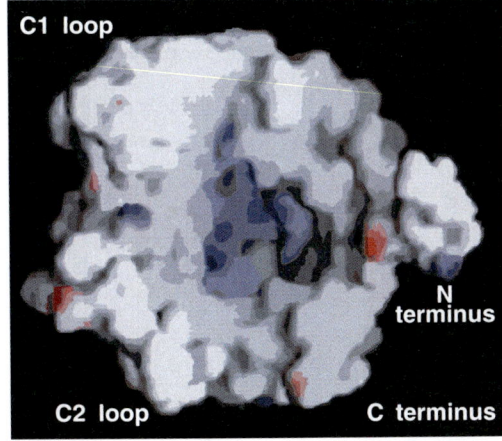

nucleotides is located at the very bottom of this positively charged cavity (i.e. around the middle of the lipid bilayer). At this site, the protein has an unusual sequence ($Arg_{234}Arg_{235}Arg_{236}Met_{237}Met_{238}Met_{239}$) that is characteristic for the primary structure of ATP/ADP-antiporters and is absent from other carriers.

The mechanism of nucleotide exchange by ATP/ADP-antiporter is probably similar to the mechanism of the above-described lactose, H^+-symporter. This mechanism could embrace substantial conformational changes of the protein during of its functioning; these changes might result in the opening of the hydrophilic cavity either in the direction of the matrix (this variant of the protein structure should have an increased affinity to ATP), or in the direction of the intermembrane space (accompanied by simultaneous decrease in affinity to ATP and increase in affinity to ADP). It is important to note that ATP/ADP-antiporter probably catalyzes nucleotide exchange only when in the form of a homodimer.

Thus, the possible mechanism of functioning of this homodimer includes cooperative change in monomer conformations with hydrophilic cavities of these monomers opening to the opposite membrane sides.

The ATP/ADP-antiport is found not only on the inner mitochondrial membrane. An indication that ATP/ADP-antiport occurs in chloroplasts was first obtained by Woldegiorgis and colleagues (1985). Nucleotide exchange in chloroplasts is sensitive to bongkrekic acid, but resistant to carboxyatractylate. The plastid antiporter was shown to function in the direction opposite to the one of mitochondria, i.e., it exchanges cytoplasmic ATP for the ADP in the organelle, thus providing for chloroplast functioning in the dark.

A special system of adenine nucleotide transfer was discovered in the cytoplasmic membrane of *Rickettsia prowazekii*, an intracellular parasitic bacterium of eukaryotic cells and epidemic typhus pathogen. The bacterium has a respiratory chain that maintains $\Delta\bar{\mu}_{H^+}$ across its membrane. However, the ATP produced by the host cell was also found to be an energy source for *Rickettsia* due to the presence of a special ATP/ADP-antiporter in the bacterium. This protein provides ATP influx into the microorganism and ADP efflux from it, i.e., the direction of nucleotide transport in this case is opposite to the one observed in mitochondria (Andersson 1998). These data explain the fact that the ATP/ADP-antiporter of this microorganism is much more similar to the antiporter of chloroplast type than to its mitochondrial analog (even though *R. prowazekii* and mitochondria are close evolutionary "relatives") and shed light on the evolutionary mechanism of the origin of mitochondria.

An ATP-import mechanism has also been described for *Bdellovibrio bacteriovorus* (Ruby and McCabe 1986). This small bacterium has the highest rate of motility among prokaryotes. Swimming very fast, it kills other microorganisms by breaking their cell wall. When it finds itself inside a cell, *B. bacteriovorus* imports the ATP of the sacrificed cell.

9.8 Role of $\Delta\bar{\mu}_{H^+}$ in Transport of Macromolecules

One of the most intriguing problems of membranology and bioenergetics is how biological macromolecules, i.e., proteins and nucleic acids, can be transported across the hydrophobic barriers of coupling membranes impermeable even to such small particles as H^+ ions.

It is firmly established that such *processes of macromolecules transport occur without membrane breakdown* since it was found that in many cases that to proceed with a normal rate they need $\Delta\bar{\mu}_{H^+}$ or one of its constituents. This rate, at least in some cases, is very high.

Two models of protein transport across the coupling membranes have been most thoroughly studied: the import of mitochondrial (chloroplast) proteins synthesized in cytosol, and the export of bacterial periplasmic or outer membrane proteins.

Another class of processes closely related to the above-mentioned ones includes insertion (into the coupling membrane) of proteins that are arranged in transmembrane fashion. To occupy the correct position in the membrane, some domains of such proteins should cross the hydrophobic barrier.

9.8.1 Transport of Mitochondrial Proteins: Biogenesis of Mitochondria

A large majority (not less than 98 %) of mitochondrial proteins are coded by the nuclear genome. This means that the proteins are usually synthesized in the cytosol and then are somehow transported to the respective mitochondrial compartment (to the matrix, intramembrane space, or to the inner or outer membrane).

It is a general rule that proteins synthesized in the cytosol and then imported by mitochondria are primarily formed as precursors containing additional N-terminal amino acid sequences (Maccecchini et al. 1979). These sequences can be viewed as an "address" that directs the protein to mitochondria. Such a *leader* sequence usually consists of 10–30 amino acid residues capable of forming an amphipathic α-helix. Half of this helix surface is hydrophobic, while the other half is hydrophilic and positively charged due to the presence of lysine and arginine residues. When reaching the destination compartment, the leader sequence is immediately removed by specific *leader peptidases*. Having lost the leader, the precursor becomes a *mature protein*.

In some cases, imported mitochondrial proteins lack the leader sequence. This was demonstrated, in particular, for many carriers of the inner mitochondrial membrane (e.g. for ATP/ADP-antiporter) and for all the proteins of the outer mitochondrial membrane (e.g. for porin). It was shown that the information to target such proteins to mitochondria is present within their primary sequence (Adrian et al. 1986).

For many cases, it was found that mitochondrial proteins can be translated on free polysomes. Let us discuss the process of mitochondrial protein transport using the example of a matrix mitochondrial enzyme with a leader sequence (Schatz 1993; Wiedemann et al. 2004). Such a protein should first of all permeate the outer mitochondrial membrane. This process is supported by the TOM complex (from "translocase of the outer mitochondrial membrane") consisting of seven different polypeptides. The TOM complex receptor subunits (Tom20, Tom22, and Tom70) bind the leader sequence, and then the precursor permeates the outer membrane. This step is mediated by the Tom40 protein, which forms a large pore in the membrane.

Protein transport across the inner mitochondrial membrane is catalyzed by the TIM complex ("translocase of the inner mitochondrial membrane") consisting of three different subunits: Tim50, Tim23, and Tim17. The leader sequence is first bound to Tim50 in the intermembrane space, and then this sequence penetrates the inner membrane through a channel formed by the Tim23 subunit. The latter process occurs due to electrophoretic transfer of positively charged amino acid residues of the leader, i.e. it is $\Delta\Psi$-driven. The absence of $\Delta\Psi$ on the inner mitochondrial membrane completely prevents the transport of mitochondrial matrix proteins.

When in the matrix, the leader sequence binds with the chaperone mtHsp70, which (together with additional proteins) fulfills two functions. The chaperone "drags" down the rest of the precursor protein into the matrix at the expense of energy of ATP hydrolysis, and then it secures the correct assembly of the mature mitochondrial protein.

Thus, protein transport to the mitochondrial matrix requires energy in the form of both $\Delta\Psi$ and ATP. However, when proteins are transported from the cytosol to inner mitochondrial membrane, it is only energy of $\Delta\Psi$ that is utilized.

There are a small number of very hydrophobic polypeptides coded by mitochondrial genome and synthesized in the matrix: seven subunits of NADH:CoQ-oxidoreductase, one subunit of $CoQH_2$:cytochrome c-oxidoreductase, three subunits of cytochrome oxidase, and two subunits of H^+-ATP-synthase.[1] All of these are synthesized as mature forms (without any precursors involved) and are incorporated into the membrane by a special complex different from the TIM complex. The mechanism of this process remains unclear.

An important conclusion from the above-discussed data is that a *mitochondrion cannot be formed by self-assembly from its constituents*. It can originate only from another mitochondrion, just as a daughter cell is formed from a mother cell. A necessary condition for mitochondrial biogenesis is a $\Delta\Psi$ across its inner membrane. Since no mitochondrial $\Delta\Psi$ generator can be formed solely from mitochondrially synthesized polypeptides, a mitochondrion that has lost its last $\Delta\Psi$-generating enzyme cannot make up for this deficiency even if the surrounding cytoplasm contains an excess of mitochondrial protein precursors. This may be the reason why yeasts growing under anaerobic conditions retain closed vesicles, promitochondria, containing no respiratory chain but competent in $\Delta\Psi$ formation mediated by reverse H^+-ATP-synthase, small amounts of which are present there.

9.8.2 Transport of Bacterial Proteins

In bacteria, proteins located in the periplasm or in the outer cell membrane must cross the inner (cytoplasmic) membrane of the cell to reach their destination. Moreover, bacteria release certain enzymes into the surrounding medium. To some degree, these processes resemble protein import by mitochondria. They require the presence of $\Delta\Psi$ on the inner bacterial membrane. In both cases, there is usually an additional signal sequence at the N-terminus of the transported polypeptide. This

---

[1] Data for higher animals.

9.8 Role of $\Delta\bar{\mu}_{H^+}$ in Transport of Macromolecules

sequence is composed of one or several positively charged amino acid residues followed by hydrophobic residues capable of transmembrane α-helix formation. When the protein transport is completed, the signal sequence can be deleted by a special peptidase (Müller 1992).

The necessity of membrane energization was studied on periplasmic proteins that bind leucine, isoleucine, valine, and maltose, and also on β-lactamase and several proteins of the outer membrane. A strict dependency of protein transport on $\Delta\Psi$ on the inner bacterial membrane has been demonstrated in all these cases.

9.8.3 Role of $\Delta\Psi$ in Protein Arrangement in the Membrane

In many cases $\Delta\Psi$ was shown to be necessary for a protein to be inserted into, rather than transported through, the membrane. Apparently, this process can be exemplified by the fact that a peptide antibiotic, alamethicin, forms channels in a planar phospholipid membrane only if $\Delta\Psi$ of the proper direction is imposed across the membrane (Ehrenstein and Lecar 1977). This effect can be explained by electrophoretic movements of charged groups of alamethicin in the membrane.

Hepatic asialoglycoprotein receptor and bee venom melittin were found to be incorporated into a planar phospholipid membrane only when $\Delta\Psi > 20$ mV was applied. Melittin is a short polypeptide (26 amino acids), and it was rather easy to imagine a scheme explaining its behavior in terms of electrophoretic movement of some parts of this molecule through the hydrophobic region of the membrane (Weinstein et al. 1982).

The presence of $\Delta\Psi$ on a lipid bilayer not only provides energy for the incorporation of a protein, but it also makes this process a directed one, i.e., it creates functional asymmetry of biological membranes. There is a rule that *positively charged domains of a membrane protein are localized on the negatively charged membrane side, while negatively charged domains are on the opposite, positively charged membrane side* (Bogdanov et al. 2009).

9.8.4 Bacterial DNA Transport

Clearly, a special molecular device must exist to transport DNA across the bacterial wall. This process can be extremely fast. For instance, the rate of DNA transport on T4-phage infection is as high as 3×10^3 nucleotide residues per second, i.e., one nucleotide residue per 330 μs. The DNA transport rates during conjugation of bacteria and their genetic transformation are somewhat slower, i.e., 600 and 55 nucleotide residues per second, respectively, but still faster, than, for example, the nucleotide flux mediated by the ATP/ADP-antiporter (25 nucleotides per second). DNA transport is often accompanied by the translocation of special ("pilot") proteins so that large molecular masses cross the hydrophobic barrier. In some cases, double-stranded DNA was found to be transported (Grinius 1986).

The role of $\Delta\bar{\mu}_{H^+}$ and ATP in DNA transport was studied by Leonas Grinius and colleagues. They found, in particular, that the *exhaustion of the intracellular ATP pool by arsenate does not inhibit either T4-phage infections or genetic transformation* (Kalasauskaite et al. 1980). On the other hand, conjugation is inhibited by arsenate. As to $\Delta\bar{\mu}_{H^+}$, *it was necessary for all three types of bacterial DNA transport processes*. This was shown for *Bacillus subtilis* transformation, *E. coli* T4-phage infection, and conjugation.

Later these observations were confirmed in other laboratories. In particular, Wout Bremer et al. found that transformation-inducing DNA transport into *B. subtilis* or *Haemophilus influenzae* cells is $\Delta\bar{\mu}_{H^+}$-dependent, both $\Delta\Psi$ and ΔpH being effective (Bremer et al. 1984). Bernard Labedan et al. confirmed that $\Delta\bar{\mu}_{H^+}$ is essential for T4-phage infection but pointed out that only $\Delta\Psi$, and not ΔpH, is effective (Labedan and Goldberg 1979; Labedan et al. 1980). Labedan's group, studying a T5 phage-induced infection, failed to observe the requirement for $\Delta\bar{\mu}_{H^+}$. The transport of the λ-phage DNA into *E. coli* was also found to be energy-independent. Apparently, bacteria possess several mechanisms for phage DNA transport across the membrane. In cases when $\Delta\bar{\mu}_{H^+}$ is necessary, it may be a driving force and/or a regulator essential for the correct orientation of the transported DNA or DNA–protein complex.

References

Abramson J, Smirnova I, Kasho V, Verner G, Iwata S, Kaback HR (2003) The lactose permease of *Escherichia coli*: overall structure, the sugar-binding site and the alternating access model for transport. FEBS Lett 555:96–101

Adrian GS, McCammon MT, Montgomery DL, Douglas MG (1986) Sequences required for delivery and localization of the ADP/ATP translocator to the mitochondrial inner membrane. Mol Cell Biol 6:626–634

Andersson SG (1998) Bioenergetics of the obligate intracellular parasite *Rickettsia prowazekii*. Biochim Biophys Acta 1365:105–111

Bogdanov M, Xie J, Dowhan W (2009) Lipid-protein interactions drive membrane protein topogenesis in accordance with the positive inside rule. J Biol Chem 284:9637–9641

Bremer J (1983) Carnitine—metabolism and functions. Physiol Rev 63:1420–1480

Bremer W, Kooistra J, Hellingwerf KJ, Konings WN (1984) Role of the electrochemical proton gradient in genetic transformation of *Haemophilus influenzae*. J Bacteriol 157:868–873

Briskin DP, Thornley WR, Wyse RE (1985) Membrane transport in isolated vesicles from sugarbeet taproot: II. evidence for a sucrose/H^+-antiport. Plant Physiol 78:871–875

Carter HE, Bhattacharyya PK, Weidman KR, Fraenkel G (1952) Chemical studies on vitamin B_T isolation and characterization as carnitine. Arch Biochem Biophys 38:405–416

Cohen GN, Rickenberg HV (1955) Direct study of the fixation of an inducer of beta-galactosidase by *Escherichia coli* cellules. C R Hebd Seances Acad Sci 240:466–468

Ehrenstein G, Lecar H (1977) Electrically gated ionic channels in lipid bilayers. Quart Rev Biophys 10:1–34

Epstein W, Laimins L (1980) Potassium transport in *Escherichia coli*: diverse systems with common control by osmotic forces. Trends Biochem Sci 5:21–23

Friedman S, Fraenkel G (1955) Reversible enzymatic acetylation of carnitine. Arch Biochem Biophys 59:491–501

Fritz I (1955) The effect of muscle extracts on the oxidation of palmitic acid by liver slices and homogenates. Acta Physiol Scand 34:367–385

Fritz IB (1963) Carnitine and its role in fatty acid metabolism. Adv Lipid Res 1:285–334

Grinius LL (1986) Bacterial transport of macromolecules. Nauka, Moscow

Gulewitsch W, Krimberg R (1905) Zur kenntnis der extraktivstoffe der muskeln. 2 Mitteilung. Über das Carnitin. Hoppe-Seyler's Z Physiol Chem 45:326–330

Kaback HR (1983) The lac carrier protein in *Escherichia coli*. J Membr Biol 76:95–112

Kaback HR, Wu J (1997) From membrane to molecule to the third amino acid from the left with a membrane transport protein. Q Rev Biophys 30:333–364

Kalasauskaite E, Grinius L, Kadisaite D, Jasaitis A (1980) Electrochemical H^+ gradient but not phosphate potential is required for *Escherichia coli* infection by phage T4. FEBS Lett 117:232–236

Klingenberg M, Rottenberg H (1977) Relation between the gradient of the ATP/ADP ratio and the membrane potential across the mitochondrial membrane. Eur J Biochem 73:125–130

Labedan B, Goldberg EB (1979) Requirement for membrane potential in injection of phage T4 DNA. Proc Natl Acad Sci U S A 76:4669–4673

Labedan B, Heller KB, Jasaitis AA, Wilson TH, Goldberg EB (1980) A membrane potential threshold for phage T4 DNA injection. Biochem Biophys Res Commun 93:625–630

Levitsky DO, Skulachev VP (1972) Carnitine: the carrier transporting fatty acyls into mitochondria by means of an electrochemical gradient of H^+. Biochim Biophys Acta 275:33–50

Maccecchini ML, Rudin Y, Blobel G, Schatz G (1979) Import of proteins into mitochondria: precursor forms of the extramitochondrially made F_1-ATPase subunits in yeast. Proc Natl Acad Sci U S A 76:343–347

Mitchell P (1963) Molecule, group and electron translocation through natural membranes. Biochem Soc Symp 22:142–169

Müller M (1992) Proteolysis in protein import and export: signal peptide processing in eu- and prokaryotes. Experientia 48:118–129

Newman MJ, Wilson TH (1980) Solubilization and reconstitution of the lactose transport system from *Escherichia coli*. J Biol Chem 255:10583–10586

Pande SV, Parvin R (1976) Characterization of carnitine acylcarnitine translocase system of heart mitochondria. J Biol Chem 251:6683–6691

Pebay-Peyroula E, Dahout-Gonzalez C, Kahn R, Trézéguet V, Lauquin GJ, Brandolin G (2003) Structure of mitochondrial ADP/ATP carrier in complex with carboxyatractyloside. Nature 426:39–44

Pivetti CD, Yen MR, Miller S, Busch W, Tseng YH, Booth IR, Saier MH Jr (2003) Two families of mechanosensitive channel proteins. Microbiol Mol Biol Rev 67:66–85

Ramsay RR, Tubbs PK (1976) The effects of temperature and some inhibitors on the carnitine exchange system of heart mitochondria. Eur J Biochem 69:299–303

Ruby EG, McCabe JB (1986) An ATP transport system in the intracellular bacterium, *Bdellovibrio bacteriovorus* 109 J. J Bacteriol 167:1066–1070

Schatz G (1993) The protein import machinery of mitochondria. Protein Sci 2:141–146

Schönheit P, Beimborn DB, Perski H-J (1984) Potassium accumulation in growing *Methanobacterium thermoautotrophicum* and its relation to the electrochemical proton gradient. Arch Microbiol 140:247–251

Severin SE, Skulachev VP, Yaguzhinskiy LS (1970) Possible role of carnitine in the transport of fatty acids through the mitochondrial membrane. Biokhimiia 35:1250–1253

Simpson SJ, Bendall MR, Egan AF, Vink R, Rogers PJ (1983) High-field phosphorus NMR studies of the stoichiometry of the lactate/proton carrier in *Streptococcus faecalis*. Eur J Biochem 136:63–69

Skulachev VP (1972) Energy transformation in biomembranes. Nauka, Moscow

Stewart LM, Bakker EP, Booth IR (1985) Energy coupling to K$^+$ uptake via the Trk system in *Escherichia coli*: the role of ATP. J Gen Microbiol 131:77–85

Teather RM, Müller-Hill B, Abrutsch U, Aichele G, Overath P (1978) Amplification of the lactose carrier protein in *Escherichia coli* using a plasmid vector. Mol Gen Genet 159:239–248

Weinstein JN, Blumenthal R, van Renswoude J, Kempf C, Klausner RD (1982) Charge clusters and the orientation of membrane proteins. J Membr Biol 66:203–212

West IC (1970) Lactose transport coupled to proton movements in *Escherichia coli*. Biochem Biophys Res Commun 41:655–661

Wiedemann N, Frazier AE, Pfanner N (2004) The protein import machinery of mitochondria. J Biol Chem 279:14473–14476

Williams AJ, Fry CH (1979) Calcium-proton exchange in cardiac and liver mitochondria. FEBS Lett 97:288–292

Woldegiorgis G, Voss S, Shrago E, Werner-Washburne M, Keegstra K (1985) Adenine nucleotide translocase-dependent anion transport in pea chloroplasts. Biochim Biophys Acta 810:340–345

Zimniak P, Barnes EM Jr (1980) Characterization of a calcium/proton antiporter and an electrogenic calcium transporter in membrane vesicles from *Azotobacter vinelandii*. J Biol Chem 255:10140–10143

Chapter 10
$\Delta\bar{\mu}_{H^+}$ as Energy Source for Heat Production

10.1 Three Ways of Converting Metabolic Energy into Heat

Certain living organisms (especially mammals and birds) are capable of maintaining a substantial temperature gradient between their body and the environment. This process requires very high energy consumption. It is obvious that oxidation of respiratory substrates by molecular oxygen is the primary energy source for thermogenesis in any animal tissues.

In the living cell, heat can be released by three main types of mechanisms: (1) by hydrolysis of ATP synthesized due to oxidative phosphorylation (respiration $\rightarrow \Delta\bar{\mu}_{H^+} \rightarrow$ ATP $\rightarrow$ heat); (2) by dissipation of $\Delta\bar{\mu}_{H^+}$ produced by proton-translocating enzymes of the respiratory chain (respiration $\rightarrow \Delta\bar{\mu}_{H^+} \rightarrow$ heat); and (3) by direct dissipation of respiration-released energy to heat with no $\Delta\bar{\mu}_{H^+}$ involved (respiration $\rightarrow$ heat). *Respiration coupled to phosphorylation* participates in mechanisms of the first type, while *uncoupled respiration* and *respiration not initially coupled to energy storage (noncoupled respiration)* are responsible for the second and third cases, respectively. To distinguish the last two mechanisms from the first, representing a much more complicated way to heat via ATP formation and utilization, it is convenient to use the term "free respiration", uniting both non- and uncoupled respiratory systems. Various organisms can use different mechanisms of thermogenesis or their combinations.

The very fact that cellular respiratory systems are somehow involved in heat regulation was already recognized in early physiological studies showing a strong increase in oxygen consumption by mammals and birds exposed to cold. Usually, such an increase, which can be as high as 2.5-fold, occurs without a corresponding nonspecific increase in the functional activity of the organism. The impression arises that under cold conditions physiological functions are performed less efficiently, which results in an increase in food and O_2 consumption and in additional heat production. An alternative possibility is that the efficiency of the functions remains the same and additional thermogenesis is caused by the activation of mechanisms specializing in heat production at low ambient temperatures.

The latter concept seems to be more reasonable since the aim of the heat regulation mechanism is to make physiological functions independent of the temperature of the medium.

10.2 Thermoregulatory Activation of Free Respiration in Animals

10.2.1 Brown Fat

Mammals have a special tissue, *brown fat*, specializing in additional heat production in the cold. Although brown fat accounts for at most 1–2 % of the total body mass, sympathetic nerve stimulation of this tissue in cold-acclimated animals can increase its heat production so strongly that it becomes an important source of thermogenesis, being responsible for up to one-third of the overall metabolic rate (Foster 1984). Under these conditions, brown fat was found to be able to produce as much as 400 W heat per kg, i.e., several orders of magnitude more than the thermogenic capacity of mammalian tissues in general (about 1 W heat per kg body mass in a resting adult person).

Brown fat is localized in the upper part of the back, where it envelops blood vessels that transport blood to the brain. Thus, the heat produced by brown fat may be of great significance for the organism to survive under extreme cooling. This tissue is mostly developed in newborns, and its share in body mass decreases rapidly with age.

Although the scientific literature on brown fat dates back to the seventeenth century, it is comparatively recently that its function was suggested to be linked to thermogenesis. In 1959, Johansson wrote the first comprehensive review on brown fat metabolism (Johansson 1959). On the basis of scarce data and preliminary evidence, he postulated that this tissue might somehow take part in thermoregulation. Thanks to studies of Robert Smith's and several other groups, it became clear that the principal physiological function of brown fat was to produce extra heat under conditions of significant increase in body heat loss (Smith and Horwitz 1969).

The brown color of this tissue comes from a very high content of mitochondria, which are well equipped for heat production. In particular, they have a significant excess of respiratory chain enzymes over H^+-ATP-synthase. Yet, most importantly the inner membrane of the brown fat mitochondria contains a special *uncoupling protein called UCP* (from *uncoupling protein*), which is also called *thermogenin*.

Thermogenin is expressed only in brown fat and it can amount to up to 15 % of the total brown fat mitochondrial protein (Nicholls and Locke 1984). UCP can cause uncoupling of oxidative phosphorylation, dissipating the proton potential energy in the form of heat (thermogenesis according to the mechanism: respiration $\to \Delta\bar{\mu}_{H^+} \to$ heat). UCP is a 32.2-kDa protein. It consists of three repeating

10.2 Thermoregulatory Activation of Free Respiration in Animals

sequences; each of these sequences contains two transmembrane α-helices connected to each other through a long hydrophilic loop. UCP is considered to function as a homodimer. The amino acid sequence of thermogenin is rather similar to that of the ATP/ADP-antiporter and several other mitochondrial anion carriers (e.g. glutamate/aspartate-antiporter and dicarboxylate carrier) (Klingenberg 1990).

The functioning of UCP is regulated at several levels. For instance, cold adaptation in animals is accompanied by an increase in the transcription of the thermogenin gene (a significant increase in thermogenin mRNA level in brown fat cells is observed already 15 min after the initial cooling). Various purine nucleotides (GDP, ADP, GTP, and ATP) were found to inhibit proton permeability of the brown fat mitochondria, while free fatty acids in micromolar concentration were shown to be necessary for increase in this permeability (Drahota et al. 1970).

Several studies on whole animals, brown adipose tissue in situ, brown adipocytes, and isolated mitochondria in vitro revealed the following chain of events involved in the thermogenic response of this tissue to cold.

1. Cold receptors in the skin are activated and send signals to the thermoregulatory center in the hypothalamus.
2. From the hypothalamus, signals are transmitted to brown fat via sympathetic neurons. Noradrenaline is released from the neuronal terminals into the intercellular spaces of brown adipose tissue.
3. Noradrenaline binds to β-adrenergic receptors that are located on the outer surface of the brown fat cell plasma membrane (this and all further effects are illustrated in Fig. 10.1).
4. The β-adrenergic receptors, combining with noradrenaline, activate adenylate cyclase. (Cyclase can also be activated by glucagon, while insulin inhibits the enzyme; in both cases, specific hormone receptors are involved).
5. Adenylate cyclase produces cAMP from ATP.
6. The cAMP switches on a protein kinase cascade that activates triglyceride lipase.
7. Lipase releases fatty acids and glycerol when triglycerides in intracellular neutral fat droplets are hydrolyzed.
8. Thus, formed fatty acids perform two functions: they serve as the respiratory substrates and the secondary cytosolic messengers of a hormonal signal. As substrates, fatty acids undergo the usual catabolic reactions. First they are activated to form fatty acyl-CoA in the outer mitochondrial membrane, then fatty acyls are transported into the mitochondrial matrix via the carnitine system, and finally they yield CO_2 and H atoms catalyzed by the β-oxidation enzymes.
9. Reducing equivalents (H atoms) supplied to the respiratory chain by β-oxidation of fatty acids, oxidative decarboxylation of pyruvate, and oxidation of acetyl groups in the Krebs cycle are transferred to O_2, the process being coupled to generation of $\Delta\bar{\mu}_{H^+}$. As a result, H^+ ions are exported from the matrix to the mitochondrial intermembrane space.

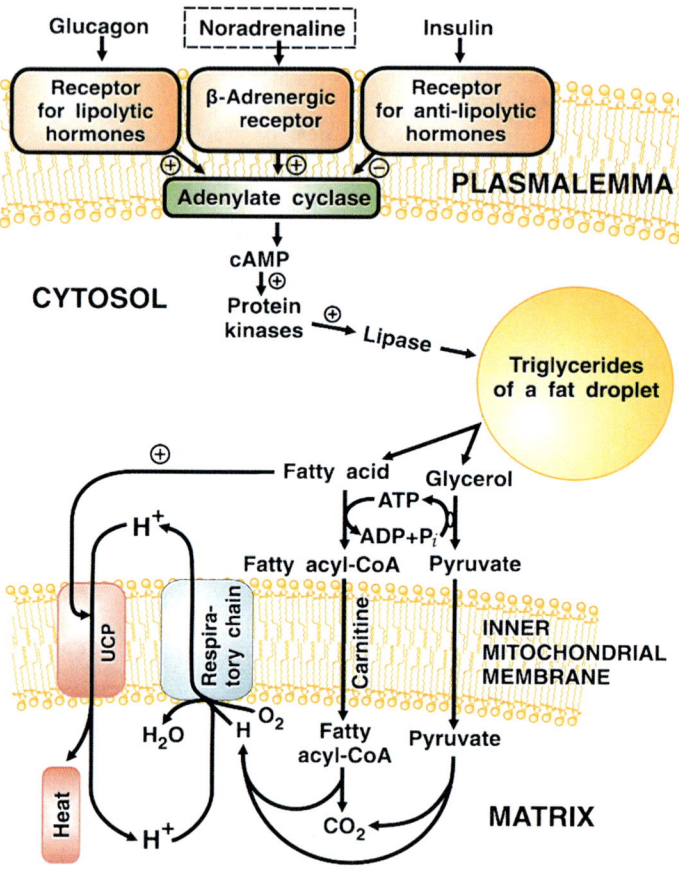

Fig. 10.1 Cold-induced thermogenesis in brown fat mitochondria. Symbols "+" and "−" stand for stimulation and inhibition, respectively. For explanation, see the text

10. The H$^+$ ions return to the matrix moving downhill. This is mediated by the fatty acid–thermogenin system. This final event results in heat production.

Two alternative hypothetical mechanisms of the uncoupling effect of UCP have been suggested.

According to one, binding of fatty acids to this protein opens a channel in it that allows transmembrane H$^+$ transfer (Klingenberg 2001).

An alternative mechanism was suggested by one of this book's authors (V.P.S.). According to this hypothesis, thermogenin is a carrier of the anion form of free fatty acids (Skulachev 1991, 1998, 1999). Facilitation of transmembrane transfer of these anions (RCOO$^-$) can play a significant role in the uncoupling effect of fatty acids, for their anion forms are stopped at the membrane/water interface while facing water with their carboxyl residue and facing lipid with the fatty "tail". That is why membranes, while being permeable for neutral (protonated)

10.2 Thermoregulatory Activation of Free Respiration in Animals

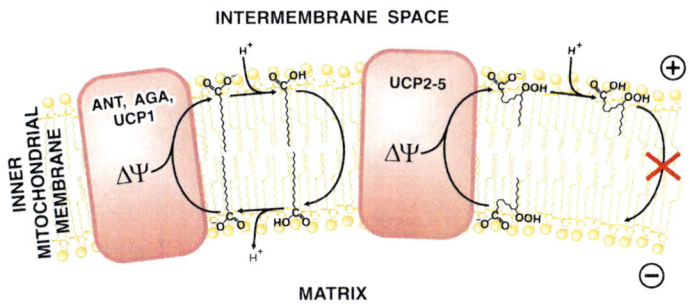

Fig. 10.2 Suggested translocation mechanism for anions of fatty acids (*left*) and their peroxides (*right*) by uncoupling proteins (UCP), ATP/ADP-antiporter (ANT), or aspartate/glutamate-antiporter (AGA). It is assumed (*left-hand* scheme) that UCP1, ANT, and AGA are competent in electrophoretic efflux of nonoxidized fatty acid anions from the inner to outer leaflet of the inner mitochondrial membrane. The exported fatty acid anions are protonated by H^+ ions from the mitochondrial intermembrane space and then return to the inner leaflet in their protonated form. As a result of such a flip-flop, the intermembrane H^+ ions are transported to the mitochondrial matrix (the fatty acid protonophorous cycle). The *right-hand* scheme illustrates another function characteristic of UCP2, 3, 4, and 5, namely purification of the inner leaflet from peroxidized fatty acids. This function is assumed to be characteristic of these minor UCPs. Protonated fatty acid peroxides cannot return to the inner leaflet since the hydrophilic peroxy group prevents their flip-flop. Electrophoretic efflux of fatty acid peroxide may preserve the matrix-located mitochondrial DNA from being oxidized by very aggressive intermediates of peroxidation of unsaturated fatty acids (from Goglia and Skulachev 2003)

fatty acids (RCOOH), have very low conductivity for fatty acid anions. Thus, fatty acids *per se* are not protonophores. However, in the presence of the $RCOO^-$-transferring UCP, the efficiency of fatty acids as uncouplers should dramatically increase (Fig. 10.2).

The transfer of $RCOO^-$ by UCP and ATP/ADP-antiporter, as well as by aspartate/glutamate-antiporter could proceed due to the interaction of this anion with positively charged Arg and Lys residues in the lower part of the funnel formed by these proteins (see previous chapter, Fig. 9.10). The $RCOO^-$-anion could reversibly bind to Arg 234, 235, or 236 of the bottom of the funnel; this would suffice for these anions to cross the hydrophobic membrane barrier.

The other very important characteristic of fatty acid-mediated uncoupling is that no special mechanism is needed for the termination of the uncoupling effect in this case. Uncoupling ceases when the rate of fatty acid influx in mitochondria becomes lower that the rate of their oxidation. This phenomenon occurs when an increase in the ambient temperature switches off the cold receptors in the skin; then the regulatory cascade, which forms fatty acids from triacylglycerides, ceases to function. When no fatty acids are transported into the mitochondria, their level is quickly lowered by the β-oxidation system. As a result, the fatty acid-mediated uncoupling disappears.

Significantly, not only cooling, but at least three other states of the organism, also related to increased oxygen consumption and heat production, are accompanied by uncoupling in brown fat tissue. One such state is the birth of a newborn. Immediately after birth, the young organism suddenly changes the comfortable isothermal conditions in the mother's body to the hypothermal outer medium. The small size of a newborn (and thus the high surface/volume ratio) additionally complicates the problem, for it substantially increases the heat loss. Brown fat was shown to play the key role in the thermal adaptation of mammal cubs after their birth. These data explain the fact of this tissue being particularly well developed in newborns. It also provides some understanding of why the share of brown fat tissue in body mass decreases during postnatal period parallel to the increase in body size and decrease in surface/volume ratio.

Another example of brown fat functioning is the *arousal of an animal after hibernation*. It was found that brown fat does not undergo postnatal involution in hibernators adapted to thermoneutral conditions. Cold acclimation increases the thermogenin concentration in brown fat mitochondria of hibernators, but less dramatically than in non-hibernators such as rats or rabbits. The uncoupling that was found to accompany the awakening of a hamster from hibernation appears to be especially important, because the brown fat warms the blood going to the brain so that the temperature rises first in the brain (Smith and Horwitz 1969).

Another model is *thermogenesis induced by excessive food consumption*. Nancy Rothwell and Michael Stock found that adult rats eating a "cafeteria" diet of varied and palatable food increase their usual dietary intake by 80 % (Rothwell and Stock 1981). But the gain in mass over a 3-week period was only 27 % greater than that for the control group. They consumed ∼25 % more oxygen than the control animals. This increase was abolished by a noradrenaline antagonist, propranolol. The mass of the brown fat after 3 weeks of "cafetering" was three times higher than that in the control rats. The UCP concentration in the brown fat mitochondria also increased. On the other hand, a mutation resulting in obesity of mice was shown to be accompanied by a decrease in the UCP content in brown fat mitochondria.

The role of UCP in thermal adaptation has been directly confirmed by molecular genetic studies. Cannon and Nedergaard obtained a mouse strain with knockout of the gene for UCP (Nedergaard et al. 1999). These mice were shown to be much more sensitive to temperature decrease than the wild type.

10.2.2 Skeletal Muscles

We have already discussed that brown fat is a specialized mammalian tissue responsible for thermal adaptation at low temperature. However, the contribution of brown fat tissue to the total body mass is rather small (especially in adult animals with high body mass), and its activity can hardly explain the greater part of additional thermogenesis observed in animals exposed to cold. It is important to note that such warm-blooded animals as birds have no brown fat at all, which means that some other tissues, especially those substantially contributing to the

10.2 Thermoregulatory Activation of Free Respiration in Animals

body mass and containing many mitochondria (e.g. skeletal muscles), should also participate in heat production in response to cooling. On the other hand, the thermogenin genes are known to be expressed only in brown fat tissue. Thus, the heat production mechanism in other tissues of warm-blooded animals may differ from the mechanism of thermogenesis in brown fat.

One of the thermogenesis mechanisms of skeletal muscles is well known to all of us—it is *muscle shivering*. It is supported (as well as any other muscle contraction) by hydrolysis of cytoplasmic ATP. The latter is restored by ATP transport from mitochondria via the ATP/ADP-antiporter. ATP is produced in mitochondria by H^+-ATP-synthase, which utilizes $\Delta\bar{\mu}_{H^+}$ generated by the respiratory chain (i.e. in this case thermogenesis is described as *respiration* $\rightarrow \Delta\bar{\mu}_{H^+} \rightarrow ATP \rightarrow heat$).

However, this mechanism of thermal adaptation is also known to be used only in case of sudden cooling. It seems to be too slow and complicated for effective long-term acclimation to a decrease in ambient temperature. Moreover, it activates muscle contraction, i.e., a specific function of this tissue (oxygen consumption in the case of abrupt cooling is similar to that of heavy muscular work). Thus, the main principle of thermoregulation, i.e., to make the functions independent of temperature, remains unresolved. Therefore, it is not surprising that muscle shivering decreases in the course of cold adaptation. This fact points to the existence of some other mechanism of heat production that is activated in muscles in the case of the thermoregulatory long-term cooling.

It is interesting that the role of mitochondria in thermogenesis was first shown for animal skeletal muscles. These experiments studying the in vivo effect of a cold exposure upon muscle mitochondria were carried out in 1960 by one of this book's authors (V.P.S.) together with Sergei Maslov in the laboratory of Sergei Severin (Skulachev and Maslov 1960). The experiments were conducted on pigeons that were shorn beforehand to exclude physical thermoregulation. The cage with the bird was put into a ventilated refrigerator at $-20\ °C$. The pigeon survived for 15–20 min when exposed to these conditions for the first time. But birds cooled for the second time the next day were shown to acquire the ability to survive for hours.

The study of oxidative phosphorylation in mitochondria isolated from the breast muscle of pigeons cooled for the first time revealed a decrease in the P/O ratio by a factor of 1.6. The second cold exposure resulted in about a sixfold lowering of the P/O ratio, i.e., almost complete uncoupling of respiration and phosphorylation. Further studies of this phenomenon revealed that development of the uncoupling effect takes ~ 15 min of cold exposure. This was shown not only in birds, but also in mammals (mice) (Skulachev 1963).

It was also found that the concentration of high-energy phosphates increases during the first cold exposure and decreases during the repeated one. This fact showed that muscles were involved in this thermoregulatory response. Measurements of in vivo respiration showed that initially it was stimulated on both the first and repeated coolings, but then decreased in the former, but not in the latter case. It looked as if respiration stimulated by the first cold exposure still formed ATP, being not sufficiently uncoupled. On repeated cooling, a system of fast uncoupling was activated so that the respiration rate was not limited by formation and utilization of ATP. If so, one might hope that the injection of an artificial uncoupler into the animal on the first cooling would save it from sudden death in the cold. Experiments undertaken by Maslov showed that this was really the case. Injection of 2, 4-dinitrophenol, which is in fact a poison for animals, significantly prolonged the survival time for mice on their first cold exposure (Skulachev et al. 1963).

In other experiments, respiration of the intact rat diaphragm muscle was shown to be stimulated by repeated short-term cold exposure of the animal, whereas the degree of the respiratory stimulation by 2, 4-dinitrophenol was strongly decreased. An effect similar to that of cold could be induced by the injection of noradrenaline. Further studies revealed that the concentration of free fatty acids increased significantly, while that of triglycerides decreased, in the muscle of an animal repeatedly subjected to short cold exposure (Levachev et al. 1965). Some other data were obtained indicating that free fatty acids were responsible for cold-induced uncoupling in muscle mitochondria. Addition of delipidated serum albumin, which binds free fatty acids, increased the P/O ratio in mitochondria from cold-treated animals. A fraction of free fatty acids isolated from the muscle of cold-treated animals and added to mitochondria from the untreated control animals was found to induce uncoupling. Moreover, palmitic acid, when added at a concentration as low as 1×10^{-5} M, caused a significant stimulation of respiration in muscle mitochondria while in the state of respiratory control.

Thus, fatty acids were shown to serve as thermoregulatory uncoupling mediators in mitochondria of both brown fat and skeletal muscles tissues. However, it has already been mentioned that muscle mitochondria possess no thermogenin. If this is the case, what is the mechanism of the uncoupling induced by fatty acids?

Elena Mokhova in our group showed that in muscle mitochondria, uncoupling induced by low concentrations of fatty acids is specifically abolished by 5×10^{-7} M carboxyatractylate, a specific inhibitor of the ATP/ADP-antiporter (Andreyev et al. 1988). ADP added with or without oligomycin also decreased the palmitate-induced stimulation of respiration. On the other hand, fatty acids were found to inhibit ATP/ADP-antiport. Thus, it was concluded that uncoupling of muscle mitochondria by fatty acids in their physiological concentration range is mediated by the ATP/ADP-antiporter. We need to stress once again that UCP is in many ways similar to the ATP/ADP-antiporter (in terms of its amino acid sequence and domain structure). The transfer of H^+ by UCP is also activated by fatty acids and inhibited by purine nucleotides (see above). Thus, fatty acids probably mediate thermoregulatory uncoupling in mitochondria of both muscle and brown fat tissues, though it is the ATP/ADP-antiporter that is the fatty acid target in the first case, while in the second one it is thermogenin, a special protein which might be a modification of an antiporter[1] (see Fig. 10.2).

---

[1] According to this hypothesis, UCP is a monofunctional version of an ATP/ADP-antiporter which has lost its main (nucleotide-transporting) function and specializes in a supplementary function, i.e., fatty acid-mediated uncoupling provided through transport of their anions.

10.2 Thermoregulatory Activation of Free Respiration in Animals

The fact that fatty acids not only increase the permeability of the inner mitochondrial membrane for H^+, but also compete with adenine nucleotides for ATP/ADP-antiporter, should insure the cell from the uncoupling-caused depletion of extramitochondrial ATP when H^+-ATP-synthase starts working in the reverse (hydrolytic direction). ATP depletion might stop oxidation of fatty acids that serve not only as a regulator of coupling, but also as the main respiration substrate in the cold. It has been already mentioned that activation of fatty acids by means of their binding to CoA at the expense of energy of ATP occurs in the outer mitochondrial membrane.

Studies on mitochondrial uncoupling in tissues other than brown fat have recently produced quite unexpected results. The mammalian genome was shown to contain not only the thermogenin (UCP) gene, but also four homologous genes. The products of these genes were given the names of uncoupling proteins UCP2, UCP3, UCP4, and UCP5, while thermogenin is now called UCP1 (Adams 2000; Skulachev 1998, 1999). Moreover, uncoupling proteins were discovered in the majority of investigated animals (both warm-blooded and cold-blooded) as well as in many plants (Ledesma et al. 2002; Vercesi et al. 1995). Moreover, some UCPs probably function also in fungi and protozoa. Such a wide distribution of these proteins suggests that they play a fundamental role in the metabolism of all eukaryotes.

The expression of genes of new mammalian uncoupling proteins is tissue-specific. UCP2 is expressed in many tissues, UCP3 in brown fat and skeletal muscles, while UCP4 and UCP5 are specific for brain. The discovery of a whole group of UCPs in mammals might suggest that the main function of these proteins is heat production in tissues other than brown fat. Certain observations seemed to support this conclusion: exposure of an animal to cold caused the increased expression of genes UCP2 and UCP3 (quite surprisingly, similar cold induction is also typical for several plant UCPs). However, it was shown later that the induction of the uncoupling proteins could be observed not only as a result of cooling, but also under many other conditions, including, e.g., starvation, when a decrease in energy storage efficiency could have no positive sense. Thus, the increased expression of the uncoupling proteins genes probably correlates with the acceleration of fat metabolism rather than to lowering of temperature.

Moreover, mutant mice with knocked-out genes of different alternative uncoupling proteins (UCP2–UCP5) demonstrate no differences in cold adaptation when compared to the control mice. Thus, UCP2–UCP5 are considered to participate in the cold-stress-induced mitochondrial uncoupling only to a minimal degree. Fernando Goglia and one of this book's authors (V.P.S.) suggested the new UCPs to be the carriers of fatty acid peroxide anions and hence to be responsible for the transfer of these dangerous compounds from the inner to the outer half-membrane layer of the inner mitochondrial membrane (see above Fig. 10.2, Goglia and Skulachev 2003). This hypothesis might explain numerous data indicating a possible role of the new UCPs in the antioxidant protection of mitochondria, and also the fact that their amount is very small when compared to that of UCP1 (similar to the quantity of fatty acid peroxides, which is always much lower than that of intact fatty acids).

10.3 Thermoregulatory Activation of Free Respiration in Plants

A widespread dogma is that thermoregulation is a function exclusively in higher animals. However, nature is not as dogmatic as its students and may bring them surprises. The only problem is that we tend to ignore such presents.

In 1778, Jean-Baptiste Lamarck noticed that the spadix of an *arum* lily is significantly warmer than other parts of the plant. It was later shown that the phenomenon of substantial warming of an inflorescence is characteristic for many species of different families of plants, such as *Araceae, Nelumbonaceae, Nymphaeaceae, Aristolochiaceae, Arecaceae, Cyclanthaceae, Annonaceae, Magnoliales,* and *Cycadaceae* (Seymour 2001). Hyperproduction of heat lasts for 12–48 h, being necessary to volatilize odoriferous compounds—usually amines or indoles—that attract the pollinators. The thermoregulatory mechanism makes volatilization independent of the ambient temperature. The temperature difference between the flower and air was found to reach up to 35 °C, and the specific heat production was up to 1000 W per kg of plant tissue (Fig. 10.3).

These observations stimulated intense studies of plant thermogenesis. In 1851, Gareau showed a close relationship between heat production and oxygen consumption in the spadices of aroids. It was later shown that a decrease in the ambient temperature causes a significant activation of respiration, a property that is usually believed to be inherent in warm-blooded animals only. For instance, in eastern skunk cabbage, oxygen consumption by the spadix was about 3.5-fold higher at + 5 °C than at + 15 °C. Flowering of thermogenic plants is accompanied by such a powerful increase in respiration that it was given the name of "metabolic burst". The maximal rate of oxygen consumption can be as high as 40 ml O_2 g^{-1} h^{-1}, which is exceptional among plants and rivals the highest respiratory rates in the tissues of higher animals. How does this transformation of energy of respiratory substrate oxidation to heat take place in thermogenic plants?

As already discussed in Sect. 5.1, the respiratory chain of plant mitochondria (in contrast to animal mitochondria) contains enzymes of alternative oxidation pathways that is not coupled to energy conservation. These are two noncoupled NADH:quinone-oxidoreductases (ND_{in} and ND_{ex} oxidizing mitochondrial and cytoplasmic NADH, respectively) and noncoupled cyanide-resistant quinol oxidase (AOX). This difference makes it possible for plants to use the pathway of direct transformation of energy of respiratory substrates oxidation to heat, while omitting the intermediate stages of proton potential generation or ATP formation.

The mitochondria of thermogenic plants have normal respiration before flowering. This respiration is sensitive to cyanide, myxothiazol, and rotenone; it was also shown to be coupled to phosphorylation (the P/O ratio for malate oxidation is about 2.7). But *the day florescence starts, respiration becomes noncoupled and loses sensitivity to all the above-mentioned inhibitors.*

Mitochondria of thermogenic plant inflorescence were shown to have very high amounts of ND_{in}, ND_{ex} (Bertsova et al. 2004), and AOX (James and Beevers

10.3 Thermoregulatory Activation of Free Respiration in Plants

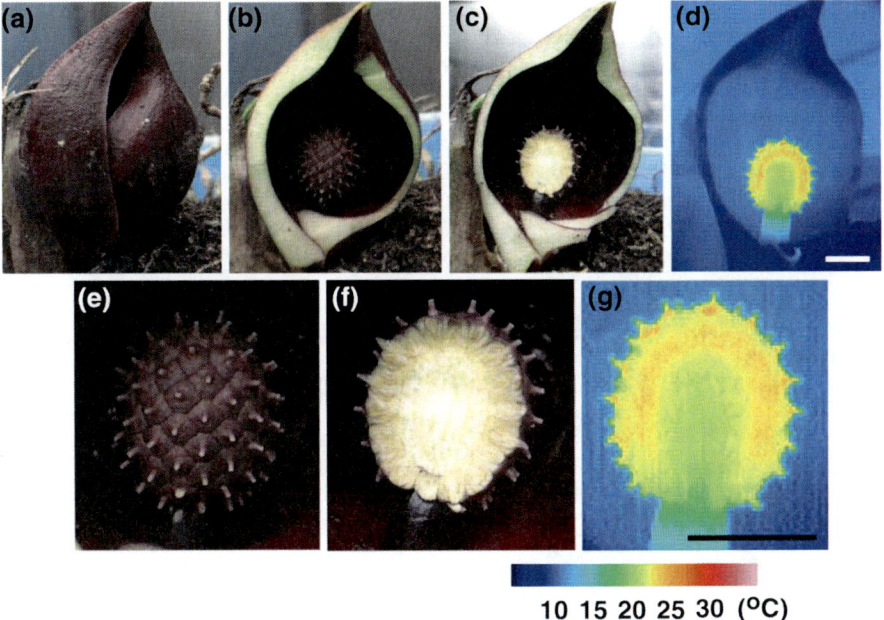

Fig. 10.3 (**a–c**)Florescence stages of the thermogenic plant *Symplocarpus renifolius*, **d** infrared picture of the **c** stage, **e, f, g** magnified versions of pictures **b, c** and **d**. Temperature scale for the pictures **d** and **g** is given in the lower right corner; bar 1 cm (Onda et al. 2008)

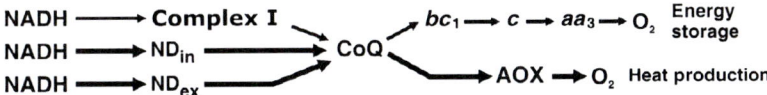

Fig. 10.4 Scheme of the mitochondrial respiratory chain of thermogenic plants. Electron transfer pathways prevalent in thermogenesis are marked with bold arrows

1950). At the same time, coupling mitochondrial enzymes (complexes I, III, and IV) were shown to provide only 5 % of the overall respiration in these inflorescences. Thus, ND_{in} and ND_{ex} shunt the first coupling site, and AOX shunts the second and the third coupling sites in these mitochondria (Fig. 10.4). This is the mechanism of direct transformation of respiratory energy to heat. Thus, heat production mechanisms are fundamentally different in plants and warm-blooded animals.

References

Adams SH (2000) Uncoupling protein homologs: emerging views of physiological function. J Nutr 130:711–714
Andreyev AYu, Bondareva TO, Dedukhova VI, Mokhova EN, Skulachev VP, Volkov NI (1988) Carboxyatractylate inhibits the uncoupling effect of free fatty acids. FEBS Lett 226:265–269

Bertsova YV, Popov VN, Bogachev AV (2004) NADH oxidation by mitochondria from the thermogenic plant *Arum orientale*. Biochemistry (Moscow) 69:580–584

Drahota Z, Alexandre A, Rossi CR, Siliprandi N (1970) Organization and regulation of fatty acid oxidation in mitochondria of brown adipose tissue. Biochim Biophys Acta 205:491–498

Foster DO (1984) Quantitative contribution of brown adipose tissue thermogenesis to overall metabolism. Can J Biochem Cell Biol 62:618–622

Goglia F, Skulachev VP (2003) A function for novel uncoupling proteins: antioxidant defense of mitochondrial matrix by translocating fatty acid peroxides from the inner to the outer membrane leaflet. FASEB J 17:1585–1591

James WO, Beevers H (1950) The respiration of *Arum* spadix. A rapid respiration, resistant to cyanide. New Phytol 49:353–374

Johansson B (1959) Brown fat: a review. Metabolism 8:221–240

Klingenberg M (1990) Mechanism and evolution of the uncoupling protein of brown adipose tissue. Trends Biochem Sci 15:108–112

Klingenberg M (2001) Uncoupling proteins—how do they work and how are they regulated. IUBMB Life 52:175–179

Ledesma A, de Lacoba MG, Rial E (2002) The mitochondrial uncoupling proteins. Genome Biol 3: REVIEWS3015

Levachev MM, Mishukova EA, Sivkova VG, Skulachev VP (1965) Energy metabolism in the pigeon during self-warming after hypothermia. Biokhimiia 30:864–874 (in Russian)

Nedergaard J, Matthias A, Golozoubova V, Jacobsson A, Cannon B (1999) UCP1: the original uncoupling protein—and perhaps the only one? New perspectives on UCP1, UCP2, and UCP3 in the light of the bioenergetics of the UCP1-ablated mice. J Bioenerg Biomembr 31:475–491

Nicholls DG, Locke RM (1984) Thermogenic mechanisms in brown fat. Physiol Rev 64:1–64

Onda Y, Kato Y, Abe Y, Ito T, Morohashi M, Ito Y, Ichikawa M, Matsukawa K, Kakizaki Y, Koiwa H, Ito K (2008) Functional coexpression of the mitochondrial alternative oxidase and uncoupling protein underlies thermoregulation in the thermogenic florets of skunk cabbage. Plant Physiol 146:636–645

Rothwell NJ, Stock MJ (1981) Influence of noradrenaline on blood flow to brown adipose tissue in rats exhibiting diet-induced thermogenesis. Pflugers Arch 389:237–242

Seymour RS (2001) Biophysics and physiology of temperature regulation in thermogenic flowers. Biosci Rep 21:223–236

Skulachev VP (1963) Regulation of coupling of oxidation and phosphorylation. Proc Int Biochem Congr 5:365–374

Skulachev VP (1991) Fatty acid circuit as physiological mechanism of uncoupling of oxidative phosphorylation. FEBS Lett 294:158–162

Skulachev VP (1998) Uncoupling: new approaches to an old problem of bioenergetics. Biochim Biophis Acta 1363:100–124

Skulachev VP (1999) Anion carriers in fatty acid-mediated physiological uncoupling. J Bioenerg Biomembr 31:431–445

Skulachev VP, Maslov SP (1960) Role of non-phosphorylating oxidation in thermoregulation. Biokhimiya 25:1058–1064 (in Russian)

Skulachev VP, Maslov SP, Sivkova VG, Kalinichenko LP, Maslova GM (1963) Cold uncoupling of oxidative phosphorylation in the muscles of white mice. Biokhimiia 28:70–79 (in Russian)

Smith RE, Horwitz BA (1969) Brown fat and thermogenesis. Physiol Rev 49:330–425

Vercesi AE, Martins IS, Silva MAP, Leite HMF, Cuccovia IM, Chaimovich H (1995) PUMPing plants. Nature 375:24

Part IV
Interaction and Regulation of Proton Potential Generators and Consumers

Chapter 11
Regulation, Transmission, and Buffering of Proton Potential

11.1 Regulation of $\Delta\bar{\mu}_{H^+}$

The $\Delta\bar{\mu}_{H^+}$-generating and $\Delta\bar{\mu}_{H^+}$-consuming systems can be regulated by an entire set of mechanisms usually used by organism to control enzymes. A number of examples of the regulatory effects have been already described in the previous chapters. In this chapter, we shall discuss specific systems of $\Delta\bar{\mu}_{H^+}$ regulation. In the preceding chapter, we considered those related to the free oxidation in thermoregulation. Below we will show that non- and uncoupled redox processes, often in combination with energy-coupled ones, are of much greater significance than heat production.

11.1.1 Alternative Functions of Respiration

When in 1932 Kazuo Okunuki described cyanide- and CO-resistant respiration in pollen of *Lilium auratum* (Okunuki 1932), it happened almost simultaneously with the discovery by Vladimir Engelhardt of phosphorylating respiration in erythrocytes of birds (Engelhardt 1930). Therefore, it is not surprising that Okunuki did not specify whether the novel respiratory pathway discovered by him is coupled to ATP synthesis or not.

The great success of bioenergetic studies initiated by Engelhardt and the obvious importance of respiration-linked phosphorylation gave birth, as is often the case, to a dogma that ATP synthesis is the *only* useful result of the oxygen-consuming process in the cell. If somebody described respiration not coupled to phosphorylation, this process was usually considered as an in vitro artifact or an in vivo pathology.

In 1958, an alternative point of view was suggested—that cellular respiration in vivo normally occurs not only in an energy-conserving manner, but also without such energy coupling (this second type of respiration was given the name *"free respiration"*). It was suggested that *free respiration takes part in thermoregulatory heat production as well as in the formation or decomposition of metabolites, detoxication of xenobiotics, and even (indirectly) in energy storage* (Skulachev 1958, 1963; Skulachev and Maslov 1960).

To test this hypothesis, we studied cold-induced thermogenesis in warm-blooded animals and found that free oxidation was really activated in skeletal muscles. Later, similar effects were observed by other groups in brown fat and in the spadices of some plants (see Chap. 10).

As to nonenergetic functions of free respiration, the most demonstrative is their role in the decomposition of metabolic end products and harmful xenobiotics.

In 1962, one of this book's authors (V.P.S.) and his colleagues discovered that oxidation of lactate by O_2 in muscle has a much lower P/O ratio than that of other NAD^+-linked substrates (Skulachev 1962). This oxidation was rotenone and antimycin-insensitive but cyanide sensitive (Agureev et al. 1981). Later, it was shown that lactate oxidation in yeast mitochondria also bypasses the initial and middle segments of the coupled respiratory chain. Here, this process is catalyzed by cytochrome b_2 in the intermembrane space. Cytochrome b_2 participated in the transport of electrons from lactate to cytochrome oxidase with no NAD^+ involved (Appleby and Morton 1959). According to Brodie et al. lactate is oxidized in *Mycobacterium phlei* much faster than other substrates, but without ATP formation (Brodie 1959).

Impressive progress has been achieved in studies on free respiration responsible for the *detoxication of xenobiotics*. In membranes of the endoplasmic reticulum, special respiratory chains were identified, composed of *cytochrome P-450* as the terminal oxidase and *NADPH:cytochrome P-450-oxidoreductase* or *NADH:cytochrome b_5-oxidoreductase* and *cytochrome b_5*. These chains are involved in the oxidation of a large number of xenobiotics. The enzymes composing these chains are present in large amounts. For example, the concentration of cytochrome *P-450* can, under certain conditions, be higher than that of any other cytochrome in the liver cell (Skulachev 1988).

Cytochrome *P-450*-mediated free respiration was found to be involved not only in destructive, but also in some constructive processes that usually occur in the endoplasmic reticulum. However, the most demonstrative example of this function of free oxidation was revealed in adrenal cortex mitochondria. It was found that the initial and final steps of formation of steroid hormones (cortisol and aldosterone) are catalyzed by two species of cytochrome *P-450* located in the inner membrane of these mitochondria. Here, there is also a special *P-450*-reducing redox chain facing the matrix space. It oxidizes $NADPH_{in}$ via NADPH:adrenodoxin-oxidoreductase, a FAD-containing flavoprotein. The latter reduces adrenodoxin, a non-heme FeS protein, which serves as an electron donor for mitochondrial cytochrome *P-450*. None of these redox reactions are coupled to generation of $\Delta\bar{\mu}_{H^+}$, which is formed by the usual phosphorylating respiratory chain localized in the same membrane (Takemori and Kominami 1984).

An interesting example of strong activation of free respiration was described in leukocytes attacking bacterial cells. In this case, special intracellular vesicles fuse

11.1 Regulation of $\Delta\bar{\mu}_{H^+}$

with the outer cell membrane of a leukocyte. This results in a burst of free respiration mediated by NADPH, flavoproteins, and the autooxidizable, cyanide-resistant b-type cytochrome. The latter mediates one-electron reduction of O_2 to superoxide anion $O_2^{\bullet-}$, which then reacts with a chloride ion. As a consequence, a chloride radical is formed, this process being catalyzed by myeloperoxidase. The chloride radical is a highly toxic compound that kills bacteria (Gabig and Babior 1981). It seems quite important that NADPH is oxidized by flavin on the inner surface of the leukocyte plasmalemma, while oxygen is reduced on its outer surface. It is due to such topography that the poisonous product is formed *outside* the leukocyte, thus minimizing the risk of the damage to the leukocyte.

The reason why the above-mentioned respiratory systems are not coupled to $\Delta\bar{\mu}_{H^+}$ generation seems to be clear enough. Energy coupling must inevitably complicate the constructive or destructive functions of these systems that are of vital importance for the organism.

Thus, we have considered the four *main functions of cellular oxidations*: (1) conservation of usable energy ($\Delta\bar{\mu}_{H^+}$ generation); (2) energy dissipation to form heat; (3) formation of useful substances; and (4) decomposition of harmful substances. One may think that the first of these functions is performed by coupled respiration, whereas the three others involve free (non- or uncoupled) respiratory systems. However, this appears to be an oversimplification.

Sometimes coupled respiration participates in functions (2)–(4). Thus, shivering thermogenesis, the first response of a warm-adapted animal to cold, requires the formation of $\Delta\bar{\mu}_{H^+}$ and then ATP to supply actomyosin ATPase with its substrate. It is also quite clear that the majority of constructive (anabolic) mechanisms utilize energy of ATP produced by coupled respiration. As to the destructive function, here ATP is usually not involved, but some important exceptions are well known. For example, removal of NH_3 via the urea cycle consumes ATP.

On the face of it, free respiration, according to the definition, cannot take part in the energy-conserving function. Paradoxically, this is not always the case.

The initial step of the coupled respiratory chain is known to be not only the slowest but also the most vulnerable segment of the respiratory chain. A wide range of hydrophobic xenobiotics arrests the coupled NADH:CoQ-oxidoreductase. Bypassing the inhibited step by free redox reactions can release respiration, which now again forms ATP but, of course, at a lower thermodynamic efficiency. This may be of significance for mitochondria of liver cells specializing in the accumulation and oxidation of hydrophobic xenobiotics by cytochrome *P-450*.

The shunting of inhibited NADH: CoQ-oxidoreductase can be achieved by a release of cytochrome *c* into the intermembrane space or by addition of vitamin K_3 (menadione) actuating the dicumarol-sensitive pathway via DT-diaphorase, a water-soluble enzyme oxidizing NAD(P)H by quinones. However, it is not clear whether such systems demonstrated in vitro are really activated under the conditions of intoxication by hydrophobic xenobiotics. On the other hand, an artificially induced shunting by injected menadione had a pronounced medicinal effect on a patient with a mutation in a respiratory chain enzyme (for details, see Sect. 4.2.4).

11.1.2 Regulation of Flows of Reducing Equivalents Between Cytosol and Mitochondria

If two oxidation pathways, one energy-coupled and the other free of coupling, coexist in the same cell, the problem is how to prevent utilization of all the reducing equivalents by the thermodynamically more favorable free respiration systems. Without any doubt, compartmentalization of metabolic processes plays a very important role here. For instance, the main substrate dehydrogenases are localized in the mitochondrial matrix so that reducing equivalents for the respiratory chain are formed inside mitochondria and, therefore, are not available for the outer, free oxidation systems. Moreover, there are several $\Delta\bar{\mu}_{H^+}$-driven substrate porters in the inner mitochondrial membrane that accumulate those substrates in the matrix for which dehydrogenases are present both in mitochondria and in cytosol. If the entire dehydrogenase pool is localized in cytosol, special shuttle mechanisms to transport cytosolic reducing equivalents into mitochondria can be employed.

In Fig. 11.1, a *malate–aspartate–glutamate shuttle* is shown. The operation of this system results in the oxidation of extramitochondrial NADH and in the reduction of intramitochondrial NAD$^+$. Two enzymes located on the two sides of the inner mitochondrial membrane are involved in the process, i.e., malate dehydrogenase and aspartate:glutamate aminotransferase. Also, two porters should be involved, i.e., the dicarboxylate antiporter and the glutamate/aspartate antiporter. The latter is $\Delta\bar{\mu}_{H^+}$-driven, since it catalyzes the exchange of aspartate^{2-} for glutamate^{2-} and H$^+$. As a result, NADH$_{out} \rightarrow$ NAD$^+_{in}$ oxidoreduction appears to be coupled to the downhill movement of one H$^+$ across the mitochondrial membrane.

Another shuttle mechanism makes use of two glycerol phosphate dehydrogenases—a cytosolic NAD$^+$-linked enzyme and another enzyme on the outer surface of the inner mitochondrial membrane that feeds electrons to the respiratory chain at the level of CoQ (Nicholls 1982). Shuttle systems show tissue specificity. For instance, the malate shuttle is very active in liver but absent from heart, where mitochondria do not have dicarboxylate antiporter. The *glycerol phosphate shuttle* is strongly activated by thyroid hormones (Cederbaum et al. 1973).

Peroxisomes are another example of the role of compartmentalization in oxidative metabolism. These organelles represent vesicles surrounded by a membrane which resembles, to some degree, the outer mitochondrial membrane—it is impermeable to proteins, but highly permeable to low molecular mass metabolites. Peroxisomes consume oxygen due to the activity of urate oxidase, D-amino acid oxidase, and α-hydroxyacid oxidase. There is no competition with coupled respiration since the substrates of these oxidases are oxidized without participation of NAD(P)$^+$ and the respiratory chain. A toxic reaction product, H$_2$O$_2$, is immediately decomposed inside peroxisomes by catalase, a major protein located in these organelles.

11.1 Regulation of $\Delta\bar{\mu}_{H^+}$

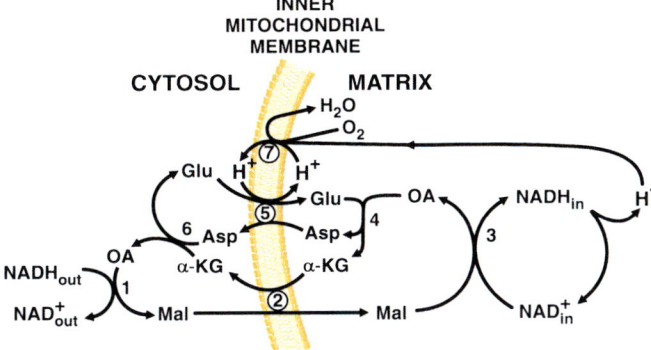

Fig. 11.1 Shuttle system to oxidize cytosolic NADH by the mitochondrial respiratory chain. Outer NADH reduces oxaloacetate (OA) by means of reversible cytosolic malate dehydrogenase (1). The formed malate (Mal) comes into the matrix in exchange for α-ketoglutarate (α-KG) via dicarboxylate-antiporter (2). In the matrix, malate reduces NAD⁺ by means of mitochondrial malate dehydrogenase (3). The resulting oxaloacetate transaminates with glutamate (Glu) to form aspartate (Asp) and α-ketoglutarate (4). Inner aspartate is exchanged for outer glutamate and H⁺ via aspartate/(glutamate + H⁺)-antiporter in a $\Delta\bar{\mu}_{H^+}$-dependent manner (5). Outer aspartate transaminates with outer α-ketoglutarate, regenerating outer oxaloacetate (6). The $\Delta\bar{\mu}_{H^+}$ needed to power reaction (5) is generated by the respiratory chain (7), which oxidizes NADH

11.1.3 Interconversion of $\Delta\Psi$ and ΔpH

Undoubtedly, *$\Delta\Psi \leftrightarrow \Delta pH$ interconversion can be of regulatory importance*. Such an effect can influence the interplay of $\Delta\Psi$- and ΔpH-driven transports, the amount of energy stored by the coupling membrane, and the pH value at least on one side of the coupling membrane.

To convert $\Delta\Psi$ into ΔpH, it is enough to electrophoretically transport any charged species other than H^+ across the membrane. In mitochondria there is only one large-scale process of this type, namely, $ATP^{4-}_{in}/ADP^{3-}_{out}$-antiport, supplying H^+-ATP-synthase with one of its substrates. However, when oxidative phosphorylation occurs, the antiport is always followed by the ΔpH-driven stoichiometric $H_2PO_4^-$, H^+-symport to supply the synthase with the other substrate necessary to form ATP from ADP. As a result, mitochondria usually maintain $\Delta\bar{\mu}_{H^+}$ mainly in the form of $\Delta\Psi$. Nevertheless, sometimes $\Delta\Psi$ may decrease and ΔpH increase. For instance, such an effect was observed by Risto Kauppinen, who measured $\Delta\Psi$ and ΔpH in mitochondria of perfused rat heart in situ (Kauppinen 1983).

Lan Bo Chen and colleagues stained mitochondria in a cell culture with methyl rhodamine (Johnson et al. 1981). The methyl rhodamine cation is accumulated in mitochondria electrophoretically, so it stains only those organelles that bear $\Delta\Psi$, and not ΔpH. Two neighboring cells of quite similar morphology were seen in a photograph. Only in one of them mitochondria accumulated rhodamine. Addition of nigericin, a K^+/H^+-antiporter, which converts ΔpH to $\Delta\Psi$, caused mitochondrial staining in the other cell. There are some indications that mitochondria have their own K$^+$/H$^+$-antiporter, which is usually in a latent state.

The uniport of Ca^{2+} into mitochondria may be one of the factors causing $\Delta\Psi \to \Delta pH$ conversion. However, this process is usually of limited importance because of a very low (10^{-7}–10^{-6} M) Ca^{2+} level in cytosol. As to the K$^+$ (cytosolic level ~ 0.1 M), its large-scale uniport must be absent from mitochondria, since K$^+$ distribution down $\Delta\Psi$ would result in the accumulation of a very high concentration of KOH in the matrix space.

The import of K$^+$ seems to be the main mechanism for the $\Delta\Psi \to \Delta pH$ transition in bacteria, which are living at a much lower outer [K$^+$] than mitochondria. In chloroplasts, $\Delta\bar{\mu}_{H^+}$ is mainly in the form of ΔpH due to the high permeability of the thylakoid membrane for Cl$^-$, K$^+$, and Mg^{2+}.

11.1.4 Relation of $\Delta\bar{\mu}_{H^+}$ Control to the Main Regulatory Systems of Eukaryotic Cells

In eukaryotic organisms, supracellular control systems, such as hormones, transmit their commands via intracellular mediators that are called *secondary messengers*. Most common among them are Ca^{2+}, cyclic nucleotides, and products of phosphatidylinositol diphosphate breakdown, i.e., inositol triphosphate and diacylglycerol. Recently, even more attention is also attracted to such a pleiotropic regulator as H$^+$.

Sometimes metabolites play the role of secondary messengers that regulate their own metabolism. Previously (see Sect. 10.2.1) we considered such a situation when discussing the mechanism of fatty acid-mediated thermoregulatory uncoupling. Fatty acids are not only the substrates of oxidation, but also uncouplers of this oxidation and inhibitors of the ATP/ADP-antiporter. Later, another regulatory effect of fatty acids was found. This is the *release of hexokinase* bound to the outer mitochondrial membrane (Klug et al. 1984). This can facilitate competition of fatty acids with glucose for ATP in formation of acyl-CoA and glucose-6-phosphate, respectively. Both these processes can occur on the surface of the outer mitochondrial membrane. Fatty acids were found to increase the negative surface charge of the membrane, resulting in the decomposition of a complex of hexokinase with porin, a function of which is to be a membrane receptor for hexokinase. Activation of certain enzymes of the Krebs cycle by Ca^{2+} is also described.

As to the "classical" secondary messengers, there are numerous publications on the effects of *Ca^{2+} ions* on various mitochondrial functions. The first is temporary $\Delta\Psi \to \Delta pH$ transition accompanying Ca^{2+} import. Moreover, Ca^{2+} activates mitochondrial phospholipase A$_2$. The latter produces free fatty acids, which cause uncoupling and other effects described above. Another product, lysophospholipid, can also influence some parameters of mitochondria.

11.1 Regulation of $\Delta\bar{\mu}_{H^+}$

Also, cAMP can apparently affect some mitochondrial functions not only via fatty acid release, but more directly as well. There is a cAMP-binding protein on the outer surface of the inner membrane of yeast mitochondria. It is a 45 kDa polypeptide synthesized in cytoplasmic ribosomes. Its affinity to cAMP is very high ($K_d = 10^{-9}$ M) (Rödel et al. 1985). One can speculate that it is this protein that plays the role of a specific receptor responsible for some in vivo effects of cAMP, namely, an increase in the respiration rate and in the synthesis of subunits *I–III* of cytochrome oxidase in mitochondria.

There is no doubt that H^+ can also be used as a secondary messenger. For instance, it is quite obvious that pH regulation is very important for the functioning of H^+-ATPase, the $\Delta\bar{\mu}_{H^+}$-generating enzyme of the plant and fungal outer cell membrane. In fact, H^+-ATPase pumps H^+ ions from the cytoplasm when cytoplasmic $[H^+]$ increases (see Sect. 7.2.3.1). The effect of auxins, plant hormones, is apparently associated with acidification of the cytoplasm resulting in the stimulation of H^+–ATPase and the increase in $\Delta\Psi$ across the outer cell membrane (Brummer and Parish 1983).

11.1.5 Control of $\Delta\bar{\mu}_{H^+}$ in Bacteria

For many bacteria, parallel pathways of respiratory chain electron transfer are typical, some of them being coupled to $\Delta\bar{\mu}_{H^+}$ formation, the others representing free oxidation systems. Moreover, free and coupled redox reactions are often consecutively included into the respiratory chain.

An interesting example of a system maintaining high $\Delta\bar{\mu}_{H^+}$ according to the self-regulation principle was revealed in motile bacteria. It was shown that artificially induced $\Delta\bar{\mu}_{H^+}$ changes can cause behavioral responses of bacterial cells (Miller and Koshland 1977). Thus, the addition of uncouplers or exhaustion of oxygen, resulting in $\Delta\bar{\mu}_{H^+}$ decrease, is sensed by bacteria as a repellent, changing the direction of their movement. On the other hand, an increase in $[O_2]$ seems to produce an attractant signal favorable for linear movement. It was noted that the effect of O_2 upon bacterial behavior (aerotaxis) can be shown only if a change in $[O_2]$ shifts the $\Delta\bar{\mu}_{H^+}$ level.

It was suggested that *bacteria have a device competent in measuring $\Delta\bar{\mu}_{H^+}$ across their membrane. By means of this system, $\Delta\bar{\mu}_{H^+}$ controls the direction of flagellar rotation*: the direction changes when $\Delta\bar{\mu}_{H^+}$ decreases and remains unchanged when it increases. As a result, the bacterium moves to a region where it can maintain higher $\Delta\bar{\mu}_{H^+}$ (Glagolev 1980). This type of hypothetical mechanism, called a *"protometer"* by one of this book's authors (V.P.S.), allows many favorable and unfavorable effects upon membrane energetics to be integrated. Functioning of a protometer has been recently directly demonstrated for the taxis system of *H. salinarium* (Bibikov et al. 1993) (see Sect. 6.5.3).

A mechanism has been described that coordinates the work of two photosystems in chloroplasts and hence optimizes $\Delta\bar{\mu}_{H^+}$ and NADPH production.

If photosystem II works too fast, a redox carrier (presumably *PQ*) included between photosystems I and II becomes completely reduced. This activates in some manner a protein kinase that phosphorylates the protein bearing light-harvesting antenna chlorophyll localized mainly in stacked thylakoids of grana. Phosphorylation adds negative charges to the antenna proteins which, due to electric repulsion, diffuse along the membrane to adjacent stroma membranes where photosystem I is mostly localized. As a result, photosystem II receives fewer, while photosystem I receives more photons from the antenna. Activation of photosystem I oxidizes PQH_2 and, hence, inhibits protein kinase. Constitutively, active protein phosphatase dephosphorylates antenna protein and prevents its further outflow from thylakoids to the stroma lamellas (Staehelin and Arntzen 1983).

11.2 Energy Transmission Along Membranes in the Form of $\Delta\bar{\mu}_{H^+}$

11.2.1 General Remarks

Biological membranes are usually regarded as structures that separate the cytoplasm from the environment or one intracellular compartment from another. No doubt, such a property is characteristic of any biomembrane. However, the separating functions always entail some integrating effects. For instance, the cytoplasmic membrane not only segregates the cytoplasm from the extracellular medium, but it also unites diverse intracellular contents into a common system.

As it is clear from the above considerations, $\Delta\bar{\mu}_{H^+}$, like ATP, is a convertible energy currency for many bacteria, plants, fungi, mitochondria, and chloroplasts. In Chaps. 12–14, it will be shown that $\Delta\bar{\mu}_{Na^+}$ plays the role of a membrane-linked energy form in the cytoplasmic membrane of some bacteria and in the outer membrane of the animal cell.

An obligatory feature of any currency must be transportability. The proton- and sodium motive forces ($\Delta\bar{\mu}_{H^+}$ and $\Delta\bar{\mu}_{Na^+}$) meet this requirement. In particular, it is obvious that both the $\Delta\Psi$ and ΔpH constituents of a proton potential, once they are formed across the membrane, immediately spread along it. The electric conductance of the media on both sides of the coupling membrane is very high since these media are aqueous solutions of electrolytes. At the same time, the conductance of the membrane can be extremely low. This means that $\Delta\Psi$, if produced by a given $\Delta\bar{\mu}_{H^+}$ generator, cannot avoid fast spreading over the membrane surface. As to ΔpH, it also must spread rather quickly due to the high rate of H^+ diffusion in water and the high concentration of mobile pH buffers that further increase the rate of H^+ movement. This means that the $\Delta\bar{\mu}_{H^+}$ produced by a $\Delta\bar{\mu}_{H^+}$ generator in a certain area of the membrane can, in principle, be transmitted as such along the membrane and transduced into work when used in another region of the same membrane. In 1969, one of this book's authors (V.P.S.) extended this line of reasoning to the hypothesis that *coupling membranes act as power-transmitting cables at the cellular (or even at the supracellular) level* (Skulachev 1969, 1971). Below, this and some other lateral transport functions of biological membranes will be considered.

11.2.2 Lateral Transmission of $\Delta\bar{\mu}_{H^+}$ Produced by Light-Dependent Generators in Halobacteria and Chloroplasts

As already discussed in Sect. 6.1, lateral $\Delta\bar{\mu}_{H^+}$ transmission should be an indispensable step in the utilization of light energy in *H. salinarium*. In this microorganism, more than 50 % of the area of the cytoplasmic membrane can be occupied by purple membrane sheets with a diameter of up to 0.5 μm. *The bacteriorhodopsin-generated $\Delta\bar{\mu}_{H^+}$ is utilized by $\Delta\bar{\mu}_{H^+}$ consumers that must be localized in membrane regions other than purple membranes* since bacteriorhodopsin is the only protein in the purple membrane.

There are at least two kinds of lateral energy transport in chloroplasts and cyanobacteria. It is well known that photons are absorbed mostly by antenna chlorophyll. Then the electron excitation energy migrates in the plane of the membrane between antenna chlorophylls until it reaches the reaction center chlorophyll.

Another lateral transport event in the thylakoid membrane is transmission of $\Delta\bar{\mu}_{H^+}$ generated in grana regions by photosystem II to stroma regions where H^+-ATP-synthase is localized.

11.2.3 Trans-Cellular Power Transmission Along Cyanobacterial Trichomes

In halobacteria and chloroplasts, energy is transmitted for distances that are not longer than micrometers. However, transmission, for example of $\Delta\Psi$, can be effective even for millimeter distances, as the calculation of energy losses accompanying the transmission process has shown (the electric cable equation was used) (Skulachev 1980; Chailakhyan et al. 1982).

The first example of such a long-distance $\Delta\bar{\mu}_{H^+}$ transmission was described by Inna Severina in the group of one of the book's authors (V.P.S.) when trichomes of filamentous cyanobacteria *Phormidium uncinatum* were studied (Skulachev 1980; Chailakhyan et al. 1982). A cyanobacterial trichome, i.e., a linear sequence of hundreds of cells, can be several millimeters long. There are indications that the trichome-forming cells are connected with microplasmodesmata, the very thin and short tubules crossing the intercellular space. If there is electric conductance via the microplasmodesmata, one might hope that the electric potential difference generated across the cytoplasmic membrane near, e.g. one of the trichome ends, can be transmitted along the trichome and utilized in its distal end to carry out work. Severina found that *illumination of the trichome end (about 5 % of the trichome length) with a small light beam-initiated motility of the trichome.* The measured amplitude and the kinetics of the $\Delta\Psi$ propagation along the trichome were in agreement with the same parameters calculated on the basis of assumption that the cyanobacterial trichome has the properties of an electric cable.

The conclusion on the cable properties of cyanobacterial trichomes can be extended to plant tissues in which the cells are connected with plasmodesmata quite permeable to ions and, in particular, to H^+. A study of the possible significance of such a power transmission for plant tissue economy seems to be an interesting subject for future investigations.

11.2.4 Structure and Functions of Filamentous Mitochondria and Mitochondrial Reticulum

The dogma of small mitochondria. Translated from Greek, the word *mitochondrion* means *thread grain*. This term was introduced by cytologists who used light microscopes. The first students of mitochondria always indicated that mitochondria can exist in two basic forms—filamentous and spherical (or ellipsoid).

The development of the electron microscope and thin-section techniques changed this opinion in such a way that filamentous mitochondria came to be regarded as a very rare exception, while the spherical shape was assumed to be canonical. This change of views stemmed from the fact that single-section electron microscopy deals with a 2D, rather than 3D picture of the cell. A clew considered this way can be incorrectly interpreted as a number of small grains if its reconstitution from many parallel sections is not carried out.

The new dogma on the shape and size of mitochondria has become especially popular among the young generation of biochemists who have never seen mitochondria under a light microscope. For this, one should possess certain essential skills since the thickness of mitochondrial filaments is usually close to the limit of resolution of light microscopy. It is easier to believe electron microscopists who use a newer physical method of much higher resolution.

It is clear that if $\Delta\bar{\mu}_{H^+}$ transmission is confined to a single spherical mitochondrion, this mechanism cannot be used to transport energy for a distance comparable to the size of an eukaryotic cell. However, if mitochondria are, at least under certain conditions, filamentous and the old students of the cell were right in putting the *"mitos"* first, then the role of mitochondria as intracellular proton cables can be discussed as a realistic hypothesis.

Giant Mitochondria and Reticulum Mitochondriale. Three approaches have shaken the dogma of spherical mitochondria: (1) the reconstitution of 3D electron

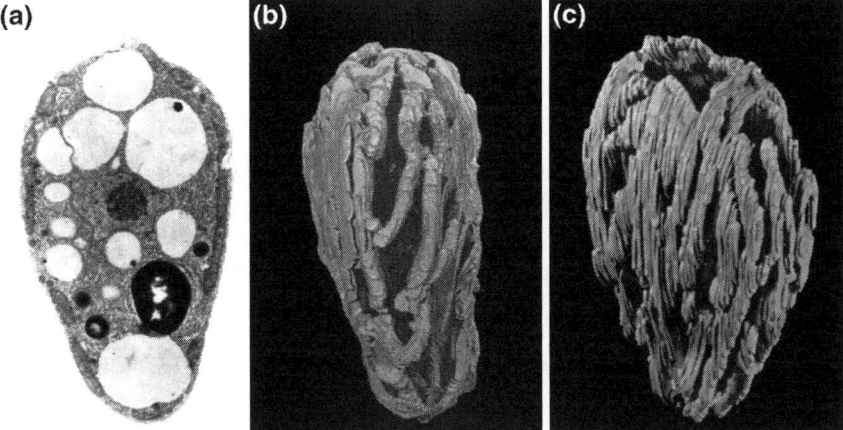

Fig. 11.2 Single mitochondrion in the cell of a flagellate, *Polytomella agilis*: **a** Single section (the picture can be wrongly interpreted as indicating the presence of many small, roundish, or ellipsoid mitochondria at the cell periphery); **b** and **c** front and side views of the model of the single giant mitochondrion of *P. agilis,* reconstituted from serial sections (Burton and Moore 1974)

micrographs of the whole cell with the aid of serial thin sections; (2) high-voltage electron microscopy allowing one to increase the thickness of the studied preparation, and (3) the staining of mitochondria with fluorescent penetrating cations, thus making it possible to return to studies of mitochondria in the living cell using the light microscope.

The serial section method was first applied in studies on unicellular eukaryotes. Here, very large and complicated mitochondrial structures were detected. For example, in a flagellate, *Polytomella agilis*, (Burton and Moore 1974) described a single (!) mitochondrion that looked like a hollow, perforated sphere arranged immediately below the outer cell membrane so that the cytoplasm and the nucleus was packed into a mitochondrial "string bag" (Fig. 11.2).

A single giant mitochondrion was described in some yeast cells, and also *Chlorella* and other unicellular algae. Several huge branched mitochondria, sometimes united into a single mitochondrion, were detected in *Chlamydomonas*, *Euglena*, *Trypanosoma*, and some fungi. Threadlike, 10-μm-long mitochondria were found in exocrine cells of the pancreas. A chain of long end-to-end arranged mitochondria were described in spermatozoa. In liver cells, mitochondria of two types were found—small spherical and the large branched ones. A similar picture was revealed in ascites tumor cells and in an algal cell, *Polytoma papillatum*. In the latter case, 228 consecutive sections were analyzed. It was found that there were 246 separate small mitochondria, two large and branched ones, and a very large one forming a perforated, hollow sphere (Smith and Ord 1983).

Muscle tissue seems to be one of the most interesting objects for studying intracellular power transmission. Very large multinuclear muscle cells (fibers) have high energy requirements. In a hard-working muscle, gradients of oxygen and substrates between the periphery and the core of the cell should arise, this effect limiting the scope of the work performed. The transmission of $\Delta\bar{\mu}_{H^+}$ from the muscle cell edge to its core along mitochondrial membranes might solve this problem. If this were the case, the dimensions of muscle mitochondria should be particularly large.

In insect flight muscle, slab-like mitochondria of about the same length as the muscle fiber radius, i.e., 10 µm, have been described.

The first indications of the existence of a mitochondrial system penetrating the muscle fibers of higher animals were obtained in the 1960s by Bubenzer (1966) and Gauthier (1969). They studied random sections of rat diaphragm muscle. In our group, Lora Bakeeva and Yuri Chentsov undertook a systematic investigation of serial sections of *rat diaphragm muscle*. It was found that in this tissue the mitochondrial material is organized into networks transpiercing the I-band regions of the muscle near the Z-disks (Bakeeva et al. 1977, 1978). These networks are connected with columns oriented perpendicular to their plane, i.e., parallel to myofibrils. Moreover, there are branches, arranged parallel to the Z-disks, connecting the networks with mitochondrial clusters at the periphery of the fiber (Fig. 11.3). Such a system, defined by one of this book's authors (V.P.S.) as *mitochondrial reticulum (Reticulum mitochondriale)*, was found to be characteristic of the diaphragm of adult animals. It is absent from the diaphragm of rat embryos and newborn rats (Bakeeva et al. 1981).

Sometimes the reticulum forming mitochondrial filaments contain dark partitions built of four membranes, the intermembrane space being filled with osmophilic material. These partitions apparently represent junctions of two branches of the mitochondrial reticulum This structure, discovered by Bakeeva and her colleagues in diaphragm (Bakeeva et al. 1978), was later studied in detail by the same group in heart muscle since here *mitochondrial junctions* were found to be especially numerous (Bakeeva et al. 1983). It was found that in this tissue that mitochondria also form a 3D system, but instead of a thin filamentous mitochondrial network found in diaphragm and skeletal muscle, the heart has a multitude of thick, poorly branched organelles. However, they are all coupled to each other by numerous junctions (Fig. 11.4).

The junction zone was found to be a disk with a diameter of 0.1–1.0 µm. In these zones, membranes and intermembrane spaces are of higher density. Two outer membranes of contacting mitochondria appear to maximally approach each other in a manner similar to that observed in tight junctions of the outer cell membranes (Fig. 11.4). Each mitochondrion was shown to be connected with its neighbors by several such junctions. Mitochondrial contacts could not be detected in the hearts of 3-day-old rats.

An even more complicated junction structure was found in the zone of the intercellular gap junction (nexus). Occasionally, one can see that two mitochondria belonging to two neighboring cells are in close contact with the outer cell membrane. As a result, a junction composed of six membranes is formed (two cellular and four mitochondrial membranes.

11.2 Energy Transmission Along Membranes in the Form of $\Delta\bar{\mu}_{H^+}$

Fig. 11.3 Mitochondrial reticulum in the diaphragm muscle of a 2-month-old rat: **a** Longitudinal section; **b** Transverse section of the isotropic region; **c–e** Reconstruction of diaphragm muscle mitochondria (**c** and **e**, *side view*; **d**, *top view*) (Bakeeva et al. 1981)

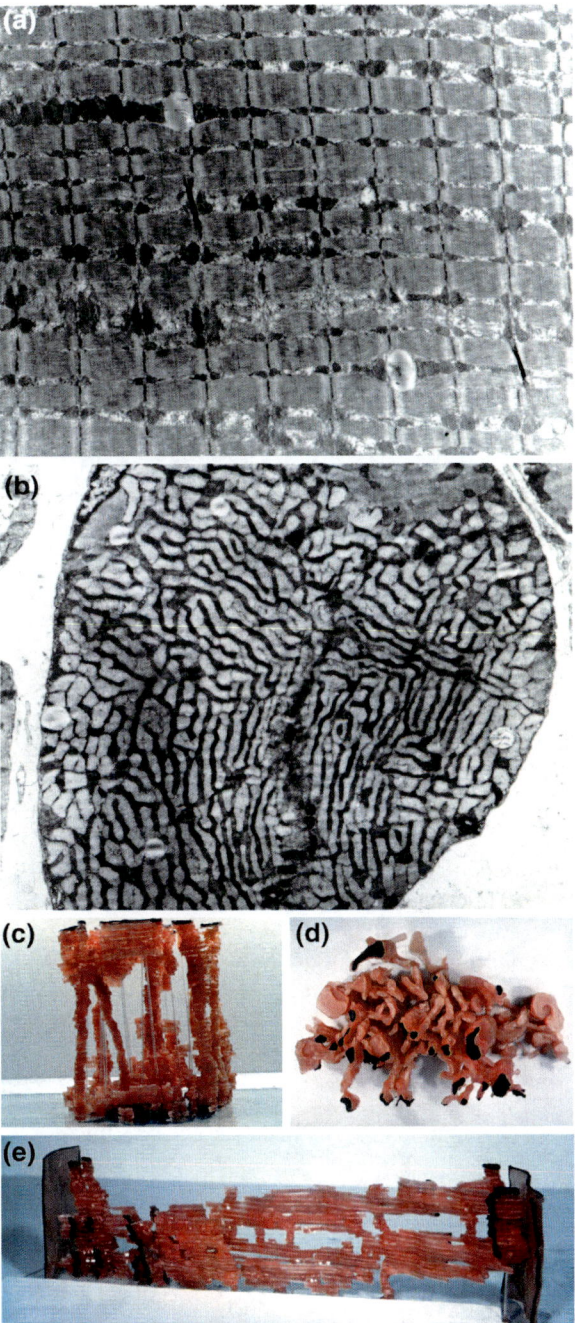

Fig. 11.4 Longitudinal section of a cardiomyocyte. Arrows indicate some of the mitochondrial junctions (Bakeeva et al. 1981)

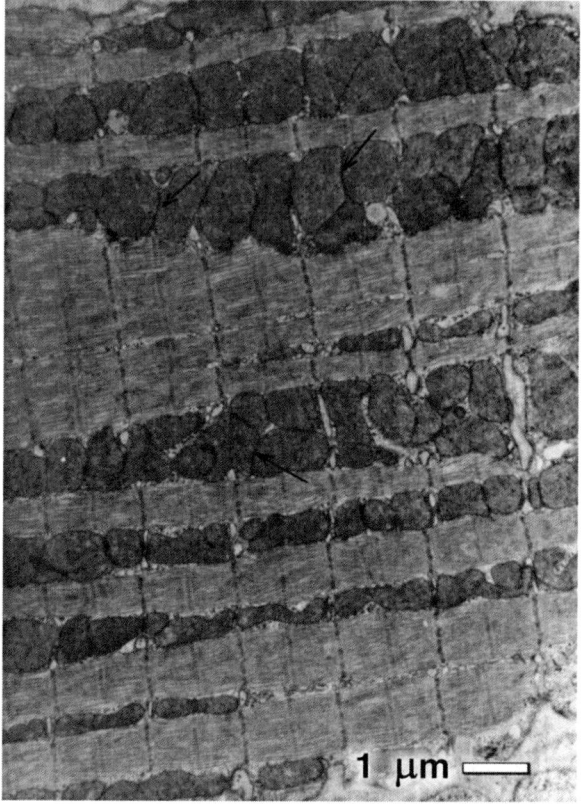

Filamentous mitochondria. If the studied tissue does not have such a high electron density as muscle, the real shape and size of mitochondria can be easily determined by high-voltage electron microscopy. Investigations of this kind revealed long filamentous mitochondria in diverse cell types.

However, two limitations of this method should be noted. First, it is impossible to observe mitochondrial junctions, and so one cannot answer the question whether it is really a single, lengthy mitochondrion or several, end-to-end joined organelles. And second, like any other electron microscopic method, it deals with fixed material, and thus it is not possible to follow directly the functioning of mitochondria. To overcome this latter limitation, we decided to return to light microscopy. Here, functional analysis is quite possible, but the great disadvantage is low resolution that is insufficient to see with certainty thin mitochondrial filaments and networks. We might overcome this limitation when dealing with light emission instead of light absorption. If the light emission is sufficiently strong, the light source will be seen in the dark even if it cannot be observed as a light-absorbing body (that is why the stars in the sky can be seen only at night).

The experiments conducted by Lan Bo Chen and colleagues in the USA (Johnson et al. 1981) and by Dmitry Zorov in our laboratory (Drachev and Zorov 1986) showed

Fig. 11.5 Mitochondria revealed in a living human fibroblast by means of the penetrating fluorescent cation ethylrhodamine (Drachev and Zorov 1986)

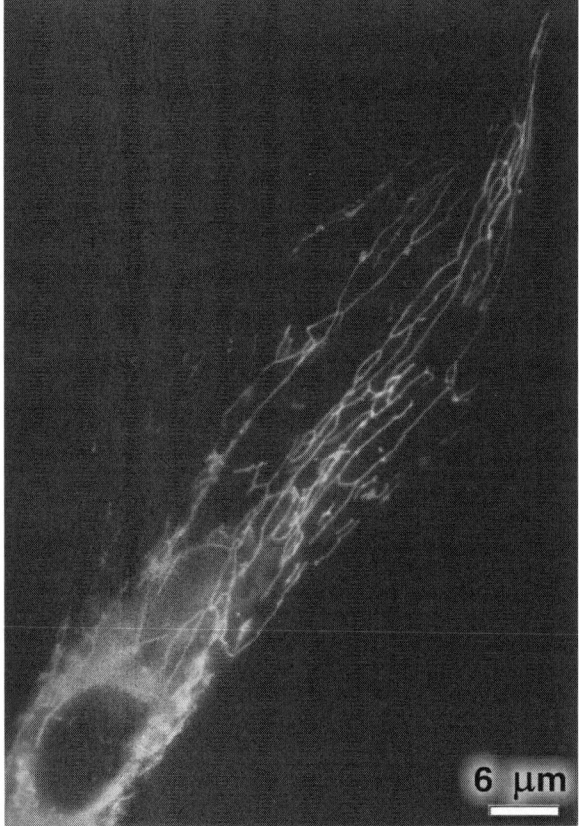

that *derivatives of the fluorescent penetrating cation rhodamine are accumulated specifically in mitochondria*. It was shown using these compounds that mitochondria in the living cell very frequently have the form of filaments tens of micrometers in length (Fig. 11.5). The filaments can be so long as to connect the cell core and the cell periphery, or even cross the cell from edge to edge. In some cases, mitochondrial reticulum was detected. A typical case in point is *that two mitochondrial populations, long filamentous and small oval*, coexist in one and the same cell.

Fluorescence studies of mitochondria have confirmed the data obtained by traditional methods, indicating that the thread–grain transition is a typical feature inherent in these organelles. A demonstrative example of such a transition was described by Ilse Foissner, who showed dramatic elongation (up to 30 μm) of mitochondria in *Nitella flexilis* cells when the photosynthesis rate was reduced by the herbicide Diuron or by a decrease in the light intensity, or when the middle or terminal steps of the mitochondrial respiratory chain were inhibited (Foissner 1983). The addition of an NADH:CoQ-oxidoreductase inhibitor rotenone, or the uncoupler 2,4-dinitrophenol, had no effect on the size of mitochondria. Elongation required protein synthesis. It is remarkable that only a few mitochondria (no more than 20 %) were elongated, indicating that some additional function requiring filamentous mitochondria, besides the small ones, might appear under the changed conditions.

Another situation was described by (Tanaka et al. 1985). Using serial sections, they showed that in the yeast *Candida albicans* at the time of entry into a bud a few mitochondria fused into a single giant one, which fragmented during mitosis, returned to a single giant form before cytokinesis, and was then partitioned into two parts.

The formation of mitochondrial filaments was described as early as 1964 by Meisel and colleagues in the yeast *Endomyces magnusii* under hypoxic conditions (Meisel et al. 1964).

The mitochondrion as an intracellular proton cable. The data discussed in the preceding section are sufficient for concluding that many types of eukaryotic cells possess very long filamentous or network-like mitochondrial structures besides, and sometimes even instead of, the numerous small spherical organelles considered in textbooks as typical mitochondria.

If a mitochondrial filament is a continuum of both inner and outer membranes, $\Delta \bar{\mu}_{H^+}$ will be transmitted throughout the entire length of the filament. Yet, if the mitochondrial filament is an assembly of many small contacting end-to-end mitochondria or mitoplasts, two possibilities should be considered: (i) the filament is similar to a cyanobacterial trichome where the constituents (individual cells) are connected to each other by junctions of high electrical conductance (this chapter, see Sect. 11.2.3); (ii) there is no electrical contact of adjacent mitochondria in the filament.

In the preceding section, we described the intermitochondrial junctions that clearly represent specific structures responsible for contact of adjacent mitochondria. It remained obscure, however, whether the electric resistance of the junctions is as high as in other regions of the inner mitochondrial membrane or alternatively is it as low as in cyanobacterial microplasmodesmata or gap junctions of two animal outer cell membranes. The latter version is identical to the mitochondrial continuum in the sense of $\Delta \bar{\mu}_{H^+}$ transmission. The former possibility requires an exchange of energy equivalents other than $\Delta \bar{\mu}_{H^+}$ between neighboring mitochondria to be postulated. These equivalents could be ATP shuttling through mitochondrial junctions.

An obvious prediction of the hypothesis considering a filamentous mitochondrion as cable is that the whole filament must be de-energized when any of its parts become leaky. This prediction was verified in the group of one of this book's author (V.P.S.). Vladimir Drachev (Fig. 11.6) constructed a special device combining a laser and a fluorescence microscope. Using this device, Dmitry Zorov succeeded in illuminating a single mitochondrial filament in a human fibroblast cell stained with ethylrhodamine. A very narrow laser beam (the diameter of the light spot was comparable with the thickness of the mitochondrial filaments) was used to cause local damage to the filament. The *laser treatment resulted in the disappearance of the rhodamine fluorescence in the entire 40-μm-long*

11.2 Energy Transmission Along Membranes in the Form of $\Delta\bar{\mu}_{H^+}$

Fig. 11.6 Vladimir Drachev

mitochondrial filament. It is significant that other filaments remained fluorescent, so that the effect of the laser was not due to nonspecific damage to the cell. Besides that, the illuminated filament was shown to retain its continuity when scrutinized under a phase-contrast or electron microscope. In fact, no detectable traces of laser-induced damage were found (Fig. 11.7) (Drachev and Zorov 1986; Amchenkova et al. 1988; Skulachev 1988).

A similar study was conducted on cardiomyocytes. It was found that the illumination of a single mitochondrion gives rise to a quenching of a cluster composed of many mitochondria. Electron microscopic analysis revealed that mitochondria, which become quenched after laser treatment, are connected with the illuminated mitochondria by intermitochondrial contacts, whereas those retaining fluorescence are not (Amchenkova et al. 1988). The simplest explanation of these data is the following. In a cardiomyocyte, there are several independent mitochondrial clusters formed by many mitochondria joined by mitochondrial junctions (*mitochondrial bunches* or *Streptio mitochondriale*). These clusters have *high electrical conductance*.

Giant mitochondria as pathways of lateral transport of compounds. It seems obvious that the above consideration of H^+ movement along the mitochondrial filaments (Fig. 11.8) can be also applied to components other than the hydrogen

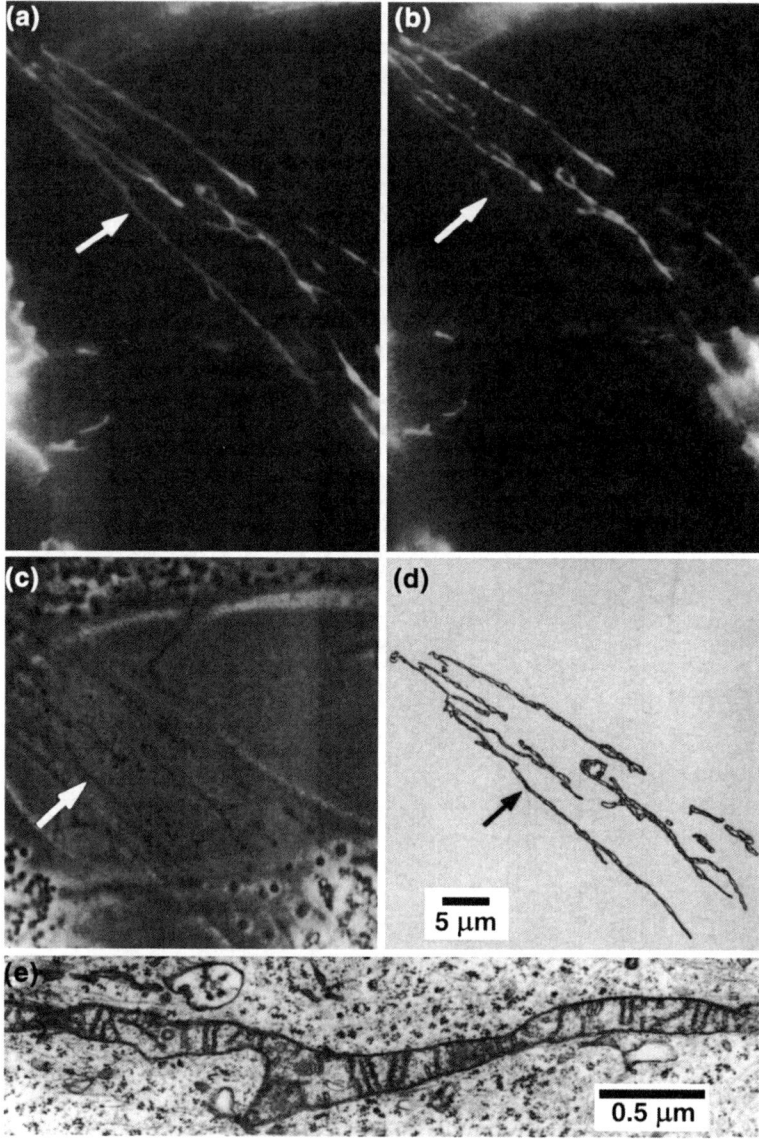

Fig. 11.7 Illumination by a narrow laser beam of a small part of the mitochondrial filament results in $\Delta\Psi$ collapse over the entire filament length. A cell of the primary culture of human fibroblasts was stained with ethylrhodamine, **a** Before and **b–e** after 100-ms laser treatment. The laser light spot was commensurate with the mitochondrial filament thickness. **a** and **b** Fluorescent microscopy. **c** Phase-contrast microscopy. **d** The top view on the model of mitochondria in the fibroblast, reconstituted with the use of the serial section technique. Arrows show the place illuminated by the laser. **e** Electron microscopy; a part of the laser-treated mitochondrion is seen (Drachev and Zorov 1986; Amchenkova et al. 1988)

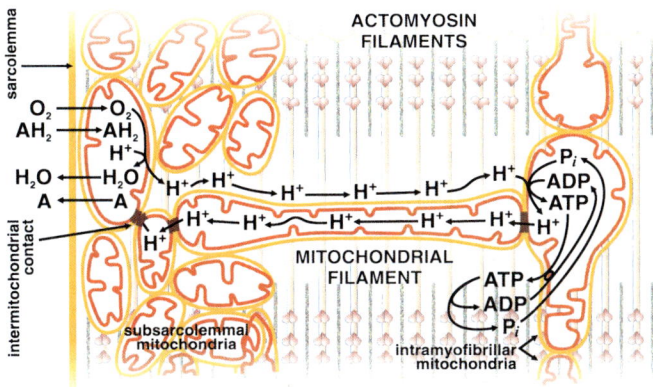

Fig. 11.8 Scheme of lateral $\Delta\bar{\mu}_{H^+}$ transmission along a mitochondrial filament of skeletal muscles

ions. In fact, the filaments might be a route to any substance that is concentrated in a mitochondrial compartment, including the membranes. For instance, if *Ca^{2+} ions* are transported from the intercellular space to cytosol at the cell periphery, they can be accumulated in the matrix space of mitochondrial filaments by means of $\Delta\Psi$-dependent Ca^{2+}-uniporter, diffuse along the filament as in a tube, and be released to cytosol in the cell core if the mitochondrial Ca^{2+}/H^+-antiporter is activated in this region of the filament. The same logic can be applied to transport of any other solute that can be reversibly accumulated in the matrix.

Lateral transport of lipid-soluble compounds can be another variation of this theme. It is generally believed that transports occurring in a membrane must be slower than in the water phase of the cell since the viscosity of the membrane is much higher than that of water, and in this respect the water route should have an advantage over the membrane route for any amphiphilic compound.

However, one must take into account that the cytosol and the mitochondrial matrix are not aqueous solutions of low molecular mass compounds, but rather colloid solutions that are crossed by insoluble structures, namely, by cytoskeleton and membranes. It is not surprising therefore that a spin-labeled synthetic compound of low molecular mass, of which the partition coefficient in a lipid/water system is close to 1, was shown to move preferentially in the lipid phase of various types of cells (Keith and Snipes 1974). Such preference must be even more pronounced for substances for which the lipid/water distribution coefficient is higher than 1, such as *free fatty acids* and their carnitine and CoA esters. For the latter, its role in the lateral transport of fatty acyls was postulated by Manfred Sumper and Hermann Trauble (Sumper and Träuble 1973). *Fatty acyl carnitine* seems adapted to this function even better since its hydrophilic moiety is much smaller than in *fatty acyl-CoA*.

The *lateral transport of molecular oxygen* poses an intriguing problem. There are at least four effects favorable for the use of the inner mitochondrial membrane as a route for intracellular oxygen transport.

(1) Solubility of O_2 in lipid is about fivefold higher than in water.
(2) The main oxygen-utilizing enzyme, cytochrome oxidase, located in the inner mitochondrial membrane, has a K_m for oxygen much lower than oxygen concentration in water under standard conditions. This should result in the formation of a local low O_2 concentration area in the inner mitochondrial membrane.
(3) The rate of diffusion of O_2 in hydrocarbons is between 10- and 100-fold more rapid than could be predicted from macroscopic viscosity data, being as high as in water (Subczynski and Hyde 1984). The rate of O_2 diffusion in phospholipids membranes is strongly decelerated by cholesterol, which is localized in membranes other than the inner mitochondrial membrane.
(4) In such tissues as muscle, the inner mitochondrial membrane comprises the bulk of the total membrane material of the cell (Bakeeva et al. 1978).

However, the inner mitochondrial membranes, even in heart muscle, occupy only a small portion of the cell volume. Further studies are required to determine whether the flow of O_2 along membranes can effectively compete with that through the cytosol.

Electron transfer along the membranes is an interesting problem. This process is probably localized in the endoplasmic reticulum and in the outer mitochondrial membrane and is provided by *NADH:cytochrome b_5—oxidoreductase* (flavoprotein fp$_5$) and *cytochrome b_5* (Nagi et al. 1983). This system has been found in liver, kidney, brain, and some other tissues. Both fp$_5$ and cytochrome b_5 are composed of two unequal parts—the larger, hydrophilic part, containing a flavin or heme group, and the smaller, hydrophobic part, requiring the protein to be anchored to the membrane. Like floats, fp$_5$ and cytochrome b_5 can move along the membrane, encountering relatively weak resistance.

Since the redox potential of fp$_5$ is close to its reductant, NADH, i.e. about − 0.3 V, it can serve as a component equilibrating the reducing equivalent at the level of NADH in different parts of the cell. As for cytochrome b_5, it might perform a similar function for electrons at redox potential level close to zero.

Another function of cytochrome b_5 was described by one of the authors of this book (V.P.S.) together with Alexander Archakov (Archakov et al. 1975). They found that rat liver endoplasmic membrane vesicles or mitochondria can reduce cytochrome b_5 in proteoliposomes containing cytochrome b_5 as the only protein species. NADH was used as the reductant. Repeated washing of endoplasmic reticulum vesicles did not diminish this effect. Thus, it was concluded that fp$_5$ and/or cytochrome b_5 are competent in *intermembrane electron transfer* without any water-soluble carriers being involved.

Further experiments showed that this property is inherent in cytochrome b_5, not fp$_5$. Endoplasmic reticulum vesicles were treated with pronase at 4 °C, which resulted in the destruction of cytochrome b_5. However, the fp$_5$ activity persisted at

11.2 Energy Transmission Along Membranes in the Form of $\Delta\bar{\mu}_{H^+}$

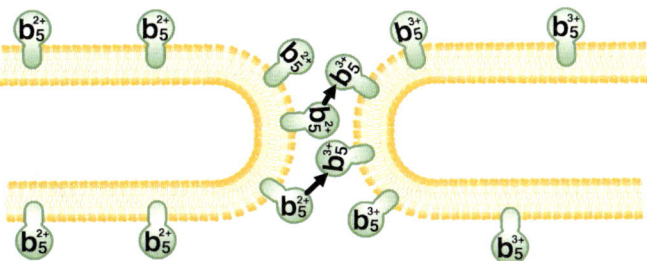

Fig. 11.9 Electron transport from the end of one cytochrome b_5-bearing membrane filament to the beginning of another filament; arrows indicate electron transport. It is assumed that cytochromes b_5 in the left and in the right filaments are reduced (b_5^{2+}) and oxidized (b_5^{3+}), respectively

a quite measurable level. This preparation completely lost the ability to catalyze the intermembrane electron transport.

An interesting finding was published by Susan Greenhut and Mark Roseman (Greenhut and Roseman 1985). When studying the amount of cytochrome b_5 in proteoliposomes differing in diameter by a factor of about 1.5 (21 and 35 nm), they found that the cytochrome preferred the smaller vesicles to the larger ones by a factor of at least 20. This means that the formation of highly curved regions of membranes in the ends, e.g., of mitochondrial filaments, may cause accumulation of cytochrome b_5.

Figure 11.9 illustrates the possible mechanism of electron transfer across the gap between the ends of two adjacent mitochondrial filaments. It has been proposed that the molecules of cytochrome b_5 diffuse on the membrane plane along the filaments by means of Brownian motion, scanning this membrane and searching for oxidants. In a region where the distance between two adjacent membranes does not exceed that of double the diameter of the cytochrome b_5 hydrophilic head, cytochrome b_5 of the more reduced filament loses an electron to cytochrome b_5 of the more oxidized one. This reaction should be facilitated by the accumulation of cytochrome b_5 at the ends of the membrane filaments, which might be an example of a "taxis" of membrane enzymes to their substrates (Skulachev 1980, 1988).

11.3 Buffering of $\Delta\bar{\mu}_{H^+}$

11.3.1 Na^+/K^+ Gradients as a $\Delta\bar{\mu}_{H^+}$ Buffer in Bacteria

To perform the role of a convertible energy currency, equivalents of $\Delta\bar{\mu}_{H^+}$ must be present in an amount sufficiently large to buffer the rate fluctuations of the $\Delta\bar{\mu}_{H^+}$-producing and $\Delta\bar{\mu}_{H^+}$-consuming processes (Mitchell 1977; Skulachev 1977).

Taking into account the electrical capacitance of the bacterial membrane, one can calculate how many H^+ ions should be exported from the bacterium to form a $\Delta\bar{\mu}_{H^+}$ of about 250 mV (the higher limit of the $\Delta\bar{\mu}_{H^+}$ value that can be maintained under energized conditions). Assuming that the capacitance is about 1 μF per cm^2, the number of exported H^+ ions is as low as 1 μmol H^+ ions per gram of protein, which is comparable to the amount of enzyme in the membrane (Mitchell 1968). To store membrane-linked energy in "substrate" (rather than in "catalytic") quantities, one must discharge the membrane by a transmembrane flux of ion(s) other than H^+. A flow of penetrating ions across the membrane discharges $\Delta\Psi$ and, hence, allows an additional portion of H^+ ions to be exported from the bacterium by $\Delta\bar{\mu}_{H^+}$ generators until ΔpH reaches the value equivalent to 250 mV. Now ΔpH, rather than $\Delta\Psi$, appears to be a factor limiting the activity of $\Delta\bar{\mu}_{H^+}$ generators. The pH buffering capacity of the cytoplasm is by about two orders of magnitude higher than the electrical capacity of the cytoplasmic membrane, so that the amount of energy stored in ΔpH appears to be much higher than that in $\Delta\Psi$ (Skulachev 1978, 1979).

A further increase in the amount of the stored membrane-linked energy can be achieved if the formed ΔpH is utilized, for example, by a cation/H^+-antiporter to substitute $\Delta p(cation)$ for ΔpH. The involvement of this antiporter exchanging, e.g. inner Na^+ for outer H^+, is equivalent to an increase in the concentration of pH buffer, as recognized already in 1968 by Mitchell (1968). Unfortunately, Mitchell ignored the role of the other monovalent cation, K^+. Accumulation of K^+ down $\Delta\Psi$ was, according to him, an unfavorable but inevitable side effect.

In 1978, one of this book's authors (V.P.S.) suggested that *influx of K^+ is used by the bacterial cell to induce $\Delta\Psi \rightarrow (\Delta pH, \Delta pK)$ transition* (Skulachev 1978). The formed ΔpK was postulated to buffer the $\Delta\Psi$ component of $\Delta\bar{\mu}_{H^+}$. As to ΔpH, it is utilized by a Na^+/H^+-antiporter so that ΔpNa buffers the ΔpH component of $\Delta\bar{\mu}_{H^+}$. According to this concept, *K^+ is accumulated in, and Na^+ is exported from, the bacterial cell when there is an excess of energy.* Energy deficiency results in some lowering of $\Delta\Psi$ and ΔpH. This causes K^+ efflux and Na^+ influx, preventing immediate dissipation of $\Delta\Psi$ and ΔpH, respectively (Fig. 11.10).

The hypothesis of a $\Delta\bar{\mu}_{H^+}$ buffer made it possible to explain a number of observations connected to bacterial energetics. It was found, in particular, that K^+ can be electrophoretically accumulated in bacterial cells. Na^+/H^+-antiport was described in 1974 by West and Mitchell in *E. coli* (1974). Since then, Na^+/H^+-antiport has been found in many species of bacteria (Krulwich 1983).

Cooperation of the $\Delta\Psi$-driven K^+ influx and ΔpH-driven Na^+ efflux seems to meet the requirements for the $\Delta\bar{\mu}_{H^+}$ buffer. Na^+ is the most common cation of the environment. Its concentration in seawater is as high as 0.5 M. Thus, a cell pumping out Na^+ ions can easily create a large ΔpNa. $[K^+]_{out}$ is usually much lower than $[Na^+]_{out}$, so a large ΔpK can be obtained via K^+-import. On the other hand, $[K^+]_{out}$ is not so low as to hinder the search for this cation in the medium. In fact, K^+ ranks second among cations in its concentration in seawater. Substitution of K^+ for Na^+ and vice versa inside the cell would hardly exert a dramatic influence upon metabolism and the cell structure owing to the similarity of many

11.3 Buffering of $\Delta\bar{\mu}_{H^+}$

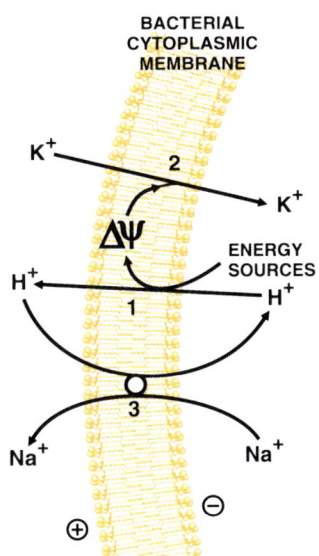

Fig. 11.10 Formation of $\Delta\bar{\mu}_{H^+}$-driven Na$^+$/K$^+$ gradients across the bacterial cytoplasmic membrane. H$^+$ is exported by a $\Delta\bar{\mu}_{H^+}$ generator (1). The formed $\Delta\Psi$ causes electrophoretic influx of K$^+$ (2), which discharges $\Delta\Psi$, converting $\Delta\Psi$ to ΔpH. The latter is utilized by a Na$^+$/H$^+$-antiporter to export Na$^+$ from the cell (3). $\Delta\Psi$ and ΔpH decrease reverse ion flows. As a result, $\Delta\Psi$ and ΔpH are formed, supporting for some time membrane-linked types of work in the absence of any external energy sources (not shown in the figure)

properties of these two monovalent cations and also because they are not directly involved in enzymatic processes.

Studies carried out in our laboratory by Brown et al. (1983) confirmed the hypothesis that *bacteria really create Na$^+$/K$^+$ gradients when there is an energy surplus, and they utilize these gradients to perform useful work when external energy sources are exhausted*. The following observations seem to be noteworthy.

(1) The Na$^+$ and K$^+$ gradients formed when external energy sources were available can sustain such a $\Delta\bar{\mu}_{H^+}$-linked function as the motility of halophilic *Halobacterium salinarium*, marine *Vibrio harveyi*, and enterobacterium *E. coli* for some time after the depletion of these energy sources. In the freshwater cyanobacterium *Phormidium uncinatum*, motility could be supported only by the K$^+$ gradient, ΔpNa being ineffective.
(2) The Na$^+$ and K$^+$ gradients stabilize the ATP level when other energy sources are absent.
(3) The capacitance of the Na$^+$/K$^+$ energy buffer is directly proportional to the ambient salt concentration, decreasing in the environment for bacteria in the series *H. salinarium* > *V. harveyi* > *E. coli* > *Ph. uncinatum*. In *H. salinarium* preincubated in the light, Na$^+$ and K$^+$ gradients are still able to support measurable motility for more than 8 h after the cessation of illumination and exhaustion of O$_2$. This means that halobacteria can invest a very large amount of energy into the Na$^+$/K$^+$ gradients during the light hours when solar energy is available, to provide it throughout the night.

It should be emphasized that cooperation of K$^+$-uniport and electroneutral Na$^+$/H$^+$-antiport is the simplest, but not the only possible mechanism for $\Delta\bar{\mu}_{H^+}$ buffering by cation gradients. For instance, *electrogenic Na$^+$/nH$^+$-antiport* apparently operates in certain bacteria at least at alkaline pH (Krulwich 1983). Moreover, K^+ influx can be carried out by the K^+, H^+-symporter (Bakker and Harold 1980). In such cases, ΔpNa buffers not only ΔpH, but also (to a lesser degree) the $\Delta\Psi$ component of $\Delta\bar{\mu}_{H^+}$. Respectively, ΔpK stabilizes both $\Delta\Psi$ and ΔpH, the former, however, more strongly than the latter.

Apparently, $\Delta\bar{\mu}_{H^+}$ buffering is necessary especially when in a single coupling membrane has are many pathways of $\Delta\bar{\mu}_{H^+}$ formation and utilization. This is the case for the cytoplasmic membranes of respiring and/or photosynthetic bacteria. These membranes contain H^+-ATP-synthase, many H^+, solute-symporters, H^+ motors, and several respiratory (photosynthetic) $\Delta\bar{\mu}_{H^+}$ generators.

On the other hand, in anaerobic, motionless species using, for example, glycolysis as the only energy-supplying mechanism, $\Delta\bar{\mu}_{H^+}$ buffering might not be essential since here there is only one type of $\Delta\bar{\mu}_{H^+}$ generator (H^+-ATPase, which utilizes glycolytic ATP) and one type of $\Delta\bar{\mu}_{H^+}$ consumer (H^+, solute-symporters responsible for the accumulation of some metabolites). It is in this group of microorganisms that one can find instances when the intracellular sodium concentration appears to be higher than the extracellular one (Sprott and Jarrell 1981). The same logic seems to be valid if we consider intracellular vesicles and organelles specializing in $\Delta\bar{\mu}_{H^+}$-driven ATP synthesis (chromatophores, mitochondria, and chloroplasts). They utilize simplified versions of the $\Delta\bar{\mu}_{H^+}$-stabilizing mechanisms.

11.3.2 Other $\Delta\bar{\mu}_{H^+}$-Buffering Systems

A biological membrane usually divides a volume into two unequal parts. Therefore, H^+ pumping across the membrane changes the pH mainly (or even exclusively) in the smaller compartment. If there are no enzymes in this compartment, the $\Delta\bar{\mu}_{H^+}$-buffering mechanism can be simplified such that its Na^+ part is eliminated. The most demonstrative example of this phenomenon is *chloroplast thylakoids*.

Here, H^+ ions are released into a narrow intrathylakoid cavity, the size of which is of the same order of magnitude with that of the membrane. This cavity is practically empty of enzymes, and thus there is no risk of enzyme inactivation by strong acidification. As mentioned previously (see Sect. 2.3.4), $\Delta\bar{\mu}_{H^+}$ buffering in thylakoids is achieved via $\Delta\Psi \to \Delta pH$ conversion accompanied by the transport of K^+, Mg^{2+}, and Cl^- across the thylakoid membrane, while Na^+/H^+-antiport is absent. Since the intrathylakoid space is very small in comparison to the chloroplast stroma, ΔpH (up to three units) is formed almost exclusively due to the increase in intrathylakoid $[H^+]$.

Some $\Delta\bar{\mu}_{H^+}$ buffering is apparently necessary for chloroplasts because fluctuations in light intensity always occur under natural conditions. This is probably why chloroplasts convert $\Delta\Psi$ to ΔpH. A similar problem exists in bacterial chromatophores. In *Rhodospirillum rubrum* it can be solved by the involvement of H^+-PP_i *synthase*. This enzyme utilizes $\Delta\bar{\mu}_{H^+}$ to synthesize PP_i from P_i. No rapid pathway of useful pyrophosphate utilization was described in *Rh. rubrum*. One of this book's authors (V.P.S.) suggested that the main way of using PP_i in this bacterium is pyrophosphate hydrolysis coupled to $\Delta\bar{\mu}_{H^+}$ formation due to the reversal of H^+-PP_i synthase (Skulachev 1979). ATP synthesis is the main function of chromatophores. It is significant that pyrophosphate stabilizes $\Delta\bar{\mu}_{H^+}$ at the level sufficient for ATP synthesis (see Sect. 7.3).

The outer cell membrane of higher plants seems to organize the $\Delta\bar{\mu}_{H^+}$ buffering in the same way as thylakoids. Protons pumped from the cell by H^+-ATPase appear in a narrow intercellular space where pH may be two units lower than in the cytosol.

In freshwater cyanobacteria, where ΔpNa of the proper direction cannot be formed because of a very low $[Na^+]_{out}$, $\Delta\bar{\mu}_{H^+}$ buffering by ion gradients is low, since only ΔpK is operative (Brown et al. 1983).

In mitochondria, the problem of $\Delta\bar{\mu}_{H^+}$ buffering is not as acute as in bacteria or chloroplasts since the level of respiratory substrates is usually more stable than that of illumination. Therefore, it is not surprising that $\Delta\bar{\mu}_{H^+}$ in mitochondria exists mainly in the form of $\Delta\Psi$. In tissues of rather constant activity like liver, some $\Delta\bar{\mu}_{H^+}$ buffering is achieved by K^+-uniport, while in tissues of varying activity, such as muscles, a similar effect can be reached due to mitochondrial reticulum formation.

References

Agureev AP, Altukhov ND, Mokhova EN, Savel'ev IA (1981) Activation of the external pathway of NADH oxidation in mitochondria at decreased pH. Biokhimiia 46:1945–1956 (in Russian)

Amchenkova AA, Bakeeva LE, Chentsov YS, Skulachev VP, Zorov DB (1988) Coupling membranes as energy-transmitting cables. I. Filamentous mitochondria in fibroblasts and mitochondrial clusters in cardiomyocytes. J Cell Biol 107:481–495

Appleby ML, Morton RK (1959) Lactic dehydrogenase and cytochrome b_2 of baker's yeast: purification and crystallization. Biochem J 71:492–499

Archakov AI, Karyakin AV, Skulachev VP (1975) Intermembrane electron transport in the absence of added water-soluble carriers. Biochim Biophys Acta 408:93–100

Bakeeva LE, Chentsov YS, Skulachev VP (1978) Mitochondrial framework (*Reticulum mitochondriale*) in rat diaphragm muscle. Biochim Biophys Acta 501:349–369

Bakeeva LE, Chentsov YS, Skulachev VP (1981) Ontogenesis of mitochondrial reticulum in rat diaphragm muscle. Eur J Cell Biol 25:175–181

Bakeeva LE, Chentsov YS, Skulachev VP (1983) Intermitochondrial contacts in myocardiocytes. J Mol Cell Cardiol 15:413–420

Bakeeva LE, Skulachev VP, Chentsov YS (1977) [Mitochondrial reticulum: organization and possible functions of a novel intracellular structure in the muscle tissue.] Vestn Mosk Univ Ser. Biol 3:23–28 (in Russian)

Bakker EP, Harold FM (1980) Energy coupling to potassium transport in *Streptococcus faecalis*. Interplay of ATP and the protonmotive force. J Biol Chem 255:433–440

Bibikov SI, Grishanin RN, Kaulen AD, Marwan W, Oesterhelt D, Skulachev VP (1993) Bacteriorhodopsin is involved in halobacterial photoreception. Proc Natl Acad Sci U S A 90:9446–9450

Brodie AF (1959) Oxidative phosphorylation in fractionated bacterial systems. I. Role of soluble factors. J Biol Chem 234:398–404

Brown II, Galperin MYu, Glagolev AN, Skulachev VP (1983) Utilization of energy stored in the form of Na^+ and K^+ ion gradients by bacterial cells. Eur J Biochem 134:345–349

Brummer B, Parish RW (1983) Mechanisms of auxin-induced plant cell elongation. FEBS Lett 161:9–13

Bubenzer HJ (1966) The thin and the thick muscular fibers of the rat diaphragm. Z Zellforsch Mikrosk Anat 69:520–550 (in German)

Burton MD, Moore J (1974) The mitochondrion of the flagellate, *Polytomella agilis*. J Ultrastruct Res 48:414–419

Cederbaum AI, Lieber CS, Beattie DS, Rubin E (1973) Characterization of shuttle mechanisms for the transport of reducing equivalents into mitochondria. Arch Biochem Biophys 158:763–781

Chailakhyan LM, Glagolev AN, Glagoleva TN, Murvanidze GV, Potapova TV, Skulachev VP (1982) Intercellular power transmission along trichomes of cyanobacteria. Biochim Biophys Acta 679:60–67

Drachev VA, Zorov DB (1986) Mitochondria as an electric cable. Experimental testing of a hypothesis. Dokl Akad Nauk SSSR 287:1237–1238 (in Russian)

Engelhardt WA (1930) Ortho- und Pyrophosphat im aeroben und anaeroben Stoffwechesel der Blutzellen. Biochem Z 251:16–21

Foissner I (1983) Inhibitor studies on formation of giant mitochondria in *Nitella flexilis*. Phyton (Austria) 28:19–29

Gabig TG, Babior BM (1981) The killing of pathogens by phagocytes. Annu Rev Med 32:313–326

Gauthier GF (1969) On the relationship of ultrastructural and cytochemical features of color in mammalian skeletal muscle. Z Zellforsch Mikrosk Anat 95:462–482

Glagolev AN (1980) Reception of the energy level in bacterial taxis. J Theor Biol 82:171–185

Greenhut SF, Roseman MA (1985) Distribution of cytochrome b_5 between sonicated phospholipid vesicles of different size. J Biol Chem 260:5883–5886

Johnson LV, Walsh ML, Bockus BJ, Chen LB (1981) Monitoring of relative mitochondrial membrane potential in living cells by fluorescence microscopy. J Cell Biol 88:526–535

Kauppinen R (1983) Proton electrochemical potential of the inner mitochondrial membrane in isolated perfused rat hearts, as measured by exogenous probes. Biochim Biophys Acta 725:131–137

Keith AD, Snipes W (1974) Viscosity of cellular protoplasm. Science 183:666–668

Klug GA, Krause J, Ostlund AK, Knoll G, Brdiczka D (1984) Alterations in liver mitochondrial function as a result of fasting and exhaustive exercise. Biochim Biophys Acta 764:272–282

Krulwich TA (1983) Na^+/H^+ antiporters. Biochim Biophys Acta 726:245–264

Meisel MN, Birjusova VI, Volkova TM, Malatjan MN, Medvedeva GA (1964) Functional morphology and cytochemistry of the mitochondrial apparatus of microorganisms. In: Kheysin EN (ed) Electron and fluorescent microcopy of the cell. Nauka, Moscow Leningrad, pp 1–15 (in Russian)

Miller JB, Koshland DE Jr (1977) Sensory electrophysiology of bacteria: relationship of the membrane potential to motility and chemotaxis in *Bacillus subtilis*. Proc Natl Acad Sci USA 74:4752–4756

Mitchell P (1968) Chemiosmotic coupling and energy transduction, Glynn Research (Bodmin)

Mitchell P (1977) A commentary on alternative hypotheses of protonic coupling in the membrane systems catalysing oxidative and photosynthetic phosphorylation. FEBS Lett 78:1–20

Nagi M, Cook L, Prasad MR, Cinti DL (1983) Site of participation of cytochrome b_5 in hepatic microsomal fatty acid chain elongation. Electron input in the first reduction step. J Biol Chem 258:14823–14828

Nicholls DG (1982) Bioenergetics, an introduction to the chemiosmotic theory. Academic Press, London

Okunuki K (1932) Gas exchange of pollen. Bot Mag (Tokyo) 46:701–721 (in Japanese)

Rödel G, Müller G, Bandlow W (1985) Cyclic AMP receptor protein from yeast mitochondria: submitochondrial localization and preliminary characterization. J Bacteriol 161:7–12

Skulachev VP (1958) [Novel aspect in studies of the mitochondrial oxidative phosphorylation.] Uspekhi Sovrem. Biol 46:241–263 (in Russian)

Skulachev VP (1962) Interrelations of the respiratory chain oxidation and phosphorylation (in Russian), The USSR Academy of Sciences Publishers, Moscow p. 156

Skulachev VP (1963) Regulation of coupling of oxidation and phosphorylation. Proc Int Biochem Congr 5:365–374

Skulachev VP (1969) Energy accumulation in the cell. Nauka, Moscow. (in Russian)

Skulachev VP (1971) Energy transformation in the respiratory chain. Curr Topics Bioenerg 4:127–190

Skulachev VP (1977) Transmembrane electrochemical H^+-potential as a convertible energy source for the living cell. FEBS Lett 74:1–9

Skulachev VP (1978) Membrane-linked energy buffering as the biological function of Na^+/K^+ gradient. FEBS Lett 87:171–179

Skulachev VP (1979) Na^+/K^+ gradient as an energy reservoir in bacteria. In: Mukohata Y, Packer L (eds) Cation flux across biomembranes, Academic Press, New York

Skulachev VP (1980a) Membrane electricity as a convertible energy currency for the cell. Can J Biochem 58:161–175

Skulachev VP (1980b) Integrating functions of biomembranes. Problems of lateral transport of energy, metabolites and electrons. Biochim Biophys Acta 604:297–310

Skulachev VP (1988) Membrane Bioenergetics. Springer, Berlin

Skulachev VP, Maslov SP (1960) The role of the non-phosphorylating oxidation in thermoregulation. Biokhimiia 25:1058–1064 (in Russian)

Smith RA, Ord MJ (1983) Mitochondrial form and function relationships in vivo: their potential in toxicology and pathology. Int Rev Cytol 83:63–134

Sprott GD, Jarrell KF (1981) K^+, Na^+, and Mg^{2+} content and permeability of *Methanospirillum hungatei* and *Methanobacterium thermoautotrophicum*. Can J Microbiol 27:444–451

Staehelin LA, Arntzen CJ (1983) Regulation of chloroplast membrane function: protein phosphorylation changes the spatial organization of membrane components. J Cell Biol 97:1327–1337

Subczynski WK, Hyde JS (1984) Diffusion of oxygen in water and hydrocarbons using an electron spin resonance spin-label technique. Biophys J 45:743–748

Sumper M, Träuble H (1973) Membranes as acceptors for palmitoyl CoA in fatty acid biosynthesis. FEBS Lett 30:29–34

Takemori S, Kominami S (1984) The role of cytochromes P-450 in adrenal steroidogenesis. Trends Biochem Sci 9:393–396

Tanaka K, Kanbe T, Kuroiwa T (1985) Three-dimensional behaviour of mitochondria during cell division and germ tube formation in the dimorphic yeast Candida albicans. J Cell Sci 73:207–220

West IC, Mitchell P (1974) Proton/sodium ion antiport in *Escherichia coli*. Biochem J 144:87–90

Part V
The Sodium World

Chapter 12
$\Delta \bar{\mu}_{Na^+}$ Generators

12.1 Na$^+$-Motive Decarboxylases

In 1980, Peter Dimroth reported the discovery of a prokaryotic, *primary* $\Delta \bar{\mu}_{Na^+}$ *generator* creating Na$^+$ potential with no $\Delta \bar{\mu}_{H^+}$ involved (Dimroth 1980). A decarboxylase from the anaerobically grown bacterium *Klebsiella aerogenes* was found to convert oxaloacetate to pyruvate and CO$_2$ only if sodium ions were present, and this *decarboxylation was coupled to uphill export* of Na$^+$ ion from the cytoplasm to the external medium:

$$HOOC - CH_2 - CO - COOH + nNa^+_{in} \rightarrow CH_3 - CO - COOH + CO_2 + nNa^+_{out}$$
(12.1)

where n appears to be two (Dimroth et al. 2001).

This process could be shown in intact cells, in inside-out subcellular vesicles, and in proteoliposomes. The biotin-containing enzyme is located in the cytoplasmic membrane of *K. aerogenes*. The reaction sequence includes two stages: (1) transfer of the carboxyl residue from oxaloacetate to the biotin prosthetic group; (2) release of free CO$_2$ from carboxylated biotin with the regeneration of the initial form of the enzyme. It is the second step that requires Na$^+$ (Dimroth 1982). The decarboxylation-dependent Na$^+$ uptake by inside-out subcellular vesicles resulted in a positive charging of the intravesicular space, this indicating that the Na$^+$ transport is electrogenic. The $\Delta\Psi$ and ΔpNa were shown to be about 65 and 50 mV, respectively, the total $\Delta \bar{\mu}_{Na^+}$ being about 115 mV.

Oxaloacetate decarboxylase from *K. aerogenes* is composed of three different subunits: α, β, and γ (65, 34, and 12 kDa, respectively). The α-subunit is a peripheral membrane protein, while the two others are integral membrane proteins that can be released only by detergent treatment. Biotin is covalently bound to a conservative lysine residue of the α-subunit (Dimroth et al. 2001).

The functioning of the Na$^+$-translocating decarboxylase makes it possible for *K. aerogenes* to ferment such an unusual for anaerobic growth substrate as citrate

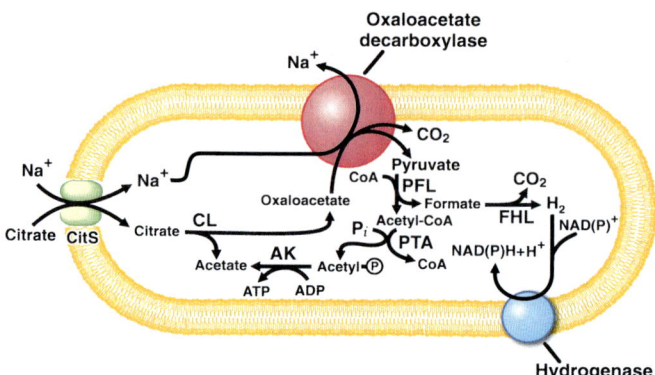

Fig. 12.1 Scheme of citrate fermentation in *K. aerogenes* and *K. pneumonia* (for explanations, see text)

(Fig. 12.1). Citrate is first imported into the cell via a Na$^+$-*motive citrate carrier* (CitS). It is then decomposed by *citrate lyase* (CL) to oxaloacetate and acetate. Oxaloacetate is decarboxylated by the respective decarboxylase, thus forming pyruvate. This process is coupled to Na$^+$ export, i.e., the energy of this reaction is stored in the form of $\Delta\bar{\mu}_{Na^+}$. *Pyruvate formate lyase* (PFL) catalyzes transformation of pyruvate to acetyl-CoA and the release of formate. The energy of the acetyl-CoA thioester bond can be used to form ATP. This process is catalyzed by *phosphotransacetylase* (PTA) and *acetate kinase* (AK). Formate, being toxic for the cell, is decomposed to CO_2 and H_2 by *formate hydrogen-lyase* (FHL), while electrons are transferred from H_2 to NAD(P)$^+$ by *hydrogenase*. Thus, substrate phosphorylation produces 1 mole of ATP per mole of substrate in the course of this fermentation. A certain amount of energy is additionally stored in the form of $\Delta\bar{\mu}_{Na^+}$ which can be used for the transport of citrate and other metabolites. The $\Delta\Psi$ component of $\Delta\bar{\mu}_{Na^+}$ might be utilized by H$^+$-ATP synthase for some additional ATP synthesis.

Further research showed that *K. aerogenes* is not the only microorganism having a Na$^+$-translocating decarboxylase. So far, four enzymes of this class dealing with different substrates have been described in various bacteria (Buckel 2001).

A Na$^+$-translocating oxaloacetate decarboxylase was discovered in enterobacteria *K. aerogenes*, *K. pneumoniae*, and *Salmonella typhimurium*.

Microorganisms, such as *Acidaminococcus fermentans*, *Peptococcus aerogenes*, *Clostridium symbiosum*, and *Fusobacterium nucleatum* have Na$^+$-translocating glutaconyl-CoA decarboxylase, which decarboxylates glutaconyl-CoA (an intermediate of the fermentation of glutamate to acetate and butyrate) to form crotonyl-CoA, the reaction being coupled to Na$^+$ export.

In the case of strictly anaerobic *Veillonella alcalescens* converting lactate into acetate and propionate and in *Propionigenium modestum*, which utilizes succinate and forms propionate, decarboxylation of methylmalonyl-CoA to propionyl-CoA is also a Na$^+$-motive process. Before decarboxylation, succinate is converted to methylmalonyl-CoA; it is decarboxylation of the latter that is coupled to export of Na^+.

Na^+-translocating malonate decarboxylase converting malonate into acetate and CO_2 was described for *Malonomonas rubra*.

Bacterial genome studies showed genes homologous to the already known genes of Na^+-translocating decarboxylases to be present in a most bacteria and archaea (Häse et al. 2001). Future research will determine substrate specificity of these putative Na^+-translocating enzymes.

12.2 Na^+-Translocating NADH:Quinone-Oxidoreductase

In 1977, Tsutomu Unemoto and coworkers discovered the specific activating effect of Na^+ ions upon NADH oxidase of marine alkalotolerant *Vibrio alginolyticus* and moderately halophilic *Vibrio costicola* (experiments were conducted on subcellular vesicles) (Unemoto et al. 1977). Other cations were ineffective as replacements for the sodium ions. Under the same conditions, NADH oxidase of *E. coli* did not require Na^+ for maximal activity. Further studies carried out in the same group revealed that the Na^+-dependent step was localized between NADH dehydrogenase and ubiquinone, whereas the cytochrome segment of the respiratory chain was Na^+-independent (Unemoto and Hayashi 1979).

Moreover, the Na^+-dependence of NADH oxidation in *V. alginolyticus* was shown to be caused by transmembrane translocation of sodium ions, and not by some allosteric regulation of NADH:quinone oxidoreductase. In 1982, Tokuda together with Unemoto demonstrated protonophore-resistant transport of Na^+ ions coupled to NADH oxidation in *V. alginolyticus* (Tokuda and Unemoto 1982). Later, they succeeded in isolating a Na^+-*motive NADH:quinone-oxidoreductase* (Na^+-NQR) and in reconstituting proteoliposomes competent in uphill Na^+ influx. In proteoliposomes, 15-fold stimulation of NADH:quinone oxidoreductase activity by Na^+ was observed. It was concluded that the NADH:quinone oxidoreductase from *V. alginolyticus* is a primary sodium pump.

12.2.1 Primary Structure of Subunits of Na^+-Translocating NADH:Quinone Oxidoreductase

In 1994, the sequence of the operon coding Na^+-NQR, was determined in the groups of Hayashi et al. (1995) and Rich et al. (1995). This operon was shown to contain six genes (*nqr*A–F) corresponding to six subunits of the Na^+-NQR complex. Three subunits (NqrB, NqrD, and NqrE) are very hydrophobic, while the other three are relatively hydrophilic (Table 12.1). Subunit NqrF is homologous to certain proteins catalyzing redox reactions. The C-terminal domain of this subunit

Table 12.1 Properties of Na$^+$-NQR subunits exemplified by the enzyme from V. harveyi, and their probable functions

Na$^+$-NQR subunit	Molecular mass (kDa)	Total number of transmembrane α-helices	Prosthetic groups and cofactors	Possible function
NqrA	48.4	0	Ubiquinone-8	Quinone-reductase center of the enzyme
NqrB	45.4	8	Covalently bound FMN residue, noncovalently bound riboflavin	Electron transfer. Possibly involved in quinone binding and/or Na$^+$ translocation
NqrC	27.5	2	Covalently bound FMN residue	Electron transfer
NqrD	22.7	6	Contains two conservative cysteine residues	Possibly involved in Na$^+$ translocation
NqrE	21.5	6	Contains two conservative cysteine residues	Possibly involved in Na$^+$ translocation
NqrF	45.2	1	Noncovalently bound FAD and [2Fe–2S] cluster	Binding site for NADH, responsible for NADH dehydrogenase activity of the enzyme

is similar to some of ferredoxin:NADP$^+$-oxidoreductases. Its primary structure contains sites responsible for NADH- and FAD-binding (Rich et al. 1995; Turk et al. 2004). The N-terminal domain of NqrF is homologous to ferredoxins containing a [2Fe–2S] cluster. Its sequence includes four conservative residues of cysteine that constitute such a cluster. Thus, NqrF is a polypeptide comprising NADH:ferredoxin-oxidoreductase and ferredoxin. The five other subunits of the Na$^+$-NQR (NqrA-E) have practically no homology to other proteins with known functions. It is important to stress that Na$^+$-NQR is very different both from the proton-translocating NADH:quinone oxidoreductases (NDH-1 or complex I) and the noncoupled NADH:quinone oxidoreductases (NDH-2). So proteins of the Na$^+$-NQR type constitute a novel class of the respiratory chain NADH:quinone oxidoreductases that developed in evolution independently from the other classes of these enzymes.

Genes homologous to *nqr*A-F from *V. alginolyticus* were later discovered in the genomes of many bacteria (Zhou et al. 1999; Häse et al. 2001). Thus, Na$^+$-NQR is widely spread among prokaryotes. In particular, pathogenic microorganisms such as *V. cholerae, K. pneumoniae, Haemophilus influenzae, Neisseria gonorrhoeae, Neisseria meningitidis, Yersinia pestis, Pseudomonas aeruginosa, Porphyromonas gingivalis*, etc., were found to contain this protein.

12.2.2 Na$^+$-NQR Prosthetic Groups

In 1986, Tokuda and Unemoto extracted Na$^+$-NQR from *V. alginolyticus* and showed this enzyme to contain only flavins as cofactors, namely one noncovalently bound FAD and one noncovalently bound FMN (Hayashi and Unemoto 1986). However, analysis of the primary structure of the Na$^+$-NQR subunits showed that only one of them (NqrF subunit) has the motif for noncovalent flavin binding. Moreover, Pfenninger-Li and Dimroth showed the purified Na$^+$-NQR to contain only one noncovalently bound flavin, FAD (Pfenninger-Li and Dimroth 1995). This contradiction was later solved, for Na$^+$-NQR was shown to contain (besides *noncovalently* bound FAD) two additional FMN molecules *covalently* bound to subunits NqrB and NqrC (Zhou et al. 1999; Hayashi et al. 2001). In both cases, FMN residues were found to be connected to threonine residues of the subunits by phosphoester bonds.[1] This type of covalent bond between flavin and protein requires a special type of primary structure of the flavin-containing domain. Apparently, the usual extraction methods should be ineffective in this case.

Studies of the flavin cofactors of Na$^+$-NQR have recently brought some unexpected results. In the group of Barquera, this protein was shown to contain (besides FAD and FMNs) a noncovalently bound riboflavin (Barquera et al. 2002). Riboflavin is usually used by living organisms only as a precursor for FMN and FAD, and there are no other known cases of its functioning as a redox-active prosthetic group of an enzyme. According to recent data, the Na$^+$-NQR contains (besides the flavins) also one [2Fe–2S] cluster and a molecule of tightly bound ubiquinone-8. *So, Na$^+$-NQR is considered now to contain the following cofactors: noncovalently bound FAD, two covalently bound FMN residues, riboflavin, [2Fe–2S] cluster, and ubiquinone-8* (Verkhovsky and Bogachev 2010).

When functioning in vivo, the Na$^+$-NQR oxidizes NADH and transfers two electrons to the ubiquinone molecule, thus forming ubiquinol. This redox reaction is coupled to a vectorial transfer of two sodium ions across the membrane. The Na$^+$/ē ratio for the Na$^+$-NQR is equal to 1 (Bogachev et al. 1997). Thus, the efficiency of energy storage by Na$^+$-translocating NADH:quinone oxidoreductases is approximately two times less than in the case of H$^+$-translocating NADH:quinone oxidoreductases, where the H$^+$/ē stoichiometry is equal to 2:

$$\mathrm{NADH} + \mathrm{H}^+_{in} + \mathrm{CoQ} + 2\ \mathrm{Na}^+_{in} \to \mathrm{NAD}^+ + \mathrm{CoQH}_2 + 2\ \mathrm{Na}^+_{out}. \qquad (12.2)$$

As translocation of sodium ions is a part of the reaction catalyzed by the Na$^+$-NQR, its rate depends on the concentration of these ions. Indeed, NADH oxidation by CoQ is practically absent in media depleted of Na$^+$. At the physiological values of ionic strength, the K_m for sodium ions are around millimolar values for Na$^+$-

---

[1] This type of covalent bond between flavin and protein is quite unique; it has not been found in any studied protein besides the Na$^+$-NQR.

NQR from different *Vibrio* species. The Na$^+$-NQR has absolute specificity for Na$^+$. As to other cations, only lithium was found to activate reaction (12.2). The Na$^+$-NQR cannot translocate protons even in the absence of sodium ions.

Translocation of sodium ions by the enzyme is accompanied by the generation of $\Delta\Psi$. The protons released from NADH upon its oxidation come to the cytoplasm, and the protons necessary for ubiquinol formation are also taken up from the cytoplasm. Thus, the Na$^+$-NQR functions as a primary electrogenic sodium pump, and not as a Na$^+$/H$^+$-antiporter (Zhou et al. 1999).

Na$^+$-NQR has a pH optimum between 8.0 and 9.0. It is specifically inhibited by very low concentrations of either Ag$^+$ ions or 2-heptyl-4-oxy-quinoline-N-oxide (*HQNO*). The antibiotic korormicin is also inhibitory (Hayashi et al. 2001).

The mechanism of sodium potential generation by the Na$^+$-NQR remains obscure. However, the stage of catalytic cycle that is specifically activated by sodium ions has been established. As found by an author of this book (A.V.B.) and Verkhovsky, this is electron transfer from FAD to covalently bound FMN residues (Bogachev et al. 2002; Verkhovsky and Bogachev 2010).

12.3 Na$^+$-Motive Methyltransferase Complex from Methanogenic Archaea

It has already been discussed in Sect. 5.3 that methanogenic archaea have a unique set of membrane proteins capable of storage of energy released on anaerobic formation of CH$_4$. It was found that methanogenic archaea require sodium ions in their growth medium (more than 1 mM) for the early stages of methanogenesis (Perski et al. 1981). Later Na$^+$ ions were shown to specifically activate the oxidative phase of methanogenesis.

When methanogens reduce various one-carbon compounds by molecular hydrogen thus forming methane and water, sodium ion-induced activation was observed only for such substrates as formaldehyde + H$_2$ or CO$_2$ + H$_2$, but not for methanol (Sect. 5.3.1). Analysis of these data showed that the transfer of the methyl group from methyltetrahydromethanopterin (CH$_3$-H$_4$MPT) to 2-thioethansulfonate (CoM) to be one of the sodium-dependent stages of methanogenesis (Gottschalk and Thauer 2001):

$$CH_3 - H_4MPT + CoM \rightarrow H_4MPT + CH_3 - CoM \quad (12.3)$$

$$\Delta G^{0\prime} = -30 \text{kJ/mol}.$$

This reaction is catalyzed by the membrane Na$^+$-translocating methyltransferase complex, and the free energy of this exergonic reaction is stored in the form of $\Delta\bar{\mu}_{Na^+}$. The methyltransferase complex consists of eight different subunits (MtrA-H) encoded by eight genes of the *mtr*A-H operon. This protein contains *5-oxybenzimidazolyl cobalamin* (*cobalamin* is a cobalt-containing cofactor, see Appendix 2) as a prosthetic group bound to MtrA subunit. During the methyl transferase reaction, the methyl group is first transferred from $CH_3 - H_4MPT$ to cobalamin:

$$CH_3 - H_4MPT + cobalamin(Co^+) \rightarrow H_4MPT + CH_3 - cobalamin(Co^{3+}). \tag{12.4}$$

The methyl group is then transferred from cobalamin to CoM:

$$CH_3 - cobalamin\left(Co^{3+}\right) + CoM \rightarrow cobalamin(Co^+) + CH_3 - CoM \tag{12.5}$$

It is reaction 12.5 that is specifically activated by Na^+ and is coupled to transmembrane electrogenic export of these ions from the cell. Two sodium ions are assumed to be transported across the membrane per each methyl group transferred. The mechanism of sodium potential generation by the methyltransferase complex has not yet been elucidated.

12.4 Na^+-Motive Formylmethanofuran Dehydrogenase from Methanogenic Archaea

Oxidation of formylmethanofuran (formyl-MFR[2]) to methanofuran (MFR) (see Appendix 2) and CO_2 is the second sodium-dependent stage of the oxidative phase of methanogenesis. This reaction is catalyzed by formylmethanofuran dehydrogenase (de Poorter et al. 2003):

$$formyl - MFR \rightarrow CO_2 + MFR + 2 H \tag{12.6}$$

$$\Delta G^{0'} = -16 kJ/mol$$

It is not clear yet which carrier of reducing equivalents functions as the electron acceptor for this reaction. Formylmethanofuran dehydrogenase is an electrogenic sodium pump; it carries out a transmembrane movement of 2–3 sodium ions per CO_2 molecule formed (Kaesler and Schönheit 1989).

Formyl-MFR oxidation occurs on conversion of formate or methanol to methane and CO_2:

$$4 HCOOH \rightarrow CH_4 + 3 CO_2 + 2 H_2O \tag{12.7}$$

$$4 CH_3OH \rightarrow 3 CH_4 + CO_2 + 2 H_2O \tag{12.8}$$

---

[2] Methanofuran (MFR) is a unique coenzyme of methanogenic archaea that plays the key role in the reactions of CO_2 fixation and reduction, and also in the reactions of oxidation of organic molecules, the latter being accompanied by CO_2 formation. The redox potential of the formyl-MFR/MFR + CO_2 couple is about -530 mV.

Methanogenic bacteria can have formylmethanofuran dehydrogenases of two types, the enzymes containing molybdenum or tungsten. All of them are multi-subunit iron–sulfur proteins. Besides FeS clusters, they have a special prosthetic group, molybdopterin guanidine dinucleotide (or tungstenopterin guanidine dinucleotide) (Hochheimer et al. 1995). The mechanism of the reaction catalyzed by formylmethanofuran dehydrogenase remains unclear, and this enzyme seems to be the most mysterious bioenergetic component of methanogenic archaea.

12.5 Secondary $\Delta\bar{\mu}_{Na^+}$ Generators: Na$^+$-Motive ATPases and Na$^+$-Pyrophosphatase

12.5.1 Bacterial Na$^+$-ATPases

In 1982, Heefner and Harold (Heefner and Harold 1982) presented convincing evidence for a Na$^+$-*motive ATPase* in the anaerobic bacterium *Enterococcus hirae* (at that time this microorganism was called *Streptococcus faecalis*). It was found that inside-out subcellular vesicles show a sodium-activated ATPase activity and that radioactive Na$^+$ is accumulated in the vesicle interior in an ATP-dependent manner.

It was later discovered that mainly a "canonical" H$^+$-translocating ATPase of F_oF_1 type functions in *E. hirae* when this bacterium grows at neutral pH and low [Na$^+$], i.e., under optimal conditions. However, increase in [Na$^+$] in the growth medium, alkalinization of the medium, or growth in the presence of protonophores result in the appearance of an alternative enzyme, Na$^+$-translocating ATPase of V_0V_1 type, which prevails under these conditions (Murata et al. 2001).

This enzyme (as well as other V-type ATPases) was found to be resistant to azide and vanadate but sensitive to nitrate and N-ethylmaleimide. It is specifically activated by sodium ions (with two K_m values of 20 μM and 4 mM); out of all other cations, only lithium ions have a similar effect (with K$_m$ of 60 μM and 3.5 mM).[3] Similar to Na$^+$-motive NADH:quinone-oxidoreductase from *V. alginolyticus*, Na$^+$-ATPase from *E. hirae* is mainly active under alkaline conditions (pH 8.0–9.0). As discussed in Sect. 7.2.2, hydrolysis of one ATP molecule by Na$^+$-translocating ATPase from *E. hirae* should be coupled to transmembrane transport of 3.3 sodium ions (Murata et al. 2005).

One may speculate that the biological function of *E. hirae* Na$^+$-motive ATPase is to energize the membrane of this glycolyzing bacterium when it is impossible to employ the usual chain of events leading to metabolite accumulation and fulfillment of other types of membrane-linked work (see Eq. 12.9):

$$\text{glucose} \rightarrow \text{ATP} \rightarrow \Delta\bar{\mu}_{H^+} \rightarrow \text{membrane} - \text{linked work.} \tag{12.9}$$

---

[3] It is interesting that in contrast to Na$^+$, increase in [Li$^+$] in the growth medium does not cause increased expression of the *E. hirae* Na$^+$-motive ATPase genes (even though Li$^+$ can be transported by the enzyme).

12.5 Secondary $\Delta\bar{\mu}_{Na^+}$ Generators: Na^+-Motive ATPases and Na^+-Pyrophosphatase

In the absence of $\Delta\bar{\mu}_{H^+}$, an alternative energy-transducing pathway seems to be induced:

$$\text{glucose} \rightarrow \text{ATP} \rightarrow \Delta\bar{\mu}_{Na^+} \rightarrow \text{membrane - linked work,} \quad (12.10)$$

where the second and the third steps are catalyzed by the Na^+-motive ATPase and the Na^+-dependent transporters, respectively.

Bacterial genome studies showed the Na^+-translocating ATPases (Na^+-ATP synthases) to be present in a substantial number of bacteria and archaea (Mulkidjanian et al. 2008).

12.5.2 Animal Na^+/K^+-ATPase and Na^+-ATPase

Na^+/K^+-ATPase was described by Skou (Fig. 12.2) in 1957, and 40 years later he was awarded the Nobel Prize in Chemistry for this work. The enzyme is located in the *outer membrane of animal cells*, being responsible for $\Delta\Psi$, ΔpNa, and ΔpK generation across this membrane. Na^+/K^+-ATPase catalyzes the exchange of 2 K^+_{out} for 3 Na^+_{in} per molecule of hydrolyzed ATP. The enzyme is composed of two subunits (α and β) with molecular masses of 100 and 40 kDa, respectively (Horisberger 2004). The β-subunit was shown to contain a short cytoplasmic

Fig. 12.2 Nobel Laureate Jens Skou

Fig. 12.3 Mechanism of functioning of Na$^+$/K$^+$-ATPase

$$E_1 + ATP + 3Na^+_{in} \longrightarrow 3Na^+ \cdot E_1 \sim P + ADP$$

$$3Na^+ \cdot E_1 \sim P \longrightarrow 3Na^+ \cdot E_2 \text{-} P$$

$$3Na^+ \cdot E_2 \text{-} P + 2K^+_{out} \longrightarrow 2K^+ \cdot E_2 \text{-} P + 3Na^+_{out}$$

$$2K^+ \cdot E_2 \text{-} P \longrightarrow E_1 + P_i + 2K^+_{in}$$

(N-terminal) domain, one transmembrane α-helix, and a large glycosylated extracellular domain. The α-subunit contains 10 transmembrane α-helices. Na$^+$/K$^+$-ATPase can be reconstituted into proteoliposomes, which are then competent in energy transduction.

The N-terminal domain of the α-subunit faces the cytoplasmic membrane side. The carboxyl of one of the aspartate residues of the conservative DKTGT sequence located in this domain is phosphorylated by ATP.

The mechanism of action of Na$^+$/K$^+$-ATPase is described in terms of the scheme for E_1E_2 ATPases (Fig. 12.3). According to this scheme, Na$^+_{in}$ is required for phosphorylation of the enzyme. The phosphoenzyme undergoes a conformational change ($E_1 \rightarrow E_2$ transition) and then decomposes to form E_2 and phosphate in a K$^+_{out}$-dependent manner. E_2 then relaxes to E_1 ready to enter the next cycle. It is the E_2P state that is fixed by ouabain. The binding sites for 3 Na$^+$ and 2 K$^+$ show positive cooperativity.

Rb$^+$, Cs$^+$, NH$_4^+$, Tl$^+$, Li$^+$, and Na$^+$ were reported to substitute to some degree for K$^+$ at the K$^+$-binding sites. On the other hand, the Na$^+$-binding sites are known to have relatively high specificity. Among the above-mentioned cations, only Li$^+$ was found to substitute for Na$^+$. The Li$^+$/K$^+$ ATPase activity is, however, very small, and the affinity for Li$^+$ is much lower than that for Na$^+$. It is interesting that H$^+$ can substitute for Na$^+$, and the affinity for H$^+$ is, in fact, at least two orders of magnitude *higher* than for Na$^+$. The only problem for H$^+$ is that at physiological pH, [Na$^+$] is much higher than [H$^+$], so that Na$^+$, rather than H$^+$, is translocated by Na$^+$/K$^+$-ATPase. However, under slightly acidic conditions (pH = 5.7), ATPase activity and H$^+$/K$^+$-antiport were observed in Na$^+$/K$^+$-ATPase proteoliposomes.

Another Na$^+$-transporting ATPase was described by Fulgencio Proverbio and colleagues in basolateral membranes from small intestinal epithelial cells and from renal proximal tubular cells. It differs from the Na$^+$/K$^+$-ATPase in several aspects. First of all, K$^+$ is unnecessary for its activity and ouabain is ineffective as an inhibitor (Proverbio et al. 1975).

12.5.3 Na$^+$-Motive Pyrophosphatase

It has already been discussed in Sect. 7.3 that in the cells of plants and some microorganisms there can be found a special enzyme, membrane-bound H$^+$-motive

12.5 Secondary $\Delta\bar{\mu}_{Na^+}$ Generators: Na$^+$-Motive ATPases and Na$^+$-Pyrophosphatase

pyrophosphatase, which catalyzes transmembrane proton transfer coupled to hydrolysis of pyrophosphate. Recently, Baykov from our institute and his colleagues from Turku University Lahti, Malinen, and Belogurov, have shown that the membrane-bound pyrophosphatases from archaea *Methanosarcina mazei* and *Moorella thermoacetica* as well as from the thermophilic bacterium *Thermotoga maritima* use Na$^+$ and not H$^+$ as the coupling ion (Malinen et al. 2007). These pyrophosphatases differ from the traditional membrane-bound H$^+$-translocating pyrophosphatases only in their substrate specificity. The catalytic activity of the membrane-bound pyrophosphatases from *Methanosarcina mazei*, *Moorella thermoacetica*, and *Thermotoga maritima* is expressed only in the presence of Na$^+$; PP$_i$ hydrolysis was shown to be coupled to electrogenic transmembrane translocation of Na$^+$ ions. Analysis of microbial genomes indicates that Na$^+$-pyrophosphatases are widely distributed among bacteria and archaea (Luoto et al. 2011).

References

Barquera B, Zhou W, Morgan JE, Gennis RB (2002) Riboflavin is a component of the Na$^+$-pumping NADH-quinone oxidoreductase from *Vibrio cholerae*. Proc Natl Acad Sci U S A 99:10322–10324

Bogachev AV, Murtazina RA, Skulachev VP (1997) The Na$^+$/e$^-$ stoichiometry of the Na$^+$-motive NADH:quinone oxidoreductase in *Vibrio alginolyticus*. FEBS Lett 409:475–477

Bogachev AV, Bertsova YV, Ruuge EK, Wikström M, Verkhovsky MI (2002) Kinetics of the spectral changes during reduction of the Na$^+$-motive NADH:quinone oxidoreductase from *Vibrio harveyi*. Biochim Biophys Acta 1556:113–120

Buckel W (2001) Sodium ion-translocating decarboxylases. Biochim Biophys Acta 1505:15–27

de Poorter LMI, Geerts WG, Theuvenet APR, Keltjens JT (2003) Bioenergetics of the formylmethanofuran dehydrogenase and heterodisulfide reductase reactions in *Methanothermobacter thermautotrophicus*. Eur J Biochem 270:66–75

Dimroth P (1980) A new sodium-transport system energized by the decarboxylation of oxaloacetate. FEBS Lett 122:234–236

Dimroth P (1982) The role of biotin and sodium in the decarboxylation of oxaloacetate by the membrane-bound oxaloacetate decarboxylase from *Klebsiella aerogenes*. Eur J Biochem 121:435–441

Dimroth P, Jockel P, Schmid M (2001) Coupling mechanism of the oxaloacetate decarboxylase Na$^+$ pump. Biochim Biophys Acta 1505:1–14

Gottschalk G, Thauer RK (2001) The Na$^+$-translocating methyltransferase complex from methanogenic archaea. Biochim Biophys Acta 1505:28–36

Häse CC, Fedorova ND, Galperin MY, Dibrov PA (2001) Sodium ion cycle in bacterial pathogens: evidence from cross-genome comparisons. Microbiol Mol Biol Rev 65:353–370

Hayashi M, Unemoto T (1986) FAD and FMN flavoproteins participate in the sodium-transport respiratory chain NADH:quinone reductase of a marine bacterium, *Vibrio alginolyticus*. FEBS Lett 202:327–330

Hayashi M, Hirai K, Unemoto T (1995) Sequencing and the alignment of structural genes in the *nqr* operon encoding the Na$^+$-translocating NADH-quinone reductase from *Vibrio alginolyticus*. FEBS Lett 363:75–77

Hayashi M, Nakayama Y, Unemoto T (2001) Recent progress in the Na$^+$-translocating NADH-quinone reductase from the marine *Vibrio alginolyticus*. Biochim Biophys Acta 1505:37–44

Heefner DL, Harold FM (1982) ATP-driven sodium pump in *Streptococcus faecalis*. Proc Natl Acad Sci U S A 79:2798–2802

Hochheimer A, Schmitz RA, Thauer RK, Hedderich R (1995) The tungsten formylmethanofuran dehydrogenase from *Methanobacterium thermoautotrophicum* contains sequence motifs characteristic for enzymes containing molybdopterin dinucleotide. Eur J Biochem 234:910–920

Horisberger JD (2004) Recent insights into the structure and mechanism of the sodium pump. Physiology 19:377–387 (Bethesda)

Kaesler B, Schönheit P (1989) The role of sodium ions in methanogenesis. Formaldehyde oxidation to CO_2 and $2H_2$ in methanogenic bacteria is coupled with primary electrogenic Na^+ translocation at a stoichiometry of 2–3 Na^+/CO_2. Eur J Biochem 184:223–232

Luoto HH, Belogurov GA, Baykov AA, Lahti R, Malinen AM (2011) Na^+-translocating membrane pyrophosphatases are widespread in the microbial world and evolutionarily precede H^+-translocating pyrophosphatases. J Biol Chem 286:21633–21642

Malinen AM, Belogurov GA, Baykov AA, Lahti R (2007) Na^+-pyrophosphatase: a novel primary sodium pump. Biochemistry 46:8872–8878

Mulkidjanian AY, Galperin MY, Makarova KS, Wolf YI, Koonin EV (2008) Evolutionary primacy of sodium bioenergetics. Biol Direct 3:13

Murata T, Kawano M, Igarashi K, Yamato I, Kakinuma Y (2001) Catalytic properties of Na^+-translocating V-ATPase in *Enterococcus hirae*. Biochim Biophys Acta 1505:75–81

Murata T, Yamato I, Kakinuma Y, Leslie AG, Walker JE (2005) Structure of the rotor of the V-type Na^+-ATPase from *Enterococcus hirae*. Science 308:654–659

Perski H-J, Moll J, Thauer RK (1981) Sodium dependence of growth and methane formation in *Methanobacterium thermoautotrophicum*. Arch Microbiol 130:319–321

Pfenninger-Li XD, Dimroth P (1995) The Na^+-translocating NADH-ubiquinone oxidoreductase from the marine bacterium *Vibrio alginolyticus* contains FAD but not FMN. FEBS Lett 369:173–176

Proverbio F, Condrescu-Guidi M, Whittembury G (1975) Ouabain-insensitive Na^+ stimulation of an Mg^{2+}-dependent ATPase in kidney tissue. Biochim Biophys Acta 394:281–292

Rich PR, Meinier B, Ward B (1995) Predicted structure and possible ion-motive mechanism of the sodium-linked NADH-quinone oxidoreductase of *Vibrio alginolyticus*. FEBS Lett 375:5–10

Skou JC (1957) The influence of some cations on an adenosine triphosphatase from peripheral nerves. Biochim Biophys Acta 23:394–401

Tokuda H, Unemoto T (1982) Characterization of the respiration-dependent Na^+ pump in the marine bacterium *Vibrio alginolyticus*. J Biol Chem 257:10007–10014

Turk K, Puhar A, Neese F, Bill E, Fritz G, Steuber J (2004) NADH oxidation by the Na^+-translocating NADH:quinone oxidoreductase from *Vibrio cholerae*: functional role of the NqrF subunit. J Biol Chem 279:21349–21355

Unemoto T, Hayashi M (1979) NADH:quinone oxidoreductase as a site of Na^+-dependent activation in the respiratory chain of marine *Vibrio alginolyticus*. J Biochem 85:1461–1467

Unemoto T, Hayashi M, Hayashi M (1977) Na^+-dependent activation of NADH oxidase in membrane fractions from halophilic *Vibrio alginolyticus* and *V. costicolus*. J Biochem 82:1389–1395

Verkhovsky MI, Bogachev AV (2010) Sodium-translocating NADH:quinone oxidoreductase as a redox-driven ion pump. Biochim Biophys Acta 1797:738–746

Zhou W, Bertsova YV, Feng B, Tsatsos P, Verkhovskaya ML, Gennis RB, Bogachev AV, Barquera B (1999) Sequencing and preliminary characterization of the Na^+-translocating NADH:ubiquinone oxidoreductase from *Vibrio harveyi*. Biochemistry 38:16246–16252

Chapter 13
Utilization of $\Delta\bar{\mu}_{Na^+}$ Produced by Primary $\Delta\bar{\mu}_{Na^+}$ Generators

13.1 Osmotic Work Supported by $\Delta\bar{\mu}_{Na^+}$

13.1.1 Na$^+$, Metabolite-Symporters

It has been shown in the alkalotolerant *Vibrio alginolyticus* possessing a Na$^+$-motive NADH:quinone-oxidoreductase that there are Na$^+$,metabolite-symporters responsible for accumulation of *19 amino acids as well as sucrose* (Kakinuma and Unemoto 1985). The Na$^+$-dependent import of nutrients into alkalophilic *Bacilli* has been described by several groups (Guffanti et al. 1981). However, it is still not clear how $\Delta\bar{\mu}_{Na^+}$ is formed in these alkalophiles.

In neutrophilic bacteria living at low or moderate NaCl concentrations, $\Delta\bar{\mu}_{H^+}$, rather than $\Delta\bar{\mu}_{Na^+}$, is used to support osmotic work (see Sect. 9.7). However, some exceptions to this rule have been reported as well. In *Mycobacterium phlei*, *Salmonella typhimurium*, and *Escherichia coli*, a Na$^+$, proline-symporter is used to accumulate proline inside the cell.

An interesting example of a "dualistic" mechanism of metabolite import was described in *E. coli*, which was shown to use H$^+$ or Na$^+$ alternatively as coupling cations for cotransport when it takes up melibiose (Bassilana et al. 1985). The import of citrate by *Klebsiella pneumoniae* is catalyzed by a carrier that symports citrate^{3-} with 2 Na$^+$ and 2 H$^+$. This means that $\Delta\Psi$, ΔpNa, and ΔpH are the driving forces of the uphill citrate import (Dimroth and Thomer 1986).

Arnold Brodie and coworkers isolated a Na$^+$, proline-symporter from *M. phlei*, which was found to be a single 20-kDa polypeptide (Lee et al. 1979). The purified symporter was reconstituted with phospholipids to form proteoliposomes, which were competent in proline accumulation driven by an artificially imposed $\Delta\Psi$. To generate $\Delta\Psi$, a valinomycin-induced downhill K$^+$ efflux was used. The accumulation of proline required Na$^+$ and was lowered by $\Delta\Psi$-discharging ionophores and by sulfhydryl reagents.

Partial purification and reconstitution of a Na$^+$, aspartate-symport system from halophilic *Halobacterium halobium* has also been described (Greene and

MacDonald 1984). Generally, *marine and halophilic bacteria, like alkaliphiles, employ Na$^+$, metabolite- rather than H$^+$, metabolite-symporters*. This is also true for the outer membrane of higher animals, the phenomenon being yet another testimony to the idea that blood is a small part of the ocean trapped inside our body. In this membrane, $\Delta\bar{\mu}_{Na^+}$ is generated due to Na$^+$ export by the Na$^+$, K$^+$-ATPase (and in some cases, by Na$^+$-ATPase). The formed $\Delta\bar{\mu}_{Na^+}$ is utilized by numerous carriers catalyzing the symport of Na$^+$ and amino acids, sugars, fatty acids, and other compounds into the cell (Hopfer 1978). Some of these Na$^+$, metabolite-symporters have been isolated and studied in detail (Krishnamurthy et al. 2009).

In some animal tissues, H$^+$-ATPase is located in the outer cell membrane (see Sect. 7.2). In these cases, H$^+$, metabolite-symporters have been described (Lee and Pritchard 1983).

13.1.2 Na$^+$ Ions and Regulation of Cytoplasmic pH

In bacteria, the pH$_{in}$ homeostasis mechanism operates very effectively. As shown by Robert Macnab's group, this homeostasis in *E. coli* is maintained at pH$_{out}$ values between 5.5 and 9.0 (Slonczewski et al. 1981). Within this limit, pH$_{in}$ could be described by the equation:

$$pH_{in} = 7.6 + 0.1 \, (pH_{out} - 7.6), \tag{13.1}$$

so that the pH$_{in}$ values appear to be equal to 7.4 and 7.8 at pH$_{out}$ 5.5 and 9.0, respectively.

When pH$_{out}$ decreases, acidification of the cytoplasm can be prevented by *K$^+$ import* discharging $\Delta\Psi$ that is produced by the $\Delta\bar{\mu}_{H^+}$ generator. As a result, $\Delta\Psi$ is converted to ΔpH, the cytoplasm becoming more alkaline than the outer medium (Fig. 13.1a).

At neutral pH$_{out}$, the only problem is how to return to the cytoplasm those H$^+$ ions that are pumped from the cell by $\Delta\bar{\mu}_{H^+}$ generators. The problem can be solved by an electroneutral Na$^+$/H$^+$-antiporter (Fig. 13.1b).

The electroneutral Na$^+$/H$^+$-antiporter is effective in the maintenance of pH homeostasis only if pH$_{out}$ ≤ pH$_{in}$. Other systems are necessary to acidify the cytoplasm when pH$_{out}$ is higher than pH$_{in}$. Bacteria confront this problem when the growth medium becomes alkaline. Now, there is no reason for the electroneutral Na$^+$/H$^+$-antiporter to carry H$^+$ from the outside to the interior of the cell since both Na$^+$ and H$^+$ gradients are in an unfavorable direction. One can overcome this difficulty, assuming that at alkaline pH$_{out}$ the Na$^+$/H$^+$-antiporter is electrogenic, transporting more than one H$^+$ per Na$^+$ (e.g. exchanging *one* Na$^+$ ion for *two* H$^+$ ions) and the $\Delta\Psi$ formed by $\Delta\bar{\mu}_{H^+}$ generators is so large that it can compensate for the unfavorable ΔpH and ΔpNa. In such a case, cooperation of a $\Delta\bar{\mu}_{H^+}$ generator and a *Na$^+$/2H$^+$-antiporter* will maintain pH$_{in}$ < pH$_{out}$ (see Fig. 13.1c).

13.1 Osmotic Work Supported by $\Delta\bar{\mu}_{Na^+}$

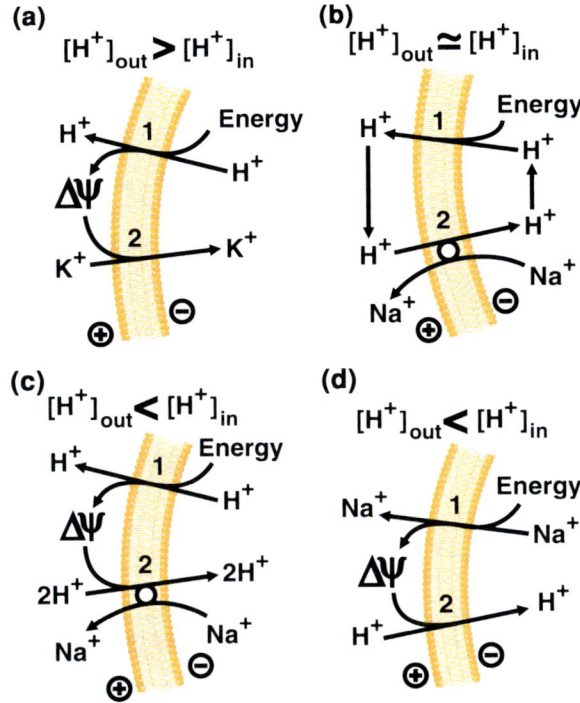

Fig. 13.1 Possible mechanisms for pH homeostasis in bacterial cytoplasm. **a** pH$_{out}$ < pH$_{in}$; a $\Delta\bar{\mu}_{H^+}$ generator pumps H$^+$ from the cell and forms $\Delta\Psi$ (reaction *1*). This $\Delta\Psi$ is discharged by the K$^+$ influx so that $\Delta\Psi \to \Delta$pH transition occurs, the cell interior being more alkaline (reaction *2*); **b** pH$_{out}$ ≈ pH$_{in}$; the H$^+$ loss due to the activity of $\Delta\bar{\mu}_{H^+}$ generators (reaction *1*) is compensated by H$^+_{out}$ influx in exchange for Na$^+_{in}$ (reaction *2*); **c** and **d** pH$_{out}$ > pH$_{in}$; **c** the H$^+$-pump exports H$^+$ and forms $\Delta\Psi$ (reaction *1*). To discharge $\Delta\Psi$, Na$^+$/2H$^+$-antiporter exchanges one Na$^+_{in}$ for two H$^+_{out}$ (reaction *2*); **d** a primary Na$^+$-pump charges the membrane, exporting Na$^+$ (reaction *1*). Uniport of H$^+$ ions into the cell discharges $\Delta\Psi$ and acidifies the cytoplasm (reaction *2*)

An alternative possibility is shown in Fig. 13.1d. Here, there is a primary Na$^+$ pump that charges the membrane and the *H$^+$-uniporter* allowing H$^+$ ions to move electrophoretically into the cytoplasm.

Na$^+$ ions are also involved in eukaryotic pH homeostasis. In animal cells, it has been firmly established that it is the Na$^+$/H$^+$-antiporter that is primarily responsible for the maintenance of stable pH$_{in}$. In plants, Na$^+$/H$^+$-antiport occurs both in the outer cell membrane and in the tonoplast.

13.2 Mechanical Work

There are many examples of bacteria performing mechanical work at the expense of $\Delta\bar{\mu}_{Na^+}$ generated by a primary Na$^+$ pump. Pavel Dibrov in our group showed (Dibrov et al. 1986) that the motility of *V. alginolyticus*: (1) occurs only in the presence of Na^+; (2) can be supported by an artificially imposed ΔpNa in a monensin-sensitive manner (under the same conditions, ΔpH was ineffective); (3) occurs at a *lowered but measurable* rate in the presence of a very high concentration of a protonophore. A 100-fold lower protonophore concentration completely arrests the motility if the medium is supplemented with monensin. Monensin added without CCCP decreases the respiration-supported motility only slightly. It was concluded that *the flagellar motor of V. alginolyticus is driven by $\Delta \bar{\mu}_{Na^+}$ rather than $\Delta \bar{\mu}_{H^+}$*. It is connected with a single giant flagellum that supports the motility of vibrios in sea water (Fig. 13.2). This sodium motor can rotate with frequency up to 1,700 revolutions per second, and it causes the cell to movie at the rate of 60 µm/s. The structure of the basal body is basically similar to that of regular lateral H^+-motive flagella (see Chap. 8). The main differences between the Na^+-motive motor and its H^+-motive analogs involve the stator elements of these organelles. The primary structures of the MotA and MotB proteins of sodium flagella are substantially different from those of the corresponding subunits of

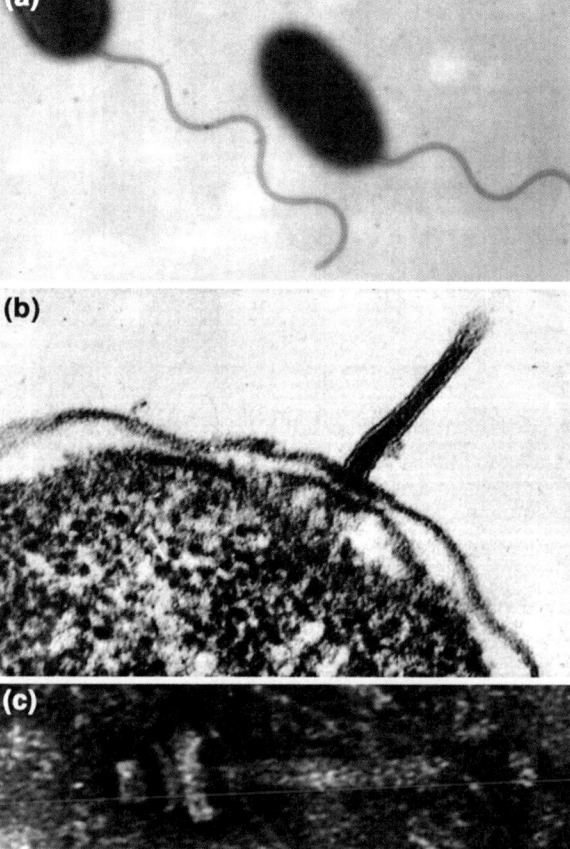

Fig. 13.2 Polar flagella of *Vibrios*, energized by $\Delta \bar{\mu}_{Na^+}$. **a** *V. parahaemolyticus;* **b** *V. alginolyticus*); **c** isolated basal body of *V. alginolyticus* (Skulachev 1988)

13.2 Mechanical Work

proton motors. Moreover, the sodium motor stator contains also two additional proteins—MotX and MotY—that are absent from H$^+$-motive flagella (Yorimitsu et al. 1999). It is noteworthy that the replacement of the *Vibrio* MotA, MotB, MotX, and MotY proteins with the proteins MotA and MotB from *E. coli* changes the motor specificity: in this case it uses the energy of $\Delta\bar{\mu}_{H^+}$ (and not $\Delta\bar{\mu}_{Na^+}$) for its rotation (Asai et al. 2003).

It was recently demonstrated that when growing in the liquid medium, *V. alginolyticus* moves due to the functioning of the above-described single polar sodium flagellum. However, growth in very viscous medium causes *V. alginolyticus* to express additional smaller flagella that are located laterally on the whole surface of the bacterium. It is the energy of $\Delta\bar{\mu}_{H^+}$ (and not $\Delta\bar{\mu}_{Na^+}$) that supports the rotation of these lateral flagella. Thus, this bacterium can use both $\Delta\bar{\mu}_{Na^+}$ and $\Delta\bar{\mu}_{H^+}$ for performing mechanical work, the choice being dependent on the growth conditions (Atsumi et al. 1992). An interesting example of energization of bacterial motility was described for the marine microorganism *Shewanella oneidensis* MR-1. It was shown that two different stator systems (H$^+$- and Na$^+$-dependent) can drive a single polar flagellum of this microorganism (Paulick et al. 2009).

Motility supported by $\Delta\bar{\mu}_{Na^+}$ was also described in other marine vibrios, in particular, in *Vibrio cholerae*. It is common knowledge that *V. cholerae* produces a toxin causing salt export from tissues to the intestinal lumen. This salt may be necessary for *V. cholerae* to support its Na$^+$ energetics.

Motility supported by $\Delta\bar{\mu}_{Na^+}$ was also demonstrated in alkalophilic *Bacilli*, for instance in *Bacillus sp. YN*-1 (Hirota and Imae 1983), *Bac. pseudofirmus* FTU (Bogachev et al. 1993), and *Bac. firmus* (Kitada et al. 1982). In the *Bacillus sp. YN*-1 studied in detail, Na$^+$ requirement for motility and partial resistance of motility to protonophores were shown. It was also found that ΔpNa and $\Delta\Psi$, generated enzymatically, are equivalent in supporting motility of this microorganism.

13.3 Chemical Work

The first indication that ΔpNa (together with ΔpK) can, in principle, be utilized to reverse an ion-transfer ATPase reaction was obtained in 1966 by Glynn and Garrahan, who described ATP synthesis by Na$^+$/K$^+$ ATPase in erythrocytes (Garrahan and Glynn 1966). The process was observed when [Na$^+$]$_{in}$ was higher than [Na$^+$]$_{out}$, and [K$^+$]$_{in}$ was lower than [K$^+$]$_{out}$. However, this reaction can hardly occur under in vivo conditions because of the opposite direction of the ion gradients.

13.3.1 $\Delta\bar{\mu}_{Na^+}$-Driven ATP Synthesis in Anaerobic Bacteria

Some evidence that $\Delta\bar{\mu}_{Na^+} \rightarrow$ ATP energy transduction really occurs in some living organisms was obtained by Peter Dimroth and colleagues (Hilpert and Dimroth

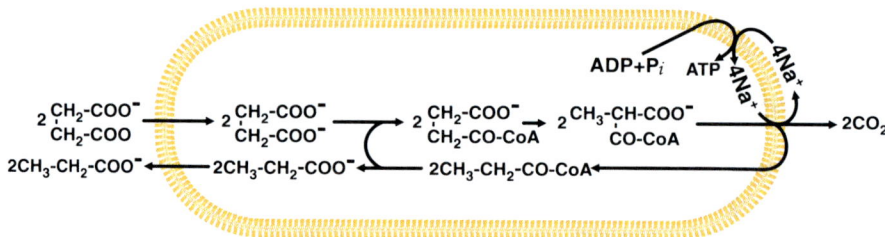

Fig. 13.3 Possible mechanism of energy supply in anaerobic bacterium *P. modestum*. Succinate enters the cell where it is converted to succinyl-CoA by transacylation with propionyl-CoA. From succinyl-CoA, methylmalonyl-CoA is formed. The latter is decarboxylated to propionyl-CoA and CO_2, this process being coupled to Na^+-pumping. Exported Na^+ ions return to the cytoplasm via Na^+-ATP synthase, which produces ATP from ADP and inorganic phosphate

1984), who studied *Propionigenium modestum*, a strictly anaerobic bacterium discovered in 1982 by (Schink and Pfennig 1982). The only biologically useful energy for *P. modestum* is gained during the decarboxylation of succinate to propionate:

$$^-OOC-CH_2-CH_2-COO^- + H_2O \rightarrow CH_3-CH_2-COO^- + HCO_3^- \quad (13.2)$$

$$\Delta G^{\circ\prime} = -21 \text{ kJ/mol}$$

The energy yield of the reaction is approximately two-fold lower than that required for ATP synthesis from ADP and inorganic phosphate under physiological conditions.[1] It seems clear that the mechanism of substrate-level phosphorylation cannot be used since it presumes the formation of one ATP molecule per molecule of the utilized substrate. A solution to the problem can be found if succinate decarboxylation is coupled to the formation of a difference in electrochemical potentials of some ion, e.g. H^+ or Na^+. In this case, succinate decarboxylation can export nH^+ or nNa^+ ions, while the formation of one ATP molecule can require the import of, e.g., $2nH^+$ or $2nNa^+$ ions.

According to Wilhelm Hilpert and Peter Dimroth, decarboxylation of methylmalonyl-CoA formed from succinyl-CoA is coupled to Na^+ export from the *Propionigenium modestum* cell (see Sect. 12.1) (Hilpert and Dimroth 1984). The formed $\Delta\bar{\mu}_{Na^+}$ is assumed to be used to synthesize ATP by a reversal of Na^+-ATPase found in large amounts in the cytoplasmic membrane of *P. modestum* (Fig. 13.3). This ATPase is similar to ATPases of F_oF_1-type in relation to its subunit composition and inhibitor sensitivity, being different only in ability to utilize $\Delta\bar{\mu}_{Na^+}$ (Kluge et al. 1992). At neutral and slightly alkaline pH, the ATPase activity of the enzyme from *P. modestum* is coupled to translocation of sodium

---

[1] This fact explains why Schink and Pfenig (1982) called this microbe *modestum*, i.e., modest with respect to its energy requirement.

ions across the membrane. Interestingly, the same enzyme can also transport protons when pH is low and Na$^+$ ions are absent from the medium.

Besides sodium ions and protons, ATPase from *P. modestum* is also able to transport Li$^+$, but its affinity for this ion is substantially lower. Ion specificity of Na$^+$-ATPase is determined by the F$_o$ factor. For instance, the chimeric enzyme composed of the F$_o$ factor from *P. modestum* and the F$_1$ factor from *E. coli* demonstrates the properties of the typical Na$^+$-pump (Gerike et al. 1995).

Besides *P. modestum*, Na$^+$-ATP-synthases have been described in such bacteria as *Acetobacterium woodii* (Müller et al. 2001) and *Ilyobacter tartaricus* (Neumann et al. 1998). These enzymes seem to be rather widespread among microorganisms, in particular in methanogenic archaea (Mulkidjanian et al. 2008).

13.3.2 $\Delta\bar{\mu}_{Na^+}$ Consumers Performing Chemical Work in Methanogenic Archaea

It has been already described that two enzymes of methanogenic archaea, i.e., methyltransferase complex and formylmethanofuran dehydrogenase, function as primary sodium pumps (Gottschalk and Thauer 2001; Poorter et al. 2003). However, under certain conditions these enzymes can catalyze the reverse reactions, i.e., they can utilize $\Delta\bar{\mu}_{Na^+}$ for performing chemical work (e.g. for methyl group transfer from methyl-CoM to H$_4$MPT and for reduction of CO$_2$ to formyl-MFR).

In the case of a methanogenic process such as:

$$CH_3COOH \rightarrow CH_4 + CO_2 \qquad (13.3)$$

$$CO_2 + 4\ H_2 \rightarrow CH_4 + 2\ H_2O \qquad (13.4)$$

$$H_2CO + 3\ H_2 \rightarrow CH_4 + 2\ H_2O, \qquad (13.5)$$

the Na$^+$-translocating methyltransferase complex functions as the sodium potential generator, but in the case of methanol molecules disproportionation to methane and carbon dioxide:

$$4\ CH_3OH \rightarrow 3\ CH_4 + CO_2 + 2\ H_2O \qquad (13.6)$$

the same enzyme catalyzes the reverse reaction and acts as the $\Delta\bar{\mu}_{Na^+}$ consumer.

Na$^+$-translocating formylmethanofuran dehydrogenase functions as a sodium potential generator when participating in conversion of formate or methanol to methane and carbon dioxide:

$$4\ HCOOH \rightarrow CH_4 + 3\ CO_2 + 2\ H_2O \qquad (13.7)$$

$$4\ CH_3OH \rightarrow 3\ CH_4 + CO_2 + 2\ H_2O. \qquad (13.8)$$

However, on reducing carbon dioxide to methane by molecular hydrogen:

$$CO_2 + 4 H_2 \rightarrow CH_4 + 2 H_2O \tag{13.9}$$

this enzyme acts as $\Delta\bar{\mu}_{Na^+}$ consumer.

It is important to note that in the case of these methanogenic processes, when one out of two sodium-dependent enzymes functions as a $\Delta\bar{\mu}_{Na^+}$ consumer, the other Na^+-translocating enzyme always acts as the sodium pump, generating the $\Delta\bar{\mu}_{Na^+}$ necessary for the consumers.

References

Asai Y, Yakushi T, Kawagishi I, Homma M (2003) Ion-coupling determinants of Na^+-driven and H^+-driven flagellar motors. J Mol Biol 327:453–463

Atsumi T, McCarter L, Imae Y (1992) Polar and lateral flagellar motors of marine *Vibrio* are driven by different ion-motive forces. Nature 355:182–184

Bassilana M, Damiano-Forano E, Leblanc G (1985) Effect of membrane potential on the kinetic parameters of the Na^+ or H^+ melibiose symport in *Escherichia coli* membrane vesicles. Biochem Biophys Res Commun 129:626–631

Bogachev AV, Murtasina RA, Shestopalov AI, Skulachev VP (1993) The role of proton and sodium potentials in motility of *Escherichia coli* and *Bacillus FTU*. Biochim Biophys Acta 1142:321–326

de Poorter LMI, Geerts WG, Theuvenet APR, Keltjens JT (2003) Bioenergetics of the formylmethanofuran dehydrogenase and heterodisulfide reductase reactions in *Methanothermobacter thermautotrophicus*. Eur J Biochem 270:66–75

Dibrov PA, Kostyrko VA, Lazarova RL, Skulachev VP, Smirnova IA (1986) The sodium cycle. I. Na^+-dependent motility and modes of membrane energization in the marine alkalotolerant *Vibrio alginolyticus*. Biochim Biophys Acta 850:449–457

Dimroth P, Thomer A (1986) Citrate transport in *Klebsiella pneumoniae*. Biol Chem Hoppe-Seyler 367:813–823

Garrahan PJ, Glynn IM (1966) Driving the sodium pump backwards to form adenosine triphosphate. Nature 211:1414–1415

Gerike U, Kaim G, Dimroth P (1995) In vivo synthesis of ATPase complexes of *Propionigenium modestum* and *Escherichia coli* and analysis of their function. Eur J Biochem 232:596–602

Gottschalk G, Thauer RK (2001) The Na^+-translocating methyltransferase complex from methanogenic archaea. Biochim Biophys Acta 1505:28–36

Greene RV, MacDonald RE (1984) Partial purification and reconstitution of the aspartate transport system from *Halobacterium halobium*. Arch Biochem Biophys 229:576–584

Guffanti AA, Cohn DE, Kaback HR, Krulwich TA (1981) Relationship between the Na^+/H^+ antiporter and Na^+/substrate symport in *Bacillus alcalophilus*. Proc Natl Acad Sci U S A 78:1481–1484

Hilpert W, Dimroth P (1984) Reconstitution of Na^+ transport from purified methylmalonyl-CoA decarboxylase and phospholipid vesicles. Eur J Biochem 138:579–583

Hirota N, Imae Y (1983) Na^+-driven flagellar motors of an alkalophilic *Bacillus* strain YN-1. J Biol Chem 258:10577–10581

Hopfer U (1978) Transport in isolated plasma membranes. Am J Physiol 234:F89–F96

Kakinuma Y, Unemoto T (1985) Sucrose uptake is driven by the Na^+ electrochemical potential in the marine bacterium *Vibrio alginolyticus*. J Bacteriol 163:1293–1295

References

Kitada M, Guffanti AA, Krulwich TA (1982) Bioenergetic properties and viability of alkalophilic *Bacillus firmus* RAB as a function of pH and Na^+ contents of the incubation medium. J Bacteriol 152:1096–1104

Kluge C, Laubinger W, Dimroth P (1992) The Na^+-translocating ATPase of *Propionigenium modestum*. Biochem Soc Trans 20:572–577

Krishnamurthy H, Piscitelli CL, Gouaux E (2009) Unlocking the molecular secrets of sodium-coupled transporters. Nature 459:347–355

Lee SH, Pritchard JB (1983) Proton-coupled L-lysine uptake by renal brush border membrane vesicles from mullet (*Mugil cephalus*). J Membr Biol 75:171–178

Lee SH, Cohen NS, Jacobs AJ, Brodie AF (1979) Isolation, purification, and reconstitution of a proline carrier protein from *Mycobacterium phlei*. Biochemistry 18:2232–2239

Mulkidjanian AY, Galperin MY, Makarova KS, Wolf YI, Koonin EV (2008) Evolutionary primacy of sodium bioenergetics. Biol Direct 3:13

Müller V, Aufurth S, Rahlfs S (2001) The Na^+ cycle in *Acetobacterium woodii*: identification and characterization of a Na^+ translocating F_1F_o-ATPase with a mixed oligomer of 8 and 16 kDa proteolipids. Biochim Biophys Acta 1505:108–120

Neumann S, Matthey U, Kaim G, Dimroth P (1998) Purification and properties of the F_1F_o ATPase of *Ilyobacter tartaricus*, a sodium ion pump. J Bacteriol 180:3312–3316

Paulick A, Koerdt A, Lassak J, Huntley S, Wilms I, Narberhaus F, Thormann KM (2009) Two different stator systems drive a single polar flagellum in *Shewanella oneidensis* MR-1. Mol Microbiol 71:836–850

Schink B, Pfennig N (1982) *Propionigenium modestum* gen. nov. sp. nov., a new strictly anaerobic, nonsporing bacterium growing on succinate. Archiv Microbiol 133:209–216

Skulachev VP (1988) Membrane Bioenergetics. Springer-Verlag, Berlin

Slonczewski JL, Rosen BP, Alger JR, Macnab RM (1981) pH homeostasis in *Escherichia coli*: measurement by ^{31}P nuclear magnetic resonance of methylphosphonate and phosphate. Proc Natl Acad Sci U S A 78:6271–6275

Yorimitsu T, Sato K, Asai Y, Kawagishi I, Homma M (1999) Functional interaction between PomA and PomB, the Na^+-driven flagellar motor components of *Vibrio alginolyticus*. J Bacteriol 181:5103–5106

Chapter 14
Relations Between the Proton and Sodium Worlds

14.1 How Often is the Na$^+$ Cycle Used by Living Cells?

Figure 14.1 shows that effective utilization of $\Delta\bar{\mu}_{H^+}$ in its usual direction is possible only at *neutral or acidic pH$_{out}$*. Here, H$^+$ pumping from the cell generates $\Delta\Psi$ and/or ΔpH, the interior negative and/or alkaline, conditions which are favorable for H$^+$ influx. The same process at *alkaline pH$_{out}$* results in the generation of $\Delta\Psi$ (interior negative) that is counterbalanced by ΔpH of the opposite direction (at high outer pH, the interior must always be maintained more acidic than the outer medium since alkalinization of the cytoplasm inactivates intracellular enzymes). Thus, the H$^+$ ion, being exported from the cell, cannot perform any work when returning to the cytoplasm since it comes to the cell down $\Delta\Psi$, but against ΔpH. Substitution of the coupling ion (H$^+$ by Na$^+$) can solve the problem as in the case of *Vibrio alginolyticus*. Other alkalotolerant and alkaliphilic bacteria also seem to energize their cytoplasmic membrane by mechanism(s) requiring no proton potential.

An *Enterococcus hirae* mutant lacking H$^+$-ATPase, as well as the wild strain of the same microorganism growing in the presence of protonophore, can serve as examples to show that an alkaline medium is not the only reason for prokaryotes to develop Na$^+$ energetics (see Sect. 12.5.1). A similar situation occurs in certain anaerobes (as well as in *Salmonella typhimurium*) employing Na$^+$-motive decarboxylases. For instance, *Propionigenium modestum*, using the Na$^+$ cycle, grows at pH 6.5–8.4 with an optimum between 7.1 and 7.7 (Schink and Pfennig 1982).

Significantly, certain partial reactions of the sodium cycle are inherent in such a neutrophile as *Escherichia coli* accumulating some nutrients in symport with Na$^+$ ions. This mechanism for osmotic work is the major one in many marine and halophilic bacteria as well as in the animal outer cell membrane (see the preceding Sect. 13.1). Again here, the reason for the use of $\Delta\bar{\mu}_{Na^+}$ must be other than alkaliphilism or alkali tolerance.

Fig. 14.1 Role of $\Delta\bar{\mu}_{H^+}$ for bacteria growing at different pH$_{out}$: $\Delta\bar{\mu}_{H^+}$ can be effectively utilized by acidophiles and neutrophiles but not by alkaliphiles, where electric potential and H$^+$ gradient are oppositely directed

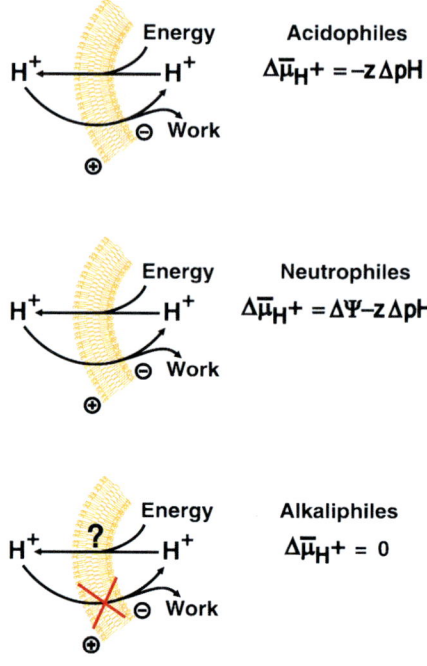

14.2 Probable Evolutionary Relationships of the Proton and Sodium Worlds

One can assume that evolutionarily, H$^+$, rather than Na$^+$, was the initial coupling ion. It is easy to imagine how a primitive cell could invent a $\Delta\bar{\mu}_{H^+}$-generator functioning as a Mitchellian redox loop (Mitchell 1961). In the redox loop mechanism, a *hydrogen atom* acceptor, e.g., quinone Q, is reduced by an *electron* donor, e.g., by inner Fe^{2+} of a cytochrome. If the process takes place near the inner surface of the bacterial membrane, *hydrogen ions* should be taken up from the inner aqueous space of the cell bathing this surface:

$$Q_{in} + 2\ Fe^{2+}_{in} + 2\ H^+_{in} \rightarrow QH_{2in} + 2\ Fe^{3+}_{in}. \qquad (14.1)$$

Then QH$_2$ diffuses down its gradient to the opposite (outer) membrane surface (14.2) and is oxidized there by an *electron* acceptor, e.g., by another outer Fe^{3+} of the cytochrome. As a result, *hydrogen ions* are released into the outer medium:

$$QH_{2in} \rightarrow QH_{2out} \qquad (14.2)$$

$$QH_{2out} + 2\ Fe^{3+}_{out} \rightarrow Q_{out} + 2\ Fe^{2+}_{out} + 2\ H^+_{out} \qquad (14.3)$$

Then electrons are transported from the outer Fe^{2+} to the inner Fe^{3+} and Q_{out} diffuse back to the inner membrane surface.

$$Fe^{2+}_{out} + Fe^{3+}_{in} \rightarrow Fe^{3+}_{out} + Fe^{2+}_{in} \qquad (14.4)$$

$$Q_{out} \rightarrow Q_{in} \qquad (14.5)$$

Thus, the very fact that it is the electrochemical potential difference of the *hydrogen* ions that is produced by the oxidative process appears to be the inevitable consequence of the chemistry of the energy-supplying oxidative reaction. Such a redox loop scheme was directly shown for certain $\Delta\bar{\mu}_{H^+}$ generators (the photosynthetic reaction centers of purple and green bacteria, for photosystems I and II in chloroplasts and cyanobacteria, Q-cycle, see Chaps. 2 and 4).

On the other hand, there are $\Delta\bar{\mu}_{H^+}$ generators operating with no H/ē-antiport involved, such as transhydrogenase (see Sect. 7.4) and bacteriorhodopsin (see Chap. 6). In these cases, $\Delta\bar{\mu}_{H^+}$ formation is not a direct consequence of the energy-releasing process, and thus the energy coupling is organized in a more sophisticated manner. Systems of this type (ion pumps) are not inevitably connected with the transport of hydrogen ions. Therefore, it is not surprising that similar, indirect coupling mechanisms can also be applied to the transport of ions other than H^+. For instance, bacteriorhodopsin pumps H^+ ions, whereas a bacteriorhodopsin analog, halorhodopsin, pumps Cl^- ions (see Sect. 6.5.1).

The sodium potential generators clearly fall into a class of indirect energy transducers that can be regarded as the latest in the evolution of membrane bioenergetics.

In cases when the uptake and release of H^+ ions are not a direct consequence of the chemistry of the energy-producing reaction, *substitution* of H^+ with Na^+ can have some advantages. As already discussed, *this makes cell energetics resistant to a decrease in the outer concentration of H^+.*

Moreover, *the conductance of the lipid membrane for H^+ is at least 10^6-fold higher than for Na^+* (Gutknecht 1984). This value should be even higher if the lipid bilayer contains inclusions of proteins that have proton-acceptor groups that can facilitate H^+ conductance through the hydrophobic membrane layer. On the other hand, H^+ concentration in the medium is usually about 10^6-fold lower than that of Na^+. The interplay of these two parameters can result in a situation in which it is easier for the cell to retain the Na^+, rather than the H^+ gradient. This is apparently not the case for extreme halophiles living in a saturating NaCl solution. This might explain why they use the H^+ cycle in spite of a very high $[Na^+]_{out}$.

The last and perhaps most significant point is that *intracellular enzymes are much more sensitive to fluctuation in H^+ than in Na^+ concentration*, especially if, e.g., a decrease in the $[Na^+]_{in}$ level is accompanied by an increase in the $[K^+]_{in}$ level. Thus, a cell pumping out Na^+, not H^+, appears more evolutionarily progressive than one pumping out H^+. In such a cell, pH_{in} becomes independent of the rate of the membrane-linked energy-coupling processes and, vice versa, these processes are not directly linked to cytoplasmic pH changes.

Proceeding from the above line of reasoning, one can explain why an evolutionarily young membrane of the most progressive kingdom of living organisms such as the animal outer cell membrane employs Na^+ as the coupling ion.

It has already been mentioned that very high $[Na^+]_{out}$ can be unfavorable for Na^+ energetics. On the other hand, $[Na^+]_{out}$ should not be too low, otherwise an Na^+ gradient of proper direction ($[Na^+]_{in} < [Na^+]_{out}$) would be impossible to maintain. Apparently, $[Na^+]_{out}$ is high enough, but not too high, in the case of animal cells and microorganisms other than freshwater and halophilic bacteria. It is noteworthy that both *V. alginolyticus* and *P. modestum*, the most demonstrative examples of Na^+-cycle studies, are marine isolates.

To reveal the taxonomic links between the living organisms employing H^+ or Na^+ as coupling ions, the search for genes that code H^+- and Na^+-motive enzymes among the large number of known bacterial genomes was conducted (Häse et al. 2001; Mulkidjanian et al. 2008). The broad distribution of the $\Delta\bar{\mu}_{Na^+}$-employing genera in various taxonomic groups indicates that the sodium cycle was discovered by living systems at a rather early stage of evolution. Barry Rosen even suggested that the Na^+ cycle had appeared before H^+ energetics, and substitution of Na^+ with H^+ occurred when marine microorganisms spread to rivers and freshwater lakes (Rosen 1986). Like many evolutionary speculations, this hypothesis is certainly difficult to exclude [it has recently been supported by Armen Mulkidjanyan and his colleagues (Mulkidjanian et al. 2008)]. However, if we proceed from the arguments given in the beginning of this section, Rosen's point of view would appear less probable than the alternative concept assuming *H^+ as the evolutionarily primary coupling ion*.

Presuming that proton energetics appeared before sodium energetics, one can visualize a *possible way of Na^+ cycle evolution*. The use of Na^+ and K^+ gradients as $\Delta\bar{\mu}_{H^+}$ buffers could be the initial step (see the preceding, Sect. 13.1). Direct utilization of $\Delta\bar{\mu}_{Na^+}$ formed by the Na^+/H^+-antiporter to support, e.g., osmotic work via Na^+, metabolite symporters is likely to have been the next step toward the sodium cycle. This cycle was then completed when biological evolution discovered primary $\Delta\bar{\mu}_{Na^+}$ generators operating with no $\Delta\bar{\mu}_{H^+}$ involved.

Most of these events apparently took place at a rather early stage of evolution, so that now some constituents of the sodium cycle can be found in quite different taxa: *Vibrio, Propionigenium, Salmonella, Clostridium, Streptococcus*, methanogenic archaea, and finally in animal cells. Perhaps even the "classical" *E. coli* has some sodium world representatives among its remote ancestors. This can explain why it uses $\Delta\bar{\mu}_{Na^+}$ to import certain metabolites into the cell.

14.3 Membrane-Linked Energy Transductions Involving Neither H^+ Nor Na^+

Considering the interrelations of the proton and sodium worlds, one can ask the question whether yet another (a third) world exists where *membrane-linked energy coupling occurs with neither H^+ nor Na^+ involved*. So far, the answer has been negative.

Yet, this does not mean that among all the diversity of membrane-linked energy transductions, it is impossible to find those that are H^+- and Na^+-independent. However, these processes are rather unique exceptions and certainly cannot be brought together into a unified system of interconversions of various forms of energy, as occurs for proton or sodium energetics.

Let us consider, for instance, the bioenergetic functions of K^+ ions. As already mentioned in Sect. 11.3.1, in bacteria ΔpK can be formed at the expense of the $\Delta\Psi$ constituent of a proton or sodium potential produced by $\Delta\bar{\mu}_{H^+}$ ($\Delta\bar{\mu}_{Na^+}$) generators. When *E. coli* grows at a very low $[K^+]_{out}$, an E_1E_2-type K^+-ATPase (*Kdp*) is induced (Epstein and Laimins 1980). This enzyme accumulates K^+ inside the cell at the expense of the energy of ATP. The functions of the formed ΔpK are limited to stabilization of $\Delta\Psi$ and osmotic pressure.

In animal tissues, the K^+ gradient is usually formed by Na^+/K^+-ATPase. The role of K^+ is limited to the functions of a substituent for Na^+ that is pumped from the cell to form $\Delta\bar{\mu}_{Na^+}$, a convertible energy currency of the animal outer cell membrane.

In, Sect. 7.2.3.2, K^+/H^+-ATPase of the gastric mucosa epithelium of higher animals was considered. Here K^+ plays the role of a substitute for H^+, which is pumped from the cell to acidify the gastric juice.

In epithelia of some insects, an electrogenic K^+-ATPase responsible for the transport of massive amounts of K^+ from the basal to apical sides of the epithelial cells was discovered. This process is essential for fluid movement through the epithelium. Unlike Na^+/K^+-ATPase or K^+/H^+-ATPase, insect K^+-ATPase pumps K^+ out of the cell, being located on the apical plasma membrane (Harvey et al. 1983). It appears to reside in 10-nm particles generally resembling factor F_1. The activity is resistant to ouabain and specifically sensitive to an insecticide, the δ-endotoxin from *Bacillus thuringiensis*.

In the midgut epithelium of the tobacco hornworm, K^+-ATPase is responsible for the generation of $\Delta\Psi$ used to import H^+ into the cell (Giordana et al. 1985). As a result, the pH in the gut lumen is maintained at the level of 10–11. In this case, the total $\Delta\bar{\mu}_{K^+}$ or one of its constituents is likely to be the driving force for the symport of amino acids and K^+ into the cytoplasm of epithelium cells.

A Ca^{2+}-ATPase is present in the endoplasmic reticulum and plasma membrane of animal cells and in the cytoplasmic membrane of some bacteria. This *E_1E_2-type ATPase pumps Ca^{2+}* from cytosol to the reticulum interior or to the extracellular medium. Most probably, the main function of the Ca^{2+}-ATPase is to regulate cytosolic $[Ca^{2+}]$.

There are several examples of $\Delta\bar{\mu}_{H^+}$ ($\Delta\bar{\mu}_{Na^+}$)-independent systems of anion transport. One of them, the light-dependent Cl^- pump of halorhodopsin of *H. salinarium*, was described in , Sect. 6.5.1. This pump returns to the cytoplasm Cl^- ions which escape the cytoplasm moving electrophoretically to the cell exterior. In a mollusk, *Aplysia californica*, a similar function is performed by the plasma membrane Cl^--ATPase, a vanadate-sensitive enzyme (Gerencser 1990).

A salt marsh plant, *Limonium vulgaris*, excretes salt through a special gland on the leaf surface. The plasmalemma of the cells composing this salt gland was shown to contain Cl^--ATPase (Auffret and Hanke 1981). It was suggested that this pump exports Cl^- from the cell and generates $\Delta\Psi$ (positive inside the cell) that is then used to electrophoretically export Na^+.

Certain nonionized compounds were shown to be accumulated in bacteria with no $\Delta\bar{\mu}_{H^+}$ ($\Delta\bar{\mu}_{Na^+}$) involved. The *phosphoenolpyruvate transferase system* is the most elaborated example of such a mechanism. It catalyzes two processes: phosphorylation of glucose by PEP and accumulation of this sugar in the form of glucose 6-phosphate inside some bacterial cells. As a result, the energy difference between phosphoryl groups in PEP and glucose 6-phosphate is utilized to obtain a large sugar gradient across the cytoplasmic membrane.

In summary, we would like to stress that all the above-mentioned systems, rather peculiar indeed, represent mechanisms of osmotic work. At the same time, the great majority of osmotic work processes are supported by $\Delta\bar{\mu}_{H^+}$ or $\Delta\bar{\mu}_{Na^+}$. As for membrane-linked chemical and mechanical work, proton and sodium potentials (or ATP) are utilized as the driving forces in all the known cases (Skulachev 1988).

References

Auffret CA, Hanke DE (1981) Improved preparation and assay and some characteristics of Cl^--ATPase activity from *Limonium vulgare*. Biochim Biophys Acta 648:186–191

Epstein W, Laimins L (1980) Potassium transport in *Escherichia coli*: diverse systems with common control by osmotic forces. Trends Biochem Sci 5:21–23

Gerencser GA (1990) Reconstitution of a chloride-translocating ATPase from *Aplysia californica* gut. Biochim Biophys Acta 1030:301–303

Giordana B, Parenti P, Hanozet GM, Sacchi VF (1985) Electrogenic K^+-basic amino-acid cotransport in the midgut of lepidopteran larvae. J Membr Biol 88:45–53

Gutknecht J (1984) Proton/hydroxide conductance through lipid bilayer membranes. J Membr Biol 82:105–112

Harvey WR, Cioffi M, Dow JA, Wolfersberger MG (1983) Potassium ion transport ATPase in insect epithelia. J Exp Biol 106:91–117

Häse CC, Fedorova ND, Galperin MY, Dibrov PA (2001) Sodium ion cycle in bacterial pathogens: evidence from cross-genome comparisons. Microbiol Mol Biol Rev 65:353–370

Mitchell P (1961) Coupling of phosphorylation to electron and hydrogen transfer by a chemi-osmotic type of mechanism. Nature 191:144–148

Mulkidjanian AY, Galperin MY, Makarova KS, Wolf YI, Koonin EV (2008) Evolutionary primacy of sodium bioenergetics. Biol Direct 3:13

Rosen BP (1986) Recent advances in bacterial ion transport. Annu Rev Microbiol 40:263–286

Schink B, Pfennig N (1982) *Propionigenium modestum* gen. nov. sp. nov. a new strictly anaerobic, nonsporing bacterium growing on succinate. Archiv Microbiol 133:209–216

Skulachev VP (1988) Membrane bioenergetics. Springer, Berlin

Part VI
Mitochondrial Reactive Oxygen Species and Mechanisms of Aging

Chapter 15
Concept of Aging as a Result of Slow Programmed Poisoning of an Organism with Mitochondrial Reactive Oxygen Species

In this part of the book, we will apply the data discussed in earlier chapters on biological membrane energetics to the solution of a quite practical problem which is of great importance, namely an attempt to retard or perhaps even eliminate human aging.

The main points of our concept are as follows:

(1) Aging of living organisms is the final stage of the program of ontogenesis.
(2) Aging is one of the mechanisms accelerating biological evolution. Minor favorable traits that are not essential for a strong young organism can be of vital importance (and hence become an object of natural selection) for organisms weakened by aging.
(3) Humans have already practically ceased to be objects of biological evolution. We do not adapt to the environment—we rather change the environment according to our purposes. Thus, in the case of human beings, aging (as well as any other evolutionary mechanism) is an atavism that we can try to eliminate because it is obviously counterproductive for our bodies.
(4) The increase in reactive oxygen species (ROS) level in mitochondria is one of the necessary steps of the aging program. ROS cause the programmed death of cells. This usually results in a decrease in cellularity of organs and tissues, which leads to the weakening of numerous physiological functions and hence to senile diseases and eventually to death.
(5) The aging program may be interrupted by removing excessive ROS from the mitochondrial interior, a place where ROS are produced. Mitochondria can be specifically loaded with cation derivatives of rechargeable antioxidants that penetrate freely the mitochondrial membrane and accumulate electrophoretically inside the negatively charged mitochondria. As such antioxidants, one can employ compounds comprising quinones and "molecular electric locomotives", i.e., synthetic cations with charge being strongly delocalized over a hydrophobic molecule (Skulachev 2003a, 2007, 2009; Longo et al. 2005; Skulachev and Longo 2005; Severin and Skulachev 2011).

Until recently, the vast majority of gerontologists have considered even the first point of the above-described concept to be completely dissident. However, the situation has been changing rapidly over the last few years, and at the end of 2005 we managed to publish the article entitled "Programmed and altruistic aging" in such a prestigious and respectable journal as *Nature* (Longo et al. 2005). At the same time, we started a broad investment project on the practical use of penetrating cationic antioxidants. The main goal of the project was to try to find a way to interrupt transmission of the aging signals from the ontogenesis-controlling "master clock" to intracellular molecular targets (Skulachev 2007).

15.1 Nature of ROS and Paths of their Formation in the Cell

Almost all the oxygen consumed by living organisms is transformed into such a harmless compound as water. This is the result of reaction of the O_2 molecule with 4 e^- and 4 H^+, catalyzed by cytochrome oxidase. However, a minor portion of the O_2 consumed (usually no more than 1–2 % of the total amount) is transformed into superoxide anion (one-electron reduction of O_2 to anion radical $O_2^{\bullet-}$) or into H_2O_2 (two-electron reduction of O_2). These processes, being seemingly insignificant in the general balance of respiration, nevertheless play a very important role in the regulation of physiological functions of an organism, and even in its final fate. The point is that superoxide anion as well as H_2O_2 can turn into ROS, in particular into hydroperoxyl ($HO_2^{\bullet}$) or $OH^{\bullet}$ radicals and also into peroxynitrite ($ONOO^-$) capable of producing $^{\bullet}ONOO$, singlet oxygen ($^1O_2$), and $OH^{\bullet}$ (Fig. 15.1). ROS in turn are known to attack different cell components including DNA, RNA, proteins, lipids, etc. Certain ROS are such aggressive oxidants that even their rather small concentration are sufficient to lead to tragic consequences for mitochondria, cells, and even for the whole organism. If only about 1 % of the 400 l of O_2 consumed per day by an adult human of average body mass is converted into $O_2^{\bullet-}$, this would mean the production of 4 l of superoxide. It is easy to imagine the consequences, especially if we take into account the fact that such a product of further superoxide transformation as hydroxyl radical can compete in its toxicity with chlorine radicals used for disinfection (Skulachev 1994; Halliwell and Gutteridge 1999).

ROS formation was once considered to be an inevitable consequence of aerobic life, for small uncharged and rather hydrophobic O_2 molecules capable of penetrating through the membranes and even accumulating in them can oxidize some membrane-localized electron donors in a nonenzymatic fashion. The list of these electron donors includes half-reduced (semiquinone) forms of ubiquinone, plastoquinone, menaquinone, flavins, and also FeS clusters of nonheme iron proteins with redox potentials close to the superoxide/oxygen couple.[1] Numerous

---

[1] The standard potential of the $O_2^{\bullet-}/O_2$ couple is about −0.3 V. The actual redox potential, being more positive due to $[O_2] > [O_2^{\bullet-}]$, is still in the negative part of the redox potential scale (Skulachev 1994, 2003a; Halliwell and Gutteridge 1999).

15.1 Nature of ROS and Paths of their Formation in the Cell

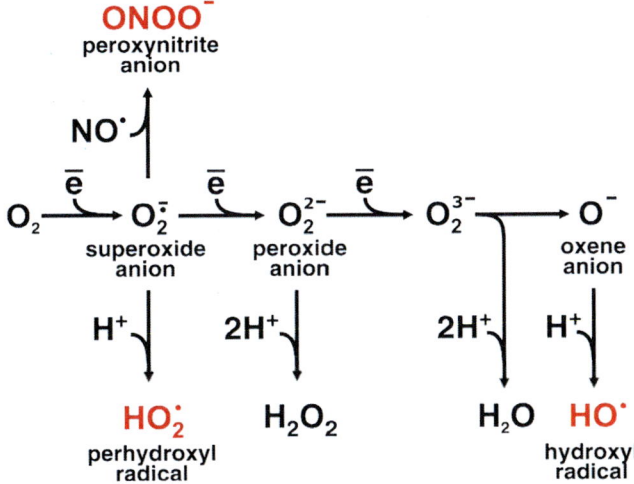

Fig. 15.1 Pathways of oxygen reduction in an organism. Upper scheme, four-electron reduction of O_2 to H_2O, which occurs in bacterial or mitochondrial membranes. Lower scheme, one-electron reduction of O_2 results in formation of superoxide anion. This anion is protonated at acidic pH values, while at neutral pH it is reduced to O_2^{2-} in a dismutation reaction ($2\ O_2^{\bullet-} \rightarrow O_2^{2-} + O_2$); $O_2^{\bullet-}$ can also combine with $NO^{\bullet}$, thus forming peroxynitrite anion. O_2^{2-} is protonated, thus forming H_2O_2, or it is reduced to O_2^{3-}, which is then transformed to H_2O and $OH^{\bullet}$ radical (Skulachev 1994; Halliwell and Gutteridge 1999)

publications in the last few years have provided data supporting the view that ROS formation and utilization are thoroughly controlled by the organism. As a result, ROS concentration in an organism can vary greatly depending on the functioning of special regulating systems (Halliwell and Gutteridge 1999; Skulachev 2003a; Tait and Green 2010; Rigoulet et al. 2011; Naik and Dixit 2011; Powers et al. 2011).

In the case of the most powerful ROS generators, such as complexes I and III (and also complex II, dihydrolipoate dehydrogenase, and cytochrome P450), $O_2^{\bullet-}$ formation appears to be a result of leakage of electrons that, instead of being transferred along the respiratory chain to cytochrome oxidase (or in case of cytochrome P450, instead of participating in substrate oxygenation), are taken by molecular oxygen by means of its one-electron reduction. Due to the huge amount

Table 15.1 ROS generators in a living cell

Enzyme	ROS type	How are ROS formed?	Intracellular localization
Complex I	$O_2^{\bullet-}$	Leakage electrons to O_2	Inner leaflet of inner mitochondrial membrane
Complex III	$O_2^{\bullet-}$	"	Outer leaflet of inner mitochondrial membrane
Complex II	$O_2^{\bullet-}$	"	Inner leaflet of inner mitochondrial membrane
Dihydrolipoate dehydrogenase	$O_2^{\bullet-}$	"	Mitochondrial matrix
Cytochrome P450	$O_2^{\bullet-}$	"	Endoplasmic reticulum of liver, inner membrane of adrenal cortex mitochondria
NADPH oxidase	$O_2^{\bullet-}$	Main product of enzymatic reaction	Outer cell membrane
Monoamine oxidase	H_2O_2	"	Outer mitochondrial membrane
Oxidases of uric acid, D- and L- amino acids	H_2O_2	"	Peroxisome
Xanthine oxidase	$O_2^{\bullet-}$, H_2O_2	"	Milk; cytosol

of oxygen consumed by the respiratory chain, the leakage of even a relatively small number of electrons would result in ROS formation in quantities substantially greater than those generated by the enzymes specialized in O_2 reduction to $O_2^{\bullet-}$ or H_2O_2, i.e., monoamine oxidase of the outer mitochondrial membrane, plasmalemma NADPH oxidase, peroxisomal oxidases, xanthine oxidase, etc. (Table 15.1). The greater input of the respiratory chain into general ROS formation corresponds to the *quantitative* prevalence of these enzymes over other oxidases and oxygenases. The respiratory chain enzymes are the main components of the inner mitochondrial membrane, the total surface of which in case of the human organism is about 14,000 m^2. Direct measurements of ROS distribution in different areas of an intact cell have shown that, e.g., in case of yeast cell suicide, an initial outburst of ROS formation takes place just in its mitochondria (see Fig. 15.9a).

15.2 How Do Living Systems Protect Themselves from ROS?

15.2.1 Antioxidant Compounds

Traditionally, antioxidant compounds neutralizing the toxic oxygen derivatives have been considered to be the main line of defense against ROS. The antioxidant is oxidized while protecting other more important molecules such as DNA, RNA, proteins, phospholipids, etc. Sometimes though there is a possibility to regenerate the oxidized antioxidant to its original active reduced form. $CoQH_2$, for instance, can reduce free-radical forms of lipids while being oxidized to $CoQ^{\bullet-}$. The formed $CoQ^{\bullet-}$ is regenerated to the original $CoQH_2$ by means of its reduction by complexes I, II, or III.

Antioxidants are usually divided into two groups: water soluble, acting in hydrophilic areas of a cell (ascorbic acid, SH-glutathione, cysteine, etc.) and lipid soluble, protecting first of all membrane structures (CoQ, plastoquinone, vitamins A, D, E, and K) (Skulachev 1994; Halliwell and Gutteridge 1999).

15.2.2 Decrease in Intracellular Oxygen Concentration

Recently, increased attention has been given to the biological mechanisms preventing ROS formation rather than to neutralizing already formed ROS.

All aerobic organisms have a multilevel system of such anti-ROS protection. Decreasing the intracellular oxygen concentration seems to be the simplest solution to the problem. The fact is that cytochrome oxidase as well as other terminal oxidases reducing O_2 to H_2O retains their maximal activity even when oxygen concentration is decreased to only several percents of its value in water under the regular oxygen pressure in air. At the same time, the rate of O_2 reduction to $O_2^{\bullet-}$ in mitochondria depends linearly on $[O_2]$ in the entire oxygen concentration range. That is why a 10–20-fold decrease in intracellular $[O_2]$ level causes dramatic inhibition of $O_2^{\bullet-}$ production without a substantial inhibition of the cytochrome oxidase reaction and hence respiration-coupled ATP synthesis. Measurements of $[O_2]$ in animal tissues have shown it to be about one order of magnitude lower than in air-saturated water (area B in Fig. 15.2). This area seems to be the optimal one, as in area A $[O_2]$ is too low to support high enough rate of cytochrome oxidase reaction, while in area C the increase in$[O_2]$ facilitates one-electron reduction of oxygen without any benefit for the cytochrome oxidase (Skulachev 1996a).

In animal tissues (except in lungs), $[O_2]$ is maintained at a level substantially lower than in equilibrium with air. The reason for this phenomenon is that in the resting state the rate of O_2 supply from the lungs to other organs is kept far lower than the maximal possible. This rate increases greatly during the rest-to-work transition. Under such a transition, lung ventilation, heartbeat, and blood flow rate increase while blood vessels dilate. Parallel to this, the oxygen consumption by cytochrome oxidase is increased because of the appearance of available ADP (see Sect. 7.1). However, even under these new conditions the ratio between the rate of oxygen consumption and supply was shown to be regulated in a way that intracellular $[O_2]$ is kept at a level no more than 10 % of its saturation in air. In addition, the organism monitors ROS concentration in the blood by special sensors, causing constriction of blood vessels in response to increased $[O_2^{\bullet-}]$ (Skulachev 1996a).

The described strategy is impossible in case of unicellular organisms. Our calculations showed that even *Azotobacter*, which has the highest respiration rate per mg protein among living organisms (see Sect. 5.2.4), cannot substantially decrease intracellular $[O_2]$ because the distance between the bacterial cytosol and the outer medium is so short. O_2 consumption by cytochrome oxidase is immediately compensated by O_2 influx from the outside. It is the polysaccharide "coat" surrounding the *Azotobacter* cell that provides the solution to the problem in this case. The "coat" prevents the convection of the liquid in the area closest to the cell wall, and this substantially slows the movement of O_2 from the medium toward the bacterium. In the case of *Azotobacter*, the problem of protection against oxygen appears to be the most crucial in media lacking ammonium salts. Under such conditions, the very survival of *Azotobacter* depends on its ability to reduce molecular nitrogen to ammonia catalyzed by a special enzyme—nitrogenase. This nitrogenase is extremely sensitive to O_2, and it is inactivated in the presence of even trace amounts of O_2. It is remarkable that initiation of nitrogen fixation is accompanied by induction of a special simplified respiratory chain in *Azotobacter*. This chain consumes oxygen much faster than the canonical one consisting of complexes I, III, and IV. The simple chain (it was given the name *the respiratory protection chain*) consists of only two enzymes: (1) noncoupled NADH dehydrogenase II with its only redox center (FAD) that catalyzes CoQ reduction by NADH and (2) *bd* quinol oxidase that oxidizes thus formed $CoQH_2$ by oxygen. The second reaction generates $\Delta\bar{\mu}_{H^+}$ with efficiency $H^+/e^- = 1$, i.e., twice less

15.2 How Do Living Systems Protect Themselves from ROS?

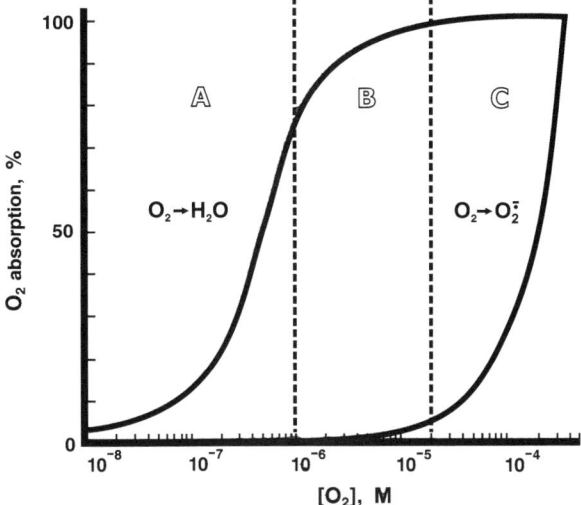

Fig. 15.2 Enzymatic (four-electron) and nonenzymatic (one-electron) reduction of O_2 as a function of oxygen concentration. The $[O_2]$ providing half-maximal rate of enzymatic reduction ($O_2 \rightarrow H_2O$) is assumed to be 3×10^{-7} M. The rate of nonenzymatic reduction ($O_2 \rightarrow O_2^{\bullet-}$) increases linearly with increase in O_2 concentration. The rate of O_2 consumption at the $[O_2]$ level corresponding to atmospheric oxygen pressure (0.22 mM) is taken as 100 %. It is important to remember that in vivo the *absolute* rate of O_2 consumption assumed to be 100 % for the reaction $O_2 \rightarrow H_2O$ is usually at least two orders of magnitude higher than that for the reaction $O_2 \rightarrow O_2^{\bullet-}$ (Skulachev 1996a)

than in the case of complex IV (see Sect. 5.2.4). As a result, the respiratory protection chain transports only 2 H^+ per NADH molecule instead of the 10 H^+ of a canonical chain. Hence, under these conditions *Azotobacter* has to reduce five times more O_2 to obtain the same amount of ATP (see Chap. 5, Fig. 5.6).

Escherichia coli adopted a different mechanism of protection against oxygen. As discovered in our laboratory by Elena Budrene, these bacteria form clusters in response to the appearance of H_2O_2 in the medium (Budrene and Berg 1991, 1995). Each of these clusters was shown to consist of thousands of individual cells. Addition of hydrogen peroxide to the medium causes bacteria to start secreting aspartic acid, which acts as the attractant. The attractant gradient serves as a signal for bacteria to start moving toward each other. As a result, many clusters of bacterial cells appear on a Petri dish with semisolid agar. These clusters can form regular structures of different, sometimes very odd and regular shapes (Fig. 15.3), which is likely to reflect uneven distribution of aspartate in the agar. One can speculate that the outer bacterial layers in each cluster absorb oxygen so as to decrease its concentration in the inner cluster layers. Thus, bacteria of the outer layers are sacrificed to protect those in the interior of the cluster.

A similar strategy is used in very large muscle cells. Here, mitochondria are accumulated right under the outer cell membrane (sarcolemma). These clusters are connected to interfibrillar mitochondria localized inside the muscle fiber (i.e. between the actomyosin filaments) via long mitochondrial cables. Oxygen is supposed to be consumed first by subsarcolemmal mitochondria, the respiratory chain of which produces $\Delta\bar{\mu}_{H^+}$. The produced $\Delta\bar{\mu}_{H^+}$ is then transferred along the mitochondrial cables to interfibrillar mitochondria, which produce ATP at the expense of the energy of $\Delta\bar{\mu}_{H^+}$ (see Chap. 11, Fig. 11.7).

15.2.3 Decrease in ROS Production by the Respiratory Chain

Mild uncoupling is probably one of the mechanisms decreasing ROS production by complexes I and III (Skulachev 1996a). Anatoly Starkov and colleagues from our laboratory showed hydrogen peroxide formation by isolated heart mitochondria oxidizing succinate to be strongly dependent on the $\Delta\Psi$ on the mitochondrial membrane (Fig. 15.4) (Korshunov et al. 1997). In these experiments, the complex I inhibitor rotenone was shown to lower H_2O_2 production by 80 %. Subsequent addition of a complex III inhibitor myxothiazol or protonophorous uncoupler SF 6847 increased the inhibition to 95 %. These relationships may be explained if one assumes that 80 % of H_2O_2 is formed by a process requiring an energy-dependent reverse electron transfer via complex I from succinate (redox potential of the fumarate/succinate couple +0.03 V) to O_2 (redox potential of $O_2/O_2^{\bullet-}$ couple −0.3 V). About 15 % more of the total amount of hydrogen peroxide is formed in the Q-cycle when the semiquinone form of ubiquinone ($CoQ^{\bullet-}$) interacts with O_2 under the conditions of respiratory control, i.e., when high $\Delta\Psi$ blocks electron transfer from heme b_L to heme b_H thus preventing $CoQ^{\bullet-}$ oxidation by the b_L heme (Fig. 15.5; see also Sect. 4.3).

Both mechanisms were switched off when $\Delta\Psi$ was somewhat lowered by an uncoupler, succinate dehydrogenase inhibitor malonate, or by addition of ADP (the latter initiated $\Delta\bar{\mu}_{H^+}$ consumption by H^+-ATP-synthase) (Fig. 15.4). Some data suggest that such a decrease in $\Delta\Psi$ in vivo can be caused by a minor uncoupling of respiration and phosphorylation by fatty acids. Such uncoupling can be mediated by ATP/ADP-antiporter, aspartate/glutamate-antiporter, and also by some other mitochondrial anion carriers that transport not only the anions of their specific substrates (nucleotides, amino acids, etc.), but also anions of fatty acids (see Sect. 10.2.2).

Inactivation of aconitase by superoxide seems to be the next mitochondrial line of defense against ROS (Skulachev 1999a). Superoxide was shown to oxidize one of four nonheme iron ions present in aconitase, the enzyme catalyzing the first reaction of the Krebs cycle. As a result of this oxidation, aconitase loses this ion and is inactivated. Such an inactivation can be reversed by binding an Fe^{2+} ion to aconitase. Inhibition of aconitase by superoxide blocks the Krebs cycle in its initial stage. As it is the Krebs cycle that plays the key role in providing electrons for respiration, its inhibition makes impossible the operation of the respiratory chain,

15.2 How Do Living Systems Protect Themselves from ROS?

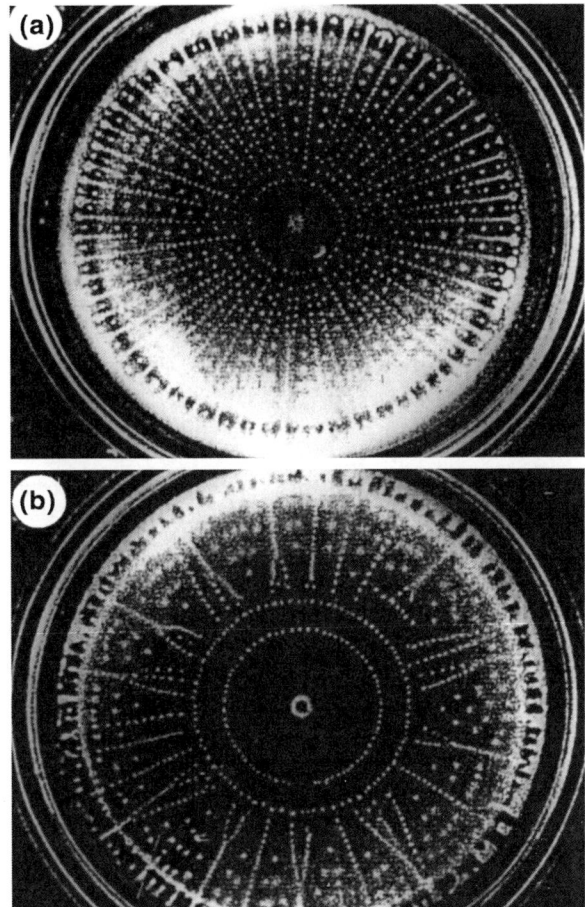

Fig. 15.3 Clusters formed by *E. coli* cells on Petri dishes with semisolid agar. Each white point corresponds to a cluster composed of many bacteria. **a** and **b** show results of two different experiments (Budrene and Berg 1991)

which now can no longer reduce O_2 either to H_2O or to $O_2^{\bullet-}$. Hence, the formation of $O_2^{\bullet-}$, the primary precursor of mitochondrial ROS (mROS), is inhibited.

Inhibition of aconitase also leads to intracellular accumulation of citrate, the substrate of this enzyme. Citrate is a triply charged anion that forms complexes with Fe^{2+} and Fe^{3+}. The citrate^{3-}–Fe^{2+} complex undergoes irreversible autooxidation by O_2, thus changing into the more stable complex citrate^{3-}–Fe^{3+}. For this reason, citrate accumulation under aerobic conditions decreases $[Fe^{2+}]$. The Fe^{2+} ions are the best electron donors for reduction of peroxide ion (O_2^{2-}). This process in return entails the formation of the radical $OH^{\bullet}$, the most dangerous ROS form (see above, Fig. 15.1). Thus, inhibition of aconitase helps mitochondria (1) to decrease $O_2^{\bullet-}$ production and, hence, to lower the formation of all ROS that are derivatives of superoxide, and (2) to decrease generation of $OH^{\bullet}$ from hydrogen peroxide (Fig. 15.6).

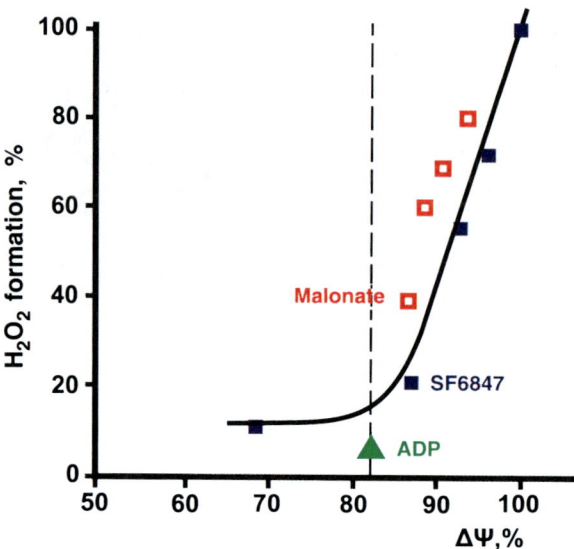

Fig. 15.4 Threshold dependence of hydrogen peroxide formation by succinate-oxidizing mitochondria on the membrane potential value ($\Delta\Psi$). Membrane potential was reduced by: (i) uncoupler SF6847, stimulating respiration by increasing H$^+$-conductivity of the mitochondrial membrane (*black squares*); (ii) malonate, inhibiting electron supply for the respiratory chain (*red squares*), or (iii) adenosine diphosphate, stimulating respiration by activating ATP synthesis (*green triangle*) (Korshunov et al. 1997)

Fig. 15.5 Two sites of ROS formation in the mitochondrial respiratory chain. Respiratory substrate, succinate

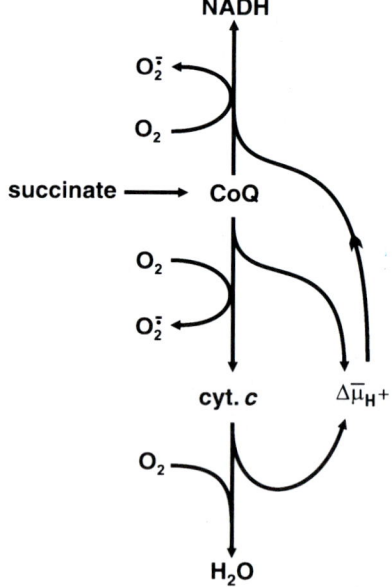

15.2.4 Mitoptosis

Excess ROS cause *mitoptosis* (or suicide of mitochondria), which appears to be the last frontier of cell protection against mitochondria that produce too much ROS (Zorov et al. 1992; Skulachev 2003a; Skulachev and Longo 2005).

Mitoptosis triggered by ROS can develop in several different ways. One includes *pore opening in the inner mitochondrial membrane*. The most likely scenario of this phenomenon is the following. ROS oxidize the matrix-facing SH-groups of ATP/ADP-antiporter (or some other mitochondrial anion carrier). This converts the antiporter into a nonspecific pore (so-called permeability transition pore) that is permeable to compounds of molecular mass <1.5 kDa. This leads to dissipation of $\Delta\bar{\mu}_{H^+}$ and all the other gradients of low molecular mass components between the matrix and intramembrane space of mitochondria. Such a mitochondrion, similar to a ship with all the Kingston valves left open, will inevitably perish. The mitochondrial death will be caused not only by the efflux of NAD^+, $NADP^+$, and other water-soluble cofactors necessary for the normal functioning of this organelle, but also by the termination of such a vitally important reparative process as the replacement of old damaged proteins by new active ones. There are about 600 mitochondrial precursor proteins. One of the stages of their import from the cytosol to mitochondria involves electrophoretic movement of their *positively* charged leader sequence toward the *negatively* charged matrix. This process cannot occur when the pores are open and $\Delta\Psi = 0$. Even those 13–15 proteins which are encoded by mitochondrial DNA cannot be properly arranged in the membrane because this process also requires $\Delta\Psi$ (Skulachev 1988, 1996b; Skulachev and Longo 2005).

There is one other consequence of excessive ROS production in mitochondria, and it is also lethal for the organelle. The point is that *ROS inactivate aconitase* (see above) that is normally bound to mitochondrial DNA. The inactivation decreases the affinity of aconitase to DNA. As a result, DNA is stripped and becomes unprotected from the damaging effects of ROS and other unfavorable influences. Mitochondria lack histones, which protect nuclear DNA, and according to the data obtained by R. Butow, aconitase fulfills in mitochondria a function similar to that of histones in the nucleus (Chen et al. 2005, 2007).

It should be emphasized that the pore opening and aconitase oxidation are direct consequences of the increase in ROS concentration in mitochondria. Both these events require no extramitochondrial factors and are carried out by those ROS that have been formed within the given mitochondrion. This means that this phenomenon (which we called *mitoptosis*) can be described in terms of suicide of the mitochondrion that could not maintain its ROS concentration at the safe low level. Thus, mitoptosis appears to be a mechanism for the purification of the mitochondrial population from mitochondria that become dangerous to the cell because of their enhanced production (or to slow removal) of ROS. Appearance of such abnormal organelles may result, for instance, from mutations in the mitochondrial DNA.

Dead mitochondria are usually removed by autophagy. Autophagosomes play the key role in this process. These organelles are surrounded by two membranes,

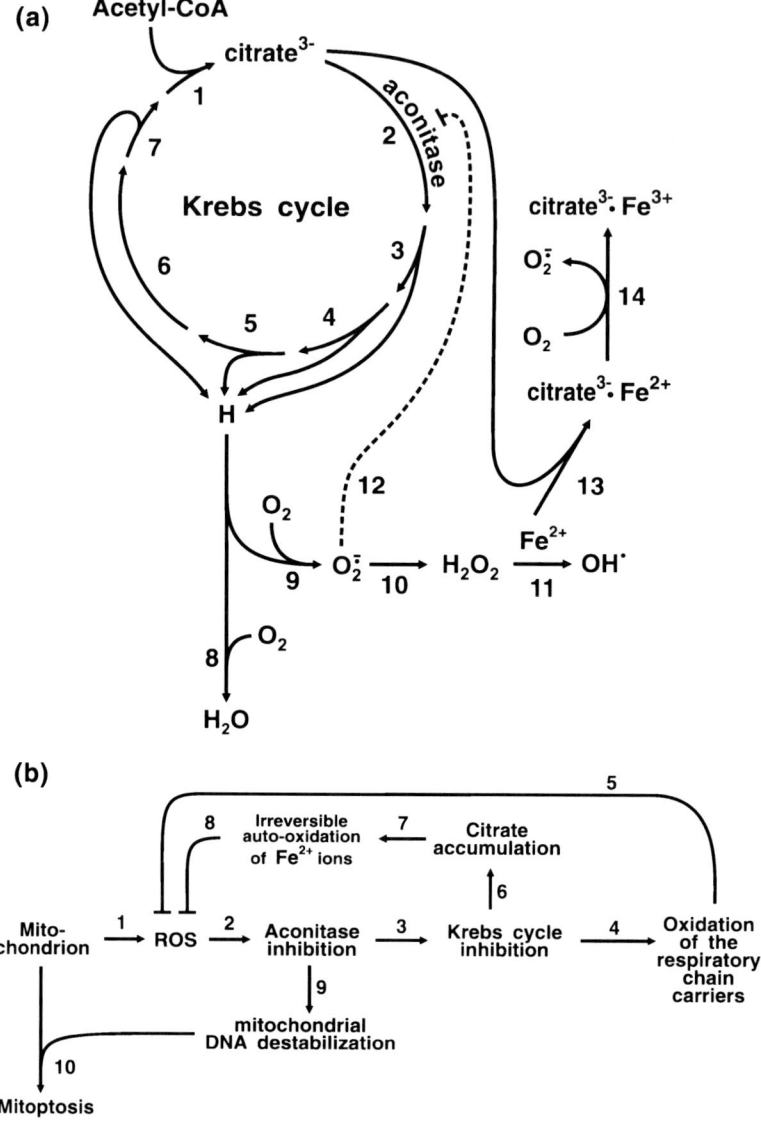

Fig. 15.6 ROS-induced inhibition of aconitase and some of its consequences. Explanations are provided in the text

but in contrast to mitochondria, their inner membrane forms no protrusions (cristae). Autophagosomes show certain specificity toward their absorbed organelles. Two types of processes are distinguished: mitophagy (removal of mitochondria) and peroxyphagy (removal of peroxisomes). Once in an autophagosome,

15.2 How Do Living Systems Protect Themselves from ROS?

the organelles are partially digested, then the autophagosome fuse with a lysosome where the digestion is completed (Narendra and Youle 2011).[2]

Mitoptosis is useful for a cell as long as it helps to remove a few damaged mitochondria from a healthy group. The situation changes dramatically when "the epidemic" of ROS superproduction spreads throughout the whole mitochondrial population of the cell or the greater part of it. In this case, death of mitochondria will inevitably lead to the death of the cell for several reasons. In the majority of non cancer cells, glycolysis is too weak to substitute effectively for respiration as the energy source. Moreover, mitochondria are also known to play a vitally important role in production of certain metabolites. However, the massive dysfunction of mitochondria leads, in fact, to cell death long before the cell is depleted of ATP and other metabolites produced by mitochondria. Here, a specific biological law seems to operate: *Living organisms seek to avoid spontaneous uncontrollable events. They do everything they can to control whatever is happening in them.*

Death is the most dramatic of such events. Thus, it is no wonder that the cell tries to control its own death, choosing to commit suicide before the damage to its structure and functions reaches a level incompatible with life. As Monsieur Bahys, a good-for-nothing physician in Moliere's comedy "L'Amour medecin" ("Love as a Healer") spoke, "Il vaut mieux mourir selon les regles, que de rechapper contre regles" ("It is better to die according to rules than to recover against the rules"). One can easily imagine the advantages that such a strategy brings to the organism. The agony of the cell dying "without any rules" might damage the functioning of the neighboring cells, and if those cells happen to release some regulatory factors, it might influence the whole organism. The consequences can be even more tragic if that agonizing cell manages to survive. For instance, if the agony was caused by a ROS excess, the cell is very unlikely to have been able to protect its DNA against ROS. As a result, the surviving cell can turn into a mutant dangerous for the organism. Cancer cells are likely examples of this phenomenon. That is why apoptosis as a form of programmed cell death is one of the main barriers for malignant transformation.

---

[2] The described mechanism of mitoptosis is probably not the only way for the cells to get rid of damaged mitochondria. Boris Chernyak, Konstantin Lyamzaev, and colleagues from our group recently discovered that a cell can gather such mitochondria in close proximity to the nucleus, surround them with a membrane, and eject the thus formed mitochondria-containing "mitoptotic body" to the extracellular space (Lyamzaev et al. 2008), see also (Lee et al. 2010; Sterky et al. 2011). This effect was demonstrated in cancer cells of HeLa line growing in the presence of an uncoupler and an inhibitor of the respiratory chain, i.e., agents that stimulate mitochondria to consume ATP instead of producing it. Cancer cells are known to successfully support ATP formation not only by respiration but also by glycolysis, and using this mechanism to destroy damaged mitochondria they can prevent the hydrolysis of the glycolytic ATP by damaged mitochondria. It is suggested that maturation of erythrocytes as well as of crystalline lens cells, accompanied by disappearance of mitochondria, may follow the same mechanism. This would explain the conversion of epithelial cells into the crystalline lens cells even in the absence of autophagy.

Fig. 15.7 Claudius Galen

15.2.5 Apoptosis

Following the reasoning discussed above, programmed cell death and, in particular, apoptosis caused by ROS, can be viewed as the next line of defense against ROS.

The term "apoptosis" was introduced by the Roman scholar and physician Claudius Galen (Fig. 15.7), who noticed that the broken branch of a tree enters winter with leaves dried up, but not fallen from it. Thus, defoliation appears to be an active process rather than the death of leaves from cold. As Greek was the scientific language in the Roman times, Galen defined his observation as "apoptosis" [from the Ancient Greek term απόπτωσις ("a falling off"), from the roots από ("away from") and πτωσις ("falling")]. When in 1972 Kerr and colleagues suggested that the cell death can be programmed by its genome, they coined this phenomenon "apoptosis" (Kerr et al. 1972). In 2002, the Nobel Prize in Physiology and Medicine was awarded to H. Robert Horwitz, John E. Sulston, and Sidney Brenner for the discovery of *genes that control ontogenesis.* The genes of programmed cell death were shown to be among them (for review, see Skulachev 2003b).

Apoptosis appears to be a result of a long chain of events where mitochondria usually play the key role. ROS are one of the factors inducing apoptosis. ROS first of all break the barrier function of the outer mitochondrial membrane[3] and let

---

[3] The ROS-induced opening of the already mentioned permeability transition pore in the inner mitochondrial membrane in one of mechanisms to eliminate the outer mitochondrial membrane as a protein-impermeable barrier. The pore opening results in disappearance of osmotic pressure at the inner membrane since K^+ and Cl^-, two major ions contributing to osmotic pressure, become permeable for the inner membrane. Now, water is distributed between mitochondrial matrix and cytosol according to the oncotic pressure, i.e., the ratio of concentrations of high molecular mass compounds (first of all, proteins) in the matrix and cytosol. The protein concentration in the matrix is much higher than that in cytosol so water goes to the matrix, which swells. Such swelling means an increase in the matrix volume due to disappearance of

mitochondrial proteins located in the intermembrane space go out into the cytosol. Some of these proteins can induce apoptosis. Among them are: cytochrome c, apoptosis-inducing factor (AIF), *smac* protein, protease *omi*, and endonuclease G. Each of these proteins stimulates apoptosis and binds to DNA. A total of 28 positively charged amino acid residues of AIF, which are located on the surface of the protein globule and thus can bind to the anions of DNA phosphate residues, seem to participate in this process. As a result, DNA loses its native conformation and becomes susceptible to nucleases. In some animals, AIF forms complexes with endonuclease G, thus increasing its activity (Susin et al. 1999, Skulachev 2003a).

Several months later (but also in 1996) Wang and colleagues (Liu et al. 1996; Yang et al. 1997) and independently Newmeyer in the beginning of 1997 (Kluck et al. 1997) demonstrated release of cytochrome c to cytosol. It was found that cytochrome c has another function besides the respiratory chain electron carrier. It triggers a cascade of processes leading to programmed cell death. Having left mitochondria, cytochrome c binds to a protein that was given the name "apoptosis protease-activating factor 1", or Apaf-1. This complex combines with ATP (or better with dATP) to form a particle called the apoptosome. The apoptosome adsorbs on its surface several molecules of the protein known as procaspase 9. When in close proximity, procaspase 9 molecules activate each other (the procaspase 9 → caspase 9 transition). Active caspase 9 molecules attack procaspase 3 and cleave off a part of its polypeptide chain. As a result, active caspase 3 is formed. This protease digests a group of intracellular proteins that occupy key positions in cell metabolism or play an essential role in the structural organization of a cell. Internucleosomal fragmentation of nuclear DNA is one of the results of the processes described above.

Two other proapoptotic proteins, smac and omi, which (similar to cytochrome c, AIF, and endonuclease G) are released from damaged mitochondria to cytosol, additionally activate the cascade triggered by cytochrome c (for reviews, see Skulachev 2003a; Bratton and Salvesen 2010; Chipuk et al. 2010; Galluzzi et al. 2010; Orrenius et al. 2011). For instance, smac was shown to neutralize inhibitors of apoptosis proteins (IAPs) that block caspases 3 and 9 (Du et al. 2000; Verhagen et al. 2000).

There are at least two more mechanisms of ROS-mediated induction of apoptosis. One deals with the modification of porin, a protein of the outer mitochondrial membrane. Normally, this protein forms pores permeable only for low molecular mass compounds. Superoxide causes porin molecules to form oligomers that become permeable also for much larger molecules of proteins. The proapoptotic protein Bax seems to enhance this effect. As a result, the proteins of the mitochondrial intermembrane space enter the cytosol, and this process no longer requires pore opening in the inner membrane.

(Footnote 3 continued)
invagination of the inner membrane (cristae). The swelling disrupts the outer mitochondrial membrane, which has no invaginations and, hence, has a much smaller area than the inner membrane (Skulachev 1996b).

Another path from ROS to the outer mitochondrial membrane is mediated by JNK protein kinase. This enzyme phosphorylates Bax, thus leading to its *activation*. In addition, JNK phosphorylates proteins Bcl-2 and Bcl-XL, which normally inhibit Bax. In the case of these proteins, phosphorylation leads to their *inactivation*. However, details of the mechanism of ROS-activating protein kinase JNK remain unclear (for reviews, see Skulachev 2003a; 2006; Bratton and Salvesen 2010; Chipuk et al. 2010; Galluzzi et al. 2010; Orrenius et al. 2011).

Figure 15.8 shows the three mechanisms of ROS-mediated triggering of apoptosis. It would be wrong though to think that this scheme embraces all the possible versions of ROS-induced induction of programmed death. There is yet another type of program lethal for cells—called necrosis.

15.2.6 Necrosis

Traditionally, necrosis has been seen as opposed to apoptosis. In contrast to apoptosis, necrosis was thought to be not programmed; it was considered to be cell death under conditions incompatible with life. The fallacy of this statement became obvious when necrosis (similar to apoptosis) was shown to require (i) energy in the form of ATP and (ii) the participation of specific proteins[4] (Galluzzi et al. 2012; Vandenabeele et al. 2010; Yuan and Kroemer 2010; Christofferson and Yuan 2010; Skulachev 2006).

Apoptosis and necrosis seem to use the same mechanisms in their early stages. In the case of ROS, it is pore opening in the inner mitochondrial membrane. There is also an alternative path of ROS-triggered necrosis mediated by JNK. The fundamental difference between necrosis and apoptosis is that it is only in the case of necrosis that the outer cell membrane is destroyed and the intracellular contents are emptied out into the extracellular space. By contrast, the outer cell membrane retains integrity in the case of apoptosis, and the death of the cell is accompanied by the decrease in its volume without intracellular macromolecules appearing in the extracellular space. The cell does not advertise its suicide resulting from apoptosis, while necrosis can be compared to a self-immolation suicide on a crowded square. A cell in the state of apoptosis attracts attention of only macrophages that recognize it because of the appearance of phosphatidylserine in the external layer of phospholipids of the outer cell membrane (that signals "Eat me!"). In contrast to that, a necrotic cell, having burst, releases to the outer medium many compounds that normally cannot be found there. These substances can provoke production of cytokines by other cells. Inflammation develops as a result. It is not a mere coincidence that one and the same stimulus can cause both apoptosis and necrosis, the result being dependent on the quantitative parameters

---

[4] When being completely depleted of ATP, the cell dies in a way different from both apoptosis and necrosis (the so-called energetic catastrophe) (Izyumov et al. 2004; Skulachev 2006).

15.2 How Do Living Systems Protect Themselves from ROS?

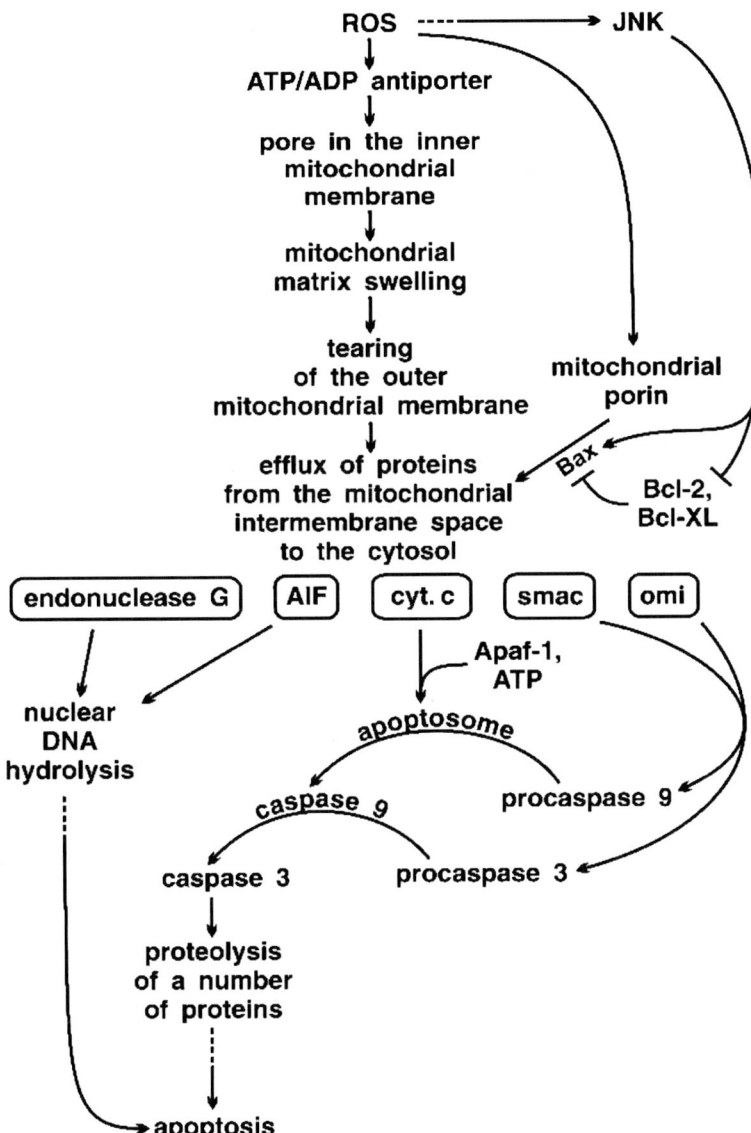

Fig. 15.8 ROS as activators of apoptosis. For explanations, see text

of the stimulus. Thus, our experiments on the HeLa cancer cells showed that a threefold reduction of intracellular ATP level led to apoptosis when applied for 3 h, and to necrosis when applied for a 5-h-long impact (Izyumov et al. 2004). It looks as if a suicidal cell chooses necrosis rather than apoptosis when a negative influence is more dangerous and lasting. An apoptotic cell might have the motto

"I could not cope, so I'm leaving", while a necrotic one—"Look at me! A catastrophe has come, and it will be lethal not only for me, but for all of us!"

15.2.7 Phenoptosis

Mitoptosis, being a programmed self-destruction of malfunctioning mitochondria, can be seen as the mechanism of *mitochondrial selection*, which discards potentially harmful mitochondria from the intracellular population of these organelles. Apoptosis and necrosis fulfill the same function on the level of a cell population in a tissue (*cell selection*). Then, we face the question whether this principle functions also at the level of the whole organism, i.e., whether there are some mechanisms of self-destruction of an individual which functions badly from the perspective of a family, community, population, or evolution of the species.

The hypothetical *mechanism of self-destruction of an organism we call as phenoptosis* (by analogy with apoptosis and mitoptosis (Skulachev 1997, 1999b). Phenoptosis might protect a community of organisms from appearance of sinister mutants or infected individuals dangerous for the community or even the whole population (Skulachev 2003a, 2006, 2012).

In the case of phenoptosis (as well as apoptosis), the main issue is to keep the stability ("purity") of the genome (Skulachev 2003a). When dealing with complex organisms of one and the same species that can exist for hundreds of thousands or even millions of years, genome stability is the main guarantee of the maintenance of the individuality of the species. Thus, it would be hardly surprising if organisms that could not cope with protection or timely repair of their genomes were eliminated due to the activation of phenoptosis, i.e., programmed biochemical suicide of an individual.

Examples of programmed death of bacteria and unicellular eukaryotes can be considered as a manifestation of phenoptosis. Here, the self-destruction of a cell is equal to the suicide of an organism, as these organisms are unicellular (Lewis 2000; Pozniakovsky et al. 2005).

The bacterial response to damage of their DNA is quite interesting. The level of damage is monitored by a special protein that first induces the enzymes responsible for DNA repair, and if they cannot cope with the situation, cell division is arrested. If the damage is still there, autolysin is activated. This enzyme hydrolyzes compounds of the cell wall, a process leading to the death of the cell. A similar mechanism has been described in yeast cells (Lewis 2000). If DNA damage is caused by ROS, the above-described mechanism of phenoptosis can be viewed as a way to protect the population of unicellular organisms from poisonous oxygen species.

Sommer (1994), Manskikh (2004, 2006), and Lichtenstein (2005) hypothesized that cancer is nothing but a phenoptosis program eliminating from the population individuals with damaged (unstable) genome. ROS can be one of the factors causing such instability.

Table 15.2 Levels of protection of living systems against ROS

(1) Antioxidants.
(2) Decrease in intracellular O_2 concentration
(3) Decrease in ROS production by the respiratory chain due to (a) mild uncoupling and (b) arresting of the Krebs cycle at aconitase level
(4) Elimination of mitochondria responsible for excessive ROS production (mitoptosis)
(5) Elimination of cells responsible for excessive ROS production (apoptosis and necrosis)
(6) Elimination of organisms that are not able to cope with DNA protection against ROS (phenoptosis)

In conclusion of our review of the mechanisms of multilevel protection of living systems against ROS (see Table 15.2), we can formulate a general biological law that might be called the "samurai" principle of biology: "*It is better to die than to be wrong*" or, in a more extended form, "*biological systems possess self-elimination programs that are activated when the system in question becomes dangerous or needless for any other system of higher position in the biological hierarchy*" (Skulachev 2000).

15.3 Biological Function of ROS

Originally, ROS were most probably only poisonous for living beings, and respiration was invented by evolution to convert the potentially poisonous O_2 into harmless water. With passing time, though, the organisms learned not only to live in the presence of small amounts of ROS (which were inevitably formed due to some leakage of electrons from respiratory enzymes and coenzymes to oxygen), but also to use the very toxicity of ROS for their own purposes. Thus, ROS turned into a kind of "*biological weapon*" used by living cells to fight pathogens, and later also into "*a samurai sword*" used by mitochondria, cells, and whole organisms for self-elimination.

NADPH oxidase of the outer cell membrane is the best example for using ROS as "biological weapons". This enzyme oxidizes NADPH on the inner side of this membrane. Electrons taken from NADPH are transferred by a flavin and a special cytochrome *b* to the opposite (outer) side of the membrane to reduce extracellular O_2 to $O_2^{\bullet-}$. The resulting superoxide is released to the outer medium where it is transformed into more aggressive ROS forms capable of killing bacteria (Halliwell and Gutteridge 1999; Skulachev 2003a).

In the meantime, ROS produced in the cell interior (first of all, in mitochondria) are powerful stimulators of mitoptosis, apoptosis, and necrosis in multicellular organisms and phenoptosis in unicellular organisms. Fedor Severin and colleagues from our group demonstrated that the pheromone-induced death of yeast requires ROS generation in mitochondria of the dying cells (Fig. 15.9a) (Pozniakovsky et al. 2005). Cytokine-induced apoptosis of HeLa carcinoma cells is another example of this phenomenon (Fig. 15.9b, tumor necrosis factor (TNF) was used as the cytokine in these experiments) (Shchepina et al. 2002). The scale of change in ROS concentration is really remarkable. In normal cells, ROS were hardly noticeable when a fluorescent ROS probe was used; the image obtained could be compared to the famous "Black Square" by Kazimir Malevich. Induction of phenoptosis or cell apoptosis resulted in cells glowing brightly. That is why ROS production in such cases is often referred to as a "burst". It is noteworthy that in the presence of excess TNF, antioxidants were found to inhibit yeast phenoptosis but showed no effect on HeLa cells apoptosis. When analyzing these results, Boris Chernyak and colleagues from our group showed that ROS, while being unnecessary for a TNF-induced suicide of any particular cell (Shchepina et al. 2002), serve as a lethal signal for other cells that have not received TNF. Hydrogen peroxide was found to be the mediator of the intercellular apoptotic signal transmission (Pletjushkina et al. 2005).

Such an effect could play an important role in tissue protection against a viral infection, when an infected cell not only undergoes apoptosis itself, but also stimulates suicide of the neighboring cells, the most likely potential recipients of the virus. As a result, a dead zone is formed around the infected cell, inhibiting the further spread of the viral infection (Skulachev 1998). Formation of such zones was revealed in a tobacco line hypersensitive to tobacco mosaic virus. Collective apoptosis was also shown to be a part of normal ontogenetic development of an organism, where it is used to eliminate tissue parts and whole organs which are no longer needed. Disappearance of the tail of the tadpole is a good example of this phenomenon. This process is triggered by thyroid hormones. It can be studied on tadpole tails that were cut and put into a medium of a particular composition. Addition of thyroxin was shown to increase significantly NO production and the level of ROS, in particular, of hydrogen peroxide. The latter causes the mass death (apoptosis) of the cells in the tadpole tail, the size of which decreases within hours (Kashiwagi et al. 1999). This phenomenon can be defined as *organoptosis* caused by ROS (Skulachev 2002).

In certain cases, *ROS act as signal molecules* of less dramatic functions than a lethal signal, first of all when mobilizing the systems of cell protection against these ROS (an example of "hormesis" when pretreatment with small amount of a poison increases resistance of an organism to subsequent treatment with much higher doses of the same poison). For instance, the increase in superoxide production is perceived by special receptors in our organism that actuate blood vessel constriction leading to decrease in aeration of tissues, lowering intracellular O_2 concentration and, hence, decreasing the rate of superoxide production (see above, Fig. 15.2). When speaking about the cell interior, the decrease in $[O_2]$ decreases hydroxylation of transcription factor HIF-1. As a result, inactivation of HIF-1 is prevented and its cleavage in proteasomes is slowed. On the other hand, the increase in activity and quantity of HIF-1 stimulates the activity of a number of genes encoding enzymes of antioxidant protection (Hirota and Semenza 2006). In this way, a cell is insured against a paradoxical effect when hypoxia causes oxidative stress. The mechanism of this stress can be explained as follows. During hypoxia, respiration slows down because of an oxygen deficiency; this leads to

15.3 Biological Function of ROS

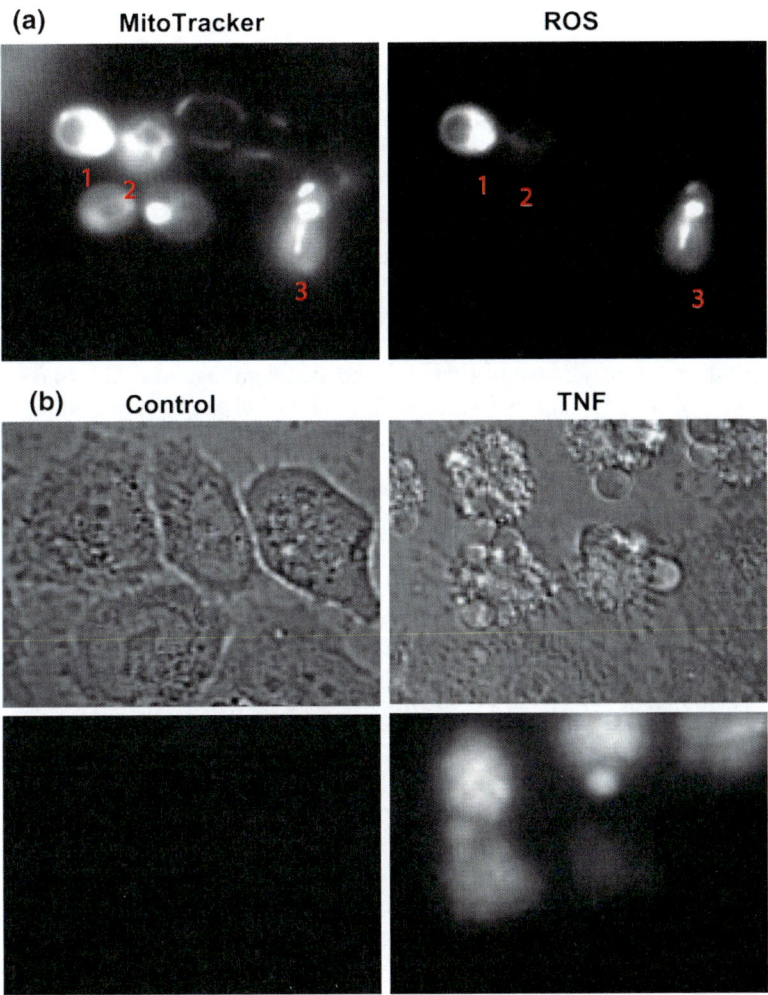

Fig. 15.9 ROS generation during yeast phenoptosis (**a**) and HeLa cell apoptosis (**b**). **a** Yeast response to amiodarone, a compound triggering the pheromone phenoptosis cascade at one of its early stages. Colocalization of mitochondria (stained with MitoTracker Orange) and ROS (stained with dichlorodihydrofluorescein diacetate, DCF) can be seen (Pozniakovsky et al. 2005). **b** Response of HeLa cells to addition of tumor necrosis factor (TNF). *Top* and *bottom* rows show the results of light and fluorescent microscopy, respectively (Shchepina et al. 2002)

activation of glycolysis, which in turn causes acidification of the tissue. Acidification leads to protonation of $O_2^{\bullet-}$, which means production of a far more aggressive $HO_2^{\bullet}$. In addition, hypoxia and especially anoxia facilitate reduction of Fe^{3+} to Fe^{2+}, which actuates the Fenton reaction ($H_2O_2 \rightarrow OH^{\bullet}$, see above Fig. 15.1). Both these effects are not so dangerous while there is no oxygen around. But hypoxia followed by reoxygenation ("oxygen stroke") immediately leads to formation of the most harmful compounds, $HO_2^{\bullet}$ and $OH^{\bullet}$.

The same logic can be applied to explain the phenomenon of ROS-stimulated angiogenesis, the growth of new blood vessels in a tissue (Gerald et al. 2004). Shortage of O_2 provokes cyclic hypoxia–reoxygenation transitions and causes an increase in ROS level following the above-described mechanisms. This increase in ROS concentration stimulates angiogenesis as one of the ways to cope with hypoxia.

One needs to add that *low ROS concentrations facilitate cell proliferation* (Giles 2006). The precise mechanism of this phenomenon remains unclear, even though ROS have already been shown to increase the level of cyclins, regulatory proteins necessary for operating of the cell cycle.

Summarizing the data discussed in this section, we may conclude that ROS are not only harmful substances, but also a necessary component of normal life. An observation of Naum Goldstein seems especially demonstrative. He showed that air completely lacking negative aeroions (i.e. $O_2^{\bullet-}$) to be lethal for animals. When breathing such air, mice died after 18 days and rats after 23 days (Goldstein 2002).

15.4 Aging as Slow Phenoptosis Caused by Increase in mROS Level

15.4.1 Definition of the Term "Aging" and a Short Historical Overview of the Problem

Aging of an organism can be defined as a slow and concerted decay of body functions with age, leading to gradually increasing probability of death (Skulachev 2009).

The key question is—what is the cause of such a decay of body functions? Over 100 years, two alternative approaches to this problem have been competing—an optimistic and a pessimistic one. The first approach suggests aging to be the final step of our ontogenetic program. If this is the case, then *aging can be canceled* if we manage to switch off this stage. The second hypothesis considers aging to be an *inevitable* result of the functioning of such a complex system as an organism due to accumulation of damage to its biomolecules, depletion of its "vitality forces", functioning of certain genes that were useful in the youth but became harmful with age, etc. (Austad 1997; Skulachev 2003a, 1997; Bredesen 2004).

It is obvious that if the pessimistic hypothesis is true, then any attempt to treat the process of aging as a disease is doomed to failure—all of us are destined to get broken like a car that is too old. One of the leading gerontologists of the twentieth century, Alex Comfort, stated though that it is difficult to believe that a horse and a carriage age in the same way (Comfort 1979). Nevertheless, the majority of gerontologists still oppose the theory of programmed aging. Only very recently, data providing direct support of the optimistic concept have been obtained. These data allow us to understand how these age-activated biochemical mechanisms of the decay of body functions might have appeared in biological evolution.

The idea that death of an organism can be organized by this organism himself apparently originates from Arthur Schopenhauer (Fig. 15.10) who wrote in 1818: "The individual is ... not only exposed to destruction in a thousand ways from the most insignificant accidents, but is even destined for this and is led toward it by nature herself, from the moment that individual has served the maintenance of the species" (Schopenhauer 1969, see Focher 2002 for discussion).

Alfred Russell Wallace (Fig. 15.10) was the next to indicate the possibility of programmed death of an organism. (He is mainly known due to the fact that—at the same time as Charles Darwin but independently from him—he postulated the idea of natural selection.) He stated in a brief note written between 1865 and 1870 that "...when one or more individuals have provided a sufficient number of successors, they themselves, as consumers of nourishment in a constantly increasing degree, are an injury to those successors. Natural selection therefore weeds them out, and in many cases favor such races as die almost immediately after they have left successors" (quoted in Weismann 1889, p. 23). Later (in 1881), this principle was independently formulated and developed by another great biologist, August Weismann (Fig. 15.10): "Worn-out individuals are not only valueless to the species, but they are even harmful, for they take the place of those which are sound... *I consider that death is not a primary necessity, but that it has been secondarily acquired as an adaptation*" (italicized by the authors) (Weismann 1889).

Weismann was immediately accused by his contemporaries of being anti-Darwinist, even though Darwin himself (Fig. 15.10) was perfectly aware of the limitations of his hypothesis (evolution follows only those directions which are favorable for an individual) in case of sexual selection and group adaptation. That is what he wrote in his second famous book *The Descent of Man*: "There can be no doubt that a tribe including many members who ...were always ready to aid one other, and to sacrifice themselves for the common good would be victorious over most other tribes; and this would be natural selection" (Darwin 1871).

Modern critics of Weismann usually quote Sir Peter Medawar, who suggested that aging, even while being useful for a population, could not have appeared as a mechanism invented by evolution. Medawar assumed that under natural conditions the great majority of animals die before they become old (Medawar 1952). Today, the falsity of this statement appears to be obvious. *Aging* starts long before it can become a direct cause of the *death* of an individual. At the same time, it can indirectly facilitate this death. For instance, the weakening of the organism that progresses with age undoubtedly increases the probability of death due to attack by predators, pathogens, etc. (Libertini 1988). Loison and colleagues (Loison et al. 1999) and independently Bonduryansky and Brassil (Bonduriansky and Brassil 2002) directly demonstrated that under natural conditions both long-lived mammals and short-lived insects suffer from aging. Such a result is not really surprising if we take into account the fact that, for example, the reduction in skeletal muscle mass in humans develops from 25 years of age (Lexellet et al.1988), and certain components of the immune system start to decay even earlier (in humans, at 15 years of age, Saule et al. 2006). Decline in visual acuity begins at 30, decrease

Fig. 15.10 *Upper row* Arthur Schopenhauer (*left*) and Alfred Russell Wallace; *lower row* August Weismann (*left*) and Charles Darwin

in the lung volume at 35, skin elasticity around 35 (Comfort 1979; Skulachev 2009).

The hypothesis of death being the result of a program encoded by the genome was first directly proven on the cell level (we speak here about apoptosis and necrosis). Only very recently, programmed death phenomena have been demonstrated also on subcellular and supracellular levels—mitoptosis, collective apoptosis, organoptosis, and phenoptosis as described above.

When speaking about the different phenomena of phenoptosis, one should distinguish between the cases of individual programmed death *dependent* and *independent* of the life cycle of an organism. Altruistic death of a unicellular or multicellular organism infected with a pathogen is an example of the latter case. It serves the purpose of preventing the spread of the infection over the whole population (Skulachev 1998, 2003a; Kirchner and Roy 1999, 2002). Similar phenomena, the molecular mechanisms of which have been already thoroughly studied in microorganisms (Lewis 2000), can be seen as a precedent of programmed death on the level of an organism and of its biological significance. But from the perspective of the problem of aging, the cases of programmed death being connected to certain stages of the individual life cycle are far more interesting.

15.4.2 Phenoptosis of Organisms that Reproduce Only Once

Wallace and Weismann were the first to stress that certain *animals reproducing only once* are constructed in a way predetermining death shortly after reproduction. For instance, the imagoes of mayflies live for only several days—they cannot eat due to the lack of a functional mouth and because their intestines are filled with air (Weismann 1989). In some invertebrates, the young hatch inside their mother's body and eat their way out (Kirkwood and Cremer 1982). The males of some squids die right after transferring their spermatophores to a female (Nesis 1997). The female of one octopus species ceases to eat and dies of starvation after her youngsters have already hatched and she no longer needs to protect her brood from marine predators. If the so-called optical glands are removed from the female's brain, the suicide program stops working and the female can reproduce several times. This operation caused fourfold increase in the female octopus's lifespan (Wodinsky 1977). The ejaculation by the male praying mantis becomes possible only after he is decapitated by the female at the end of sexual intercourse (Dawkins 1976); and the male Australian marsupial mouse dies a couple of weeks after rut under the influence of its own pheromones, which are originally produced to attract the female (Bradley et al. 1980) (compare to the yeast phenoptosis caused by its pheromone (Pozniakovsky et al. 2005)).

Quite amazing observations were made when salmon were studied. The pacific salmon *Oncorhynchus keta* dies soon after spawning. It is accelerated aging (progeria) that causes such a death. This process is activated when the fish leaves the ocean and starts swimming up the river to its headwaters. According to the traditional explanation of this phenomenon, the eventual death is caused by the animal spending too much energy while moving against the water current. This point of view proved to be wrong, as (1) aging and death do not occur if the gonads or adrenal glands are removed and (2) progeria develops even if the spawning takes place in the headwaters of very short rivers. Two populations of pacific salmon were compared: one spawning in Amur headwaters (a huge river over 2,000 km long) and the other spawning on Sakhalin Island, in a small lake connected to the ocean via a brook only 200 m long. In both cases, the spawning fish had all the signs of premature aging that ultimately caused death. It is the transfer from the sea to freshwater that seems to be the signal for aging in this case (Skulachev 2009).

To what extent can the principles of accelerated aging of salmon be applied to the slow aging of other animals? Austad believes these two phenomena are completely different in their nature and mechanism: progeria appears to be a program, while slow aging seems to be mere accumulation of occasional damage (Austad 1997). This explanation seems to be wrong since salmon progeria is accompanied by multiple signs of the usual aging process—osteoporosis, reduction of immune defense, appearance of tumors, thinning of the skin, reduction in skeletal muscle mass (sarcopenia), etc. Peptides of amyloid plaques have been recently discovered in the brain of spawning salmon, just as in the brain of old rats or humans.

The increase in the diversity of offspring seems to be one of the functions of the above-described phenoptosis phenomena. For example, fulfillment of the phenoptosis program in the case of a female octopus or male marsupial mouse leads to the situation when she or he can be a mother or father only once in a lifetime. Especially impressive and well-studied cases of phenoptosis were described in *plants reproducing only once in their life*. It has been known for a long time that the death of soybean plants occurring soon after maturation of seeds can be prevented by depodding or deseeding. In both cases, the lifespan of the plant is greatly increased. A demonstrative experiment was carried out by Nooden and Murray (1982). When a soybean plant was depodded except for a single pod cluster in the center of the plant, the pod cluster induced yellowing of the nearest leafs only, whereas the rest of the plant remained green. The yellowing occurred even if the petiole contained a zone treated with a jet of steam killing the phloem. This treatment inactivated transport of compounds from leaf to pods occurring via phloem (a living tissue) but not from pods to leaf, which occurs via xylem (an already dead tissue at functional maturity of plants). This indicates that pods induce senescence by producing a dead signal or a poison killing leaves. This observation is in obvious contrast with a statement in a recent review by well-known gerontologists–pessimists Kirkwood and Melov (2011): "There is a little evidence that semelparous (capable of only single reproduction) organisms are *actively* destroyed once reproduction is complete".

Senescence of soybeans is a fast process that takes about 3 weeks. The lifespan of this plant is about 90 days, which is greatly increased by depodding (deseeding). A similar increase was revealed in many other semelparous plants including *Arabidopsis thaliana*, the classical model species for plant physiologists and geneticists. Just this organism was recently used by Melzer and coworkers from Ghent, who disprove one more point in Kirkwood and Melov's reasoning against programmed aging: "Yet among the many gene mutations that have been discovered that affect lifespan, often increasing it significantly, none has yet been found that abolishes aging altogether" (Kirkwood and Melov 2011). The Belgian authors reported in *Nature Genetics* (Melzer et al. 2008) that a plant having mutations in the two genes (*soc1* and *ful*) switches from sexual to vegetative reproduction. In the mutant, formation of seeds is inhibited. As a result, the plant does not die due to seed-induced senescence. The lifespan of the mutants increases from 90 days to

15.4 Aging as Slow Phenoptosis Caused by Increase in mROS Level

Fig. 15.11 Double mutant of *Arabidopsis thaliana* in *soc1* and *ful* genes switches from sexual to vegetative reproduction. In the mutant, formation of seeds (which kill the plant at age 2.5 month) is inhibited, so the plant lives longer than 14 month, forming a woody stem. **a** Wild-type 2-month-old plant (*left*) and an 8-month-old mutant (*right*); **b** 14-month-old mutant (from Melzer et al. 2008)

practically infinity as in perennial trees or bushes reproducing with rhizomes. In fact, the *soc1 ful* mutant forms woody stems and rootstocks; it becomes much larger than the wild-type *A. thaliana*, changing from a grass with a single basal rosette of small leaves to a highly branched shrub with many aerial rosettes formed by rather large and thick leaves (Fig. 15.11). Inflorescence meristems are reversed to vegetative meristems. Secondary growth appears, being mediated by cambial activity absent from the wild type. The authors speculate that originally the species *A. thaliana* appeared as a perennial shrub vegetatively reproducing with rootstocks and competing with other shrubs and trees. The modern version of the species became a small short-lived grass reproducing with very light seeds transmitted by wind to open soil. If an opening is fresh, the seeds quickly germinate to give tiny plants growing with no competition with other plant species, which are still absent from the opening. The grass-type *A. thaliana* is short-lived, being killed by its own seeds. Such death might be necessary to accelerate succession of new generations and, hence, their evolution. Apparently, the modern form of *A. thaliana* appeared quite recently, since the program for its preceding (immortal) form is still conserved in the genome of this plant. Another advantage of short lifespan of the modern *A. thaliana* might be that it guarantees for this organism a life under comfortable conditions when competition with other species for the niche is excluded. Small *Arabidopsis* can hardly complete with other grasses and, if long-lived, will inevitably fail in the struggle for existence. In any case, the study of *A. thaliana* mutants clearly shows that senescence and death of a higher organism can be canceled by means of inactivation of a few genes in its genome.[5]

---

[5] Remarkably, the very fact that the death of semelparous plants is caused by their seeds literally confirms a famous maxim of Weismann (1889) that highly organized living organisms contain "seeds of death".

The *A. thaliana* case is hardly an exception. Melzer et al. (2008) mentioned that "in angiosperms, the perennial woody habit is believed to be the ancestral condition from which annual herbaceous lineages have evolved several times independently. Conversely, evolution from annual herbaceous ancestors to perennial woody taxa has also repeatedly occurred. For example, in various annual herbaceous lineages, such as *Sonchus* and *Echium*, woody perennial species evolved on isolated islands from their continental annual ancestors (Groover 2005; Kim et al. 1996; Bohle et al. 1996)".

Among perennial plants, there are examples of organisms vegetatively reproducing for many years, then switching over to the sexual reproduction and dying when their seeds mature.[6] Several species of bamboo (Fig. 15.12) have fixed lifespans determined by the time of inflorescence (Weismann 1889). This time is species specific. It varies from 6 to 120 years in species that belong to a single genus. A similar situation was described for some other perennial plants, e.g., *Agave* and the Madagascar palm *Ravenala madagascariensis*, flourishing in the 10th and 100th year of life, respectively, and dying after maturation of seeds (Skulachev 2011a).

Recently, progress was made in understanding the mechanism of the senescence-induced effect of seeds. It was shown that they somehow change the hormone balance in leaves. Level of abscisic acid strongly increases whereas that of auxins, gibberellins, salicylic acid, and cytokinins decreases (He et al. 2005; Shan et al. 2012; Ross et al. 2011) (see Addendum 4 for formulas of phytohormones). In the case cytokinins, it was shown that their synthesis is inhibited (Guo and Gan 2011) whereas their oxidative decomposition is stimulated (Price et al. 2008). It was also found that an inhibitor of cytokinin oxidase, 6-methyl purine, retards senescence (Price et al. 2008). For many years, it has been known that plants produce ethylene causing defoliation (Neljubow 1901, Burg 1968). Now it is well established that this compound activates genes involved in plant senescence (John et al. 1995; Jing et al. 2005; Shan et al. 2012; Ross et al. 2011).

There are indications that ROS induce expression of the plant senescence-inducing genes, and antioxidants inhibit such an effect (Navabpour et al. 2003; He et al. 2005; Barth et al. 2006). Most probably, induction of senescence, like responses to various stress-causing factors, first initiates a burst ROS production and then production of ethylene (Wi et al. 2010).

Thus the above-summarized observations clearly indicate at least in plants aging can be programmed. Paradoxically, at present the majority of plant and animal gerontologists occupy quite opposite positions concerning the aging program. The plant and animal students are sure that aging is programmed and nonprogrammed, respectively (Libertini 2012).

---

[6] Such a death seems to be needed for success of sexual reproduction. Bamboo plants reproducing vegetatively are growing so dense that seeds have little chance to find a place to give one more young plant.

Fig. 15.12 Bamboo phenoptosis. Botanical garden in Göteborg in the beginning of July

15.4.3 Can Aging be a Slow Form of Phenoptosis?

If we assume that not only senescence of annual plants but also aging of perennial plants and animals is programmed, we need to answer two questions: (1) what is the possible biological significance of aging seen as a slow decay of biological functions and (2) what is the mechanism of the aging program that would not be eliminated by natural selection—in spite of the fact that nonaging organisms lacking this aging program would seem to possess undeniable advantages over the aging ones?

When looking for an answer to the first question, one can suggest the slow aging process to be one of the mechanisms accelerating evolution. Let us consider the following situation. Two young hares, a clever one and the second not as clever, upon meeting a fox, have almost equal chances to escape from the enemy since they run much faster than a fox. The way Aesop put it, a hare would always run away from a fox, as it is a matter of life or death for it, while for the fox it is but a question of dinner. However, with age, the Aesop principle ceases to operate; the clever hare will gain advantage over the stupid one since they will run slower due to sarcopenia. Now the clever hare, as soon as it notices a fox, will immediately run and will have much more chances to save itself than the stupid one that will be delayed at the start. This means that only the clever hare will proceed with producing leverets. As a result, the hare population will grow wiser (Skulachev 2003a).

Essentially, the most actively reproducing young part of the population will not be involved in such an experiment, being a guarantor of the stability of what has already been reached by evolution. At the same time, the aging part of the population can afford to slightly change the genotype of the species by selecting certain new properties. If these properties prove to be useful, they will eventually be fixed in the offspring. However, if a new property has some unfavorable side effects, it will not pass through the sieve of selection, and the experiment *per se* will have no serious negative consequences for the species as the number of old individuals in any given population is rather limited, they do not reproduce as actively as the young ones, and they are destined to die out.

The above-described aging effect favorable for evolution suggests the brain to age more slowly than muscles, and muscle strength weakens while reproduction is still possible. These conditions seem to be met at least in the case of humans—the first signs of muscle atrophy can be noted as early as at the age of 25, even though at the beginning this process progresses slowly. Now substantial weakening of muscle strength starts at the age of 50 for Swedish men (Larsson et al. 1979) and at the age of 40 for men from Saudi Arabia (Al-Abdulwahab 1999). On the other hand, at the beginning of the nineteenth century, the decrease in this parameter was noticeable already in 25-year-old men (the studies were conducted on Belgian men) (Larsson et al. 1979). Such dynamics can be easily explained if we take into account the improvement of living standards that took place over the last two centuries. In any case, the way Goldsmith puts it, "due to the fact that even a minor deterioration causes a statistically significant increase in mortality, one should expect the evolutionary effect of aging to start in a relatively young age in the case of wild animals" (Goldsmith 2006).

Within certain limits, shortening of life can have a positive impact on evolution rate due to the accompanying effect of the acceleration of generational change. That is why it appears to be advantageous to regulate the "rheostat" determining the rate of the aging program, so that to shorten lifespan in case of deterioration of external conditions. And vice versa, having accomplished a major breakthrough, i.e., having achieved a better accommodation to external conditions, a population can allow itself to reduce the rate of evolution, i.e., to increase lifespan. That is exactly what happened to vertebrates when they managed to conquer the airspace: birds and bats live several times longer than terrestrial mammals of the same size, while the intensity of metabolism in birds is 2.5-fold higher than for terrestrial mammals of the same mass (Holmes et al. 2001). The lifespan of a bat is 17 times longer than that of a shrew which is similar to a bat in terms of its mass and diet (it also eats insects) (Austad 1997). In the case of birds, one can find species that live up to 100 years of age and even longer while having no visible signs of aging (Austad 1997).[7] When turning to terrestrial vertebrates, we will see that it is those animals that have practically no enemies that have the longest lifespan. That is the case for giant tortoises. One of them, having been marked by Charles Darwin, was recently still alive being fed by the zoo attendants. The problem is that the

---

[7] An elegant experiment was done on wild albatrosses, birds living longer than 50 years. Foraging behavior was studied using satellite tracking. It was found that old males foraged in remote Antarctic waters, whereas young and middle-age males never foraged south of the Polar Front (Lecomte et al. 2010).

15.4 Aging as Slow Phenoptosis Caused by Increase in mROS Level

tortoise's shell has become too heavy for this animal's muscles. As a result, the tortoise has lost its mobility and would have died in the wild because of thirst. No signs of aging have been described also for a bowhead whale. The maximal lifespan of this animal is over two centuries. An Aleut ruff, a northern species of a very large marine perch, can live just as long. It is interesting to note that there are also short-lived species among this kind of predatory fish (with maximal lifespan of 12 years) (Austad 1997; Skulachev 2005). In any case, the fact that the lifespan of the species of one genus can differ 100-fold indicates the presence of the program of aging, which can hardly be explained here by mere accumulation of random injuries. The other option seems to be more probable—the species better adapted to the environment can afford to switch off the aging program and to evolve more slowly.

An interesting example illustrating the presence of an aging program of major biological significance is provided by recent studies conducted on the African fish *Nothobranchius* (Terzibasi et al. 2007). The lifespan of different species of this genus varies up to fivefold depending on their habitat in the wild. For instance, *N. furzeri*, which lives in Zimbabwe in puddles formed during the rainy season and dried after this period, was shown to have the lifespan of 3–4 months, which corresponds to the length of the rainy season in this country. *Nothobranchius rachovii* and *N. kuhntae* from Mozambique, where the rainfall is four times more, live for about 9.5 months, while *N. guentheri* from Zanzibar (having a humid climate with two rainy seasons) lives for over 16 months. In case of the shortest-lived species, *N. furzeri*, growth and sexual maturation occur very rapidly, within the first month of life. Then the fish mates many times and lays eggs during 2–3 months, till the puddle dries up. The next year fingerlings will hatch from the eggs that have survived the drought. It seems to be quite remarkable that over the short period determined by the habitat conditions (2 months), the fish manages to age, manifesting by the end of its life a whole set of signs of senescence (slowing of movements, loss of a "research"-type of behavior in an open space, osteoporosis, accumulation of lipofuscin granules in liver, a strong increase of β-galactosidase activity in skin fibroblasts, etc.). In the case of the longer living species, completion of growth, sexual maturation, and senescence occur much later. It seems to be significant that the described properties remain even in the case of aquarium breeding, which means that they are determined genetically and are no longer dependent on the current habitat conditions. It looks like the fish of the short-lived species seek to age within that short lifespan given to them. It is also important that the aging program has similar signs even in such distant vertebrate classes as fish and mammals. This conservatism presents clear evidence of operation of a program.

It was Nature Herself that conducted an interesting experiment in the headwater of the rivers Oropouche and Yarra (Trinidad). Reznik and colleagues (Reznick et al. 2004) discovered the headwater of these rivers to be cut from their downstream by waterfalls impassable for predatory fish (in particular, for pikes) feeding on guppies. As a result, guppies living in the headwater had practically no natural enemies, while those from the downstream were constantly facing the danger of being eaten by pikes. The researchers caught guppies from the different areas and started to breed them in aquariums with no enemies present. The fish of the second aquarium generation (from the group that used to coexist with pikes) were shown: (1) to live 35 % longer, solely due to the increase in the reproductive life period; (2) to reproduce twice as fast, and (3) to respond to danger 1.5 times faster (similar to that clever hare in the example discussed above). An increase in the first two parameters is favorable for higher fertility of the brood under conditions of constant loss of fish due to the attacks of the predator, while the third decreases such losses. Overall, these data contradict one of the key postulates of classical gerontology that fertility can be increased only if the lifespan is reduced.[8]

When analyzing the above relationships, one can conclude fertility and lifespan to be regulated by, so to say, two "rheostats", which sometimes react differently to the changes of the habitat conditions. A population uses these types of regulations not only for survival, but also for changes in the rate of evolution—under unfavorable conditions evolution is accelerated due to the increase in fertility and hence, the diversity of offspring. Parallel to that, lifespan can be shortened under the influence of habitat pressure as well as internal factors of an organism. It is important to stress that this phenomenon cannot be explained by an assumption that the organism spends too many of its resources for reproduction and, hence, becomes short-lived (as suggested by Kirkwood and Austad 2000). Within certain limits, the reduction of lifespan can favorably affect the pace of evolution due to acceleration of alternation of generations. That is why it appears to be advantageous to regulate the "rheostat" that determines the pace of the aging program for a shorter life under worsening external conditions. Having attained new frontiers, i.e., having reached better accommodation to external conditions, a population can reduce its evolutionary pace. The above-described case of two populations of guppies seems to be quite peculiar in this respect. When facing the threat of immediate extinction, the population had to use all the available resources by increasing both fertility and duration of reproductive lifespan (Skulachev 2005).

15.4.4 Mutations that Prolong Lifespan

If there is a genetic program of aging, there should be mutations that disrupt thus program and prolong lifespan. In this chapter (see Sect. 15.4.2), we already described elegant experiments of Melzer et al. (2008) on a plant, *Arabidopsis thaliana*, showing that mutations in two genes result in substituting vegetative for sexual reproduction, an event converting this short-lived organism into a very

---

[8] Reznik was so sure about the lifespan "in hell" being much shorter than "in heaven", that, when starting his long-term project, he published his prediction of this result (Reznick 1997). In their recent paper, Reznik and colleagues no longer quote their erroneous predictions (Reznick et al. 2004).

15.4 Aging as Slow Phenoptosis Caused by Increase in mROS Level

long-lived creature. In this section, we shall consider some examples of mutations strongly increasing the lifespan of fungi and animals.

In the case of the filamentous fungus *Podospora anserina*, a point mutation in a gene encoding a subunit of the cytochrome *c* oxidase inactivates this enzyme, thus inducing an alternative oxidase that oxidizes $CoQH_2$ by oxygen without any ROS produced. This leads to (1) sharp reduction of ROS level and (2) switch from sexual to vegetative reproduction by means of some still unknown mechanism. As a result, signs of aging disappear and the lifespan of the fungus is prolonged more than 20-fold (from 25 days to many months). A quite similar result can be achieved by other mutations (or adding respective inhibitors) inactivating the main mitochondrial respiratory chain (Fig. 15.13), or alternatively by mutation in the gene encoding the dynamin-related protein. This protein is essential for one of the events accompanying an early stage of apoptosis mediated by mitochondria, namely, fragmentation of extended mitochondria into small ones. In the latter case too, the author observed not only a strong retardation of the stage of the life cycle, when ROS superproduction takes place, but also a decrease in sensitivity of *P. anserina* cells to the apoptogenic effect of H_2O_2 (Osiewacz 2003). It should be stressed that the switch of a program from short-life expectancy to one of long-life expectancy can also be achieved without any mutations or respiratory chain inhibitors, namely, through the cultivation of *Podospora* in a liquid medium.

Hekimi and coworkers (Lakowski and Hekimi 1996; Hekimi and Guarente 2003) managed to prolong the lifespan of the nematode *Caenorhabditis elegans* 5.5-fold. This phenomenon resulted from a mutation in the genes encoding insulin receptor Daf-2 and the enzyme catalyzing the final stage of CoQ biosynthesis. Kenyon and colleagues (Arantes-Oliveira et al. 2003) achieved an even more spectacular result (sevenfold increase in lifespan) when arresting the synthesis of Daf-2 and the proteins responsible for sexual reproduction. No differences in the other vital functions of the long-lived animals were found.

Pelicci and coworkers (Migliaccio et al. 1999; Trinei et al. 2002; Napoli et al. 2003; Giorgio et al. 2004) found that mice with a mutant *p66shc* gene live 30 % longer, and the fibroblasts obtained from these animals did not activate the apoptosis program in response to treatment with hydrogen peroxide.[9] The wild-type mice showed the following chain of events: ROS → DNA damage → p53 → p66shc stabilization → [p66shc]↑ → formation of a complex between cytochrome *c* and p66shc → $O_2^{\bullet-}$ → pore opening in the inner mitochondrial membrane → apoptosis. Formation of the complex of p66shc and cytochrome *c* is of the most interest for us, as it probably provides this cytochrome with the ability to react with O_2, thus forming superoxide.

Under in vivo conditions, knockout of gene p66shc decreases the degree of oxidative damage to both mitochondrial and nuclear DNA in lungs, liver, spleen,

---

[9] Compare with the data of Kirkwood and coworkers on the positive correlation between the lifespan of different mammals and the resistance of their fibroblasts to oxidative stress (Kapahi et al. 1999).

Fig. 15.13 Respiratory chains of a wild-type *Podospora* (*left*) and its long-lived mutant (*right*). For explanation, see text

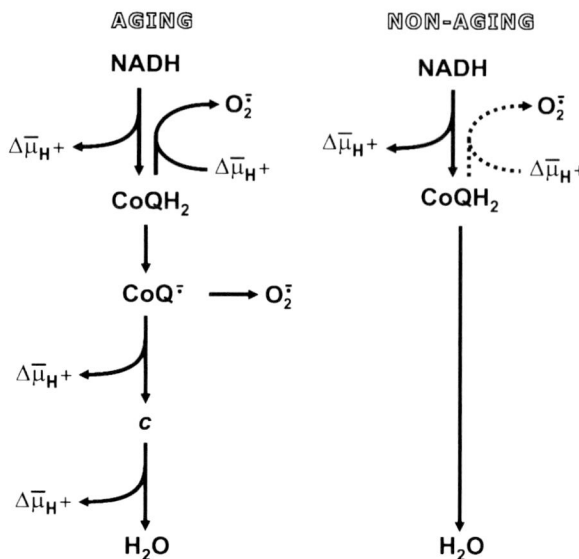

skin, skeletal muscles, and kidneys, but not in brain and heart. This corresponds to the content of p66shc in different organs, which is the lowest in brain and heart, i.e., in those organs that are the last to age.

In 1971, the Russian gerontologist Vladimir Dilman (Fig. 15.14; Dilman 1971; 1978; Dilman and Anisimov 1979) postulated that aging is controlled hormonally, insulin being the factor that stimulates aging. Later, this hypothesis was confirmed by research carried out on eukaryotes of different taxonomic groups. Blücher and colleagues (Bluher et al. 2003) showed the knockout of insulin receptor in adipocytes prolonged the lifespan of mice by 18 %. Independently, Holzenberger and colleagues (Holzenberger et al. 2003) showed that mice heterozygous for the receptor of insulin-like growth factor I lived 26 % longer than the wild type. The effect was accompanied by a decrease in [p66Shc] by 50 %. The Ras protein and Akt/PKB-protein kinase phosphorylating the serine and threonine residues of proteins were shown to be part of the system responsible for the shortening of the lifespan of the mice (Longo and Finch 2003).

Some data indicating insulin-like factors shorten lifespan were also obtained with flies and nematodes. In the case of *C. elegans*, a mutation of the *daf-2* gene prolonged life 2–3-fold (Lakowski and Hekimi 1996). NAD^+-dependent histone deacetylase (SIR2.1 protein) was shown to suppress one of the regulatory cascades of insulin and to prolong the lifespan of these animals. In case of yeast, the increase in *sir2* gene dose prolongs their life. Longo and Finch showed that the deletion of gene *Ras2* and an *Akt* gene analog cause 2–3-fold increase in the yeast lifespan (Longo and Finch 2003).

There is much evidence for the participation of mitochondrial DNA damage in the regulatory cascade that causes aging of both yeast and animals. This statement has been recently directly proved in elegant experiments carried out in the

Fig. 15.14 *Upper row* Denham Harman (*left*) and Nikolay Emanuel'; *lower row* Vladimir Dilman (*left*) and Alex Comfort

laboratories of Zassenhaus (Mott et al. 2004), Larsson (Trifunovic et al. 2004), and Prolla (Kujoth et al. 2005). These researchers found that the expression of a mutant mitochondrial DNA-polymerase, which can still synthesize DNA but cannot correct the mistakes appearing during this synthesis, leads to a substantial increase in the mutation frequency in mitochondrial DNA, especially in the cytochrome *b* area. It also causes a substantial shortening of lifespan and *development of many typical signs of aging*. Injection of cyclosporin A, an inhibitor of the pores in the inner mitochondrial membrane, was shown to prevent the effect of the mutation (these studies were conducted in Zassenhaus's group where only DNA-polymerase from the heart muscle was modified) (Mott et al. 2004).

15.4.5 ROS and Aging

About 50 years ago, Denham Harman in the USA (Harman 1956; Fig. 15.14) and then Nikolay Emanuel' in Russia (Emanuel' 1975; Fig. 15.14) suggested that aging is a result of biopolymers (especially DNA) being damaged by ROS. Over the years, much evidence has appeared supporting this postulate, and it is mitochondrial DNA that seems to be an important target of ROS during the aging process.

There is no doubt that aging leads an increase in both the level of ROS and the ROS-induced damage. This situation inevitably leads to gradual increase in the number of cells that are sent by ROS to apoptosis. Not all the cells that die are replaced by new ones, and as a result, *the overall number of cells in particular organs and tissues is reduced* (Szilard 1959; Severin and Skulachev 2009). This effect increases mortality, the main sign of aging. In such a case, decline of physiological functions occurs even if those cells that survive are still operating efficiently.[10]

It is noteworthy that H_2O_2 generation by mitochondria of different tissues is much lower in birds than in mammals of similar mass. According to Austad and coworker (Brunet-Rossinni and Austad 2004), similar differences can also be observed between flying and terrestrial mammals, a bat and a shrew.

Research conducted independently in the groups of Sohal (Ku et al. 1993), Barja (Barja 1998; Barja and Herrero 2000), and Brand (Lambert et al. 2007) showed the lifespan of various species of warm-blooded animals to shorten in parallel with the increase in ROS generation during *reverse* electron transfer via complex I of the mitochondrial respiratory chain. This correlation could not be observed in the case of ROS generation during *direct* electron transfer via complex I. The study of the Brand group was the most thorough: 12 different species of mammals (including mice, baboons, and cattle) and birds (quails, pigeons) were studied. A total of 11 species fit perfectly well the pattern describing their lifespan as an inverse function of the ROS generation rate. Only one species was an exception to the rule. This was the so-called naked mole-rat, an African rodent of mouse size that lives 10 times longer than a mouse while producing ROS much faster. Remarkably, the naked mole-rat was the only "*non-aging*" animal among the studied species (see the next section).

15.4.6 Naked Mole-Rat

The naked mole-rat, *Heterocephalus glaber* (*Rodentia: Bathyergidae*), is a small rodent species. They live under laboratory conditions for up to 28 years. These animals do not suffer from cancer, cardiovascular diseases, immunodeficiency, or atherosclerosis. In the wild they live in large colonies of about 100–250

---

[10] It is interesting that this feature of aging was noticed by ancient Greeks, who attributed the thought that "aging is the damage to the whole body when its parts remain intact" to the god of healing Asclepius (who turned into Aesculapius in Rome). This does not mean that particular parts of the body do not ultimately decay, but such events seem to occur at a terminal stage which develops long after the aging program had been activated.

"soldiers", a "queen", and her several breeding partners. Only the queen and her "husbands" breed. Under laboratory conditions, the naked mole-rats die at the age of 25–28 years for unknown reasons (Buffenstein 2005). *The level of their mortality does not depend on age.* An aging program seems to be not operative in these mammals. The program is most likely to be interrupted downstream of ROS formation, because the level of ROS production during reverse electron transfer and the degree of oxidation of biopolymers in naked mole-rats are *higher* than in mice (Labinskyy et al. 2006; Buffenstein 2005; Lambert et al. 2007). These observations are not really surprising if we take into account the fact that superoxide dismutase and catalase activities are similar in these two species, while the activity of the third most important antioxidant enzyme, glutathione peroxidase, is 70 times lower in a naked mole-rat. The fact that the cells of the naked mole-rat are extremely resistant to H_2O_2-induced apoptosis can provide an explanation of this paradoxical situation (Labinskyy et al. 2006). In this respect, the naked mole-rat is like the long-lived mice carrying the mutations in the genes encoding the p66shc protein: their cells are also resistant to apoptosis induced by ROS (in this chapter, see Sect. 15.4.4 above).

The absence of some pain sensors is another specific feature of naked mole-rat. Park and colleagues (Park et al. 2008) recently showed that these rodents lack a small peptide of 11 amino acid residues called the P substance; in other animals, it participates in the transmission of the pain signal induced by capsaicin, a potent inhibitor of complex I of the respiratory chain (see Sect. 4.2.4 and Appendix 3). It seems possible that insensitivity of naked mole-rats to capsaicin is related to their ROS resistance since capsaicin, like other complex I inhibitors, should increase ROS production by mitochondria oxidizing NAD-linked substrates (Skulachev 1996a). Sensitivity to capsaicin can be restored by injecting DNA encoding the P substance into a naked mole-rat's leg. This treatment does not influence the other legs, which were still insensitive to the pain inducer. Apparently, the naked mole-rat "queen" and her "husbands" do not interact with external pain sources. If this is the case, then the mutation interfering with the process of signal transfer from skin pain receptors will be neutral and will avoid being eliminated by natural selection. The aging program, which according to our hypothesis provides evolutionary advantages under the pressure of natural selection only in the presence of enemies, might be switched off in a similar manner. We need to remember that the queen, the only individual bearing offspring, and hence, participating in the evolutionary process, has practically no enemies. She is protected from them by living in the vast underground labyrinths guarded by the army of numerous "soldiers" who, when in the wild, live not longer than 3 years and die because of fighting with other colonies of naked mole-rats or with snakes that creep into their holes. Quite recently, Gladyshev and his colleagues (Kim et al. 2011) published the complete sequence of the naked mole-rat genome and some data on age-related changes in its expression. In particular, expression of 54 genes of brain was investigated. In humans, 33 genes among these 54 genes were underexpressed and 21 overexpressed during aging. Of these, 32 genes did not show consistent expression changes with aging in naked mole-rat, including 30 genes that had stable expression and two genes that changed in the opposite direction to human brain (4-year-old and 20-year-old mole-rats were compared). These results clearly show that expression of a large group of genes in the naked mole-rat brain is, in fact, age-independent in contrast to that in human brain.

15.4.7 Aging Program: Working Hypothesis

So, we can suppose the aging program is triggered by an enhancement of ROS level in mitochondria, the effect being programmed in the genome.

The question arises—why is it that ROS were chosen by evolution as an instrument of aging? There is no doubt that ROS still present a problem for current aerobic forms of life. Perhaps that is the reason for aging, while being a specialized evolutionary mechanism, to function so as to stimulate an improvement of the antioxidant system of organisms. From that point of view, ROS can be compared to the fox in the example discussed above (see Sect. 15.4.3), but here the selection is directed to the better system of protection against oxygen rather than to higher intellect. Such an effect is the consequence of participation of ROS in programs of self-destruction of mitochondria, cells, and organisms.

What is it that triggers the aging program? There is no doubt that its actuation starts relatively early, most likely when an organism stops growing, or even earlier, at the completion of puberty. The simplest version is that the "master clock" that counts time and gives the signal for the cessation of growth not only switches off the growth program, but also triggers the aging program as postulated by Vladimir Dilman (1978, 1982) (Fig. 15.14) and Comfort (1979) (Fig. 15.14). It is noteworthy that such nonaging long-lived animals as pearl oyster, pike, large crabs, turtles, and whales never stop growing (Comfort 1979; Finch 1990; Austad 1997).

The next question is where the "master clock" is located? If our organism counts years by the same molecular clock which counts hours, the master clock operates in the suprachiasmatic nucleus of the hypothalamus, which is responsible for the maintenance of the circadian rhythm of mammals. In the case of birds, it is the pineal gland (epiphysis) that fulfills the same function. This gland participates in the circadian rhythm of mammals as well, but here it is governed by the suprachiasmatic nucleus.

Periodic biochemical reactions were discovered in the cells of suprachiasmatic nucleus. Neurons transmit these cycles to the pineal gland, which responds to this stimulus by fluctuations in production of melatonin, the hormone that is excreted by pineal gland cells and is transported with blood to all organs and tissues. Circadian melatonin fluctuations in blood are the mediator of the circadian rhythm. Melatonin happens to be an antioxidant, and, what seems to be even more important, an inducer of the synthesis of many of the enzymes of the antioxidant cascades. It is interesting to note that the magnitude of circadian fluctuations of melatonin concentration in blood strongly decreases with age, and it is this decrease that seems to be the earliest sign of aging, as in the case of humans it can be observed starting from 7 years of age (Austad 1997).

On the whole, the hypothetical aging program of mammals can be presented as follows (Fig. 15.15). It is assumed that aging represents the final step of ontogenesis, resulting from a slow controlled poisoning of the organism by endogenous toxins, namely ROS. This intoxication increases with age and is initiated by a signal generated by the master clock, being mediated by a decrease in the level of juvenile hormone(s) and by an increase in level of age-inducing hormone(s) (step 1). Such a change in hormonal balance somehow increases ROS level specifically in mitochondria (step 2), which in turn entails stimulation of apoptosis whose rate becomes higher than the rate of proliferation if we deal with proliferating tissues (step 3). This situation leads to a decrease in number of cells in the tissue (cellularity) (step 4), inevitably resulting in a decay of functions of organs (step 5) (Skulachev 2003a, 2009).

15.4.8 Paradox of Protein p53

If the aging program really exists, what might be a way to interrupt it? We might knock out the genes encoding this program, but how can we find them and how can we avoid possible adverse effects of such an intervention? The case of p53 protein seems to be illustrative in this respect. Donehower and coworkers (Tyner et al. 2002) discovered that due to the mutation increasing protein p53 activity, cancer ceased to be the cause of animal mortality. Because usually at least half of old mice die because of cancer, one could expect this mutation to increase significantly the lifespan of these animals. Thus, it came as a big surprise that the lifespan of mutants was in fact reduced by 20 % instead of being prolonged. Later, Serrano and colleagues (Garcia-Cao et al. 2002), who also obtained p53 protein activation (that was not as significant as the one obtained by Donehower though), showed the number of malignant tumors to decrease from 45 to 17 %, but could observe no statistically reliable change in lifespan. This result is basically consistent with the data of Donehower, as such a considerable reduction in incidence of cancer should have extended the lifespan, which was not the case. Recently, Scrable and coworkers succeeded in activating p53 in probably the most effective way, and as a result the lifespan decreased threefold (Scrable et al. 2009).

The p53 protein is known to be the "guard of the genome". It is required for the stimulation of DNA repair, cell cycle arrest, and apoptosis in response to increasing levels of DNA damage. Besides that, the experiments of Donehower and others proved that p53 somehow stimulates aging. The simplest explanation of the progeric effect of p53 seems to be the following. This protein is likely to become too active in discarding cells with certain genome damage. As a result, it is not only malignant cells (whose genome was seriously damaged) that die because of apoptosis, but also those whose genome was damaged but slightly. Hence, in

Fig. 15.15 Hypothetical scheme of an aging program (for explanation, see text)

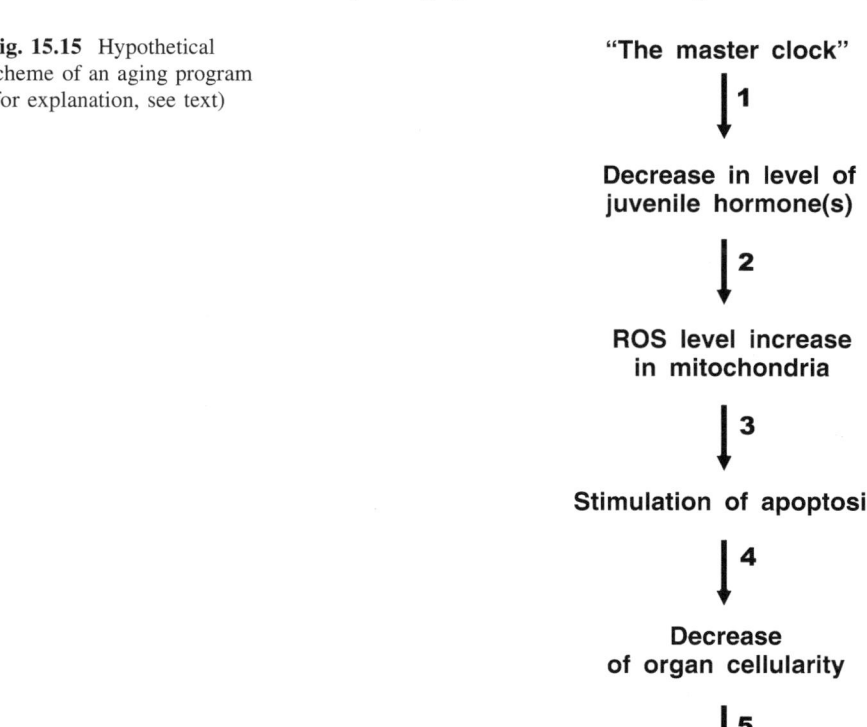

this case aging seems to be stimulated at stage 3 (see Fig. 15.15) (Skulachev 2003a).

15.4.9 Arrest of Age-Dependent Increase of Mitochondrial ROS as a Possible Way to Slow the Aging Program

One approach that could interrupt the aging program shown in Fig. 15.15 seems to be quite obvious. If the appearance of excess ROS in mitochondria with age is an essential step of this program, then let us prevent such an increase in the mROS level. On the face of it, natural antioxidants, such as vitamin E that is accumulated in biomembranes, look like good candidates for this task. However, so far nobody has succeeded in significant slowing of human or animal aging using antioxidants. An attempt to cancel aging, or some other biological program as well, results in a situation when an organism tries to fulfill this program by any means. For instance, increase in vitamin E consumption was shown to induce a special form of cytochrome P450 in liver, this protein being responsible for the destruction of excess

15.4 Aging as Slow Phenoptosis Caused by Increase in mROS Level

Fig. 15.16 Efim Liberman

vitamin E. That is why synthetic, rather than natural, antioxidants that seem to be better candidates for a tool to cancel the aging program. These antioxidants need to be specifically addressed to mitochondria, as their primary targets should be intramitochondrial ROS. Thus, the question is how to provide addressed targeted delivery of antioxidants to mitochondria.

In the late 1960s, hydrophobic cations and anions with their charge strongly delocalized over a relatively large hydrophobic molecule were described in our group in cooperation with the group of Efim Liberman (Fig. 15.16) (Liberman et al. 1969; Liberman and Skulachev 1970). Later these compounds were named *Skulachev ions* (Sk^+ and Sk^-) by the well-known American biochemist David E. Green (Green 1974). Due to low level of their hydration (water dipoles cannot form a "water coat" around an ion if its charge is delocalized), Sk^+ and Sk^- can easily penetrate membranes. As a result, their distribution between two compartments separated by a membrane follows the Nernst rule (10-fold gradient of ion concentration per 60 mV of $\Delta\Psi$) (Liberman and Skulachev 1970). As any ion gradient depends logarithmically on $\Delta\Psi$, a mitochondrion with $\Delta\Psi = 180$ mV (negative inside) is capable of maintaining a 1000-fold gradient of Sk^+ or Sk^-. The tetraphenylphosphonium cation and tetraphenylboron anion were studied as examples of Sk^+ and Sk^-, respectively. Later, Sergey Severin, Lev Yaguzhinsky, and one of this book's authors (V.P.S) (Severin et al. 1970) suggested that ions of Sk^+-type can serve as "molecular electric locomotives": if a compound X needs to be delivered to mitochondria, it would be enough to have it bound to Sk^+, for mitochondrial matrix is the only intracellular compartment that is charged negatively in comparison with cytosol (Liberman and Skulachev 1970).

Such logic was applied when we tried to explain the role of the carnitine cation group in the transport of fatty acid residues into mitochondria (Severin et al. 1970; Skulachev 1988). It was also used by Murphy and Smith for the construction of mitochondria-targeted antioxidants (Burns et al. 1995). Later, Murphy mainly used for his experiments the compound called MitoQ with ubiquinone serving as the antioxidant, and decyl triphenylphosphonium cation as the Sk$^+$ (Smith et al. 1999; Kelso et al. 2001, 2002; James et al. 2005; Saretzki et al. 2003; Jauslin et al. 2003; Murphy and Smith 2007).

We confirmed Murphy's data that micromolar concentrations of MitoQ are able to accumulate in mitochondria and to protect them from oxidative stress. But at the same time, Mikhail Vysokikh in our group (Skulachev 2007; Antonenko et al. 2008; Skulachev et al. 2009) showed that even a small overdose of this compound causes the opposite effect: MitoQ was shown to turn into a powerful prooxidant catalyzing H_2O_2 generation in mitochondria, the rate of this process being so high that it becomes comparable with the respiration rate of mitochondria in the State 4. Similar observations were reported in three other laboratories including Murphy's group (James et al. 2005; O'Malley et al. 2006; Doughan and Dikalov 2007). This is why we attempted to find an antioxidant much better than ubiquinone and combine it with Sk$^+$. These experiments proved to be successful (see Chap. 16).

References

Al-Abdulwahab SS (1999) The effects of aging on muscle strength and functional ability of healthy Saudi Arabian males. Ann Saudi Med 19:211–215

Antonenko YN, Avetisyan AV, Bakeeva LE, Chernyak BV, Chertkov VA, Domnina LV, Ivanova OY, Izyumov DS, Khailova LS, Klishin SS, Korshunova GA, Lyamzaev KG, Muntyan MS, Nepryakhina OK, Pashkovskaya AA, Pletjushkina OY, Pustovidko AV, Roginsky VA, Rokitskaya TI, Ruuge EK, Saprunova VB, Severina II, Simonyan RA, Skulachev IV, Skulachev MV, Sumbatyan NV, Sviryaeva IV, Tashlitsky VN, Vassiliev JM, Vyssokikh MY, Yaguzhinsky LS, Zamyatnin AA Jr, Skulachev VP (2008) Mitochondria-targeted plastoquinone derivatives as tools to interrupt execution of the aging program. 1. Cationic plastoquinone derivatives: synthesis and in vitro studies. Biochemistry (Moscow) 73:1273–1287

Arantes-Oliveira N, Berman JR, Kenyon C (2003) Healthy animals with extreme longevity. Science 302:611

Austad SN (1997) Why we age?. Wiley, New York

Barja G (1998) Mitochondrial free radical production and aging in mammals and birds. Ann NY Acad Sci 854:224–238

Barja G, Herrero A (2000) Oxidative damage to mitochondrial DNA is inversely related to maximum life span in the heart and brain of mammals. FASEB J 14:312–318

Barth C, De Tullio M, Conklin PL (2006) The role of ascorbic acid in the control of flowering time and the onset of senescence. J Exp Bot 57:1657–1665

Bluher M, Kahn BB, Kahn CR (2003) Extended longevity in mice lacking the insulin receptor in adipose tissue. Science 299:572–574

Bohle UR, Hilger HH, Martin WF (1996) Island colonization and evolution of the insular woody habit in *Echium L* (*Boraginaceae*). Proc Natl Acad Sci U S A 93:11740–11745

Bonduriansky R, Brassil CE (2002) Senescence: rapid and costly ageing in wild male flies. Nature 420:377

Bradley AJ, McDonald IR, Lee AK (1980) Stress and mortality in a small marsupial (*Antechinus stuartii, Macleay*). Gen Comp Endocrinol 40:188–200

Bratton SB, Salvesen GS (2010) Regulation of the Apaf-1-caspase-9 apoptosome. J Cell Sci 123:3209–3214

Bredesen DE (2004) The non-existent aging program: how does it work? Aging Cell 3:255–259

References

Brunet-Rossinni AK, Austad SN (2004) Ageing studies on bats: a review. Biogerontology 5:211–222

Budrene EO, Berg HC (1991) Complex patterns formed by motile cells of *Escherichia coli*. Nature 349:630–633

Budrene EO, Berg HC (1995) Dynamics of formation of symmetrical patterns by chemotactic bacteria. Nature 376:49–53

Buffenstein R (2005) The naked mole-rat: a new long-living model for human aging research. J Gerontol A Biol Sci Med Sci 60:1369–1377

Burg SP (1968) Ethylene, plant senescence and abscission. Plant Physiol 43:1503–1511

Burns RJ, Smith RA, Murphy MP (1995) Synthesis and characterization of thiobutyltriphenylphosphonium bromide, a novel thiol reagent targeted to the mitochondrial matrix. Arch Biochem Biophys 322:60–68

Chen XJ, Wang X, Kaufman BA, Butow RA (2005) Aconitase couples metabolic regulation to mitochondrial DNA maintenance. Science 307:714–717

Chen XJ, Wang X, Butow RA (2007) Yeast aconitase binds and provides metabolically coupled protection to mitochondrial DNA. Proc Natl Acad Sci U S A 104:13738–13743

Chipuk JE, Moldoveanu T, Llambi F, Parsons MJ, Green DR (2010) The BCL-2 family reunion. Mol Cell 37:299–310

Christofferson DE, Yuan J (2010) Necroptosis as an alternative form of programmed cell death. Curr Opin Cell Biol 22:263–268

Comfort A (1979) The biology of senescence, 3rd edn. Churchill Livingston, Edinburgh

Darwin C (1871) The Descent of Man. John Murray, London

Dilman VM (1971) Age-associated elevation of hypothalamic threshold to feedback control, and its role in development, ageing, and disease. Lancet 1:1211–1219

Dilman VM (1978) Ageing, metabolic immunodepression and carcinogenesis. Mech Ageing Dev 8:153–173

Dilman VM, Anisimov VN (1979) Hypothalmic mechanisms of ageing and of specific age pathology—I. Sensitivity threshold of hypothalamo-pituitary complex to homeostatic stimuli in the reproductive system. Exp Gerontol 14:161–174

Dilman VM (1982) Master biological clock. Znanie (in Russian), Moscow

Doughan AK, Dikalov SI (2007) Mitochondrial redox cycling of mitoquinone leads to superoxide production and cellular apoptosis. Antioxid Redox Signal 9:1825–1836

Dawkins R (1976) The selfish gene. Oxford University Press, Oxford

Du C, Fang M, Li Y, Li L, Wang X (2000) Smac, a mitochondrial protein that promotes cytochrome *c*-dependent caspase activation by eliminating IAP inhibition. Cell 102:33–42

Emanuel' NM (1975) Certain molecular mechanisms and perspectives for prevention of aging. Izv Akad Nauk SSSR 4:785–794 (in Russian)

Finch CE (1990) Longevity, senescence, and the genome. University of Chicago Press, Chicago

Focher F (2002) "Programmed death" in Schopenhauer's World? Apoptosis 7:285

Galluzzi L, Morselli E, Kepp O, Vitale I, Rigoni A, Vacchelli E, Michaud M, Zischka H, Castedo M, Kroemer G (2010) Mitochondrial gateways to cancer. Mol Aspects Med 31:1–20

Galluzzi L, Vitale I, Abrams JM, Alnemri ES, Baehrecke EH, Blagosklonny MV, Dawson TM, Dawson VL, El-Deiry WS, Fulda S, Gottlieb E, Green DR, Hengartner MO, Kepp O, Knight RA, Kumar S, Lipton SA, Lu X, Madeo F, Malorni W, Mehlen P, Nunez G, Peter ME, Piacentini M, Rubinsztein DC, Shi Y, Simon HU, Vandenabeele P, White E, Yuan J, Zhivotovsky B, Melino G, Kroemer G (2012) Molecular definitions of cell death subroutines: recommendations of the Nomenclature Committee on Cell Death 2012. Cell Death Differ 19:107–120

Garcia-Cao I, Garcia-Cao M, Martin-Caballero J, Criado LM, Klatt P, Flores JM, Weill JC, Blasco MA, Serrano M (2002) "Super p53" mice exhibit enhanced DNA damage response, are tumor resistant and age normally. EMBO J 21:6225–6235

Gerald D, Berra E, Frapart YM, Chan DA, Giaccia AJ, Mansuy D, Pouyssegur J, Yaniv M, Mechta-Grigoriou F (2004) JunD reduces tumor angiogenesis by protecting cells from oxidative stress. Cell 118:781–794

Giles GI (2006) The redox regulation of thiol dependent signaling pathways in cancer. Curr Pharm Des 12:4427–4443

Giorgio M, Migliaccio E, Paolucci D (2004) p66shc is a signal transduction red-ox enzyme. In: 13th EBEC Meeting Abstract: 27

Goldsmith T (2006) Evolution of aging. Azinet Press, Annapolis

Goldstein N (2002) Reactive oxygen species as essential components of ambient air. Biochemistry (Moscow) 67:161–170

Green DE (1974) The electromechanochemical model for energy coupling in mitochondria. Biochim Biophys Acta 346:27–78

Groover AT (2005) What genes make a tree a tree? Trends Plant Sci 10:210–214

Guo Y, Gan S (2011) AtMYB2 regulates whole plant senescence by inhibiting cytokinin-mediated branching at late stages of development in *Arabidopsis*. Plant Physiol 156:1612–1619

Halliwell B, Gutteridge JMC (1999) Free radicals in biology and medicine, 3rd edn. Oxford University Press, Oxford

Harman D (1956) Aging: a theory based on free radical and radiation chemistry. J Gerontol 11:298–300

He P, Osaki M, Takebe M, Shinano T, Wasaki J (2005) Endogenous hormones and expression of senescence-related genes in different senescent types of maize. J Exp Bot 56:1117–1128

Hekimi S, Guarente L (2003) Genetics and the specificity of the aging process. Science 299:1351–1354

Hirota K, Semenza GL (2006) Regulation of angiogenesis by hypoxia-inducible factor 1. Crit Rev Oncol Hematol 59:15–26

Holmes DJ, Fluckiger R, Austad SN (2001) Comparative biology of aging in birds: an update. Exp Gerontol 36:869–883

Holzenberger M, Dupont J, Ducos B, Leneuve P, Geloen A, Even PC, Cervera P, Le Bouc Y (2003) IGF-1 receptor regulates lifespan and resistance to oxidative stress in mice. Nature 421:182–187

Izyumov DS, Avetisyan AV, Pletjushkina OY, Sakharov DV, Wirtz KW, Chernyak BV, Skulachev VP (2004) "Wages of fear": transient threefold decrease in intracellular ATP level imposes apoptosis. Biochim Biophys Acta 1658:141–147

James AM, Cocheme HM, Smith RA, Murphy MP (2005) Interactions of mitochondria-targeted and untargeted ubiquinones with the mitochondrial respiratory chain and reactive oxygen species. Implications for the use of exogenous ubiquinones as therapies and experimental tools. J Biol Chem 280:21295–21312

Jauslin ML, Meier T, Smith RA, Murphy MP (2003) Mitochondria-targeted antioxidants protect Friedreich ataxia fibroblasts from endogenous oxidative stress more effectively than untargeted antioxidants. FASEB J 17:1972–1974

Jing HC, Schippers JH, Hille J, Dijkwel PP (2005) Ethylene-induced leaf senescence depends on age-related changes and OLD genes in Arabidopsis. J Exp Bot 56:2915–2923

John I, Drake R, Farrell A, Cooper W, Lee P, Horton P, Grierson D (1995) Delayed leaf senescence in ethylene-deficient ACC-oxidase antisense tomato plants: molecular and physiological analysis. Plant J 7:483–490

Kapahi P, Boulton ME, Kirkwood TB (1999) Positive correlation between mammalian life span and cellular resistance to stress. Free Radic Biol Med 26:495–500

Kashiwagi A, Hanada H, Yabuki M, Kanno T, Ishisaka R, Sasaki J, Inoue M, Utsumi K (1999) Thyroxine enhancement and the role of reactive oxygen species in tadpole tail apoptosis. Free Radic Biol Med 26:1001–1009

Kelso GF, Porteous CM, Coulter CV, Hughes G, Porteous WK, Ledgerwood EC, Smith RA, Murphy MP (2001) Selective targeting of a redox-active ubiquinone to mitochondria within cells: antioxidant and antiapoptotic properties. J Biol Chem 276:4588–4596

Kelso GF, Porteous CM, Hughes G, Ledgerwood EC, Gane AM, Smith RA, Murphy MP (2002) Prevention of mitochondrial oxidative damage using targeted antioxidants. Ann NY Acad Sci 959:263–274

Kerr JF, Wyllie AH, Currie AR (1972) Apoptosis: a basic biological phenomenon with wide-ranging implications in tissue kinetics. Br J Cancer 26:239–257

Kim EB, Fang X, Fushan AA, Huang Z, Lobanov AV, Han L, Marino SM, Sun X, Turanov AA, Yang P, Yim SH, Zhao X, Kasaikina MV, Stoletzki N, Peng C, Polak P, Xiong Z, Kiezun A, Zhu Y, Chen Y, Kryukov GV, Zhang Q, Peshkin L, Yang L, Bronson RT, Buffenstein R, Wang B, Han C, Li Q, Chen L, Zhao W, Sunyaev SR, Park TJ, Zhang G, Wang J, Gladyshev VN (2011) Genome sequencing reveals insights into physiology and longevity of the naked mole rat. Nature 479:223–227

Kim SC, Crawford DJ, Francisco Ortega J, Santos Guerra A (1996) A common origin for woody *Sonchus* and five related genera in the Macaronesian islands: molecular evidence for extensive radiation. Proc Natl Acad Sci U S A 93:7743–7748

Kirchner IW, Roy BA (1999) The evolutionary advantages of dying young: epidemiological implications of longevity in metapopulations. Am Nat 145:140–159

Kirchner IW, Roy BA (2002) Evolutionary implications of host-pathogen specificity: fitness consequences of pathogen virulence traits. Evol Ecol Res 4:27–40

Kirkwood TB, Austad SN (2000) Why do we age? Nature 408:233–238

Kirkwood TB, Cremer T (1982) Cytogerontology since 1881: a reappraisal of August Weismann and a review of modern progress. Hum Genet 60:101–121

Kirkwood TB, Melov S (2011) On the programmed/non-programmed nature of ageing within the life history. Curr Biol 21:R701–R707

Kluck RM, Bossy-Wetzel E, Green DR, Newmeyer DD (1997) The release of cytochrome *c* from mitochondria: a primary site for Bcl-2 regulation of apoptosis. Science 275:1132–1136

Korshunov SS, Skulachev VP, Starkov AA (1997) High protonic potential actuates a mechanism of production of reactive oxygen species in mitochondria. FEBS Lett 416:15–18

Ku HH, Brunk UT, Sohal RS (1993) Relationship between mitochondrial superoxide and hydrogen peroxide production and longevity of mammalian species. Free Radic Biol Med 15:621–627

Kujoth GC, Hiona A, Pugh TD, Someya S, Panzer K, Wohlgemuth SE, Hofer T, Seo AY, Sullivan R, Jobling WA, Morrow JD, Van Remmen H, Sedivy JM, Yamasoba T, Tanokura M, Weindruch R, Leeuwenburgh C, Prolla TA (2005) Mitochondrial DNA mutations, oxidative stress, and apoptosis in mammalian aging. Science 309:481–484

Labinskyy N, Csiszar A, Orosz Z, Smith K, Rivera A, Buffenstein R, Ungvari Z (2006) Comparison of endothelial function, $O_2^{\bullet-}$ and H_2O_2 production, and vascular oxidative stress resistance between the longest-living rodent, the naked mole rat, and mice. Am J Physiol Heart Circ Physiol 291:H2698–H2704

Lakowski B, Hekimi S (1996) Determination of life-span in *Caenorhabditis elegans* by four clock genes. Science 272:1010–1013

Lambert AJ, Boysen HM, Buckingham JA, Yang T, Podlutsky A, Austad SN, Kunz TH, Buffenstein R, Brand MD (2007) Low rates of hydrogen peroxide production by isolated heart mitochondria associate with long maximum lifespan in vertebrate homeotherms. Aging Cell 6:607–618

Larsson L, Grimby G, Karlsson J (1979) Muscle strength and speed of movement in relation to age and muscle morphology. J Appl Physiol 46:451–456

Lecomte VJ, Sorci G, Cornet S, Jaeger A, Faivre B, Arnoux E, Gaillard M, Trouvé C, Besson D, Chastel O, Weimerskirch H (2010) Patterns of aging in the long-lived wandering albatross. Proc Natl Acad Sci U S A 107:6370–6375

Lee JY, Nagano Y, Taylor JP, Lim KL, Yao TP (2010) Disease-causing mutations in parkin impair mitochondrial ubiquitination, aggregation, and HDAC6-dependent mitophagy. J Cell Biol 189:671–679

Lewis K (2000) Programmed death in bacteria. Microbiol Mol Biol Rev 64:503–514

Lexell J, Taylor CC, Sjostrom M (1988) What is the cause of the ageing atrophy? Total number, size and proportion of different fiber types studied in whole *vastus lateralis* muscle from 15- to 83-year-old men. J Neurol Sci 84:275–294

Liberman EA, Skulachev VP (1970) Conversion of biomembrane-produced energy into electric form. IV. General discussion. Biochim Biophys Acta 216:30–42

Liberman EA, Topaly VP, Tsofina LM, Jasaitis AA, Skulachev VP (1969) Mechanism of coupling of oxidative phosphorylation and the membrane potential of mitochondria. Nature 222:1076–1078

Libertini G (1988) An adaptive theory of increasing mortality with increasing chronological age in populations in the wild. J Theor Biol 132:145–162

Libertini G (2012) Classification of phenoptotic phenomena. Biochemistry (Moscow) 77:707–715

Lichtenstein AV (2005) Cancer as a programmed death of an organism. Biochemistry (Moscow) 70:1055–1064

Liu X, Kim CN, Yang J, Jemmerson R, Wang X (1996) Induction of apoptotic program in cell-free extracts: requirement for dATP and cytochrome c. Cell 86:147–157

Loison A, Festa-Blanchet M, Gaillard J-M, Jorgenson JT, Jullien J-M (1999) Age-specific survival in five populations of ungulates: evidence of senescence. Ecology 80:2539–2554

Longo VD, Finch CE (2003) Evolutionary medicine: from dwarf model systems to healthy centenarians? Science 299:1342–1346

Longo VD, Mitteldorf J, Skulachev VP (2005) Programmed and altruistic ageing. Nat Rev Genet 6(11):866–872

Lyamzaev KG, Nepryakhina OK, Saprunova VB, Bakeeva LE, Pletjushkina OY, Chernyak BV, Skulachev VP (2008) Novel mechanism of elimination of malfunctioning mitochondria (mitoptosis): formation of mitoptotic bodies and extrusion of mitochondrial material from the cell. Biochim Biophys Acta 1777:817–825

Manskikh VN (2004) Studies in evolutionary oncology. Siberian Medical University Publisher, Tomsk

Manskikh VN (2006) Tumor growth as a regulatory mechanism of mutational burden in populations. Biochemistry (Moscow) 71:933–936

Medawar PB (1952) An unsolved problem of biology. H. K. Lewis, London

Melzer S, Lens F, Gennen J, Vanneste S, Rohde A, Beeckman T (2008) Flowering-time genes modulate meristem determinacy and growth form in *Arabidopsis thaliana*. Nat Genet 40:1489–1492

Migliaccio E, Giorgio M, Mele S, Pelicci G, Reboldi P, Pandolfi PP, Lanfrancone L, Pelicci PG (1999) The p66shc adaptor protein controls oxidative stress response and life span in mammals. Nature 402:309–313

Mott JL, Zhang D, Freeman JC, Mikolajczak P, Chang SW, Zassenhaus HP (2004) Cardiac disease due to random mitochondrial DNA mutations is prevented by cyclosporin A. Biochem Biophys Res Commun 319:1210–1215

Murphy MP, Smith RA (2007) Targeting antioxidants to mitochondria by conjugation to lipophilic cations. Annu Rev Pharmacol Toxicol 47:629–656

Naik E, Dixit VM (2011) Mitochondrial reactive oxygen species drive proinflammatory cytokine production. J Exp Med 208:417–420

Napoli C, Martin-Padura I, de Nigris F, Giorgio M, Mansueto G, Somma P, Condorelli M, Sica G, De Rosa G, Pelicci P (2003) Deletion of the p66Shc longevity gene reduces systemic and tissue oxidative stress, vascular cell apoptosis, and early atherogenesis in mice fed a high-fat diet. Proc Natl Acad Sci U S A 100:2112–2116

Narendra DP, Youle RJ (2011) Targeting mitochondrial dysfunction: role for PINK1 and Parkin in mitochondrial quality control. Antioxid Redox Signal 14:1929–1938

Navabpour S, Morris K, Allen R, Harrison E, A-H-Mackerness S, Buchanan-Wollaston V (2003) Expression of senescence-enhanced genes in response to oxidative stress. J Exp Bot 54:2285–2292

Neljubow D (1901) Über die Nutation der Stengel von *Pisum sativum* und einiger anderen Pflanzen. Bot Cetrabl Beihefte 10:128–139

Nesis KN (1997) Cruel love among the squids. In: Byalko AV (ed) Russian science: withstand and revive. Nauka-Physmatlit, Moscow, pp 358–365 (in Russian)

Nooden LD, Murray BJ (1982) Transmission of the monocarpic senescence signal via the xylem in soybean. Plant Physiol 69:754–756

O'Malley Y, Fink BD, Ross NC, Prisinzano TE, Sivitz WI (2006) Reactive oxygen and targeted antioxidant administration in endothelial cell mitochondria. J Biol Chem 281:39766–39775

Orrenius S, Nicotera P, Zhivotovsky B (2011) Cell death mechanisms and their implications in toxicology. Toxicol Sci 119:3–19

Osiewacz HD (2003) Aging and mitochondrial dysfunction in the filamentous fungus *Podospora anserina*. In: Nyström T, Osiewacz HD (eds) Topics in current genetics, vol 3. Springer, Berlin, pp 17–38

Park TJ, Lu Y, Juttner R, Smith ES, Hu J, Brand A, Wetzel C, Milenkovic N, Erdmann B, Heppenstall PA, Laurito CE, Wilson SP, Lewin GR (2008) Selective inflammatory pain insensitivity in the African naked mole-rat (*Heterocephalus glaber*). PLoS Biol 6:e13

Pletjushkina OY, Fetisova EK, Lyamzaev KG, Ivanova OY, Domnina LV, Vyssokikh MY, Pustovidko AV, Vasiliev JM, Murphy MP, Chernyak BV, Skulachev VP (2005) Long-distance apoptotic killing of cells is mediated by hydrogen peroxide in a mitochondrial ROS-dependent fashion. Cell Death Differ 12:1442–1444

Powers SK, Talbert EE, Adhihetty PJ (2011) Reactive oxygen and nitrogen species as intracellular signals in skeletal muscle. J Physiol 589:2129–2138

Pozniakovsky AI, Knorre DA, Markova OV, Hyman AA, Skulachev VP, Severin FF (2005) Role of mitochondria in the pheromone- and amiodarone-induced programmed death of yeast. J Cell Biol 168:257–269

Price AM, Aros Orellana DF, Salleh FM, Stevens R, Acock R, Buchanan-Wollaston V, Stead AD, Rogers HJ (2008) A comparison of leaf and petal senescence in wallflower reveals common and distinct patterns of gene expression and physiology. Plant Physiol 147:1898–1912

Reznick DN (1997) Life history evolution in guppies (*Poecilia reticulata*): guppies as a model for studying the evolutionary biology of aging. Exp Gerontol 32:245–258

Reznick DN, Bryant MJ, Roff D, Ghalambor CK, Ghalambor DE (2004) Effect of extrinsic mortality on the evolution of senescence in guppies. Nature 431:1095–1099

Rigoulet M, Yoboue ED, Devin A (2011) Mitochondrial ROS generation and its regulation: mechanisms involved in H_2O_2 signaling. Antioxid Redox Signal 14:459–468

Ross JJ, Weston DE, Davidson SE, Reid JB (2011) Plant hormone interactions: how complex are they? Physiol Plant 141:299–309

Saule P, Trauet J, Dutriez V, Lekeux V, Dessaint JP, Labalette M (2006) Accumulation of memory T cells from childhood to old age: central and effector memory cells in CD4(+) versus effector memory and terminally differentiated memory cells in CD8(+) compartment. Mech Ageing Dev 127:274–281

Schopenhauer A (1969) The world as will and representation. Dover Publications, New York

Saretzki G, Murphy MP, von Zglinicki T (2003) MitoQ counteracts telomere shortening and elongates lifespan of fibroblasts under mild oxidative stress. Aging Cell 2:141–143

Scrable H, Burns-Cusato M, Medrano S (2009) Anxiety and the aging brain: stressed out over p53? Biochim Biophys Acta 1790:1587–1591

Severin FF, Skulachev VP (2011) Programmed cell death as a target to interrupt the aging program. Adv Gerontol 1:16–27

Severin SE, Skulachev VP, Yaguzhinskii LS (1970) Possible role of carnitine in the transport of fatty acids through the mitochondrial membrane. Biokhimia 35:1250–1253 (in Russian)

Shan X, Yan J, Xie D (2012) Comparison of phytohormone signaling mechanisms. Curr Opin Plant Biol 15:84–91

Shchepina LA, Pletjushkina OY, Avetisyan AV, Bakeeva LE, Fetisova EK, Izyumov DS, Saprunova VB, Vyssokikh MY, Chernyak BV, Skulachev VP (2002) Oligomycin, inhibitor of the F_o part of H^+-ATP-synthase, suppresses the TNF-induced apoptosis. Oncogene 21:8149–8157

Skulachev VP (1988) Membrane bioenergetics. Springer, Berlin

Skulachev VP (1994) Lowering of intracellular O_2 concentration as a special function of respiratory systems of cells. Biochemistry (Moscow) 59:1433–1434 (in Russian)

Skulachev VP (1996a) Role of uncoupled and non-coupled oxidations in maintenance of safely low levels of oxygen and its one-electron reductants. Q Rev Biophys 29:169–202

Skulachev VP (1996b) Why are mitochondria involved in apoptosis? Permeability transition pores and apoptosis as selective mechanisms to eliminate superoxide-producing mitochondria and cell. FEBS Lett 397:7–10

Skulachev VP (1997) Aging is a specific biological function rather than the result of a disorder in complex living systems: biochemical evidence in support of Weismann's hypothesis. Biochemistry (Moscow) 62:1191–1195

Skulachev VP (1998) Possible role of reactive oxygen species in antiviral defense. Biochemistry (Moscow) 63:1438–1440

Skulachev VP (1999a) Mitochondrial physiology and pathology; concepts of programmed death of organelles, cells and organisms. Mol Aspects Med 20:139–184

Skulachev VP (1999b) Phenoptosis: programmed death of an organism. Biochemistry (Moscow) 64:1418–1426

Skulachev VP (2000) Mitochondria in the programmed death phenomena; a principle of biology: "It is better to die than to be wrong". IUBMB Life 49:365–373

Skulachev VP (2002) Programmed death phenomena: from organelle to organism. Ann NY Acad Sci 959:214–237

Skulachev VP (2003a) Aging and programmed death phenomena. In: Nystrom T, Osiewacz HD (eds) Topics in current cenetics. Model systems in ageing, vol 3. Springer, Berlin, pp 191–238

Skulachev VP (2003b) A Nobel Prize for studies on programmed cell death. Biochemistry (Moscow) 68:290–291 (in Russian)

Skulachev VP (2005) Aging as an atavistic program which might be switched off? Vestnik Rus Acad Sci 75:831–843 (in Russian)

Skulachev VP (2006) Bioenergetic aspects of apoptosis, necrosis and mitoptosis. Apoptosis 11:473–485

Skulachev VP (2007) A biochemical approach to the problem of aging: "megaproject" on membrane-penetrating ions. The first results and prospects. Biochemistry (Moscow) 72:1385–1396

Skulachev VP (2009) How to cancel the aging program of organism? Rus Khim Zh 53(3):125–140 (in Russian)

Skulachev VP (2011) Aging as a particular case of phenoptosis, the programmed death of an organism (a response to Kirkwood and Melov "On the programmed/non-programmed nature of ageing within the life history"). Aging 3:1120–1123

Skulachev VP (2012) What is "phenoptosis" and how to fight it? Biochemistry (Moscow) 77:689–706

Skulachev VP, Longo VD (2005) Aging as a mitochondria-mediated atavistic program: can aging be switched off? Ann NY Acad Sci 1057:145–164

Skulachev VP, Anisimov VN, Antonenko YN, Bakeeva LE, Chernyak BV, Erichev VP, Filenko OF, Kalinina NI, Kapelko VI, Kolosova NG, Kopnin BP, Korshunova GA, Lichinitser MR, Obukhova LA, Pasyukova EG, Pisarenko OI, Roginsky VA, Ruuge EK, Senin II, Severina II, Skulachev MV, Spivak IM, Tashlitsky VN, Tkachuk VA, Vyssokikh MY, Yaguzhinsky LS, Zorov DB (2009) An attempt to prevent senescence: a mitochondrial approach. Biochim Biophys Acta 1787:437–461

Smith RA, Porteous CM, Coulter CV, Murphy MP (1999) Selective targeting of an antioxidant to mitochondria. Eur J Biochem 263:709–716

Sommer SS (1994) Does cancer kill the individual and save the species? Hum Mutat 3:166–169

Sterky FH, Lee S, Wibom R, Olson L, Larsson NG (2011) Impaired mitochondrial transport and Parkin-independent degeneration of respiratory chain-deficient dopamine neurons in vivo. Proc Natl Acad Sci U S A 108:12937–12942

Susin SA, Zamzami N, Castedo M, Hirsch T, Marchetti P, Macho A, Daugas E, Geuskens M, Kroemer G (1996) Bcl-2 inhibits the mitochondrial release of an apoptogenic protease. J Exp Med 184:1331–1341

Susin SA, Lorenzo HK, Zamzami N, Marzo I, Snow BE, Brothers GM, Mangion J, Jacotot E, Costantini P, Loeffler M, Larochette N, Goodlett DR, Aebersold R, Siderovski DP, Penninger JM, Kroemer G (1999) Molecular characterization of mitochondrial apoptosis-inducing factor. Nature 397:441–446

Szilard L (1959) On the nature of the aging process. Proc Natl Acad Sci U S A 45:30–45

Tait SW, Green DR (2010) Mitochondria and cell death: outer membrane permeabilization and beyond. Nat Rev Mol Cell Biol 11:621–632

Terzibasi E, Valenzano DR, Cellerino A (2007) The short-lived fish *Nothobranchius furzeri* as a new model system for aging studies. Exp Gerontol 42:81–89

Trifunovic A, Wredenberg A, Falkenberg M, Spelbrink JN, Rovio AT, Bruder CE, Bohlooly YM, Gidlof S, Oldfors A, Wibom R, Tornell J, Jacobs HT, Larsson NG (2004) Premature ageing in mice expressing defective mitochondrial DNA polymerase. Nature 429:417–423

Trinei M, Giorgio M, Cicalese A, Barozzi S, Ventura A, Migliaccio E, Milia E, Padura IM, Raker VA, Maccarana M, Petronilli V, Minucci S, Bernardi P, Lanfrancone L, Pelicci PG (2002) A p53–p66Shc signalling pathway controls intracellular redox status, levels of oxidation-damaged DNA and oxidative stress-induced apoptosis. Oncogene 21:3872–3878

Tyner SD, Venkatachalam S, Choi J, Jones S, Ghebranious N, Igelmann H, Lu X, Soron G, Cooper B, Brayton C, Hee Park S, Thompson T, Karsenty G, Bradley A, Donehower LA (2002) p53 mutant mice that display early ageing-associated phenotypes. Nature 415:45–53

Vandenabeele P, Galluzzi L, Vanden Berghe T, Kroemer G (2010) Molecular mechanisms of necroptosis: an ordered cellular explosion. Nat Rev Mol Cell Biol 11:700–714

Verhagen AM, Ekert PG, Pakusch M, Silke J, Connolly LM, Reid GE, Moritz RL, Simpson RJ, Vaux DL (2000) Identification of DIABLO, a mammalian protein that promotes apoptosis by binding to and antagonizing IAP proteins. Cell 102:43–53

Weismann A (1889) Essays upon heredity and kindred biological problems. Claderon Press, Oxford

Wi SJ, Jang SJ, Park KY (2010) Inhibition of biphasic ethylene production enhances tolerance to abiotic stress by reducing the accumulation of reactive oxygen species in *Nicitiana tabacum*. Mol Cells 30:37–49

Wodinsky J (1977) Hormonal inhibition of feeding and death in octopus: control by optic gland secretion. Science 198:948–951

Yuan J, Kroemer G (2010) Alternative cell death mechanisms in development and beyond. Genes Dev 24:2592–2602

Yang J, Liu X, Bhalla K, Kim CN, Ibrado AM, Cai J, Peng TI, Jones DP, Wang X (1997) Prevention of apoptosis by Bcl-2: release of cytochrome *c* from mitochondria blocked. Science 275:1129–1132

Zorov DB, Kinnally KW, Tedeschi H (1992) Voltage activation of heart inner mitochondrial membrane channels. J Bioenerg Biomembr 24:119–124

Chapter 16
Possible Medical Applications of Membrane Bioenergetics: Mitochondria-Targeted Antioxidants as Geroprotectors

16.1 SkQ Decelerate the Aging Program

When it became obvious that mitochondria-targeted antioxidant MitoQ is not capable of decelerating the aging program, we turned to mitochondria-targeted derivatives of *plastoquinone*, an electron carrier found in the photosynthetic electron transport chains of plant chloroplasts and cyanobacteria. It is the much better antioxidant properties of plastoquinone (in comparison to ubiquinone) (Kruk et al. 1997; Roginsky et al. 2003) that might have been the reason for biological evolution to replace ubiquinone of the mitochondrial respiratory chain by plastoquinone in the chloroplast photosynthetic chain of the same plant cell. An important point is that *oxygen-producing* chloroplasts constantly experience much stronger oxidative stress than *oxygen-consuming* mitochondria. In plastoquinone, two methoxy groups of ubiquinone are replaced by methyl groups, and the methyl group of ubiquinone is replaced by a hydrogen atom (see Fig. 16.1, Appendix 5). These substitutions greatly increase the antioxidant activity of the compound in isolated mitochondria. In the case of MitoQ, the concentrations responsible for 20 % anti- and prooxidant effects differ by less than a factor of two (300 and 500 nM), while for the plastoquinone derivative of decyltriphenylphosphonium, which was given the name SkQ1, this difference increased to 400-fold (2 and 800 nM, Fig. 16.2) (Antonenko et al. 2008a; Skulachev et al. 2009). With such promising results, we realized that SkQ1 is a remarkably efficient mitochondria-targeted antioxidant with little risk of prooxidant side effect. At this stage (at the very end of 2003), a study, originally initiated as a series of grant-supported scientific experiments, was transformed to an investment project dubbed "Practical application of penetrating ions" (Prof. Vladimir Skulachev, supervisor). The project was supported since that time by private funds delivered by Mr. Oleg Deripaska ("Basic Element"), then by Mr. Alexander Chikunov ("Rostok") and now by "Agro-BUS" Company. Since 2010, the state corporation RUSNANO (Mr. Anatoly Chubays, Director) contributes 50 % to the funding of the project.

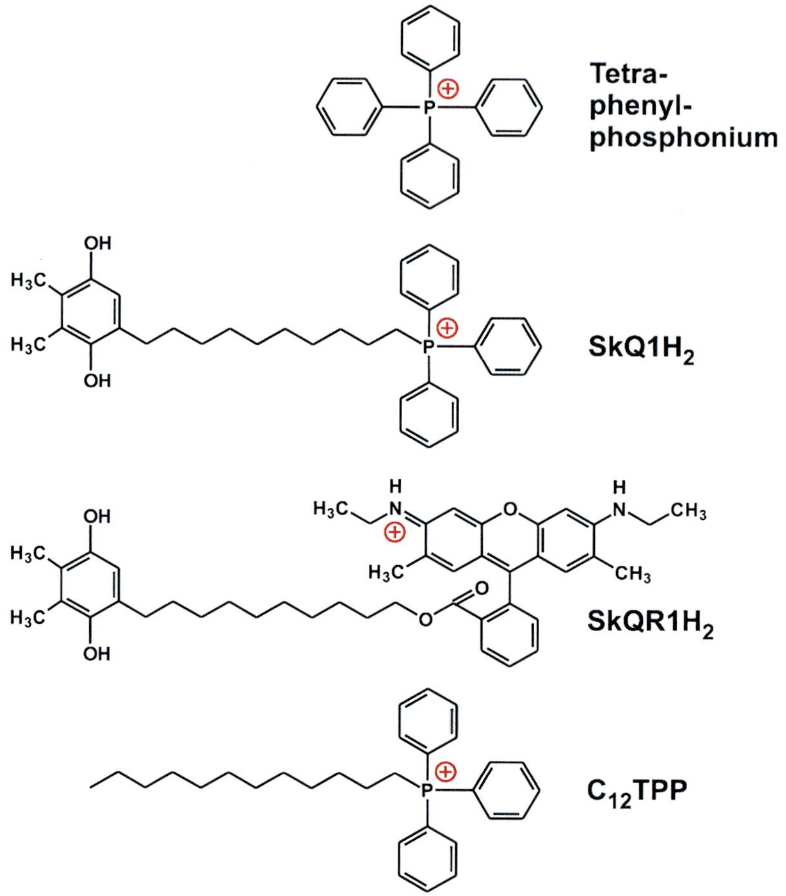

Fig. 16.1 Structural formulas of tetraphenylphosphonium, SkQ1H$_2$, SkQR1H$_2$, and C$_{12}$TPP

Researchers of the Belozersky Institute of Physico-Chemical Biology and a number of departments of Moscow State University, as well as over 30 other research groups in Russia, Sweden, Germany, and the United States have been involved in the project (major publications include: Skulachev 2007, 2012; Skulachev et al. 2009, 2011; Antonenko et al. 2008a, b, 2011; Bakeeva et al. 2008; Agapova et al. 2008; Anisimov et al. 2008, 2011; Rokitskaya et al. 2008; 2010; Plotnikov et al. 2007; 2011; Neroev et al. 2008; Roginsky et al. 2009; Severin et al. 2010; Shipounova et al. 2010; Fetisova et al. 2011; Stefanova et al. 2010; Lyamzaev et al., 2011; Kapay et al. 2011). The present state of the project can be summarized as follows.

Series of small (about 2-nm size) "molecular electric locomotives" (penetrating cations) conjugated with different quinones were synthesized by Galina Korshunova and her coworkers. Decyl- or amyltriphenylphosphonium (in SkQ1 and SkQ5, respectively), decylrhodamine 19 (in SkQR1), decylmethylcarnitine (in SkQ2),

16.1 SkQ Decelerates the Aging Program

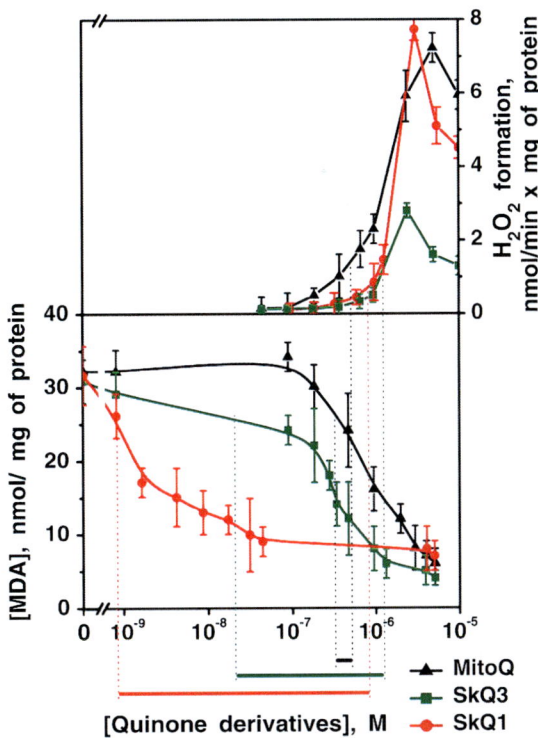

Fig. 16.2 Prooxidant (*top panel*) and antioxidant (*bottom panel*) activity of SkQ and its analogs in isolated rat heart mitochondria. Prooxidant activity was estimated as stimulation of H_2O_2 formation by mitochondria in State 4. Antioxidant activity was measured as inhibition of malondialdehyde accumulation in mitochondria incubated with ascorbate and Fe^{2+}. *Colored bars* below the abscissa indicate windows between concentrations of given quinone derivative causing 20 % anti- and 20 % prooxidant effects (Antonenko et al. 2008a; Skulachev et al. 2009)

decyltributylammonium (in SkQ4), and also similar derivatives of methylplastoquinone (in SkQ3), of ubiquinone (in MitoQ), desmethoxyubiquinone (DMQ), and a number of others have been used as Sk^+ in these compounds. (Antonenko et al. 2008a; Skulachev et al. 2011). The penetrating ability of these compounds was studied by Inna Severina, Yury Antonenko, and coworkers in artificial bilayer phospholipid membrane (BLM). This ability was shown to decrease is the series: SkQR1 > SkQ1, SkQ3, MitoQ > SkQ5 ≫ SkQ4 (Antonenko et al. 2008a; Rokitskaya et al. 2008). Antioxidant properties tested in aqueous solutions, BLMs, liposomes, linoleic acid micelles, and mitochondria followed the pattern SkQR1, SkQ1 > SkQ3, DMQ > MitoQ (Antonenko et al. 2008a, b). Antonenko et al. found that the prooxidant properties of these followed the opposite pattern, reaching their maximal value in MitoQ (Antonenko et al. 2008a). On the basis of these data, SkQ1 and its fluorescent derivative SkQR1 were chosen for further studies

In in vitro experiments on heart mitochondria, Vyssokikh et al. showed that the respiratory chain was able to reduce SkQ1. This means that this compound can act as a *rechargeable antioxidant*. The b_H heme of complex III was found to be the site of reduction of SkQ1. This heme is located close to the inner surface of the inner mitochondrial membrane. The reduction of SkQ1 was completely inhibited by antimycin (Antonenko et al. 2008a; Skulachev et al. 2009).

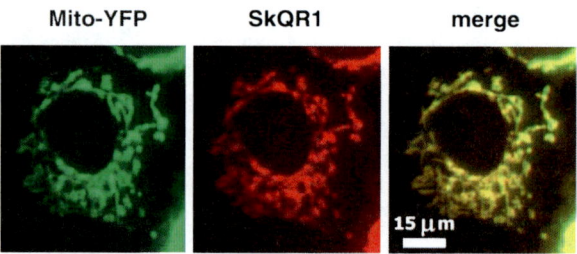

Fig. 16.3 Colocalization of SkQR1 (a fluorescent SkQ derivative) and mitochondria-targeted jellyfish yellow fluorescent protein fused with the leader sequence of cytochrome oxidase subunit VIII (Mito-YFP). HeLa cells were transfected with Mito-YFP and incubated for 15 min with 100 nM SkQR1. (Skulachev et al. 2009)

When using SkQR1, Chernyak and his colleagues showed the SkQ-type compounds added to the cell culture are selectively accumulated in mitochondria (Figs. 16.3, 16.4a) (Antonenko et al. 2008a; Skulachev et al. 2009). Then, the studied compounds were shown to inhibit the H_2O_2-induced apoptosis of HeLa cells and human fibroblasts (activity: SkQR1 > SkQ1 > MitoQ; Fig. 16.4c) (Antonenko et al. 2008a; Skulachev et al. 2009). The uncoupler FCCP abolishing $\Delta\psi$ and preventing SkQ1 accumulation by mitochondria (Fig. 16.4a) inhibited the antiapoptotic effect of SkQ1 (Fig. 16.4b). Induction of apoptosis by H_2O_2 was found to have two consequences soon after its addition: strong increase in production of ROS by cells (see DCF probe for ROS, Fig. 16.4d) and decomposition of mitochondrial filaments to small roundish mitochondria (Fig. 16.4e). Both these effects were prevented by SkQ1. $C_{12}TPP$, an SkQ1 analog without plastoquinone, was ineffective (Fig. 14e). Chernyak et al. showed that ROS-induced necrosis was also sensitive to SkQ1 (Antonenko et al. 2008a; Skulachev et al. 2009). Decylplastoquinone (DPQ) and tetraphenylphosphonium failed to substitute for SkQ1, whereas antioxidants N-acetylcysteine (NAC) and Trolox were effective but at very much higher concentrations (Fig. 16.4f). SkQ1 was shown to prolong the median lifespan of the mycelial fungus *Podospora anserina*, the crustacean *Ceriodaphnia affinis*, the insect *Drosophila melanogaster*, the fish *Nothobranchius furzeri*, and mice (in case of mice the median lifespan was doubled; Fig. 16.5). Vyssokikh, Filenko, Pasyukova, Shidlovsky, Anisimov et al. took part in these studies (Skulachev et al. 2009; Anisimov et al. 2008, 2011). In all the studied cases, SkQ1 was the most effective in reduction of mortality at early stages of aging (the first 20 % of animals to die), but it showed little effect on the maximal lifespan and the last 10 % of deaths. Anisimov et al. showed that in the case of mice in a nonsterile vivarium, SkQ1 added to drinking water (0.5–5 nmol/kg per day) strongly lowered mortality caused by infectious diseases but showed little effect on mortality caused by cancer (Anisimov et al. 2008; 2011; Skulachev et al. 2009). In studies on mice and rats, the groups of Kolosova, Anisimov, Drize, Cannon, and some others showed that SkQ1 decelerates the development of such

16.1 SkQ Decelerates the Aging Program

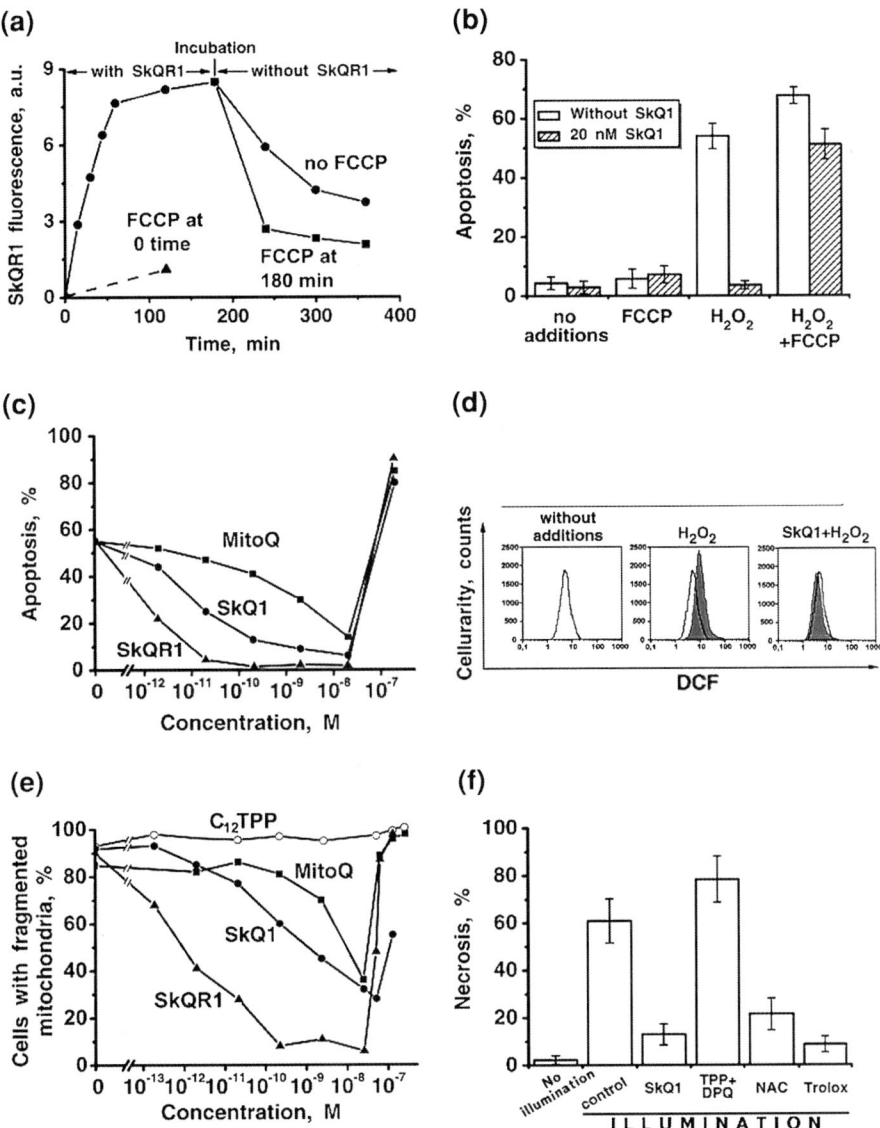

◀ **Fig. 16.4** Effects of SkQ1, SkQR1, and MitoQ on human cell cultures. **a** Kinetics of SkQR1 accumulation in HeLa cells and its efflux from cells. Cells were treated with 50 nM SkQR1 at zero time. Uncoupler (10 μM FCCP) was added where indicated. SkQR1 accumulation and efflux were measured fluorimetrically in a flow cytometer. **b** Incubation of human fibroblasts for 7 days with 20 nM SkQ1 prevents H_2O_2-induced apoptosis. Apoptotic cells were counted 24 h after addition of H_2O_2. Additions: 1 μM FCCP, 400 μM H_2O_2. **c** SkQ1 and SkQR1 are more efficient than MitoQ in arresting H_2O_2-induced apoptosis in human fibroblasts; for the experimental conditions, see **b**. **d** SkQ1 (20 nM, 7 days) prevents accumulation of H_2O_2 by H_2O_2-treated HeLa cells (200 μM, 45 min treatment). The cells were stained with 4 μM DCF for 15 min before being analyzed in a flow cytometer. **e** Preventive effects of SkQR1, SkQ1, and MitoQ and ineffectiveness of C_{12}TPP (an SkQ1 analog without the quinone moiety) on H_2O_2-induced fission of elongated mitochondria. Human fibroblasts were preincubated with SkQR1, SkQ1, MitoQ, or C_{12}TPP for 2 h and then treated with 400 μM H_2O_2 for 3 h. **f** SkQ1 prevents necrosis caused by illumination of HeLa cells stained with Mitotracker Red. The cells were illuminated for 15 min with the green light (545 nm) so as to deliver the light energy of 34.8 J/cm². Necrosis was measured 5 h after the illumination. The cells were treated with 1 μM SkQ1, the mixture of 1 μM tetraphenylphosphonium (TPP) and 1 μM decylplastoquinone (DPQ), 20 mM N-acetylcysteine (NAC), or 1 mM Trolox 1 h before illumination (Skulachev et al. 2009)

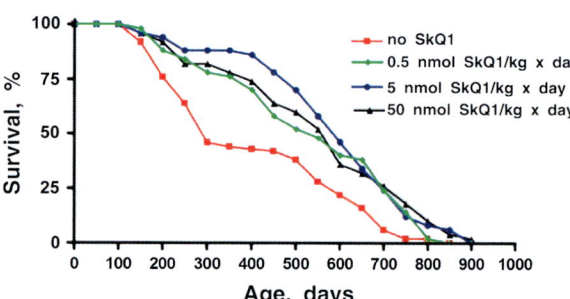

Fig. 16.5 SkQ1 increases lifespan of SHR mice. The figure summarizes results of two experiments on lifespan measurement of 200 females. In each experiment, four groups of females (25 mice in each group) were studied. (Skulachev et al. 2009)

traits of aging as involution of the thymus[1] and follicular region of spleen, reduction in the ratio of lymphocytes and neutrophils in blood, osteoporosis, lordokyphosis, cataract, retinopathy, glaucoma, alopecia, achromotrichia, canities, disappearance of regular estrous cycles in females and libido in males, an increase in the left vertical volume, slow wound healing, hypothermia, depression, and some other age-linked changed in behavior, etc. (Figs. 16.6, 16.7, 16.8) (Skulachev et al. 2009; Stefanova et al. 2010; Shipounova et al. 2010; Neroev et al. 2008; Anisimov et al. 2008, 2011). It is remarkable that effective prevention of age-related eye diseases in OXYS rats (a line of rodents suffering from permanent

---

[1] Age-related involution of the thymus is the most demonstrative example of aging being programmed. This effect surely cannot be explained by a mere accumulation of injuries in an aging organism. Involution of the thymus, the organ producing T-lymphocytes, one of the types of key cells of the immune system, undoubtedly leads to weakening of immunity. These changes are aggravated by the parallel involution of the spleen follicles, where the other very important type of immune cells is produced, B-lymphocytes. SkQ decelerates (but does not cancel) both of these processes (Obukhova et al. 2009).

Fig. 16.6 SHR mice (630-days old) receiving (*top* photo) and not receiving (*bottom* photo) 0.5 nmol SkQ1/kg per day. Traits of senescence, such as lordokyphosis and alopecia (loss of whiskers and hair on cheeks and side) are seen in the control mouse. The SkQ1-receiving mouse lacks these traits. (Skulachev et al. 2009)

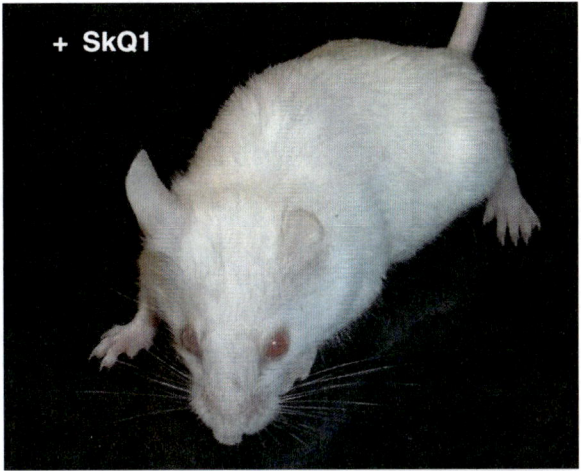

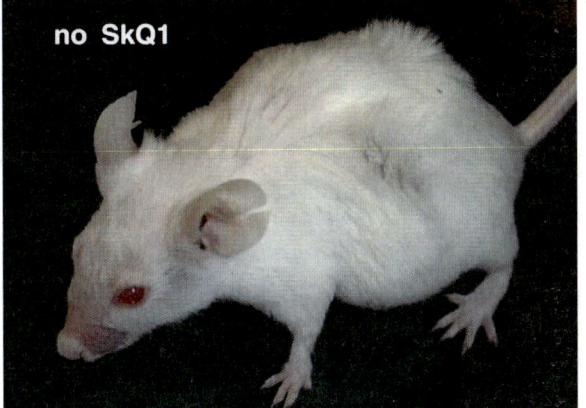

oxidative stress) was observed with SkQ1 but not with the same dose of MitoQ or even with much larger dose of α-tocopherol (Fig. 16.7a, b).

In fibroblasts obtained from control and life-long SkQ-fed mice, SkQ1 was shown to prevent the age-related appearance of β-glucosidase activity and phosphorylation of H2AX histone (Anisimov et al. 2008; Skulachev et al. 2009). Studies carried out on the same model by Spivak and Mikhelson demonstrated the age-related increase in spontaneous apoptosis and its stimulation by the addition of hydrogen peroxide. Both effects were canceled by SkQ1 treatment (Anisimov et al. 2008; Skulachev et al. 2009). It is important that SkQ1 did not arrest apoptosis completely, but prevented its threefold increase in old mice. Exactly, this type of effect might be expected within the framework of our scheme (see Fig. 15.15). A demonstrative effect of in vivo treatment with SkQ1 was observed in Kolosova's group in Novosibirsk, where age-dependent behavioral changes were studied in Wistar rats (Stefanova et al. 2010). Rats of 3- and 14-month age were investigated. An elevated plus maze with two open and two closed arms was used. Young

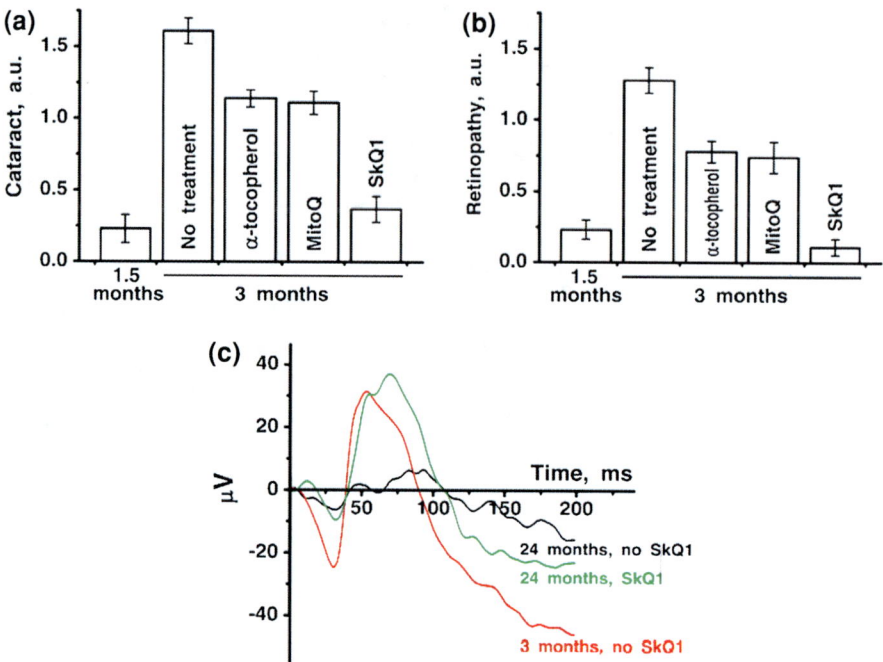

Fig. 16.7 SkQ1 prevents development of cataract and retinopathy in OXYS rats. SkQ1 and other compounds were administered *per os*. **a, b** 3-month-old rats; ∝-tocopherol (670 μmol/kg body mass/day), MitoQ (250 μmol/kg/day), or SkQ1 (250 μmol/kg/day) was administrated during the last 1.5 months (Zinovkin et al. 2013). Electroretinograms of three OXYS rats: *1* 3 months, no SkQ1 (*red*); *2* 24 months, no SkQ1 (*black*), and *3* 24 months, 250 nmol SkQ1/kg per day with food (*green*). (Skulachev et al. 2009)

animals placed in the maze center entered both types of arms with equal probability. However, the 14-month-old rats preferred to enter the closed arms only, the probability of entering the open arms being negligible. Addition of a very small amount of SkQ1 to the food (250 nmol/kg body mass daily) for the last 10 weeks completely reversed the age effect. With SkQ1, the probability of entering an open arm for the 14-month-old rats was as high as for young rats and, when entering an open arm, the SkQ1-treated old rats spent in it a time which was almost as long as for the young rodents (Fig. 16.8). The number of squares crossed by the animals in the open field test was slightly (by 25 %) smaller in the 14-month-old rats than for the 3-month-old animals. This difference was also abolished by SkQ1. Rearings in the maze or in the open field and head dips in the maze were age independent and SkQ1 insensitive.

In a number of cases, SkQ was shown to retard the development of accelerated aging—progeria. Such results were obtained on OXYS rats. SkQ1 addition to food (250 nmol/kg) was shown not only to prevent (Fig. 16.7) the development of early cataract and retinopathy, but also prevented peroxidation of cardiolipin, accumulation of malondialdehyde and carbonylated proteins in tissues (Kolosova,

16.1 SkQ Decelerates the Aging Program

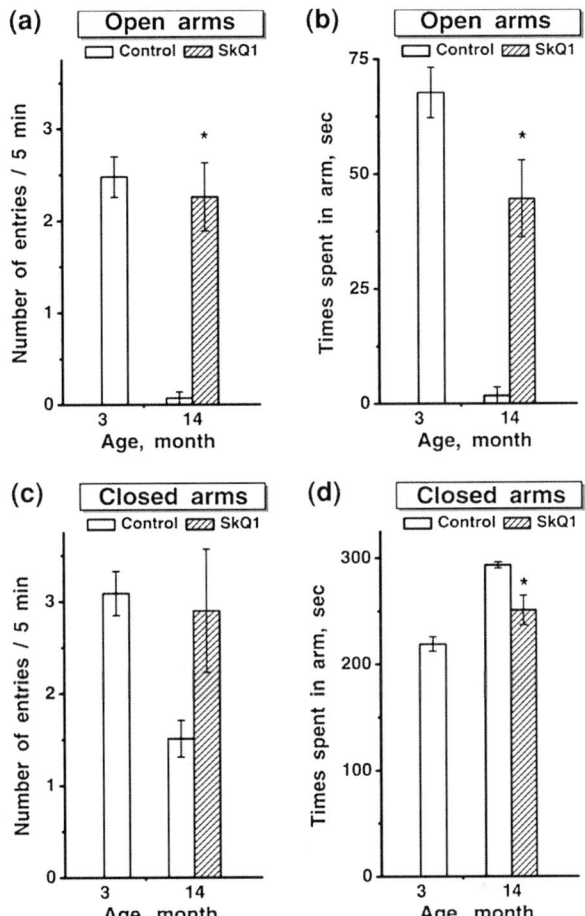

Fig. 16.8 SkQ1 reverses age-dependent behavioral changes in Wistar rats. The elevated plus maze was used. Where indicated, SkQ1 (250 nmol/kg body mass daily) was added to the food of old rats during the final 10 weeks (Reprinted from Stefanova et al. (2010), with permission from IOS Press)

Zinovkin et al.) (Antonenko et al. 2008a; Skulachev et al. 2009; 2010; Skulachev et al. 2011). In case of cataract and retinopathy, eye drops containing 2.5×10^{-7} M SkQ1 cured already developed disease within 1.5 months (Antonenko et al. 2008a; Skulachev et al. 2009). This therapy was also effective in the case of cataract in normal Wistar rats. However, Wistar rats hardly ever live long enough to experience retinopathy, so in this case we had to change the object of research. The experiment carried out by Kopenkin, Sotnikova, and their coworkers (Antonenko et al. 2008a; Skulachev et al. 2009, 2011) from the Skryabin Veterinary Academy, Moscow showed the SkQ1 eye drops to be effective in treating retinopathy in dogs, cats, and horses. A total of 271 animals, for which all the traditional methods failed to have positive effect, received SkQ1. The most striking positive effects were achieved in cases of congenital dysplasia of the retina (dramatic improvement in 67 % of cases) and its secondary degeneration (54 %). In the case of progressive retinal degeneration, SkQ1 was less effective

(29 %). In 89 cases, animals that had recently become blind were the object of research. Of these, 67 animals regained their sight after being treated with SkQ1 eye drops for 2–6 weeks.

These positive results are in contrast to the failure of Murphy's group, where congenital retinopathy in mice was treated with MitoQ (Vlachantoni et al. 2006). The likely reason for this failure is the effective MitoQ concentrations (i.e. those having antioxidant effect) being too close to those causing a prooxidant effect. The authors, however, assumed that the studied types of retinopathy might have been resistant to antioxidants, which is also possible.

When completing this series of experiments, we studied artificially imposed eye diseases, namely, experimental uveitis (Senin et al., see Neroev et al. 2008; Skulachev et al. 2009) and glaucoma in rabbits (Erichev and Iomdina, see Neroev et al. 2008; Skulachev et al. 2009). Both of these diseases are known to be closely connected to very strong oxidative stress, and in both cases SkQ1 eye drops were shown to have therapeutic effect. Experiments on treatment of uveitis were especially demonstrative. Immunization of rabbits with arrestin, a protein specific for photoreceptor cells, caused the development of this autoimmune disease, which eventually led to blindness. Then, one eye was treated four times daily with drops of 2.5×10^{-7} M SkQ1. After several days, the animals started to regain their sight, but only in the eye that was treated with SkQ1. The same procedure prevented the development of uveitis if the eye drops were used during the immunization. The treatment was successful in 100 % of cases, both in prevention of uveitis and in treatment of the already developed disease.

Obukhova et al. studied progeria in OXYS rats, where SkQ1 prevented the age-dependent acceleration of decrease in number of cells in thymus and spleen primary follicles that are responsible for the formation of T- and B-lymphocytes, respectively (Obukhova et al. 2009). Ryazanov in the USA studied radiation-induced progeria, and SkQ1 canceled such aging sign as graying of hair (black mice were used for these experiments).

Results obtained on mutant mice with alanine substituted for aspartate-257 in "the proofreading" domain of mitochondrial DNA-polymerase γ are interesting. (These mutants have been already discussed above in connection to the role of mitochondria in aging, see Chap. 15). Addition of SkQ1 to drinking water increased not only in lifespan but also "healthspan" of such animals. SkQ-treated "mutator" mice died without the whole set of senile traits that could be observed in "mutator" mice that did not receive SkQ1. Among these traits were osteoporosis (lordokyphosis), loss of hair on the cheeks and body, depression, decrease in body temperature, disappearance of estrous cycles, etc. The effect of SkQ on mortality was developed in spite of the fact that mice were kept under sterile conditions (Cannon and coworkers, Sweden).

Experiments performed in the groups of Kapelko and Pisarenko (Bakeeva et al. 2008; Skulachev et al. 2009) and the group of Zorov (Bakeeva et al. 2008; Skulachev et al. 2009; Plotnikov et al. 2010) showed that SkQ1 or SkQR1 treatment reduced considerably an unfavorable effect of ischemia and subsequent reperfusion on heart rhythm in rats; SkQ1 also reduced the areas of myocardial

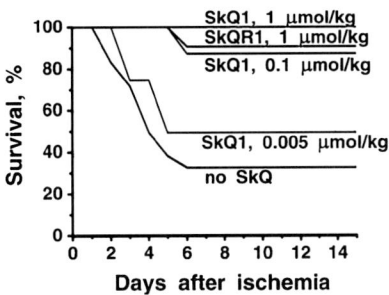

Fig. 16.9 Survival of single-kidney rats after 90 min ischemia with subsequent reperfusion. SkQ1 or SkQR1 were injected intraperitoneally a day before the ischemia (Skulachev et al. 2009)

infarction and brain stroke. Both of these compounds prevented the death of rats who had one kidney removed and the other one having undergone 90-min-long ischemia (Skulachev et al. 2009; Plotnikov et al. 2011) (Fig. 16.9). SkQR1 was also effective in the case of pyelonephritis and rhabdomyolysis (Plotnikov et al. 2010, 2011). In experiments on p53$^{-/-}$ mice performed in the group of Boris Kopnin, SkQ1 as well as nontargeted antioxidant N-acetylcysteine (NAC) slowed development of lymphoma, reduced ROS level in spleen, and prolonged lifespan, the effective SkQ1 concentration being as low as 5 nmol/kg per day. To reach a similar effect, 6 mmol NAC/kg per day was needed, i.e., *1.2 million times more than SkQ1* (Agapova et al. 2008; Skulachev et al. 2009). SkQ1 was even more effective in the case of rat cardiac arrhythmia and in prolongation of the lifespan of mice: SkQ1 concentrations of 0.05 and 0.5 nmol/kg per day caused the maximal effect (Bakeeva et al. 2008; Skulachev et al. 2009).

There are several reasons for such high efficiency of SkQ. SkQ1 can accumulate electrophoretically in the cell cytosol reaching a level about ten times higher than outside the cell. It is the electric potential difference on the outer cell membrane that causes this accumulation ($\Delta\Psi$ of about 60 mV). Even much more significant (1000-fold) accumulation takes place in mitochondria ($\Delta\Psi$ of about 180 mV). The distribution coefficient in a membrane/water system is about 10,000 for SkQ1. Because of this, SkQ1 concentration in the inner half-membrane layer of the inner mitochondrial membrane should be $10 \times 10^3 \times 10^4 = 10^8$ times higher than in the extracellular space (Skulachev et al. 2009; 2010).

The mechanism of the antioxidant effect of SkQ can involve two types of phenomena, i.e., (1) quenching of radical intermediates of lipid peroxidation and (2) mild uncoupling. SkQH$_2$, when accumulated in the inner half-membrane layer of inner mitochondrial membrane, can quench peroxyl radicals of polyunsaturated fatty acids (LO$_2^{\bullet}$) of phospholipids forming this layer (especially of cardiolipin) (Fig. 16.10). Most probably, LO$_2^{\bullet}$ is reduced by SkQH$_2$ to fatty acid peroxide (LOOH), and a chain reaction of oxidative damage to mitochondrial lipids is interrupted:

$$LO_2^{\bullet} + SkQH_2 \rightarrow LOOH + SkQ^{\bullet-} + H^+. \tag{16.1}$$

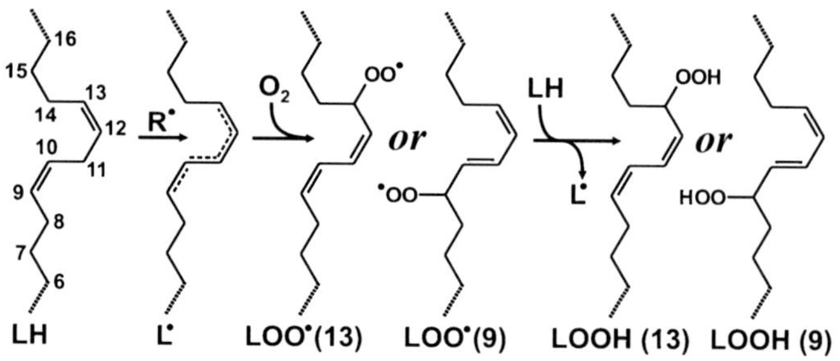

Fig. 16.10 Conversion of linoleate residue (LH) into its primary radical form (L•), peroxyl radicals (LOO•), and peroxides (LOOH). R•, a radical initiating the chain reaction of lipid peroxidation (OH•, HO$_2^•$, L•, etc.)

Then, obtained SkQ•⁻ is reduced by cytochrome b_H of complex III of the respiratory chain close to the inner membrane surface. As a result, SkQH$_2$ is regenerated:

$$\text{SkQ}^{•-} + \text{cytochrome } b_H^{2+} + 2\text{ H}^+ \rightarrow \text{SkQH}_2 + \text{cytochrome } b_H^{3+}. \quad (16.2)$$

As shown in Fig. 16.10, a linoleate residue of cardiolipin is assumed to be attacked by a free radical R• to initiate the chain reaction of lipid peroxidation. This assumption is based upon two facts: (1) linoleate is the main (in some tissues the only) type of fatty acid residues found in cardiolipin; (2) both in vitro and in vivo, the initial step of oxidative stress results in peroxidation of cardiolipin, the other phospholipids remaining intact (Antonenko et al. 2008a, b; Skulachev et al. 2009, 2010; Shabalina et al. in preparation).

The linoleate peroxide residue (LOOH) formed as a result of cardiolipin peroxidation is split from cardiolipin and transferred from the inner to the outer half-membrane layer. Here, it is decomposed by cytochrome *c* or a special glutathione peroxidase capable of using lipid peroxides as substrates. Minor uncoupling proteins (UCP2, UCP3, UCP4, and UCP5, see above, Sect. 10.2.2) or SkQH$_2$ might function as carriers of linoleate peroxides. Severina, Severin, Antonenko, Vyssokikh, and their coworkers from our group (Severin et al. 2010; Rokitskaya et al. 2010; Skulachev et al. 2011) have recently shown SkQ1 and its analog containing no plastoquinone (C$_{12}$TPP) to be able to act as carriers of fatty acid anions in artificial and mitochondrial membranes. This additional activity of SkQ1 allows one to define it as a mitochondria-targeted uncoupler. It is quite remarkable that SkQ1, in contrast to classic protonophores of dinitrophenol type, should act as a *mild uncoupler*. Through $\Delta\Psi$ decrease, SkQ1 restricts its own accumulation in mitochondria, preventing complete dissipation of $\Delta\Psi$. That is why the maximal protonophoric effect is expected only at high $\Delta\Psi$ values. It is such a mild uncoupling that seems to be an effective mechanism to lower ROS generation in mitochondria: a minor decrease in $\Delta\Psi$ was shown to cause strong inhibition of ROS production (Korshunov et al. 1997), but it does not hinder ATP synthesis (Skulachev 1996; Skulachev et al. 2010; Severin et al. 2010; Antonenko et al.

2011). Our data can be compared to the recent results obtained by Padalko and Kowaltowski (Padalko 2005; Caldeira da Silva et al. 2008) using low dinitrophenol concentrations, which increased the lifespan of *Drosophila* and mice as well as prevented oxidation of DNA and proteins during aging. Mild uncoupling may be the second antioxidant mechanism of SkQs. It requires somewhat higher SkQ1 concentrations and differs from direct antioxidant effect of SkQ1 described by Eq. (16.1) in that $C_{12}TPP$ is also active.

Table 16.1 summarizes our results obtained in studies of effects of SkQ1 and SkQR1 on animals in vivo. SkQ-type compounds were shown to decelerate 26 different traits of senescence, this showing that our compounds are powerful geroprotectors. It should be emphasized that all effects of SkQs occur at their nanomolar and in some cases even subnanomolar concentrations per kg body mass daily, which is much lower than for other known geroprotectors. It is noteworthy that they not only decrease mortality at the early and middle stages of aging, but also prevent the development of a large number of senile defects. In other words, SkQs can prolong healthspan.[2] It is clinical trials that will determine to what an extent a similar effect will be observed in humans, and these trials have already started.

Clinical trials of SkQ1 are already completed for dry-eye syndrome. 3-week administration of SkQ1-based eye drops were shown to result in significant increase in (1) amount of tears produced, (2) tear meniscus height, (3) visual acuity, etc. In the double-blind trial, SkQ1-based eye drop efficacy was compared to that of Tears Naturale (Alcon), a standard medicine recommended for light and moderate dry-eye syndrome treatment. The SkQ1-based drops proved to be much better than Tears Naturale. The dry-eye traits completely disappeared in 60 and 20 % patients after 3 weeks treatment with SkQ1 and Tears Naturale, respectively (Fig. 16.11) (Yani et al. 2012). In December, 2011, the Ministry of Health of Russia issued permission to produce and sell in drugstores "Visomitin" eye drops containing 250 nM SkQ1. The new drug is available in Russian drugstores since 2012.

---

[2] When speaking about prolongation of youth, we would like to stress that canceling of an aging program does not mean personal immortality of an individual. If we look at the aging program as a biological mechanism accelerating evolution, then increase in the maximal lifespan is rather unlikely to be an indispensable consequence of canceling of such a program. The input of the most long-lived individuals into the general productivity of the population is rather small because of their paucity. Moreover, age-dependent genome damage may accumulate not only because of the programmed increase in ROS, but also for other reasons. This process is particularly dangerous for females: DNA of their oolytes has to remain intact through the whole life of an individual. Thus, it is not surprising that at a certain age females lose their ability to reproduce. Menopause begins, which is particularly long in the case of humans and whales, but it is present even in *Drosophila*. It is no coincidence that in the early and middle stages of aging the programmed death input in mortality rate is rather large, and its SkQ-induced canceling causes substantial increase in lifespan. Further aging brings forward other causes of death that do not depend on the aging program and hence cannot be canceled by SkQs. Cancer is one of the main causes of death of this type. It seems to be also a phenoptotic program, but of a different function, i.e. elimination of individuals with overloaded damage to the genome.

Table 16.1 SkQ decelerates programs of aging (slow phenoptosis) and death after crisis (acute phenoptosis) in animals

	Result	Species	Studied treatment	References
1	Increase in median lifespan	Fungus *Podospora anserina*	Addition of SkQ1 to growth medium	(Anisimov et al. 2008; Skulachev et al. 2009)
2	Increase in median lifespan	Crustacean *Ceriodaphnia affinis*	Life-long addition of SkQ1 to aquarium water	(Anisimov et al. 2008; Skulachev et al. 2009)
3	Increase in median lifespan	Insect *Drosophila melanogaster*	Life-long addition of SkQ1 to food	(Anisimov et al. 2008; Skulachev et al. 2009)
4	Increase in median and maximal lifespan	Fish *Nothobranchius furzeri*	Life-long addition of SkQ1 to aquarium water	(Skulachev et al. 2011)
5	Increase in median lifespan	Mice	Life-long administration of SkQ1 with drinking water	(Anisimov et al. 2008; Skulachev et al. 2009)
6	Increase in median lifespan	Dwarf hamster	Life-long administration of SkQ1 with food	(Anisimov et al. 2011)
7	Increase in median and maximal lifespan	Mole-vole	Life-long administration of SkQ1 with food	(Anisimov et al. 2011)
8	Increase in median and maximal lifespan	"Mutator" mice	Life-long administration of SkQ1 with drinking water	(Shabalina et al. in preparation)
9	Increase in median and maximal lifespan	$p53^{-/-}$ mice or thymus-lacking mice with inoculated cervical carcinoma SiHa	Life-long administration of SkQ1 with food	(Agapova et al. 2008; Skulachev et al. 2009)
10	Increase in survival of animals treated with benzpyrene	Mice	Life-long administration of SkQ1 with drinking water	(Anikin IV et al. in press)
11	Reduction of level of peroxides in blood	Rats	Daily infusion of SkQ1 solution into cheek pouches for 14 days	(Skulachev et al. 2011)
12	Reduction of 8-oxo-2′-deoxyguanosine level in blood	Rats	Daily infusion of SkQ1 solution into cheek pouches for 14 days	(Skulachev et al. 2011)
13	Decrease in chromosomal aberrations	Rats	Daily infusion of SkQ1 solution into cheek pouches for 14 days	(Skulachev et al. 2011)
14	Decrease in lipid peroxidation and protein carbonylation	Rats of OXYS-line	Life-long administration of SkQ1 with food	(Neroev et al. 2008; Skulachev et al. 2009)

(continued)

16.1 SkQ Decelerates the Aging Program

Table 16.1 (continued)

Result	Species	Studied treatment	References
15 Prevention of decrease in cardiolipin level	"Mutator" mice	Life-long administration of SkQ1 with drinking water	(Shabalina et al. in preparation)
16 Prevention of decrease in unsaturated/saturated fatty acid ratio in cardiolipin	"Mutator" mice	Life-long administration of SkQ1 with drinking water	(Shabalina et al. in preparation)
17 Prevention of age-related oxidative stress-induced apoptosis of fibroblasts	Mice	Life-long administration of SkQ1 with drinking water	(Anisimov et al. 2008; Skulachev et al. 2009)
18 Preventing of age-related increase in β-galactosidase in fibroblasts	Mice	Life-long administration of SkQ1 with drinking water	(Anisimov et al. 2008; Skulachev et al. 2009)
19 Preventing of age-related increase in histone H2AX phosphorylation in fibroblasts	Mice	Life-long administration of SkQ1 with drinking water	(Anisimov et al. 2008; Skulachev et al. 2009)
20 Preventing of age-related myeloid shift of blood formula	Mice	Life-long administration of SkQ1 with drinking water	(Shipounova et al. 2010)
21 Deceleration of age-related damage of hematopoietic system	Mice	Life-long administration of SkQ1 with drinking water	(Shipounova et al. 2010)
22 Decrease in infection-caused mortality	Mice	Life-long administration of SkQ1 with drinking water	(Anisimov et al. 2008; 2011; Skulachev et al. 2009)
23 Deceleration of age-related involution of thymus and spleen follicles	Rats of OXYS-line	Life-long administration of SkQ1 with food	(Obukhova et al. 2009)
24 Deceleration of age-related osteoporosis	Rats of OXYS-line	Life-long administration of SkQ1 with food	(Neroev et al. 2008)
25 Prevention of age-related lordokyphosis	Mice	Life-long administration of SkQ1 with drinking water	(Anisimov et al. 2011; Shabalina et al., in preparation)
26 Prevention of age-related deterioration of wound healing	Rats	Life-long administration of SkQ1 with drinking water	(Demianenko et al. 2010)
27 Prevention of age-related disappearance of estrous cycles in females	Mice	Life-long administration of SkQ1 with drinking water	(Anisimov et al. 2008; Skulachev et al. 2009)
28 Prevention of reduction of libido in males	Rats of OXYS-line	Life-long administration of SkQ1 with food	(Neroev et al. 2008)

(continued)

Table 16.1 (continued)

	Result	Species	Studied treatment	References
29	Prevention of age-related behavioral changes	Rats	Administration of SkQ1 with drinking water for last 10 weeks	(Stefanova et al. 2010)
30	Prevention of age-related alopecia and loss of whiskers	Mice	Life-long administration of SkQ1 with drinking water	(Anisimov et al. 2008; Skulachev et al. 2009)
31	Prevention of the development of age-related cataract and retinopathies	Rats of OXYS-line	Life-long administration of SkQ1 with food	(Neroev et al. 2008; Skulachev et al. 2009; Skulachev et al. 2011)
32	Reversal of development of age-related cataract and retinopathies	Rats of OXYS-line	Administration of SkQ1 with food or as eye drops for 1.5 months	(Neroev et al. 2008; Skulachev et al. 2009; Skulachev et al. 2011)
33	Reversal of development of age-related retinopathies	Dogs, cats, horses	Using SkQ1-eye drops for 1–3 months	(Neroev et al. 2008; Skulachev et al. 2011)
34	Partial prevention of age-dependent sarcopenia	Rats of OXYS-line, Wistar rats	Life-long SkQ1 administration with food	(Kolosova et al. in preparation)
35	Partial reversal of decay of the grow hormone and IGF-1 levels	19-months-old rats	Administration of SkQ1 with food for 4 months	(Kolosova et al. in preparation)
36	Prevention of canities caused by γ-radiation	Black mice	Administration of SkQ1 with drinking water for one month	(Skulachev MV et al. 2011)
37	Prevention of death from kidney ischemia	Rats with one kidney	Single injection of SkQ1 or SkQR1 one day before ischemia	(Plotnikov et al. 2007; Bakeeva et al. 2008; Skulachev et al. 2011)
38	Partial normalization of creatinine level in blood, diuresis, and Ca^{2+} resorption after kidney ischemia	Rats with one kidney	Single injection of SkQ1 or SkQR1 one day before ischemia	(Bakeeva et al. 2008; Skulachev et al. 2009)
39	Normalization of kidney functioning after rhabdomyolysis	Rats	SkQR1 injection 1, 12, 24, and 36 h after induction of rhabdomyolysis	(Plotnikov et al. 2011)

(continued)

16.1 SkQ Decelerates the Aging Program

Table 16.1 (continued)

	Result	Species	Studied treatment	References
40	Decrease in brain damage zone and prevention of locomotor dysfunction after ischemia	Rats	Single injection of SkQR1 before the brain ischemia	(Bakeeva et al. 2008; Skulachev et al. 2009; Skulachev et al. 2011)
41	Decrease in heart damage zone after ischemia	Rats	Administration of SkQ1 with drinking water for 2–3 weeks before the ischemia	(Bakeeva et al. 2008; Skulachev et al. 2009)
42	Decrease in arrhythmia after heart ischemia	Rats	Administration of SkQ1 with drinking water for 2–3 weeks before the ischemia	(Bakeeva et al. 2008; Skulachev et al. 2009; Skulachev 2011)
43	Prevention or reversal of development of experimental uveitis	Rabbits	Administration of SkQ1-eye drops for 1–2 months	(Neroev et al. 2008; Skulachev et al. 2009)
44	Partial prevention of experimental glaucoma	Rabbits	Administration of SkQ1-eye drops for 1–2 months	(Neroev et al. 2008; Skulachev et al. 2009)

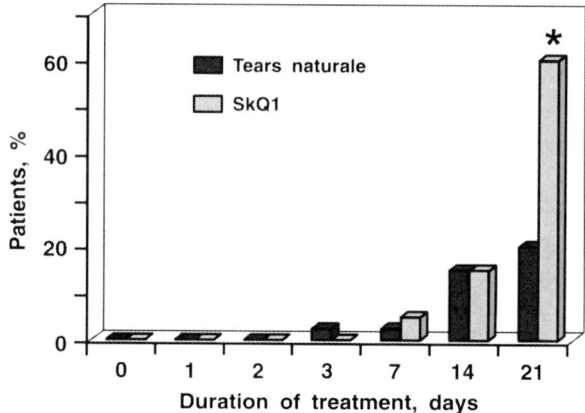

Fig. 16.11 Comparison of efficiency of SkQ1-based eye drops and "Tears Naturale" (Alcon) in clinical trials on patients suffering from dry-eye syndrome. Percentage of successful cases is indicated. *$p < 0.001$. A case was considered successful if the dry-eye traits disappeared completely (Zinovkin et al. 2013)

16.2 Comparison of Effects of Food Restriction and SkQ

The best evidence favoring the hypothesis of programmed aging could be provided by a successful attempt to cancel it. There are some reasons to suggest that such an attempt gave a positive result as early as 1934, when McCay and coworkers (MacCay and Crowell 1934; MacCay et al. 1935, 1943) showed extension the lifespan of rats by restricting their diet. This restriction was introduced at early stages of life and initially resulted in retardation of growth. When the food restriction was canceled, the animals rapidly increased in size to reach the norm but lived longer by 70 % (males) and 48 % (females) than rats that were fed ad libitum over their entire life. As to the reasons for deaths, a strong decrease in death rates from pulmonary diseases and tumors was observed. The researchers noted that long-lived animals looked mobile and young irrespective of their age. Their pulse rate was 340 beats/min, which is about 100 beats below the control. Later, the positive effect of a certain dietary restriction on lifespan was demonstrated on a great variety of organisms—from yeast to rhesus monkeys and humans (Austad 1997; Comfort 1979; Holloszy and Fontana 2007; Colman et al. 2009). With the appearance of the Harman's hypothesis on the role of ROS in aging (Harman 1956), this effect came to be explained in terms of decreased volume of food oxidized by oxygen and, as a consequence, of decreased production of ROS. The inconsistency of this assumption was evident already in early works on dietary restriction, when the same group of researchers (Will and MacCay 1943) found that the heat production per kg body mass in the food-restricted rat cohort was *higher* than in the control. Further research gave direct evidence against the commonly accepted viewpoint.

First, it was found that it is enough for *Drosophila* to have food restriction for only 2 days to prolong its life in the same degree as food restriction over its entire life (Mair et al. 2003). Second, it turned out that the *smell* of food attenuated the geroprotective effect provided by dietary restriction. (Libert and Pletcher 2007; Libert et al. 2007). As long ago as 1934, Robertson et al. (1934) discovered in experiments on albino mice that a 2-day long "fasting" every week is sufficient to extend their lifespan by 50–60 %. Carr et al. (1949) observed, again on mice, that the reproductive ability (which was lost by the end of the first year in the case of unrestricted feeding) was preserved until at least the 21 months of life when the mice were limited in food over the first 11–15 months, and then received food ad libitum. According to the data of Stuchlikova et al. (1975), rats, mice, and golden hamsters experiencing food restriction by 50 % over 2 years lived 20 % longer than the control. Dietary restriction during the first year of life only prolonged the lifespan by 40–60 %, and during the second year by 30–40 %.

Further research showed that the effect of dietary restriction involves both the carbohydrate and protein components of food. The effect of proteins is associated with only one amino acid—methionine (Richie et al. 1994; Miller et al. 2005; Sanz et al. 2006; Caro et al. 2009). Methionine belongs to the group of essential amino acids, so food is the only source of this compound. It was found that a diet in which proteins were replaced by a mixture of amino acids containing no methionine not only favors a longer life, but also decreases mitochondrial ROS generation and oxidative damage to *mitochondrial* DNA (Sanz et al. 2006; Caro et al. 2009). Interestingly, dietary restriction has no effect on the oxidation of *nuclear* DNA (Edman et al. 2009), which might be expected by the generally accepted view on aging as a nonspecific age-related ROS damage to DNA and other cellular biopolymers.

In our opinion, calorie restriction is perceived by an organism as a worrying signal of food shortage (Skulachev 2011). Even partial starvation may entail decreased fecundity. And this, in turn, jeopardizes the very existence of the population. To prevent this situation, at least partially, it is sufficient to cancel the aging program, thereby prolonging the reproductive period of the individual, i.e., increasing the possible total number of its descendants. In other words, the impact of dietary restriction on lifespan is only indirectly related to ROS—it is a regulatory effect, being first of all biology, and not chemistry. That is why such evident signaling effects as short-term fasting (or, vice versa, smell of food), rather than life-long food shortage (or excess) exert a powerful effect on the life-cycle parameters. It is remarkable that a temporary dietary restriction (fasting) is better than constant restriction. The signal can be given within a fairly short time, whereas long-term starvation is clearly harmful for an organism.[3]

The signaling nature of the effect of dietary restriction provides a good rationale for the experiments with methionine. Apparently, the organism determines the amount of available food and, first of all, essential amino acids required for protein synthesis, by monitoring the level of only one of them—methionine.

It seems quite significant that dietary restriction not only extends lifespan, but also prolongs youth, as already mentioned by the discoverer of this phenomenon— C. M. McCay. In the food-restricted animals, age-related diseases are retarded and

---

[3] Religious fasting probably prolongs life by short-term dietary restrictions. Believers are known to live longer than atheists (Eng et al. 2002).

old animals become indistinguishable from young ones in behavior and even exterior appearance. Quite illustrative in this respect are the recently reported results of research of Weindruch and his group (Colman et al. 2009) on primates. 20-year-long experiments on 76 macaques (adult animals from 7 to 14 years old) showed that a long-term 30 % dietary restriction gives the following effects: (1) sharp decrease in age-related death rate (for over 30-year-old animals, 20 % in the food-restricted group against 50 % in the control group fed ad libitum); (2) absence of diabetes from the causes of death; (3) halving the death rate from cancer (in macaques, this is primarily intestine adenocarcinoma); (4) decrease in death from cardiovascular diseases; (5) decrease in osteoporosis; (6) arrest of the development of such age-related traits as sarcopenia, decline in brain gray matter, alopecia, canities, etc. By the age of 30 years, 80 % of the surviving control macaques showed some traits of aging, whereas in the experimental group such traits were observed in as few as 20 % of the animals.

The experiment on monkeys is still far from completion, and, therefore, we can say nothing about the effect of dietary restriction on the maximum lifespan of primates. However, some evidence on this subject is available for rodents (MacCay et al. 1943; Stuchlikova et al. 1975). This evidence shows that the median lifespan (50 % death rate) in mice and hamsters increases much more markedly than the maximum lifespan. The simplest explanation of the mechanism of this phenomenon is an arrest (or a strong retardation) of the aging program. Therewith, other ontogenetic programs, first of all body growth, can also be retarded. These phenomena are observed with a rather strong and long-term fasting (MacCay and Crowell 1934). However, a more moderate dietary restriction can prolong life while not causing inhibition of growth (Berg and Simms 1960).

The effects of dietary restriction and SkQ antioxidants have much in common. Rectangularization of survival curves and decrease in early mortality rates are observed with either treatment and the median lifespan increases much more than the maximal lifespan (Anisimov et al. 2008, 2011; Skulachev et al. 2009). Both of these factors operate not only, and not so much, as to prolong life as such, as to prolong a healthy young life. Both are very effective in living beings quite different in their systematic position (dietary restriction works well in yeast, insects, and mammals, and SkQ, in fungi, crustaceans, insects, fish, and mammals). The effect of both factors is clearly pleiotropic in nature, i.e., they influence a variety of physiological functions, and the lists of those functions are similar. Cardiovascular diseases, osteoporosis, vision disorders, and certain types of cancer are decreased, graying and loss of hair as well as age-related depression (torpor) do not occur. Contradictory data concerning effects of dietary restriction on sarcopenia and immune responses have been reported. Some authors state that such exposure adversely affects both the muscle system and immunity.[4] On the other hand,

---

[4] Fernandes et al. (Sun et al. 2001) noted that dietary restriction, especially at a young age, inhibits interleukin secretion by macrophages as a response to bacterial lipopolysaccharides. According to Gardner (2005), partial starvation reduces resistance to influenza virus.

Weindruch et al. (Colman et al. 2009) reported the lack of sarcopenia in monkeys, and McCay et al. (MacCay and Crowell 1934), resistance to pulmonary diseases in rats subjected to dietary restriction (cf. decrease in age-dependent involution of the thymus and follicular spleen compartments OXYS in rats treated with SkQ1 (Anisimov et al. 2008; Skulachev et al. 2009; Obukhova et al. 2009)). At the same time, the effects of dietary restriction and SkQ1 on wound healing are opposite: starvation inhibited, while the antioxidant SkQ1 stimulated healing (Demianenko et al. 2010). Dietary restriction decreases body temperature and inhibits growth, which was not observed with SkQ1 (Skulachev et al. 2011). These data rule out such a trivial interpretation of our data as the assumption that SkQ1 decreases animal food intake, say, by decreasing appetite. Direct measurements on mice treated with SkQ1 revealed no decrease in food intake at the low SkQ1 concentrations used in our experiments.

The fact that dietary restriction adversely affects some vitally important parameters is not surprising. In fact, animals do not tend to overeat even if they are not restricted in food. Usually, they eat as much as it is needed for their organism. Therefore, long-term dietary restriction entails some disorders in vital functions. It is also clear that such disorders are more probable the longer starvation continues. We already mentioned that long-term and continuous dietary restriction is not necessary for the geroprotecting effect. This explains the controversy in data on the effects of dietary restriction. In cases when dietary restriction was not too severe and not too long, positive effects were observed, whereas when gerontologists overdid the restriction, unfavorable influence occurred. Thus, it is commonly accepted that long-term dietary restriction decreases the frequency of estrous cycles (sometimes until they cease completely) (Hopkin 2003), but already in 1949 Carr et al. (1949) showed the temporary dietary restriction with its subsequent cancelation prolonged estrous cycles, which could be observed until extreme old age. Let us remember that the same effect was found with SkQ1 (Anisimov et al. 2008, 2011; Skulachev et al. 2009). In principle, there is no need to starve through the whole life if starvation is a signal to switch off the aging program. However, there is a probability that too weak or delayed dietary restriction will only partially retard the program, and the geroprotective effect will be weak. The use of SkQ1 appears not to pose such a threat, since SkQ1 acts at such low concentrations that no adverse effects have been observed so far.

Another circumstance should be taken into account when considering restriction in eating as a geroprotector for humans. Actually, if dietary restriction is a signal to warn against starvation, then the organism should respond not only by prolonging life to compensate for the decay of birth rates in lean years. Other responses also seem quite reasonable, and some of them are not as attractive as an increase in a healthy life. For example, it was noted that a hungry mouse, once finding itself in a squirrel wheel, does not want to leave it and runs from 6 to 8 km overnight (with normal feeding, this distance is always shorter than 1 km) (Hopkin 2003). Obviously, this effect cannot be explained by starvation-induced exhaustion and muscle weakness. More likely, we deal here with another response to the starvation signal—extreme anxiety and attempt to scan as large of territory as possible in a search for food. Paraphrasing Strugatsky's novel "Monday Begins on Sunday", we deal here with a mouse "unsatisfied stomachically". Had this effect been characteristic of SkQ1, we would observe enhanced food intake by SkQ1-receiving animals, which is not the case. An impression arises that the use of SkQ1 is a "purer" way to switch off the aging program, not overburdened with undesirable effects typical for food restriction (Skulachev 2011).

16.3 From *Homo sapiens* to *Homo sapiens liberatus*

There is at least one more aspect where dietary restriction is unlikely to replace SkQ. As a matter of fact, SkQs act not only as a geroprotector (see Table 16.1, items 1–32), but also helps young animals suffering from experimental cardiac arrhythmia, cardiac and renal infarctions, stroke, and some other experimental diseases studied in young animals (see Table 16.1, items 33–40). Surely, we might suggest that SkQs merely "cleans the dirtiest place in a cell" (mitochondrial interior) and thus exerts a certain nonspecific favorable effects. However, this explanation is unlikely to be true. In this case, it remains unclear why during their evolution plants that synthesize separately plastoquinone and the penetrating cation berberine have not formed their combination capable of removing ROS from the mitochondrial interior (Lyamzaev et al. 2011). Such an effect should be especially desirable for plant mitochondria experiencing a danger of "oxygen threat" than animal mitochondria, since during daytime, the cytosol of plant cells is always saturated with oxygen formed by chloroplasts.

An alternative possibility is that the program triggering aging as slow phenoptosis is also used in other deadly programs which trigger rapid, or acute, phenoptosis. Here, we come to a fundamental question of biology—how could genetic programs counterproductive for an individual come into existence? Charles Darwin, Alfred Russell Wallace, and August Weismann suggested as early as in the nineteenth century that the death of an individual can be altruistic, i.e., serving a family or a community (Weismann 1989; Darwin 1871). In 1964, Hamilton wrote a series of two papers entitled "The Genetic Evolution of Social Behavior" (Hamilton 1964a, b). And in 1976, Richard Dawkins published the book "The Selfish Gene" (Dawkins 1976), in which he developed and popularized Hamilton's concept, concluding that "the main selection unit is a gene rather than an individual". In essence, the issue here is already not social well-being, but the "tyranny" of the genome, the only self-reproducing biological structure whose preservation, development, and expansion have taken priority over the well-being of an individual or a group of individuals. In terms of this concept, an organism is nothing but a machine serving the genome's interests (Skulachev 2009a, b). The individual always follows a principle called the "samurai" law of biology: "It is better to die than to be wrong" (Skulachev 2000). Together with the "genome tyranny" concept, this law implies that any critical state of an organism, when it no longer guarantees safety of its genome and, in case of convalescence, can generate offspring with a changed genome, should be a signal for the organism's self-elimination, i.e., acute phenoptosis.

It seems quite possible that the mechanism of acute phenoptosis, like that of slow phenoptosis (aging), is mediated by intramitochondrial ROS at early stages of the process. If this assumption is valid, then the positive effect of SkQs not only on aging, but also on a variety of acute pathologies both in young and old organisms can be explained in terms of the neutralization of these ROS.

Thus, there is a chance to use SkQs as a tool in the "rise of the machines", an attempt of *Homo sapiens* to overcome "genome tyranny" and to cancel those genome-dictated programs that are useful for genome evolution, but unfavorable for an individual. Their cancelation would symbolize human conversion into *Homo sapiens liberatus*,[5] which would be the highest achievement of medicine in the twenty-first century (Skulachev 2009a).

16.4 Conclusions

To prevent age-dependent increase in mitochondrial reactive oxygen species (mROS), cationic derivatives of plastoquinone (PQ, a photosynthetic electron carrier and very active antioxidant) were synthesized in our group. In the majority of experiments, the compound called SkQ1 was used. It is composed of PQ conjugated with decyltriphenylphosphonium cation that easily penetrates through membranes due to delocalization of its positive charge over the bulky hydrophobic phenyl residues. The following properties of SkQ1 were revealed. (1) SkQ1 has low solubility in either water or hydrocarbons, but it is highly soluble in octanol (1:10,000 water/octanol distribution coefficient) that means it has extremely high affinity for membrane structures. (2) Experiments on bilayer planar phospholipid membrane showed that SkQ1 is good penetrant for such a membrane. (3) SkQ1 accumulates in mitochondria in vitro in a $\Delta\Psi$-dependent manner. (4) In various in vitro systems, SkQ1 is a very potent antioxidant. (5) In mitochondria, SkQ1 is also a prooxidant if added in micromolar concentrations. However, these concentrations are very much higher than those required for the antioxidant effect of SkQ1, which is quite measurable at nanomolar concentrations. (6) In respiring mitochondria, SkQ1 operates as a rechargeable antioxidant that is reduced at center i (heme b_H) of respiratory chain complex III. (7) In human cell cultures, SkQR1, a fluorescing SkQ1 analog, was shown to stain all the mitochondria and only the mitochondria, which can be easily explained by the fact that the mitochondrial interior is the only negatively charged intracellular compartment. (8) In the same cell cultures, SkQ1 and SkQR1 are very effective inhibitors of H_2O_2-induced apoptosis; the SkQ1 analog lacking PQ is ineffective. (9) In vivo treatment with SkQ1 increased the median lifespan of some tested fungi, crustaceans, insects, fish,

---

[5] *Liberatus* means *liberated* in Latin.

and mammals. (10) In animals, appearance of typical traits of aging were retarded, stopped, or even reversed by SkQ1. These traits included osteoporosis, lordokyphosis, achromotrichia, balding, slow wound healing, involution of thymus and spleen follicles, sarcopenia, increase in left ventricle volume of the heart, retinopathy, glaucoma, cataract, dry-eye syndrome, disappearance of estrous cycles in females and libido in males, age-dependant changes in behavior, etc. In the cases of age-related eye diseases, drops of very diluted (250 nM) SkQ1 solution were efficient in rats, rabbits, dogs, cats, and horses. (11) Double-blind clinical trials on humans suffering from dry-eye syndrome clearly showed that drops of SkQ1 effectively treated this age-related disease regarded as incurable. In 2012, these drops became available in drugstores in Russia. (12) A single injection of SkQ1 was shown to prevent the death of animals after a crisis such as kidney ischemia, rhabdomyolysis, pyelonephritis, etc. It decreases infarct area in heart and brain after experimental ischemia in rats. It was concluded that SkQ1 is competent in inhibiting a programmed death of organisms ("phenoptosis"), being effective in both slow (aging) and fast (sudden death after crisis) version of this phenomenon which is counter-productive for the individual but apparently favorable for evolution of the species. (13) The majority of the effects of SkQ1 are similar to those of partial food restriction. This indicates that food restriction, like SkQ1 treatment, is a way to downregulate the aging program. However, in contrast to SkQ1, food restriction has undesirable side effects revealed in rodents, such as decrease in fecundity, elevated anxiety, and a tendency to scan as large area as possible. Apparently, food restriction stimulates a forager activity of the individual, being a signal of starvation, and downregulation of the aging program prevents the age-linked decrease in efficiency of functioning of the organism under adverse conditions. (14) It is hypothesized that SkQ-type treatments might be used to switch off the aging program as well as some other programs counter-productive for an individual but useful for evolution.

Acknowledgments The authors would like to express their gratitude to their families for patience and support expressed over several years of work on this book. Excellent work of Drs. Anna Brzyska, Richard Lozier, Konstantin Lyamzaev, and Antonina Pustovidko in preparing the English version of the book is greatly appreciated. The original illustrations and the general graphic style of the book were created by one of the authors (F.O.K.) in LLC "MASTER-MULTIMEDIA" (www.master-multimedia.ru).

References

Agapova LS, Chernyak BV, Domnina LV, Dugina VB, Efimenko AY, Fetisova EK, Ivanova OY, Kalinina NI, Khromova NV, Kopnin BP, Kopnin PB, Korotetskaya MV, Lichinitser MR, Lukashev AL, Pletjushkina OY, Popova EN, Skulachev MV, Shagieva GS, Stepanova EV, Titova EV, Tkachuk VA, Vasiliev JM, Skulachev VP (2008) Mitochondria-targeted plastoquinone derivatives as tools to interrupt execution of the aging program. 3. Inhibitory effect of SkQ1 on tumor development from p53-deficient cells. Biochemistry (Moscow) 73:1300–1316

References

Anikin IV, Popovich IG, Tyndyk ML, Zabezhinsky MA, Yurova MN, Skulachev VP, Anisimov VN (in press) Effect of a mitochondria-targeted plastoquinone derivative SkQ1 on cancerogenesis of soft tissues, induced by benzapyren in mice. Voprosy onkologii (In Russian)

Anisimov VN, Bakeeva LE, Egormin PA, Filenko OF, Isakova EF, Manskikh VN, Mikhelson VM, Panteleeva AA, Pasyukova EG, Pilipenko DI, Piskunova TS, Popovich IG, Roshchina NV, Rybina OY, Saprunova VB, Samoylova TA, Semenchenko AV, Skulachev MV, Spivak IM, Tsybul'ko EA, Tyndyk ML, Vyssokikh MY, Yurova MN, Zabezhinsky MA, Skulachev VP (2008) Mitochondria-targeted plastoquinone derivatives as tools to interrupt execution of the aging program. 5. SkQ1 prolongs lifespan and prevents development of traits of senescence. Biochemistry (Moscow) 73:1329–1342

Anisimov VN, Egorov MV, Krasilshchikova MS, Manskikh VN, Moshkin MP, Novikov EA, Rogovin KA, Shabalina IG, Shekarova ON, Skulachev MV, Titova TV, Vyssokikh MY, Yurova MN, Zabezhinsky MA, Skulachev VP (2011) Effects of mitochondria-targeted antioxidant SkQ1 on lifespan of rodents. Aging (Albany NY) 3:1110–1119

Antonenko YN, Avetisyan AV, Bakeeva LE, Chernyak BV, Chertkov VA, Domnina LV, Ivanova OY, Izyumov DS, Khailova LS, Klishin SS, Korshunova GA, Lyamzaev KG, Muntyan MS, Nepryakhina OK, Pashkovskaya AA, Pletjushkina OY, Pustovidko AV, Roginsky VA, Rokitskaya TI, Ruuge EK, Saprunova VB, Severina II, Simonyan RA, Skulachev IV, Skulachev MV, Sumbatyan NV, Sviryaeva IV, Tashlitsky VN, Vassiliev JM, Vyssokikh MY, Yaguzhinsky LS, Zamyatnin AA Jr, Skulachev VP (2008a) Mitochondria-targeted plastoquinone derivatives as tools to interrupt execution of the aging program. 1. Cationic plastoquinone derivatives: synthesis and in vitro studies. Biochemistry (Moscow) 73:1273–1287

Antonenko YN, Avetisyan AV, Cherepanov DA, Knorre DA, Korshunova GA, Markova OV, Ojovan SM, Perevoshchikova IV, Pustovidko AV, Rokitskaya TI, Severina II, Simonyan RA, Smirnova EA, Sobko AA, Sumbatyan NV, Severin FF, Skulachev VP (2011) Derivatives of rhodamine 19 as mild mitochondria-targeted cationic uncouplers. J Biol Chem 286:17831–17840

Antonenko YN, Roginsky VA, Pashkovskaya AA, Rokitskaya TI, Kotova EA, Zaspa AA, Chernyak BV, Skulachev VP (2008b) Protective effects of mitochondria-targeted antioxidant SkQ in aqueous and lipid membrane environments. J Membr Biol 222:141–149

Austad SN (1997) Why we age?. John Wiley and Sons, New York

Bakeeva LE, Barskov IV, Egorov MV, Isaev NK, Kapelko VI, Kazachenko AV, Kirpatovsky VI, Kozlovsky SV, Lakomkin VL, Levina SB, Pisarenko OI, Plotnikov EY, Saprunova VB, Serebryakova LI, Skulachev MV, Stelmashook EV, Studneva IM, Tskitishvili OV, Vasilyeva AK, Victorov IV, Zorov DB, Skulachev VP (2008) Mitochondria-targeted plastoquinone derivatives as tools to interrupt execution of the aging program. 2. Treatment of some ROS- and age-related diseases (heart arrhythmia, heart infarctions, kidney ischemia, and stroke). Biochemistry (Moscow) 73:1288–1299

Berg BN, Simms HS (1960) Nutrition and longevity in the rat. II. Longevity and onset of disease with different levels of food intake. J Nutr 71:255–263

Caldeira da Silva CC, Cerqueira FM, Barbosa LF, Medeiros MH, Kowaltowski AJ (2008) Mild mitochondrial uncoupling in mice affects energy metabolism, redox balance and longevity. Aging Cell 7:552–560

Caro P, Gomez J, Sanchez I, Garcia R, Lopez-Torres M, Naudi A, Portero-Otin M, Pamplona R, Barja G (2009) Effect of 40% restriction of dietary amino acids (except methionine) on mitochondrial oxidative stress and biogenesis, AIF and SIRT1 in rat liver. Biogerontology 10:579–592

Carr CJ, King JT, Visscher B (1949) Delay of senescence infertility by dietary restriction. Proc Fed Am Soc Exp Biol 8:22

Colman RJ, Anderson RM, Johnson SC, Kastman EK, Kosmatka KJ, Beasley TM, Allison DB, Cruzen C, Simmons HA, Kemnitz JW, Weindruch R (2009) Caloric restriction delays disease onset and mortality in rhesus monkeys. Science 325:201–204

Comfort A (1979) The biology of senescence. Churchill Livingston, Edinburgh
Darwin C (1871) The descent of man. John Murray, London
Demianenko IA, Vasilieva TV, Domnina LV, Dugina VB, Egorov MV, Ivanova OY, Ilinskaya OP, Pletjushkina OY, Popova EN, Sakharov IY, Fedorov AV, Chernyak BV (2010) Novel mitochondria-targeted antioxidants, "Skulachev-ion" derivatives, accelerate dermal wound healing in animals. Biochemistry (Moscow) 75:274–280
Dawkins R (1976) The selfish gene. Oxford Univ Publ, Oxford
Eng PM, Rimm EB, Fitzmaurice G, Kawachi I (2002) Social ties and change in social ties in relation to subsequent total and cause-specific mortality and coronary heart disease incidence in men. Am J Epidemiol 155:700–709
Edman U, Garcia AM, Busuttil RA, Sorensen D, Lundell M, Kapahi P, Vijg J (2009) Lifespan extension by dietary restriction is not linked to protection against somatic DNA damage in *Drosophila melanogaster*. Aging Cell 8:331–338
Fetisova EK, Avetisian AV, Iziumov DS, Korotetskaia MV, Tashlitskiū VN, Skulachev VP, Cherniak BV (2011) Multidrug resistance P-glycoprotein inhibits the antiapoptotic action of mitochondria-targeted antioxidant SkQR1. Cell Tissue Biol 1:37–46
Gardner EM (2005) Caloric restriction decreases survival of aged mice in response to primary influenza infection. J Gerontol A Biol Sci Med Sci 60:688–694
Hamilton WD (1964a) The genetical evolution of social behaviour. I. J Theor Biol 7:1–16
Hamilton WD (1964b) The genetical evolution of social behaviour. II. J Theor Biol 7:17–52
Harman D (1956) Aging: a theory based on free radical and radiation chemistry. J Gerontol 11:298–300
Holloszy JO, Fontana L (2007) Caloric restriction in humans. Exp Gerontol 42:709–712
Hopkin K (2003) Dietary drawbacks. Sci Aging Knowl Environ 2003 (8):NS4
Kapay NA, Isaev NK, Stelmashook EV, Popova OV, Zorov DB, Skrebitsky VG, Skulachev VP (2011) In vivo injected mitochondria-targeted plastoquinone antioxidant SkQR1 prevents β-amyloid-induced decay of long-term potentiation in rat hippocampal slices. Biochemistry (Moscow) 76:1367–1370
Korshunov SS, Skulachev VP, Starkov AA (1997) High protonic potential actuates a mechanism of production of reactive oxygen species in mitochondria. FEBS Lett 416:15–18
Kruk J, Jemiola-Rzeminska M, Strzalka K (1997) Plastoquinol and alpha-tocopherol quinol are more active than ubiquinol and alpha-tocopherol in inhibition of lipid peroxidation. Chem Phys Lipids 87:73–80
Libert S, Pletcher SD (2007) Modulation of longevity by environmental sensing. Cell 131:1231–1234
Libert S, Zwiener J, Chu X, Vanvoorhies W, Roman G, Pletcher SD (2007) Regulation of *Drosophila* life span by olfaction and food-derived odors. Science 315:1133–1137
Lyamzaev KG, Pustovidko AV, Simonyan RA, Rokitskaya TI, Domnina LV, Ivanova OY, Severina II, Sumbatyan NV, Korshunova GA, Tashlitsky VN, Roginsky VA, Antonenko YN, Skulachev MV, Chernyak BV, Skulachev VP (2011) Novel mitochondria-targeted antioxidants: plastoquinone conjugated with cationic plant alkaloids berberine and palmatine. Pharm Res 28:2883–2895
MacCay CM, Crowell MF (1934) Prolonging the life span. Sci Mon 39:405–414
MacCay CM, Crowell MF, Maynard LA (1935) The effect of retarded growth upon the length of life span and upon the ultimate body size. J Nutr 10:63–79
MacCay CM, Maynard LA, Barnes LL (1943) Growth, aging, chronic disease and lifespan in rats. Arch Biochem 2:469
Mair W, Goymer P, Pletcher SD, Partridge L (2003) Demography of dietary restriction and death in *Drosophila*. Science 301:1731–1733
Miller RA, Buehner G, Chang Y, Harper JM, Sigler R, Smith-Wheelock M (2005) Methionine-deficient diet extends mouse lifespan, slows immune and lens aging, alters glucose, T4, IGF-I and insulin levels, and increases hepatocyte MIF levels and stress resistance. Aging Cell 4:119–125

Neroev VV, Archipova MM, Bakeeva LE, Fursova A, Grigorian EN, Grishanova AY, Iomdina EN, Ivashchenko ZhN, Katargina LA, Khoroshilova-Maslova IP, Kilina OV, Kolosova NG, Kopenkin EP, Korshunov SS, Kovaleva NA, Novikova YP, Philippov PP, Pilipenko DI, Robustova OV, Saprunova VB, Senin II, Skulachev MV, Sotnikova LF, Stefanova NA, Tikhomirova NK, Tsapenko IV, Shchipanova AI, Zinovkin RA, Skulachev VP (2008) Mitochondria-targeted plastoquinone derivatives as tools to interrupt execution of the aging program. 4. Age-related eye disease. SkQ1 returns vision to blind animals. Biochemistry (Moscow) 73:1317–1328

Obukhova LA, Skulachev VP, Kolosova NG (2009) Mitochondria-targeted antioxidant SkQ1 inhibits age-dependent involution of the thymus in normal and senescence-prone rats. Aging (Albany NY) 1:389–401

Padalko VI (2005) Uncoupler of oxidative phosphorylation prolongs the lifespan of *Drosophila*. Biochemistry (Moscow) 70:986–989

Plotnikov EY, Kazachenko AV, Vyssokikh MY, Vasileva AK, Tcvirkun DV, Isaev NK, Kirpatovsky VI, Zorov DB (2007) The role of mitochondria in oxidative and nitrosative stress during ischemia/reperfusion in the rat kidney. Kidney Int 72:1493–1502

Plotnikov EY, Silachev DN, Chupyrkina AA, Danshina MI, Jankauskas SS, Morosanova MA, Stelmashook EV, Vasileva AK, Goryacheva ES, Pirogov YA, Isaev NK, Zorov DB (2010) New-generation Skulachev ions exhibiting nephroprotective and neuroprotective properties. Biochemistry (Moscow) 75:145–150

Plotnikov EY, Chupyrkina AA, Jankauskas SS, Pevzner IB, Silachev DN, Skulachev VP, Zorov DB (2011) Mechanisms of nephroprotective effect of mitochondria-targeted antioxidants under rhabdomyolysis and ischemia/reperfusion. Biochim Biophys Acta 1812:77–86

Richie JP Jr, Leutzinger Y, Parthasarathy S, Malloy V, Orentreich N, Zimmerman JA (1994) Methionine restriction increases blood glutathione and longevity in F344 rats. FASEB J 8:1302–1307

Robertson TB, Marston R, Walters JW (1934) Influence of intermittent starvation and of intermittent starvation plus nucleic acid on growth and longevity in white mice. Aust J Exp Biol Med Sci 12:33

Roginsky V, Barsukova T, Loshadkin D, Pliss E (2003) Substituted p-hydroquinones as inhibitors of lipid peroxidation. Chem Phys Lipids 125:49–58

Roginsky VA, Tashlitsky VN, Skulachev VP (2009) Chain-breaking antioxidant activity of reduced forms of mitochondria-targeted quinones, a novel type of geroprotectors. Aging (Albany NY) 1:481–489

Rokitskaya TI, Klishin SS, Severina II, Skulachev VP, Antonenko YN (2008) Kinetic analysis of permeation of mitochondria-targeted antioxidants across bilayer lipid membranes. J Membr Biol 224:9–19

Rokitskaya TI, Sumbatyan NV, Tashlitsky VN, Korshunova GA, Antonenko YN, Skulachev VP (2010) Mitochondria-targeted penetrating cations as carriers of hydrophobic anions through lipid membranes. Biochim Biophys Acta 1798:1698–1706

Sanz A, Caro P, Ayala V, Portero-Otin M, Pamplona R, Barja G (2006) Methionine restriction decreases mitochondrial oxygen radical generation and leak as well as oxidative damage to mitochondrial DNA and proteins. FASEB J 20:1064–1073

Severin FF, Severina II, Antonenko YN, Rokitskaya TI, Cherepanov DA, Mokhova EN, Vyssokikh MY, Pustovidko AV, Markova OV, Yaguzhinsky LS, Korshunova GA, Sumbatyan NV, Skulachev MV, Skulachev VP (2010) Penetrating cation/fatty acid anion pair as a mitochondria-targeted protonophore. Proc Natl Acad Sci U S A 107:663–668

Shabalina IG et al. (in preparation)

Shipounova IN, Svinareva DA, Petrova TV, Lyamzaev KG, Chernyak BV, Drize NI, Skulachev VP (2010) Reactive oxygen species produced in mitochondria are involved in age-dependent changes of hematopoietic and mesenchymal progenitor cells in mice. A study with the novel mitochondria-targeted antioxidant SkQ1. Mech Ageing Dev 131:415–421

Skulachev MV, Antonenko YN, Anisimov VN, Chernyak BV, Cherepanov DA, Chistyakov VA, Egorov MV, Kolosova NG, Korshunova GA, Lyamzaev KG, Plotnikov EY, Roginsky VA,

Savchenko AY, Severina II, Severin FF, Shkurat TP, Tashlitsky VN, Shidlovsky KM, Vyssokikh MY, Zamyatnin AA Jr, Zorov DB, Skulachev VP (2011) Mitochondrial-targeted plastoquinone derivatives. Effect on senescence and acute age-related pathologies. Curr Drug Targ 12:800–826

Skulachev VP (1996) Role of uncoupled and non-coupled oxidations in maintenance of safely low levels of oxygen and its one-electron reductants. Q Rev Biophys 29:169–202

Skulachev VP (2000) Mitochondria in the programmed death phenomena; a principle of biology: "It is better to die than to be wrong". IUBMB Life 49:365–373

Skulachev VP (2007) A biochemical approach to the problem of aging: "megaproject" on membrane-penetrating ions. The first results and prospects. Biochemistry (Moscow) 72:1385–1396

Skulachev VP (2009a) How to cancel the aging program of organism? Rus Khim Zh LIII:125–140 (in Russian)

Skulachev VP (2009b) New data on biochemical mechanism of programmed senescence of organisms and antioxidant defense of mitochondria. Biochemistry (Moscow) 74:1400–1403

Skulachev VP (2011) SkQ1 treatment and food restriction, two mechanisms inhibiting an aging program of organism. Aging (Albany NY) 3:1045–1050

Skulachev VP (2012) Mitochondria-targeted antioxidants as promising drugs for treatment of age-related brain diseases. J Alzheimers Dis 28:283–289

Skulachev VP, Anisimov VN, Antonenko YN, Bakeeva LE, Chernyak BV, Erichev VP, Filenko OF, Kalinina NI, Kapelko VI, Kolosova NG, Kopnin BP, Korshunova GA, Lichinitser MR, Obukhova LA, Pasyukova EG, Pisarenko OI, Roginsky VA, Ruuge EK, Senin II, Severina II, Skulachev MV, Spivak IM, Tashlitsky VN, Tkachuk VA, Vyssokikh MY, Yaguzhinsky LS, Zorov DB (2009) An attempt to prevent senescence: a mitochondrial approach. Biochim Biophys Acta 1787:437–461

Skulachev VP, Antonenko YN, Cherepanov DA, Chernyak BV, Izyumov DS, Khailova LS, Klishin SS, Korshunova GA, Lyamzaev KG, Pletjushkina OY, Roginsky VA, Rokitskaya TI, Severin FF, Severina II, Simonyan RA, Skulachev MV, Sumbatyan NV, Sukhanova EI, Tashlitsky VN, Trendeleva TA, Vyssokikh MY, Zvyagilskaya RA, Skulachev VP (2010) Prevention of cardiolipin oxidation and fatty acid cycling as two antioxidant mechanisms of cationic derivatives of plastoquinone (SkQs). Biochim Biophys Acta 1797:878–889

Stefanova NA, Fursova A, Kolosova NG (2010) Behavioral effects induced by mitochondria-targeted antioxidant SkQ1 in Wistar and senescence-accelerated OXYS rats. J Alzheimers Dis 21:479–491

Stuchlikova E, Juricova-Horakova M, Deyl Z (1975) New aspects of the dietary effect of life prolongation in rodents. What is the role of obesity in aging? Exp Gerontol 10:141–144

Sun D, Muthukumar AR, Lawrence RA, Fernandes G (2001) Effects of calorie restriction on polymicrobial peritonitis induced by cecum ligation and puncture in young C57BL/6 mice. Clin Diagn Lab Immunol 8:1003–1011

Vlachantoni D, Tulloch B, Taylor RW, Turnbull DM, Murphy MP, Wright AF (2006) Effects of MitoQ and Sod2 on rates of retinal degeneration in Rd1, Atrd1, Rho$^{-/-}$ and Rds mutant mice. Invest Ophthalmol Vis Sci 47:E-5773

Weismann A (1989) Essays upon heredity and kindred biological problems. Claderon Press, Oxford

Will LC, MacCay CM (1943) Ageing, basal metabolism and retarded growth. Arch Biochem 2:481

Yani EV, Katargina LA, Chesnokova NB, Beznos OV, Savchenko AY, Vygodin VA, Gudkova EY, Zamyatnin AA(Jr), Skulachev MV (2012) The first experience of using the drug Vizomitin in the treatment of "dry eyes". Prakticheskaya Meditsina 59:134–137

Zinovkin RA, Bakeeva LE, Chernyak BV, Egorov MV, Isaev NK, Kolosova NG, Korshunova GA, Manskikh VN, Moshkin MP, Plotnikov EY, Rogovin KA, Savchenko AY, Zamyatnin AA Jr, Zorov DB, Skulachev MV, Skulachev VP (2013) Penetrating cations as mitochondria-targeted antioxidants. Systems biology of oxidative stress. Springer, Heidelberg

Appendices

Appendix 1
Energy, Work, and Laws of Thermodynamics

A.1 Introduction

Energy (from Greek ενέργειας—action, activity) can be generally defined as the property of a material system that characterizes the ability of this system to produce some changes in the environment (work, **W**). Such a work can consist of growth of the total amount of matter in a system, the uneven distribution of matter particles in space, and/or movement of these particles. Physical interactions between matter particles in a system are accompanied by the mutual transformation of different types of energy (mechanical, thermal, electromagnetic, gravitational, nuclear, and chemical). How is life ensured by energy? The laws of thermodynamics provide us with a quantitative answer to this question. Energy transductions in living self-replicating material systems (organisms) follow the laws of thermodynamics. It seems quite remarkable that the first law of thermodynamics (the law of conservation and transformation of energy) was first formulated in 1841 by the German physicist and physician Julius Robert von Mayer as the result of his studies of the energy processes in the human organism; it was only some time later that this law was applied to the systems of technical energetics.

Thermodynamics deals with three types of systems: isolated, closed, and open. Hypothetical isolated (adiabatic) systems are completely self-contained, i.e., there is no exchange of matter, energy, and information with the environment. Closed systems are materially self-sufficient, but they are open energetically and informationally. In other words, there is exchange of energy and information, but not matter through the borders of closed systems. Many nonliving systems of technical energetics function in this way. Living organisms are open systems that constantly exchange matter, energy, and information with the environment.

The term *heat* (Q) is used in thermodynamics as the measure of energy passing from one body to another. A system uses received energy to perform an external work and also to carry out transitions between the alternative states of the components of the system. Changes in the amount of work can be calculated as the product of pressure (*P*) and an infinitesimal increment of volume (ΔV):

$$\Delta W = P \Delta V \qquad (A1.1)$$

Entropy (S) which is often assumed to be as a measure of disorder means the number of possible states of a thermodynamic system. That is why this parameter is a related degree of disorder in the studied system. The term entropy (from Greek *entropia*—turn, transformation), introduced in 1865 by the German physicist Rudolf Clausius, is now widely used in physics, chemistry, biology, and information theory. The relationship between the changes in the general amounts of heat and entropy during reversible processes follows the equation:

$$\Delta Q = T \Delta S \qquad (A1.2)$$

where *T* stands for absolute temperature.

The second law of thermodynamics, formulated in 1851 by the English physicist William Thomson (Lord Kelvin), allows to predict the possibility of performance of external work. It connects entropy with directions of transfer of energy (heat) and of information across the border of a system. Reversible processes in isolated systems do not lead to growth in entropy; neither do they result in performance of useful work. Non equilibrium transformations necessary for performing work in isolated systems are inextricably linked to the approach of thermodynamic equilibrium with entropy reaching its maximal value. Once this equilibrium state is reached, no useful work can be performed in isolated systems. Entropy in closed and open systems increases when they perform external work; this can also result from their receiving information from the outside. Increase in the number of possible alternative states of a system reduces the probability of its being in the functionally operative (native) conformation; this in turn decreases the efficiency of its work. When using closed systems of technical energetics, one faces the necessity of their regular reparation. Living systems strive to minimize growth of their entropy, which might lead to an equilibrium state fatal for such systems. They maintain steady-state organization of its components at the expense of an external entropy increase. The existence of steady-state open systems, which correspond to minimal entropy production, was postulated by the Nobel Prize Laureate in Chemistry Ilya Prigogine, who in 1947 proved the fundamental theorem of thermodynamics of non equilibrium processes.

It seems quite obvious that the complexity of thermodynamic descriptions of systems increases as the level of their isolation decreases. Simple equations of classical equilibrium thermodynamics cannot be directly applied to open systems, because matter flow across their borders prevents true equilibrium from being established. However, a living organism can be considered as an aggregate of interacting subsystems (organs, tissues, cells, organelles, and molecules), an assumption that makes it easier to understand the essence of energetic processes. In the majority of cases, individual reactions between the subsystem components

Appendix 1: Energy, Work, and Laws of Thermodynamics

can be viewed as closed systems, which means that the equations of equilibrium thermodynamics can be applied to these processes. Thus, we conclude (on the basis of these equations) that for any given system the potential ability to perform work is determined by the degree of deviation of its components from the equilibrium state. To use in practice the approximate quantitative estimate of open systems, one should understand the nature of the equilibrium state of a particular chemical reaction and estimate equilibrium shift of the real reaction with respect to the equilibrium state predicted by the equations of classical equilibrium thermodynamics. It is this shift in equilibrium that determines the quantitative ability of the reaction to support performance useful work. Furthermore, the mathematical apparatus of equilibrium thermodynamics can be effectively used for the identification of "forbidden" reactions. Only processes leading to increase in entropy and reduction in energy reserve can proceed spontaneously (without any energy input into the system).

A.2 Thermodynamic Potentials

According to the second law of thermodynamics, only a portion of energy of a system can be "useful", i.e., can be employed to carry out a work. In 1878, the American physicist and mathematician Josiah Willard Gibbs developed the theory of "fundamental functions" to estimate the amount of useful energy. These functions characterized the state of a thermodynamic system depending on its main properties: number of particles (N), volume (V), pressure (P), temperature (T), entropy (S), and chemical potential of the components (μ). These "fundamental functions" are now known as "thermodynamic potentials". The latter term was suggested in 1884 by the French philosopher and physicist Pierre Duhem, and it superseded the previous one due to its greater accuracy. Potential (from Greek *potentia* – power) corresponds to the sources, possibilities, means, and reserves that can be used to solve some problem, to reach a certain goal. Different thermodynamic potentials are optimized so as to facilitate characterization of systems with different sets of constant and variable properties (macroscopic parameters):

- internal energy (U), a function of entropy (S) and volume (V);
- enthalpy (H), a function of entropy (S) and pressure (P);
- Helmholtz free energy (F), a function of temperature (T) and volume (V);
- Gibbs free energy, also known as free enthalpy (G), a function of temperature (T) and pressure (P);
- Landau potential, or grand thermodynamic potential (Ω), a function of temperature (T), volume (V), and chemical potential of transferred particles (μ).

The internal energy of a system (U) is defined as the total energy of the given system minus the kinetic energy of the system as a whole (energy of motion in the environment) and the potential energy of the system in external force fields (when relating to other systems). When applied to bioenergetics, internal energy can be defined as the sum of kinetic energy of chaotic motion, potential energy of interaction, and intramolecular energy of its molecules. According to *the first law of thermodynamics*, the total energy of a system is equal to the difference between the amount of heat (Q) transferred to the system and the performed external work (W):

$$U = Q - W \qquad (A1.3)$$

For idealized processes comprising consecutive equilibrium states, changes of internal energy of a system with a constant number of particles during performance of work (ΔU) are described by the equation:

$$\Delta U = \Delta Q - \Delta W = T\Delta S - P\Delta V \qquad (A1.4)$$

Thermodynamic potential H, or enthalpy (from Greek *enthálpō*—heating), is used to characterize the amount of heat transferred to a system for the performance of work. Changes in enthalpy are connected to the internal energy of a system as follow:

$$\Delta H = \Delta U + \Delta(PV) = T\Delta S - P\Delta V + P\Delta V + V\Delta P = T\Delta S + V\Delta P \qquad (A1.5)$$

For constant pressure, enthalpy change is equal to the amount of heat transferred to the system for the performance of useful work; that is why enthalpy is often referred to as heat function, or heat content. It is convenient to use enthalpy of a system when pressure P and temperature T are chosen as independent variables. In studies on biological systems, enthalpy can be measured directly by calorimetric measurements of the heat effect of a reaction:

$$-\Delta H = \Delta Q \qquad (A1.6)$$

The second law of thermodynamics prohibits the complete conversion of internal energy into work, for a certain portion of inner energy is inevitably spent for entropy (S) increase. The fraction of total energy that a system can use for the performance of the greatest amount of external work is called the Helmholtz free energy (F), or isochoric–isothermal potential:

$$\Delta F = \Delta U - \Delta(TS) = T\Delta S - P\Delta V - T\Delta S - S\Delta T = -P\Delta V - S\Delta T \qquad (A1.7)$$

This equation leads us to the conclusion that Helmholtz free energy decreases with increase in volume and temperature. However, a certain fraction of total work is spent on useless change of the environment due to changes in the volume of the system.

Work of a system on external bodies that does not cause changes in the outer medium is called useful work. The maximal useful work of a system is quantitatively equal to the decrease in Gibbs energy (G), or isobaric–isothermal potential, which is calculated as follows:

$$\Delta G = \Delta H - \Delta(TS) = \Delta U + \Delta(PV) - \Delta(TS)$$
$$= T\Delta S + V\Delta P - T\Delta S - S\Delta T = V\Delta P - S\Delta T \quad (A1.8)$$

According to this equation, Gibbs energy decreases with increasing temperature, but it increases when the pressure in the system increases. Helmholtz free energy is used to characterize physical processes, while Gibbs potential is used when describing chemical and physicochemical processes. The difference of Helmholtz and Gibbs potentials characterizes the amount of energy spent by a system for nonproductive change in the outer medium:

$$\Delta F - \Delta G = -P\Delta V - V\Delta P \quad (A1.9)$$

It has to be emphasized that in the biological literature Gibbs potential is often called "Gibbs free energy" or simply "free energy", which may be misleading, as it can be confused with Helmholtz potential. In the case of biological systems, the differences between these two potentials are often ignored, for volume changes are insignificant in the majority of such cases.

Biochemical reactions proceed under conditions of constant pressure and temperature, hence the formula for calculation of the Gibbs potential changes can be simplified as follows:

$$\Delta G = \Delta H - T\Delta S \quad (A1.10)$$

The type of Gibbs energy change makes it possible to estimate the maximal theoretical possibility of a process. When $\Delta G < 0$, a process can proceed spontaneously if no kinetic factors interfere with it. For instance, a negative value of Gibbs energy change does not hinder reduction of O_2 to $2H_2O$ by NADH with no mediators involved. However, an NADH molecule, when oxidized, releases two electrons while an oxygen molecule, when reduced to water, accepts four electrons. As a result, such a process is forbidden because of its not matching the Schaeffer parity rule, which presumes the equal number of electrons released and accepts in one and the same reaction. Respiratory chain electron carriers mediating electron transfer from NADH to O_2 help to outflank kinetic limitations of the thermodynamically possible process of NADH oxidation by oxygen. When $\Delta G = 0$, the system is in the state of chemical equilibrium, and hence no gross transformations of the system components occur. When $\Delta G > 0$, a process can proceed only if there is an input of energy from the outside. For instance, formation of glucose from water and carbon dioxide in photosynthesis is coupled to Gibbs energy increase, and it can proceed only due to the input of the energy of photons into the system. Change in Gibbs energy during biochemical reactions can be determined in several ways: on the basis of affinity constant, values of standard redox potentials, calorimetric measurement of heat produced in a reaction, and also by summing standard values ΔH_0 and $T\Delta S_0$ of the formation of the reaction components from the respective simple elements (the latter are known from tables).

Thermodynamic potentials describing homogeneous systems with no transfer of particles (solutions) seem to be simpler than potentials for calculating energetics of membrane-containing heterogeneous systems. When energetic processes in a system are coupled to matter (particle) transfer across its border (membrane) to an external medium, an extra item ($+\mu \Delta N$) is added to all the equations of thermodynamic potentials. This item is quantitatively equal to Gibbs energy change in the system due to the addition of N molecules with chemical potential μ.

Chemical potential (μ) of a component is equal to the energy that needs to be spent on the addition of an infinitesimal molar quantity of this component (one particle) to the system. In other words, chemical potential can be defined as specific Gibbs potential (per particle or per mole of molecules if N stands for the number of moles of molecules):

$$\mu = \Delta G / \Delta N \tag{A1.11}$$

Estimation of chemical potential helps to characterize reactions connected to redistribution of particles (molecules) between the phases or compartments divided by a membrane. Particles can move spontaneously from a compartment (phase) with a higher chemical potential to a compartment (phase) with a lower chemical potential. This reduces the general free energy of the system and brings it closer to the equilibrium state, when chemical potentials of the particles of both compartments (phases) become the same. It is important to note that due to the course of historical development, specialists from different sciences have different understanding of the term "gradient". When speaking about "gradient" (from Latin *gradiens*—walking) biochemists often mean the direction of decrease in particles concentration, while mathematicians and physicists mean the direction of increase in a certain value which changes from one point in space to the other.

When particles moving across an interphase boundary have electric charge (e.g. ions), this creates conditions for the transmembrane difference of electric potentials ($\Delta \Psi$) to be formed. Electrochemical potential ($\bar{\mu}$) is used to estimate the probable redistribution of ions (this was thoroughly discussed in Sect. 1.3).

The "grand thermodynamic potential", or Landau potential (Ω), can be determined for a statistical ensemble of states of a system with a variable number (N) of particles. This potential shows the connection between the Helmholtz free energy and chemical potential:

$$\Omega = A - \mu N = - PV \tag{A1.12}$$

$$\Delta \Omega = -S \Delta T - P \Delta V - N \Delta \mu \tag{A1.13}$$

A.3 Gibbs Free Energy and Metabolism in Biological Systems

How do living organisms get Gibbs free energy? As biological evolution has been directed toward increase in reliability of energy supply rather than its efficiency, living systems do not use nuclear, thermal, and mechanical energy sources. It is the energy of chemical reactions, light, and ion gradients that is used by organisms to perform work.

External resources are absorbed by an organism and become part of metabolism. The constructive branch of metabolism (anabolism or assimilation) comprises reactions of synthesis of organic polymers (proteins, carbohydrates, lipids, and nucleic acids) that become parts of an organism. In the case of autotrophs, it is carbon dioxide that is the source of carbon, while for heterotrophs it is organic compounds. Organic polymers are cleaved to low molecular mass compounds (monomers) in the preliminary reactions of the exergonic branch of metabolism (catabolism or dissimilation). These compounds later become substrates for anabolic reactions. Exergonic catabolic reactions are accompanied by release of Gibbs energy, which is used for the performance of energetically unfavorable anabolic reactions. In the great majority of cases, Gibbs energy storage is connected to the transfer of one or two electrons from a donor compound to an acceptor compound having a higher affinity for electrons; in this process, the originally reduced donor is oxidized, while the oxidized acceptor is reduced. In the case of lithotrophs, it is inorganic compounds (sulfur derivatives, ferrous iron, ammonia and its salts, nitrites, hydrogen, and carbon monoxide) that act as primary electron donors, while for organotrophs it is organic compounds (proteins, carbohydrates, lipids, and products of their partial cleavage) that play the same role. Oxygen acts as electron acceptor in aerobic organisms, and nitrates, nitrites, sulfates, carbon dioxide, and certain organic compounds are electron acceptors in anaerobic organisms.

The scale of standard redox potentials measured in volts (V) is used for the quantitative evaluation of the affinity of compounds for electrons. Donors have more negative redox potential than acceptors. The values of redox potentials of low molecular mass compounds participating in bioenergetic transformations are in the range from -0.7 (α-ketoglutarate/succinate) to $+0.8$ V (oxygen/H_2O). If the redox potential difference of electron donor/acceptor pare is ≥ 0.2 V, this makes it possible to convert free energy of the oxidative process to unified convertible form of energy, ATP (see Chap. 7), which can be used to perform different types of work. The discussed quantitative restriction entails the quantum principle of free energy transfer in biological systems.

A.4 Assimilatory Force and Its Sources

Electrons received from donors with high reducing potential can be used for binding of molecules in anabolic reactions. Such electrons are often called *reducing equivalents*. The total of reducing equivalents and Gibbs free energy obtained in catabolic reactions is called *assimilation force*. What are the sources of assimilation force for biological systems?

Vital functions of chemotrophs depend on the presence of strong inorganic or organic reductants in the environment. Phototrophs reduce the molecules of their own acceptors (see Chap. 2) using electrons received from weak reductants due to the energy of light quanta. The fight for energy resources might be the driving force of evolution during the mastering of new ecological niches. Distribution of external carbon sources and assimilation force between organisms led to the appearance of eight types of bioenergetics. Chemotrophic bioenergetics (see Chap. 5) is used by chemoorganoautotrophs (certain methylotrophic prokaryotes), chemoorganoheterotrophs (majority of prokaryotes and two groups of eukaryotes: animals and fungi), chemolithoautotrophs, and chemolithoheterotrophs (many prokaryotes). Light-dependent energetics (see Chaps. 2 and 6) is used by photoorganoautotrophs and photoorganoheterotrophs (prokaryotes), photolithoautotrophs (plant eukaryotes), and photolithoheterotrophs (prokaryotes). In ecological systems, waste products of some organisms can be resources for others. Biogenic migration of chemical elements is the consequence of donor–acceptor interactions in the biosphere and its biogeocenoses. For instance, photolithoautotrophic cyanobacteria and plants, when consuming carbon dioxide and water, produce glucose and oxygen, which are later transformed back to carbon dioxide and water by aerobic chemoorganoheterotrophs. Only 5–20 % of Gibbs energy is accumulated in the produced biomass of heterotrophs, while the rest is dissipated as heat and is spent on maintenance of vital functions. Photolithoautotrophs, being the main biomass producers in the majority of ecosystems, use only about 1 % of the available light energy for anabolism.

In multicellular heterotrophs, the preliminary reactions of catabolism occur mainly in the cavities of their digestive systems and in the cells of storage tissues, while oxidative reactions are localized in mitochondria. Electron donors and acceptors enter the multicellular organism through the body surface or through some special structures (stomata of plants, gills, digestive system, and lungs of animals); later they are transported to cells via special conductive systems (sieve tubes, tracheids, and trachea of plants; blood–vascular system of animals), and then via intercellular spaces. The efficiency of transport of O_2, the most important electron acceptor to the sites of its use is increased by means of respiratory pigments (e.g., hemoglobin) and special cellular carriers (erythrocytes).

Oxidoreductase enzymes serve as intracellular mediators between donors and acceptors. Oxidoreductases use compounds of nonprotein nature (cofactors and prosthetic groups) for the transport of reducing equivalents. These are nicotinamide and flavin nucleotides, quinones, metal porphyrins, and iron–sulfur clusters. Components of electron transport chains are arranged in order of increasing affinity to an electron, thus helping to reduce the activation barrier of the overall oxidoreductase reaction. In solutions, reducing equivalents are transferred by nicotinamide coenzymes, their structure preventing them from being directly oxidized by oxygen. NAD^+ mainly participates in catabolic transformations, while its phosphorylated analog $NADP^+$ participates in anabolic reactions. Oxidation of donors (or reduction of acceptors) is often accompanied by the release (or binding) of 1–2 hydrogen ions (H^+).

Gibbs free energy can be accumulated and transmitted in the form of ion gradients (see Chaps. 9 and 11) as well as by means of molecules containing groups with high transfer potential (the so-called "macroergs", a term attributed to these groups by V. A. Engelhardt in 1945). ATP, being the "water-soluble energy currency" of the cell (see Chap. 7), is the best known molecule of this type. Under physiological conditions, the ATP structure provides combination of high transfer potential of each of the two terminal phosphoryl groups to the acceptor molecule (thermodynamic instability) with kinetic stability (ATP is not spontaneously hydrolyzed at body temperature and neutral pH). ATP was shown to be utilized for the performance of different types of work: chemical (biosyntheses), electrical (formation of electric potential difference on biomembranes), osmotic (creation of concentration gradients of uncharged compounds), and mechanical (contraction of actomyosin complexes in muscles). Gibbs free energy stored in the transfer potential of ATP groups (phosphoryl potential) can be redistributed between different nucleoside tri- and diphosphates via special enzymes (nucleoside di- and monophosphate kinases), thus driving specific biosyntheses. In the majority of animals, phosphoryl potential is stabilized by means of equilibrium reaction of reversible phosphoryl transfer to creatine. In certain crustaceans, it is phospho-arginine that serves as phosphoryl group buffer, while polyphosphates play the same role in fungi.

Besides "water-soluble energy currency" (ATP), cells always have at least one, and often two "membrane-linked energy currencies", transmembrane differences of electrochemical potentials of H^+ and Na^+ ions ($\Delta \bar{\mu}_{H^+}$ and $\Delta \bar{\mu}_{Na^+}$, respectively) that can support various types of chemical, osmotic, and mechanical work. Moreover, $\Delta \bar{\mu}_{H^+}$ is used as a transportable form of energy, transmitted among extended mitochondrial profiles operating as intracellular electric cables (Chap. 11).

Carbohydrate polymers (starch in plants, glycogen in animals and fungi), lipids (oils in bacteria, plants, and fungi, fats in animals), proteins (vitellin in oocytes, etc.) and polyphosphates (in certain bacteria and fungi) are used for long-term storage of assimilation force. A high rate of mobilization is characteristic for carbohydrates, and the largest energy storage capacity for lipids. Reserve carbohydrates are located in stocking organs in plants (tubers et al.) and in liver, muscles, brain, and some other tissues in animals. Lipids are stored in special fatty tissue in animals. Availability of electron acceptors for the cells of chemoheterotrophs is often the limiting ecological factor. It is because of this that such an electron acceptor as O_2 can be accumulated by a special muscle respiratory pigment (myoglobin). Some organisms can outwait periods of limited availability of external resources in the state of anabiosis or hibernation. Certain biological assimilation reserves can be used by technical energetics. Oils of fossil diatoms are supposed to have become the source of petroleum formation; cellulose of ancient Lycopodium and Equisetum turned into black and brown coal, and methanogenic Archaea contributed to the formation of natural gas reserves.

Appendix 2
Prosthetic Groups and Cofactors

Figs. A2.1, A2.2, A2.3, A2.4, A2.5, A2.6, A2.7, A2.8, A2.9, A2.10, A2.11, A2.12, A2.13

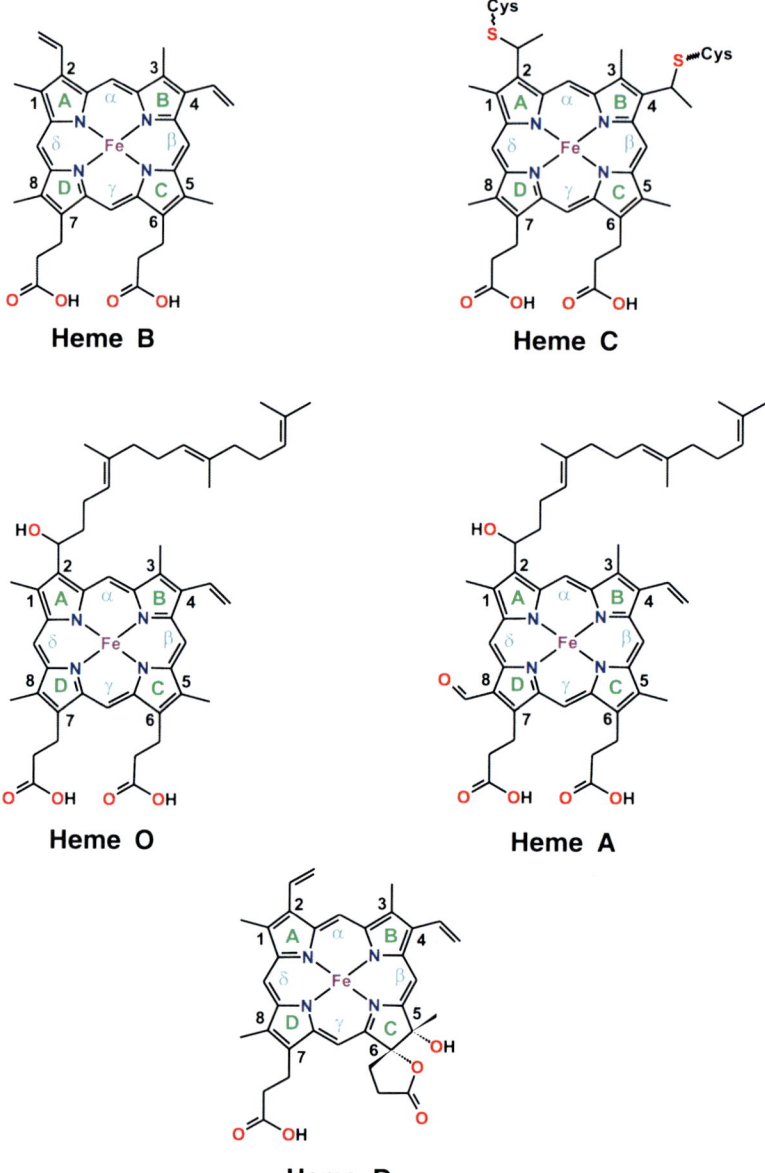

Fig. A2.1 Structural formulas of various hemes of the enzymes of electron transport chains

Appendix 2: Prosthetic Groups and Cofactors

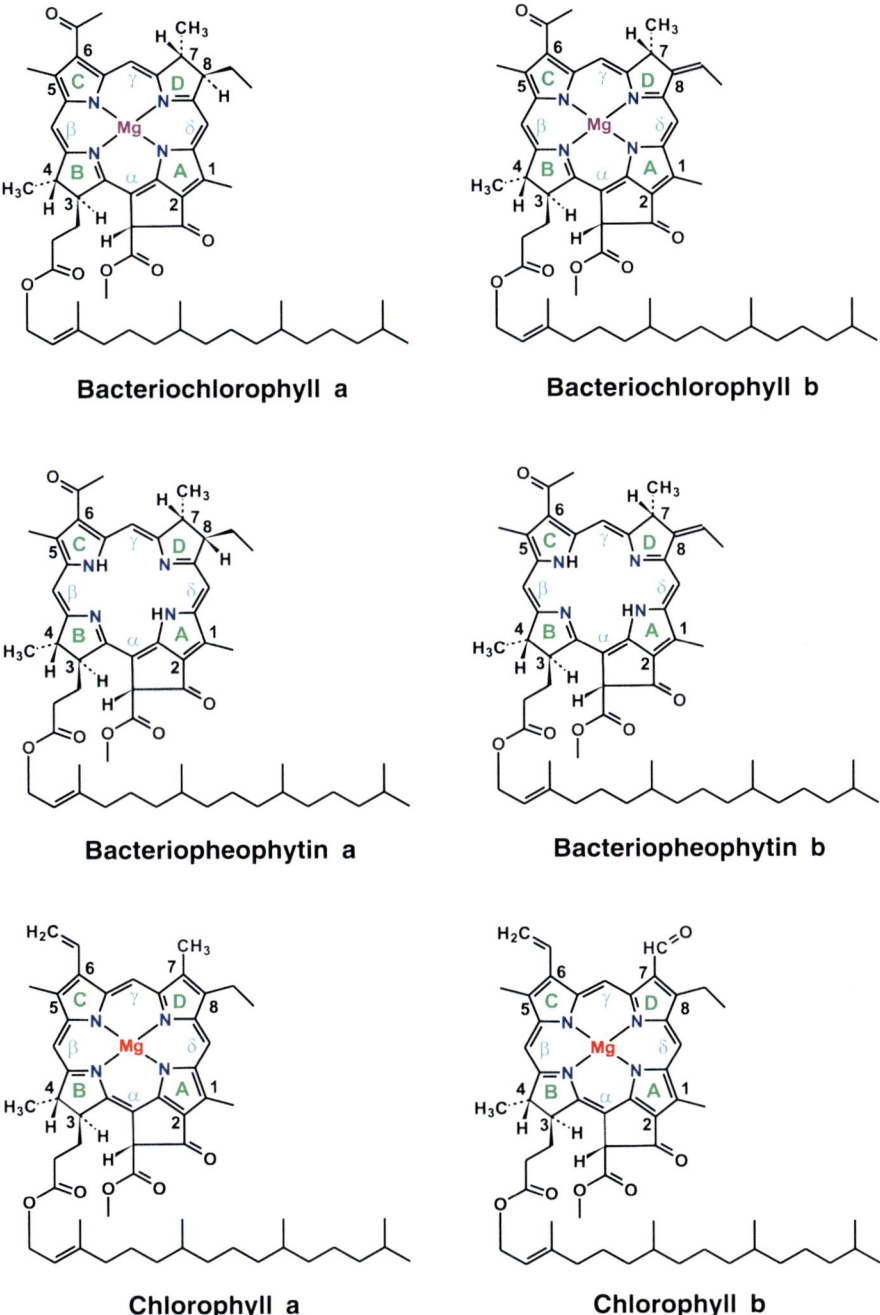

Fig. A2.2 Chlorophylls of photosynthetic electron transport chains of prokaryotes and plants

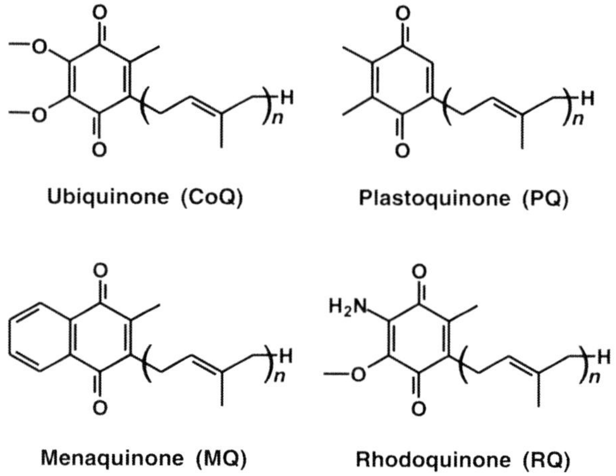

Fig. A2.3 Quinone derivatives participating in enzymatic redox processes: the number of isoprene residues (n) varies in different species from 6 to 10; ubiquinones have been found in bacterial and mitochondrial membranes, menaquinones in bacterial membranes, plastoquinone in the membranes of chloroplast thylakoids and cyanobacteria, and rhodoquinone in the respiratory chain of *Ascaris* imago

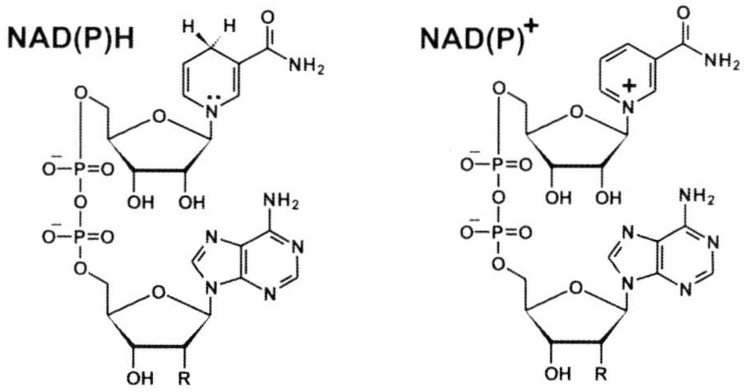

Ubiquinone (CoQ)

Plastoquinone (PQ)

Menaquinone (MQ)

Rhodoquinone (RQ)

Fig. A2.4 Structural formulas of nicotinamide cofactors. In NAD(H), R is a hydroxyl group, and in NADP(H) it is a phosphate group

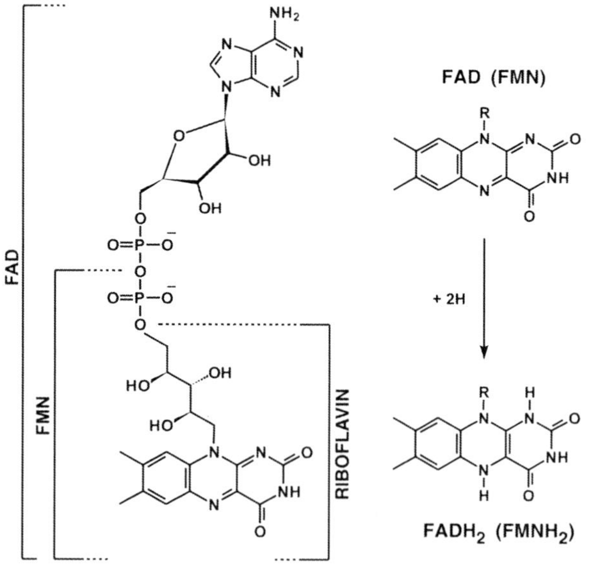

NAD(P)H

NAD(P)$^+$

Fig. A2.5 Structural formulas of riboflavin, FMN, and FAD; the figure also shows the mechanism of their two-electron reduction

FAD (FMN)

+ 2H

FADH$_2$ (FMNH$_2$)

Fig. A2.6 Main iron–sulfur clusters of FeS proteins: FeS-center, [2Fe–2S]–cluster, and [4Fe–4S]–cluster. A center containing one Fe atom has been described in different rubredoxins and related proteins. However, there are usually more than one atom of Fe present—two, three, four, etc. In rubredoxin, all the sulfur atoms belong to cysteine residues. In the case of other clusters, sulfur atoms of sulfide (S^{2-}) as well as cysteine residues participate in the formation of the complex

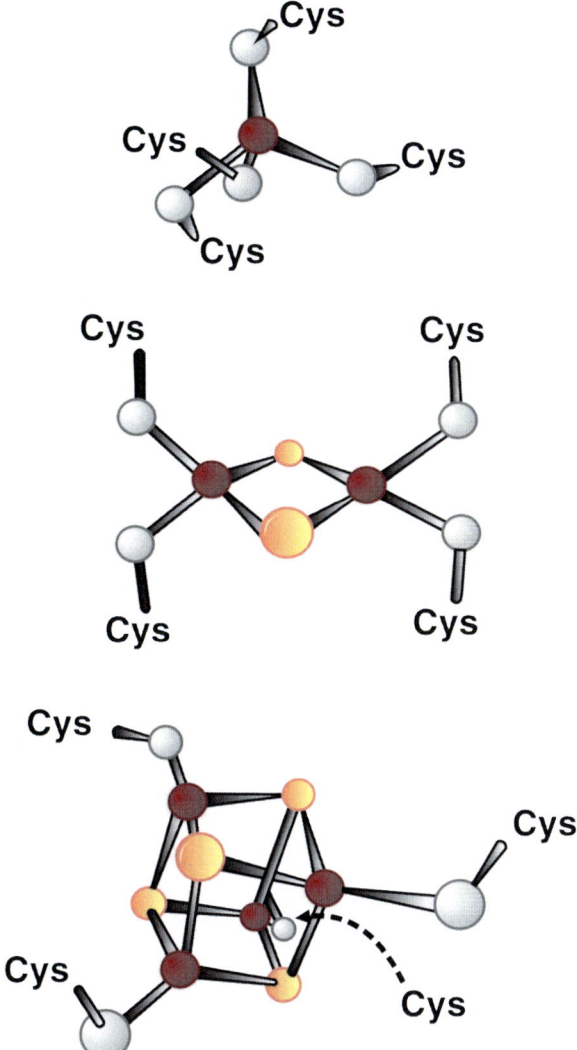

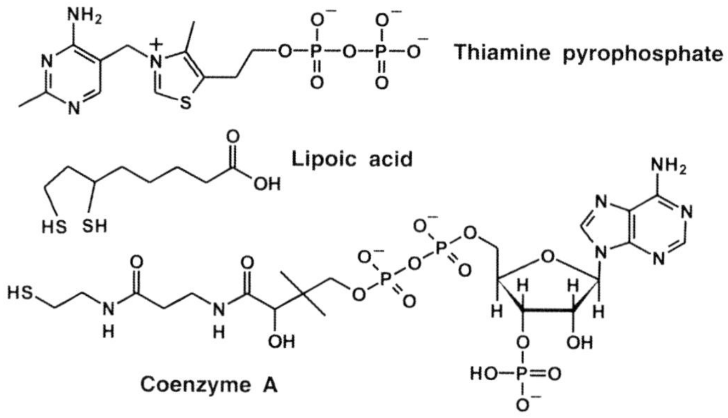

Fig. A2.7 Structural formulas of thiamine pyrophosphate, dihydrolipoic acid, and coenzyme A

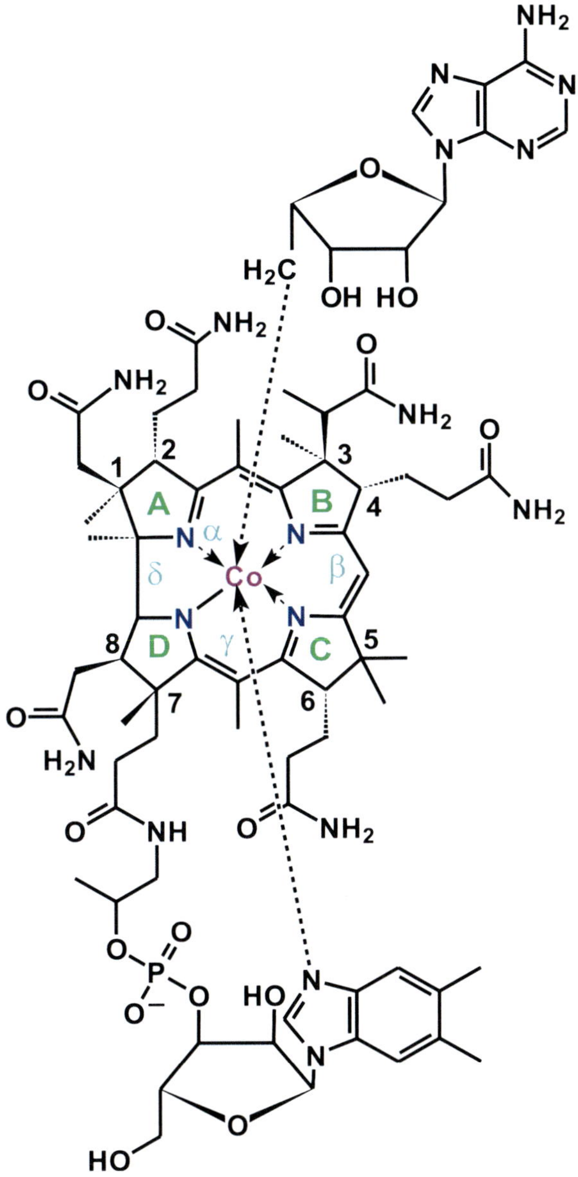

Fig. A2.8 Cobalamin (vitamin B_{12})

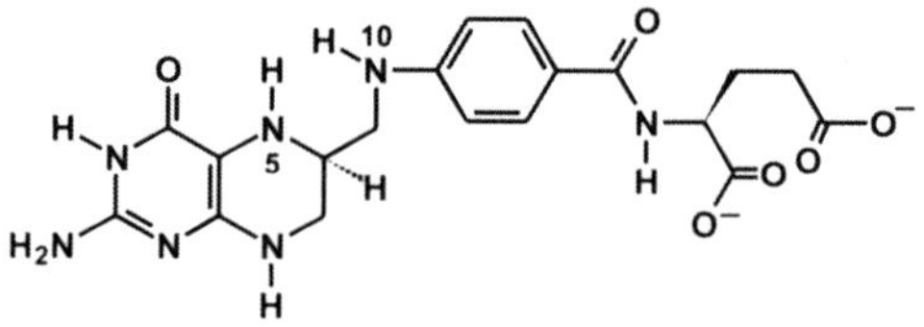

Tetrahydrofolate (H_4F)

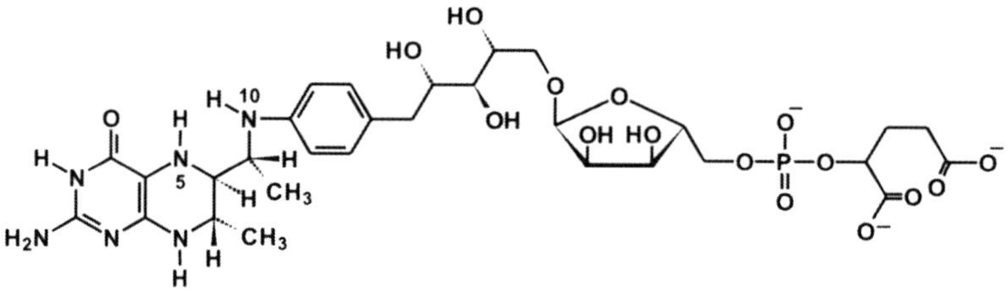

Tetrahydromethanopterin (H_4MPT)

Fig. A2.9 Tetrahydrofolate and tetrahydromethanopterin (methanogenic archaea)

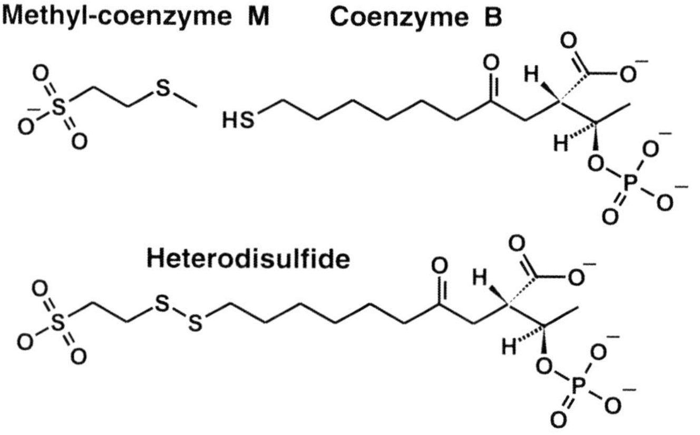

Fig. A2.10 Structural formulas of methyl-coenzyme M, coenzyme B, and CoM–S–S–CoB heterodisulfide (methanogenic archaea)

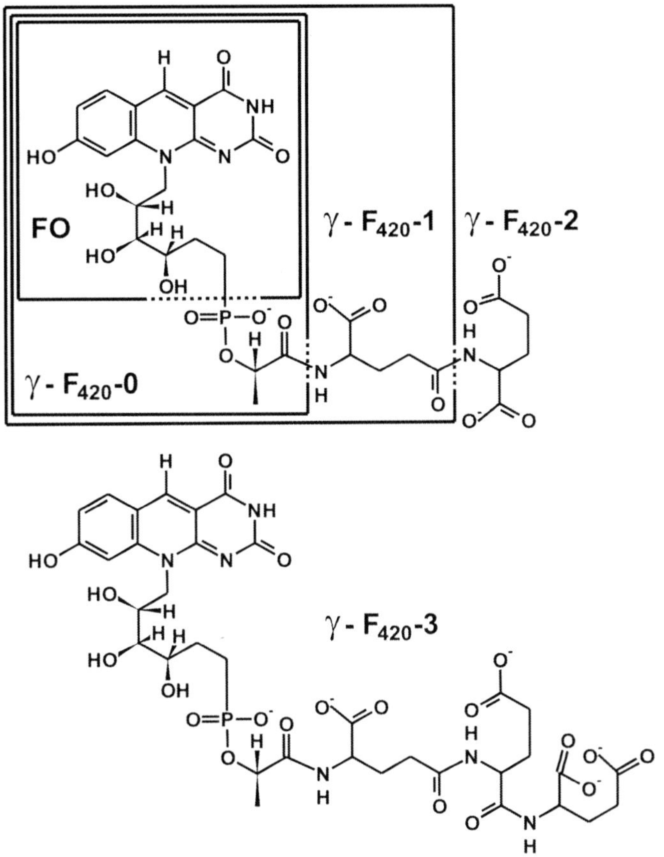

Appendix 2: Prosthetic Groups and Cofactors

Fig. A2.11 Structural formulas of different versions of coenzyme F_{420} (methanogenic archaea)

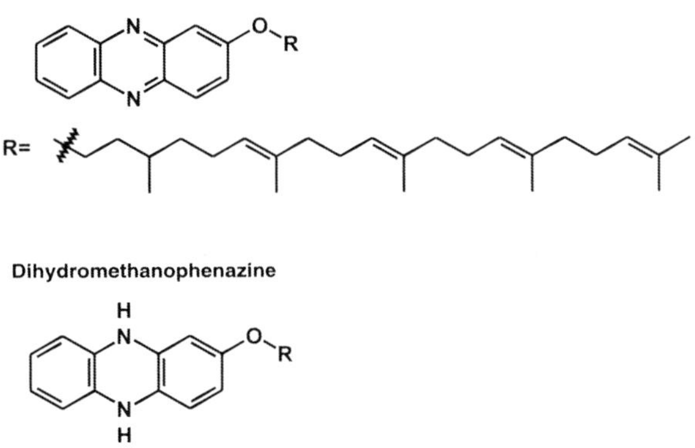

Fig. A2.12 Methanophenazine and dihydromethanophenazine of methanogenic archaea

Fig. A2.13 Structural formulas of methanofuran (MFR), formylmethanofuran, and carboxymethanofuran (methanogenic archaea)

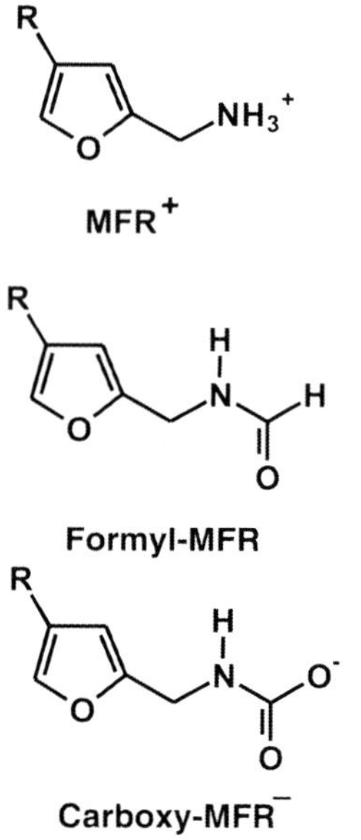

Appendix 3
Inhibitors of Oxidative Phosphorylation

Figs. A3.1, A3.2, A3.3, A3.4

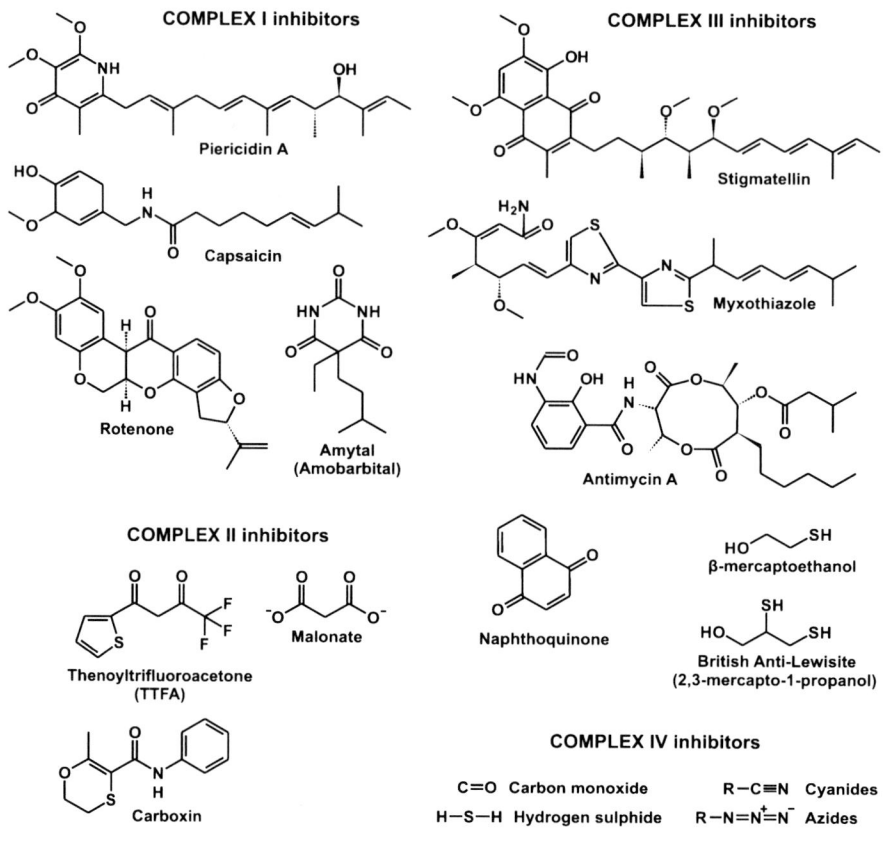

Fig. A3.1 Inhibitors of the respiratory chain complexes

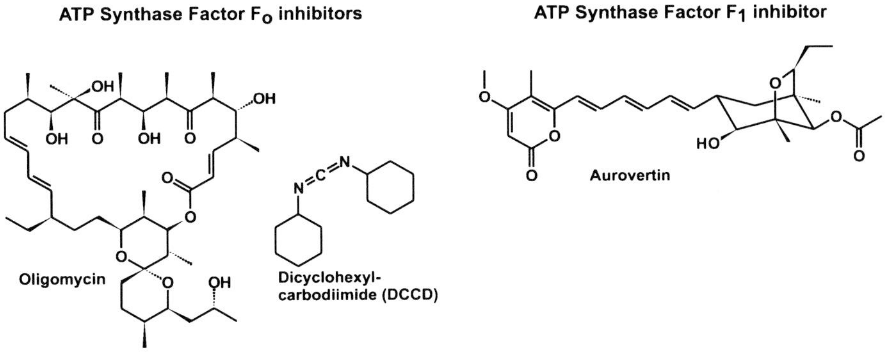

Appendix 3: Inhibitors of Oxidative Phosphorylation

ATP Synthase Factor F_O inhibitors

Oligomycin

Dicyclohexyl-carbodiimide (DCCD)

ATP Synthase Factor F_1 inhibitor

Aurovertin

Fig. A3.2 Inhibitors of H^+-ATP synthase

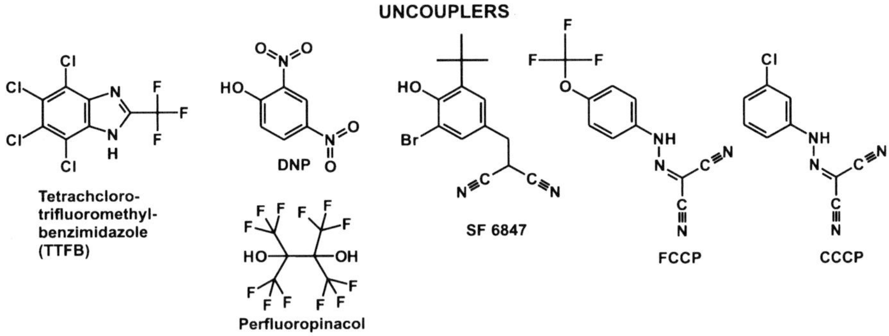

UNCOUPLERS

Tetrachloro-trifluoromethyl-benzimidazole (TTFB)

DNP

Perfluoropinacol

SF 6847

FCCP

CCCP

Fig. A3.3 Protonophorous uncouplers

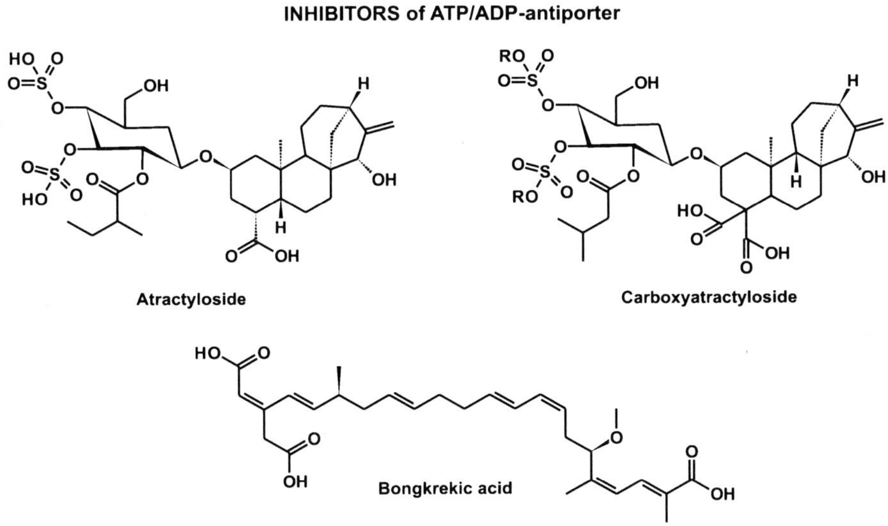

INHIBITORS of ATP/ADP-antiporter

Atractyloside

Carboxyatractyloside

Bongkrekic acid

Fig. A3.4 Inhibitors of ATP/ADP-antiporter

Appendix 4
Plant Hormones

Fig. A4.1

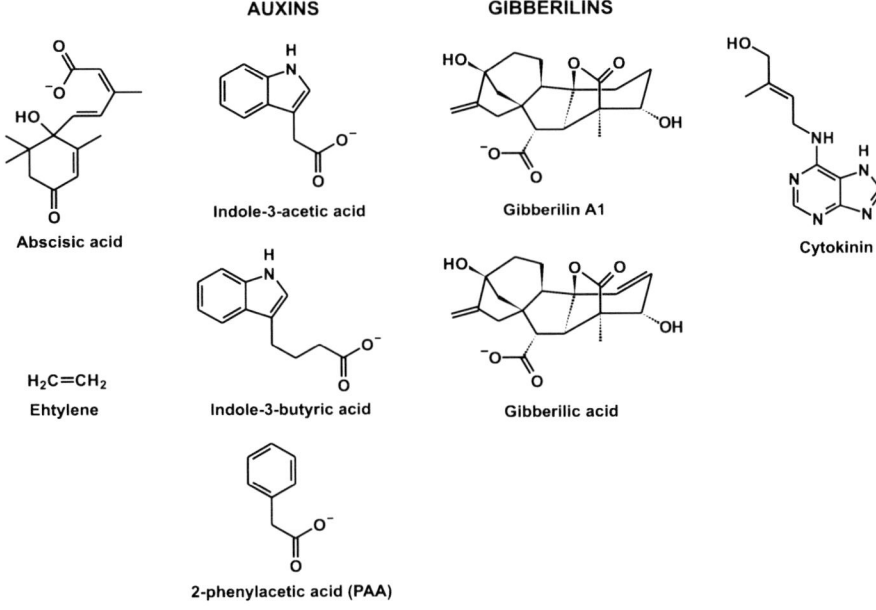

Fig. A4.1 Plant hormones

Appendix 5
Mitochondria-Targeted Antioxidants and Related Penetrating Compounds

Fig. A5.1

410   Appendix 5: Mitochondria-Targeted Antioxidants and Related Penetrating Compounds

ANTIOXIDANTS

(1) SkQ1H2

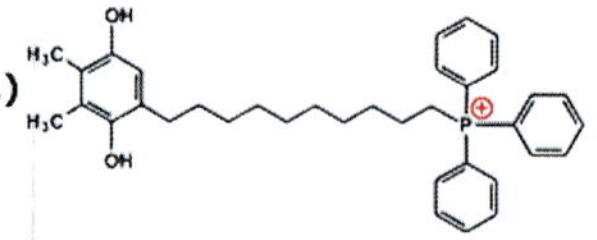

(2) SkQ2MH2

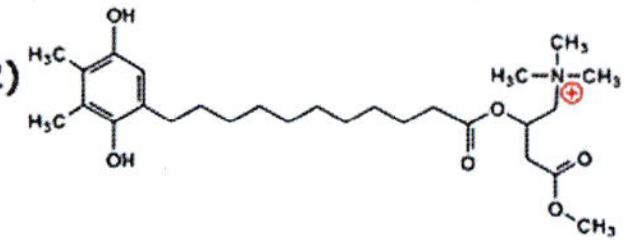

(3) SkQ3H2

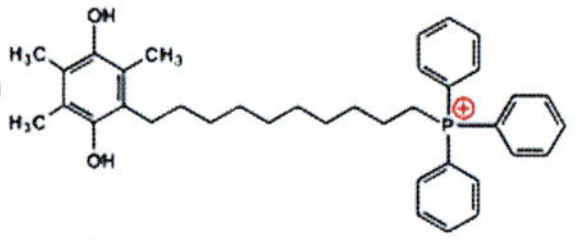

(4) SkQ4H2

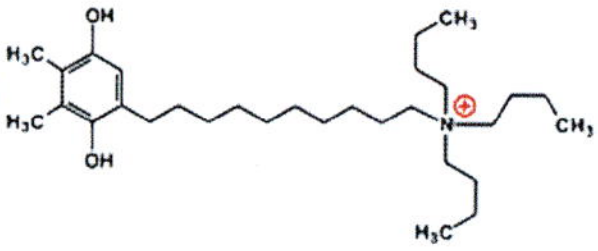

(5) DPQH2

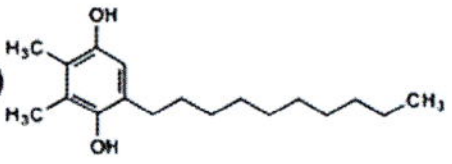

(6) SkQ5H2

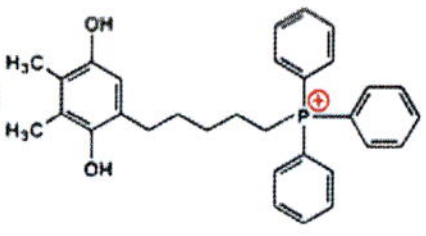

(7) MitoQH2

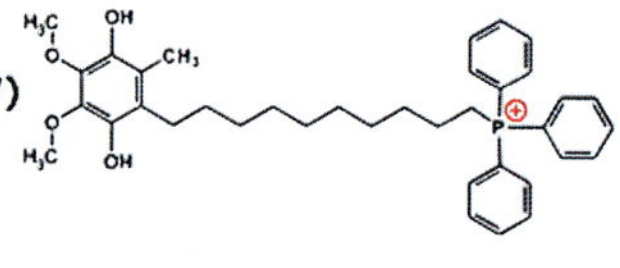

(8) DMQH2

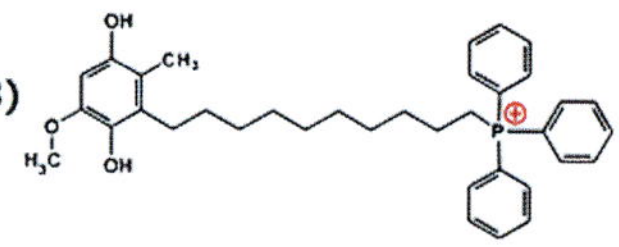

(9) Reduced Ibedenone

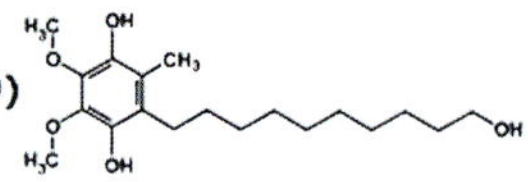

(10) SkQR1H2

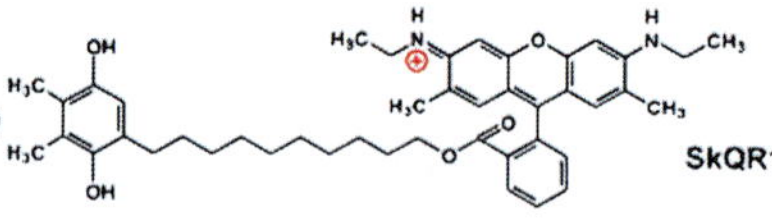

(11) SkQR2H2

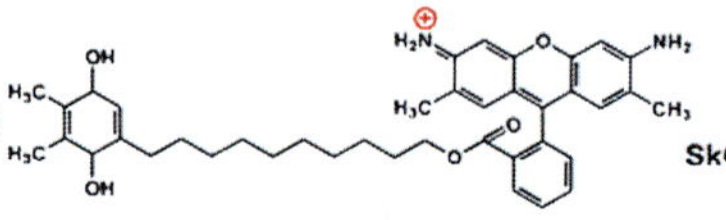

(12) SkQR3H2

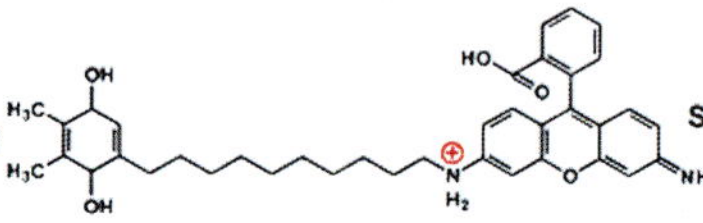

(13) SkQR4H2

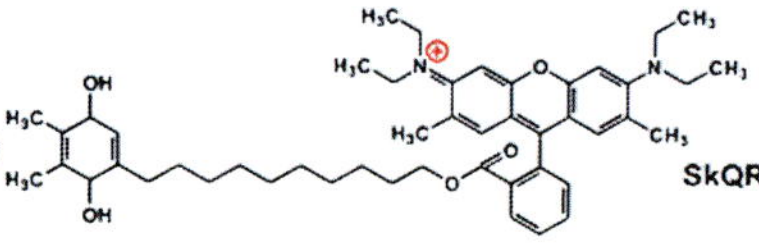

(14) SkQBH2

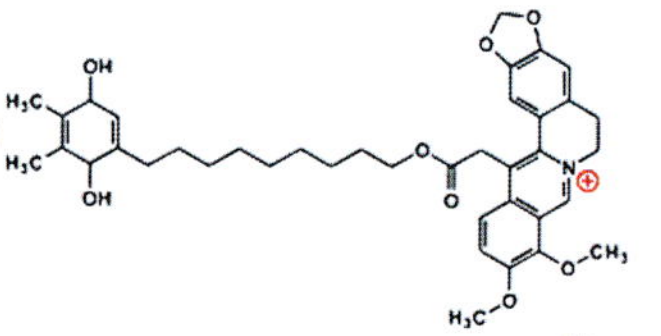

(15) SkQPH2

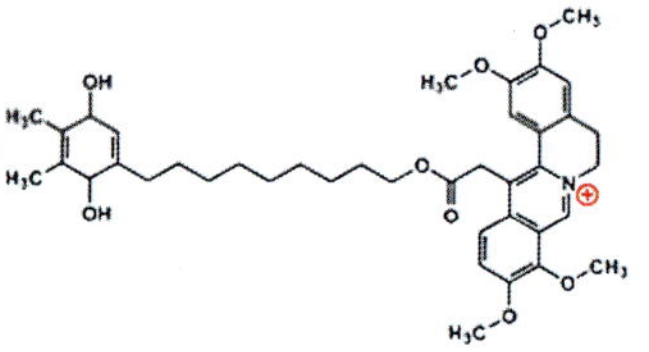

(16) $C_{12}R1$

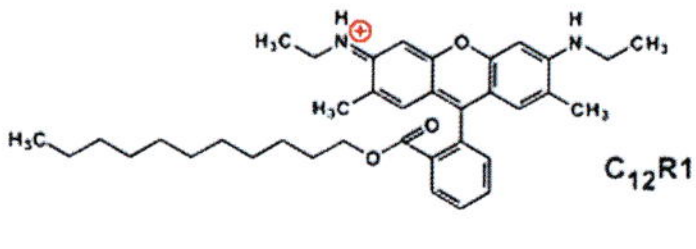

(17) $C_{12}R4$

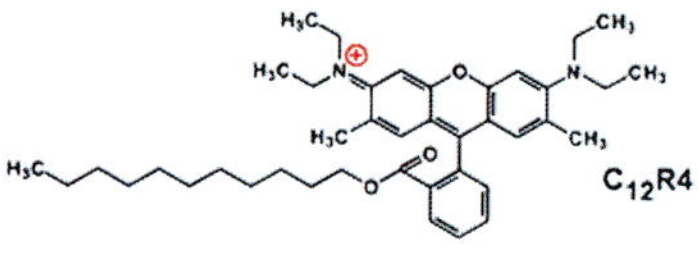

(18) $C_{12}TPP$

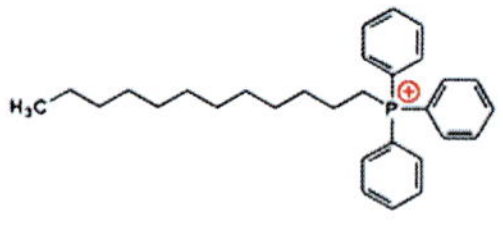

Appendix 5: Mitochondria-Targeted Antioxidants and Related Penetrating Compounds

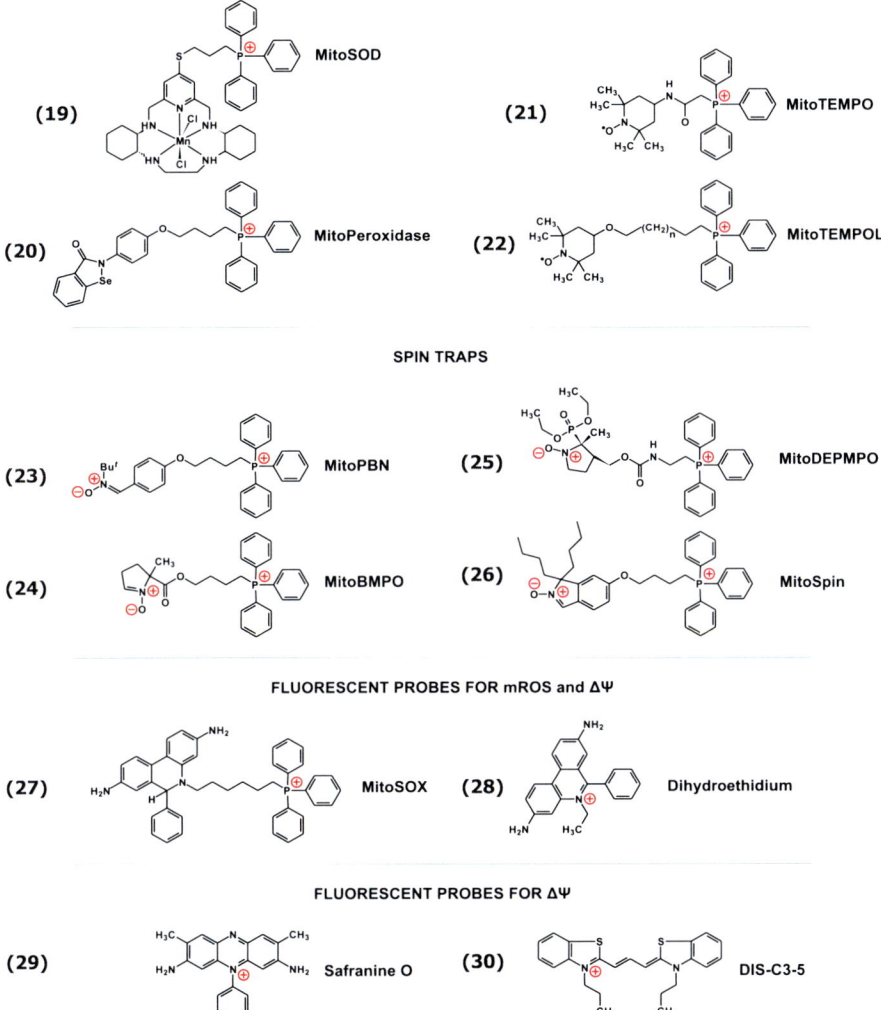

Fig. A5.1 Structures of mitochondria-targeted antioxidants and related compounds. Compounds (1–8) and (10–18) were synthesized in our group by G.A. Korshunova and N.V. Sumbatyan (Antonenko et al. 2008, 2011; Lyamzaev et al. 2011) (for the first synthesis of MitoQ, see (Kelso et al. 2001). For compound (19), see (Salvemini et al. 1999); for compound (20), see (Filipovska et al. 2005); for compounds (21, 22), see (Trnka et al. 2008); for compound (23), see (Murphy et al. 2003); for compound (24), see (Xu and Kalyanaraman 2007); for compound (25), see (Hardy et al. 2007); for compound (26), see (Quin et al. 2009); for compound (27), see (Robinson et al. 2006); for compound (28), see (Zhao et al. 2003); for compound (29), see http://pubchem.ncbi.nlm.nih.gov/summary/summary.cgi?sid=24899820 {Safranin O}; and for compound (30), see http://pubchem.ncbi.nlm.nih.gov/summary/summary.cgi?sid=135074750&loc=es_rss {diSC3(5)}

Appendix 6
Mitochondria-Targeted Natural Rechargeable Antioxidant

In Chap. 16 we discussed SkQ, *a synthetic* rechargeable antioxidant specifically targeted to the inner mitochondrial membrane. In 2008, Bender et al. (2008) published an article about a *natural* rechargeable antioxidant located in the same membrane. It is quite amazing that mitochondria chose to change their genetic code so as to achieve this goal. The authors of the cited work showed that in many species of eukaryotes, proteins encoded by mitochondrial DNA contain many more methionine residues than their analogs from other eukaryotic species. It was also shown that for species poor in methionine, this amino acid is encoded by one mitochondrial codon (AUG), while in species rich in methionine it is encoded by both AUG and AUA (in the first case, AUA is one of three isoleucine codons). As a rule, codon AUA encodes methionine instead of isoleucine in species with high respiration rate (vertebrates, insects, crustaceans, and nematodes), while in species with low respiration rate (echinoderms, sponges, flatworms, and cnidarians), AUA encodes isoleucine in mitochondria just as in the case of proteins encoded by nuclear DNA. Fungi demonstrate remarkable diversity. Some representatives of the group *Saccharomycotina* (e.g. *Saccharomyces cerevisiae*) use AUA for encoding methionine, while others use it for encoding isoleucine (*Schizosaccharomyces pombe* also belongs to the latter group).

It seems quite significant that both isoleucines and methionines encoded by AUA are usually located on the surface of the proteins of the inner mitochondrial membrane; some of them face the membrane/water interface, while others are located in the membrane core, i.e., in its hydrophobic area. The following example is rather demonstrative in this respect. The number of surface methionines in mitochondrial cytochrome *b* of the bee *Melipona bicolor* is 10 times higher than that of the starfish *Florometra serratissima* (33 residues in the former case to only 3 in the latter). As a result, the whole surface of the insect cytochrome *b* molecule is covered with methionines (Fig. A6.1).

According to Bender et al. (2008), it is the antioxidant activity of methionine that is the key to understanding this phenomenon. Methionine is the only amino acid capable of interacting with practically all the natural ROS, even the relatively inert H_2O_2. In these interactions, reactive oxygen species (ROS) are neutralized

Fig. A6.1 Methionine residues (*red*) in cytochrome *b* from the mitochondria of starfish (*Florometra serratissima*) and bee (*Melipona bicolor*) (Bender et al. 2008)

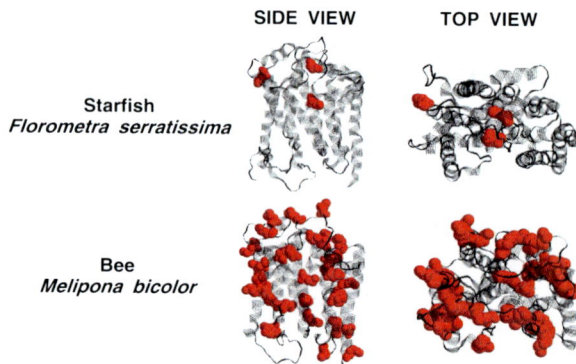

and methionine is transformed into the stable and harmless methionine sulfoxide (in contrast, another sulfur-containing amino acid, cysteine, when interacting with ROS, forms a very aggressive radical[1]). In addition, it seems to be very important that methionine sulfoxide can be reduced by electrons from the respiratory chain, this leading to its regeneration to the original methionine:

NADPH → thioredoxin reductase → thioredoxin → methionine sulfoxide reductase → methionine sulfoxide

Thus, methionine of the proteins encoded by mitochondrial DNA is an analog of SkQs, for it is also a rechargeable antioxidant located in the inner mitochondrial membrane (Fig. A6.2). It seems quite significant that additional methionines resulting from changes in the mitochondrial genetic code are located on the periphery of the protein molecule, for they meet ROS on their way toward the active sites of the key proteins of oxidative phosphorylation. This location of methionine residues originates from the fact that the same sites were formerly occupied by the residues of the hydrophobic amino acid isoleucine. The discussed proteins are integrated into the membrane as bricks into the wall, and surface hydrophobic amino acids strengthen their bonds with the intramembrane partners.

The antioxidant role of methionine residues has been recently confirmed by Luo and Levine in experiments on *E. coli* (Luo and Levine 2009). The bacterium was grown in the medium containing no methionine (it was replaced by norleucine). Because the protein biosynthesis system of *E. coli* can use norleucine instead of methionine, one could observe gradual accumulation of cells with methionine being replaced by norleucine in their proteins. When the level of replacement reached 40 %, the cells were washed from free norleucine and methionine was added. As a result, the amount of free methionine and S-adenosylmethionine in *E. coli* cells became normal, while the amount of methionine in proteins remained reduced. These cells were found to be much more sensitive to toxic effect of H_2O_2 than the control cells (Fig. A6.1).

---

[1] It is not surprising that there exists an inverse correlation between the organism lifespan and the number of cysteine residues in proteins encoded in mitochondria (Moosmann 2011).

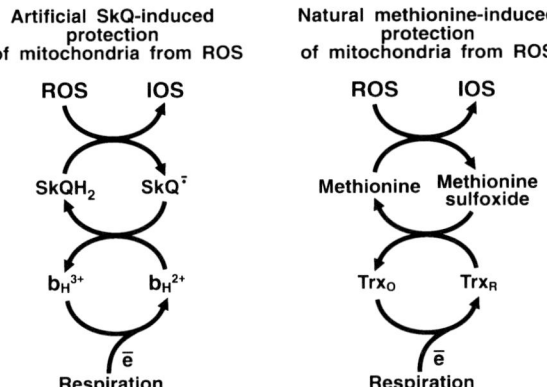

Fig. A6.2 SkQ and methionine as rechargeable antioxidants protecting mitochondria from ROS. ROS, reactive oxygen species; IOS, inactive oxygen species; SkQH$_2$, reduced form of SkQ; SkQ$^-$, anion-radical of SkQ; b_H^{2+}, reduced cytochrome b_H; b_H^{3+}, oxidized cytochrome b_H; methionine and methionine sulfoxide, respective amino acid residues in proteins encoded by mitochondrial DNA; Trx$_r$, reduced thioredoxin; and Trx$_o$, oxidized thioredoxin. Respiration regenerates b_H^{2+} and Trx$_r$ by reducing b_H^{3+} and Trx$_o$

Even though there are many similarities in antioxidant effects of methionines in proteins and SkQ1, there is also a significant difference connected to the possibility of using these two mechanisms for therapy against senescence. We cannot increase the number of methionine residues in mitochondrial proteins (thus enhancing the methionine-based protective system) without changing the mitochondrial genome. However, a physician can easily vary the impact of SkQ1 by changing the amount of this compound given a patient. And one more factor also seems to be important: methionine is known to be the most toxic amino acid; excess of it in one's diet leads to many pathological consequences for an organism (Gomez et al. 2009) (Fig. A6.2).

Appendix 7
Key Participants of the Project "Practical Application of Penetrating Cations"

Fig. A.7.1

(1) Prof. V. A. Sadovnichii, Chairman of the Supervisory Board, Leader of the Mathematics Group, Rector of Moscow State University, Moscow
(2) Prof. G. Blobel, Nobel Laureate, Member of the Supervisory Board, New York
(3) Prof. B. Cannon, President of Swedish Academy of Sciences, Member of the Supervisory Board, Stockholm
(4) Prof. A. Ciechanover, Nobel Laureate, Member of the Supervisory Board, Haifa
(5) Dr. M. V. Skulachev, Deputy Chairman of Scientific Board, Moscow
(6) Dr. B. V. Chernyak, Deputy Chairman of Scientific Board, Leader of Cell Biology Group, Moscow
(7) Prof. V. N. Anisimov, President of Russian Society of Gerontology, Leader of Mammalian Gerontology Group, St. Petersburg
(8) Dr. Yu. N. Antonenko, Leader of Model Membrane Group, Moscow
(9) Dr. L. E. Bakeeva, Leader of Electron Microscopy Group, Moscow
(10) Dr. D. A. Cherepanov, Leader of Molecular Dynamics Group, Moscow
(11) Dr. V.A. Chistyakov, Leader of Physiological Group, Rostov
(12) Dr. M. V. Egorov, Leader of Preclinical Study Group, Moscow
(13) Dr. B.A. Feniouk, Leader of ATP-synthase Group, Moscow
(14) Dr. N.K. Isaev, Leader of Brain Study Group, Moscow
(15) Prof. V.I. Kapel'ko, Leader of Cardiology Group, Moscow
(16) Prof. N. G. Kolosova, Leader of the Progeria Studies Group, Novosibirsk
(17) Dr. G.A. Korshunova, Leader of the Group of Chemical Synthesis, Moscow

Fig. A.7.1 Some key participants of the project "Practical application of penetrating cations" (V. P. Skulachev, Chairman of Scientific Board, Moscow)

Appendix 7: Key Participants of the Project

(18) Dr. A. Mulkidjanian, Leader of Biophysical Group, Osnabruck
(19) Dr. K. G. Lyamzaev, Leader of Respiratory Chain Group, Moscow
(20) Dr. V.N. Manskikh, Leader of Animal Pathologoanatomy Group, Moscow
(21) Prof. J. Nedergaard, Leader of Uncoupling Protein Group, Stockholm
(22) Dr. E.G. Pasyukova, Leader of Drosophila Study Group, Moscow
(23) Dr. P.P. Philippov, Leader of Ophthalmology Group, Moscow
(24) Dr. E.Y. Plotnikov, Leader of Nephrology Group, Moscow
(25) Dr. F. F. Severin, Leader of the Yeast Bioenergetics Group, Moscow
(26) Dr. I. I. Severina, Leader of the Planar Membrane Group, Moscow
(27) Dr. V.N. Tashlitsky, Leader of the Chemical Analysis Group, Moscow
(28) Dr. L. S. Yaguzhinsky, Leader of the Mitochondriology Group, Moscow
(29) Dr. A.A. Zamyatnin, Jr, Leader of Clinical Study Group, Moscow
(30) Dr. D. B. Zorov, Leader of Acute Phenoptosis Group, Moscow

References

Antonenko YN, Avetisyan AV, Bakeeva LE, Chernyak BV, Chertkov VA, Domnina LV, Ivanova OY, Izyumov DS, Khailova LS, Klishin SS, Korshunova GA, Lyamzaev KG, Muntyan MS, Nepryakhina OK, Pashkovskaya AA, Pletjushkina OY, Pustovidko AV, Roginsky VA, Rokitskaya TI, Ruuge EK, Saprunova VB, Severina II, Simonyan RA, Skulachev IV, Skulachev MV, Sumbatyan NV, Sviryaeva IV, Tashlitsky VN, Vassiliev JM, Vyssokikh MY, Yaguzhinsky LS, Zamyatnin AA Jr, Skulachev VP (2008) Mitochondria-targeted plastoquinone derivatives as tools to interrupt execution of the aging program. 1. Cationic plastoquinone derivatives: synthesis and in vitro studies. Biochemistry (Mosc) 73:1273–1287

Antonenko YN, Avetisyan AV, Cherepanov DA, Knorre DA, Korshunova GA, Markova OV, Ojovan SM, Perevoshchikova IV, Pustovidko AV, Rokitskaya TI, Severina II, Simonyan RA, Smirnova EA, Sobko AA, Sumbatyan NV, Severin FF, Skulachev VP (2011) Derivatives of rhodamine 19 as mild mitochondria-targeted cationic uncouplers. J Biol Chem 286:17831–17840

Bender A, Hajieva P, Moosmann B (2008) Adaptive antioxidant methionine accumulation in respiratory chain complexes explains the use of a deviant genetic code in mitochondria. Proc Natl Acad Sci USA 105:16496–16501

Filipovska A, Kelso GF, Brown SE, Beer SM, Smith RA, Murphy MP (2005) Synthesis and characterization of a triphenylphosphonium-conjugated peroxidase mimetic. Insights into the interaction of ebselen with mitochondria. J Biol Chem 280:24113–24126

Gomez J, Caro P, Sanchez I, Naudi A, Jove M, Portero-Otin M, Lopez-Torres M, Pamplona R, Barja G (2009) Effect of methionine dietary supplementation on mitochondrial oxygen radical generation and oxidative DNA damage in rat liver and heart. J Bioenerg Biomembr 41:309–321

Hardy M, Chalier F, Ouari O, Finet JP, Rockenbauer A, Kalyanaraman B, Tordo P (2007) Mito-DEPMPO synthesized from a novel NH_2-reactive DEPMPO spin trap: a new and improved trap for the detection of superoxide. Chem Commun (Camb) 10:1083–1085

Kelso GF, Porteous CM, Coulter CV, Hughes G, Porteous WK, Ledgerwood EC, Smith RA, Murphy MP (2001) Selective targeting of a redox-active ubiquinone to mitochondria within cells: antioxidant and antiapoptotic properties. J Biol Chem 276:4588–4596

Luo S, Levine RL (2009) Methionine in proteins defends against oxidative stress. FASEB J 23:464–472

Lyamzaev KG, Pustovidko AV, Simonyan RA, Rokitskaya TI, Domnina LV, Ivanova OY, Severina II, Sumbatyan NV, Korshunova GA, Tashlitsky VN, Roginsky VA, Antonenko YN, Skulachev MV, Chernyak BV, Skulachev VP (2011) Novel mitochondria-targeted antioxidants: plastoquinone conjugated with cationic plant alkaloids berberine and palmatine. Pharm Res 28:2883–2895

Moosmann B (2011) Respiratory chain cysteine and methionine usage indicate a causal role for thiyl radicals in aging. Exp Gerontol 46:164–169

Murphy MP, Echtay KS, Blaikie FH, Asin-Cayuela J, Cocheme HM, Green K, Buckingham JA, Taylor ER, Hurrell F, Hughes G, Miwa S, Cooper CE, Svistunenko DA, Smith RA, Brand MD (2003) Superoxide activates uncoupling proteins by generating carbon-centered radicals and initiating lipid peroxidation: studies using a mitochondria-targeted spin trap derived from alpha-phenyl-N-tert-butylnitrone. J Biol Chem 278:48534–48545

Quin C, Trnka J, Hay A, Murphy MP, Hartley RC (2009) Synthesis of a mitochondria-targeted spin trap using a novel Parham-type cyclization. Tetrahedron 65:8154–8160

Robinson KM, Janes MS, Pehar M, Monette JS, Ross MF, Hagen TM, Murphy MP, Beckman JS (2006) Selective fluorescent imaging of superoxide in vivo using ethidium-based probes. Proc Natl Acad Sci USA 103:15038–15043

Salvemini D, Wang ZQ, Zweier JL, Samouilov A, Macarthur H, Misko TP, Currie MG, Cuzzocrea S, Sikorski JA, Riley DP (1999) A nonpeptidyl mimic of superoxide dismutase with therapeutic activity in rats. Science 286:304–306

Trnka J, Blaikie FH, Smith RA, Murphy MP (2008) A mitochondria-targeted nitroxide is reduced to its hydroxylamine by ubiquinol in mitochondria. Free Radic Biol Med 44:1406–1419

Xu Y, Kalyanaraman B (2007) Synthesis and ESR studies of a novel cyclic nitrone spin trap attached to a phosphonium group—a suitable trap for mitochondria-generated ROS? Free Radic Res 41:1–7

Zhao H, Kalivendi S, Zhang H, Joseph J, Nithipatikom K, Vásquez-Vivar J, Kalyanaraman B (2003) Superoxide reacts with hydroethidine but forms a fluorescent product that is distinctly different from ethidium: potential implications in intracellular fluorescence detection of superoxide. Free Radic Biol Med 34:1359–1368

Author Index

A
Abdulaev, N., 141
Adler, J., 196, 197
Agre, P., 5
Aleem, M., 130
Anisimov, V., 338, 356, 358, 360, 361, 375
Antonenko, Yu., 346, 356–358, 363, 366
Archakov, A., 264
Asadov, A., 43
Austad, S., 326, 330, 334–336, 343, 372

B
Bakeeva, L., 256, 264, 356, 364, 365
Baltscheffsky, M., 184, 185
Barja, G., 340
Barquera, B., 279
Bate Smith, E., 84
Baykov, A., 185, 285
Belitser, V., 3, 174
Belogurov, G., 285
Berg, H., 197, 200, 201, 311, 374
Bibikov, S., 152, 153, 251
Blobel, G., 417
Blumenfeld, L., 16
Bogomolni, R., 145, 151
Bonduryansky, R., 327
Boyer, P., 5, 167, 169, 172
Brand, M., 340
Brassil, C., 327
Bremer, B., 215, 228
Brenner, S., 318
Brodie, A., 246, 287
Brown, I., 267, 269
Bubenzer, H., 256
Budrene, E., 311, 313
Burton, M., 255

C
Cannon, B., 236, 358, 364
Carr, C., 373
Chance, B., 103, 108
Chen, L., 249, 258, 315
Chentsov, Y., 256
Cherepanov, D., 417
Chernyak, B., 317, 324, 358
Chikunov, A., 355
Chubays, A., 355
Ciechanover, A., 417
Cleveland, L., 204
Cohen, G., 218
Comfort, A., 326, 328, 342, 372
Crick, F., 4

D
Dalton, H., 128
van Dam, K., 4
Danielson, L., 187
Darwin, C., 327, 334, 376
Dawkins, R., 329, 376
Deisenhofer, J., 5, 37–39, 41, 42, 45
Deripaska, O., 355
Dibrov, P., 289
Dilman, V., 338, 342
Dimroth, P., 275, 279, 291, 292
Donehower, L., 343
Drachev, L., 44, 45
Drachev, V., 260
Drize, N., 358

E
Egorov, M., 417
Eisenbacher, M., 197

E (*cont.*)
Emanuel, N., 339
Engelhardt, V., 3, 87, 245
Epstein, W., 209, 301
Erichev, V., 364
Ernster, L., 4, 187

F
Feldman, R., 169
Feniouk, B., 417
Fernandes, J., 374
Filenko, O., 358
Finch, C., 338, 342
Foissner, I., 259
Fritz, I., 9, 214, 215

G
Galen, C., 318
Gardner, E., 374
Garrahan, P., 291
Gauthier, G., 256
Glynn, I., 291
Goglia, F., 235, 239
Goldsmith, T., 334
Goldstein, N., 326
Green, D., 89, 307, 345
Greenhut, S., 265
Grinius, L., 228
Gulevitch, V., 82, 214

H
Haldane, J. B. S., 3
Hamilton, W., 376
Harman, D., 339, 372
Harold, F., 268, 282
Heefner, D., 282
Hekimi, S., 337, 338
van Helmont, J. B., 3
Henderson, R., 39, 141, 143
Holzenberger, M., 338
Horwitz, H., 232, 236, 318
Huber, R., 5, 37, 38

I
Iomdina, E., 364

J
Jasaitis, A., 139, 140
Johansson, B., 232

K
Kapel'ko, V., 364
Kaulen, A., 45, 145, 146
Kauppinen, R., 249
Keilin, D., 13
Kenyon, C., 337
Kerr, J., 318
Khorana, H., 141
Klingenberg, M., 221, 233, 234
Kolosova, N., 358, 361, 362
Kopenkin, Ye., 363
Kopnin, B., 365
Korshunova, G., 312, 314, 356, 366
Krasnovsky, A., 32
Krebs, H., 4, 73
Krimberg, R., 214
Kroemer, G., 5, 320

L
Labedan, B., 228
Lahti, R., 285
Laimins, L., 209, 301
Lamarck, J.-B., 240
Larsson, N., 334
Lewis, K., 152, 196, 200, 322, 329
Liberman, E., 187, 345
Lichtenstein, A., 322
Lipmann, F., 9
Loison, A., 327
Longo, V., 305, 306, 315, 338
Lozier, R., 145, 146, 378
Lyamzaev, K., 317, 356, 376, 378

M
MacKinnon, R., 5
Macnab, R., 288
Malinen, A., 285
Manskikh, V., 322
Maslov, S., 87, 237, 238, 246
McCay, C., 372, 373, 375
Medawar, P., 327
Meisel, M., 260
Michel, H., 5, 37, 112
Mikhelson, V., 361
Mitchell, P., 4, 5, 10, 48, 51, 98, 106, 215, 219, 265
Mokhova, E., 238
Moore, J., 255
Moosmann, B.
Mozenok, E., 203
Mulkidjanyan, A., 293, 300
Murphy, M., 345, 346

Author Index

N
Nedergaard, J., 236
Newmeyer, D., 5, 319

O
Obukhova, L., 364, 369, 375
Oesterhelt, D., 139, 140, 144
Okunuki, K., 245
Ovchinnikov, Y., 141, 142, 153

P
Pande, S., 216
Papa, S., 98
Park, T., 341
Parvin, R., 216
Pasyukova, Ye., 358
Pelicci, P., 337
Pfenig, N., 292
Pfenninger-Li, X., 279
Ponnamperuma, C., 16, 17
Postgate, J., 128
Prolla, T., 339
Proverbio, F., 284

R
Racker, E., 161, 162
Ramsay, R., 216
Reznik, D., 335, 336
Rich, P., 36, 49, 51, 52, 107, 277, 278
Rickenberg, H., 218
Robertson, T., 372
Roginsky, V., 355, 356
Roseman, M., 265
Rosen, B., 300
Rothwell, N., 236
Rottenberg, H., 221
Ruuge, E., 417
Ryazanov, A., 364

S
Sadovnichii, V., 417
Sagan, C., 16
Sazanov, L., 93, 94, 96, 98, 99, 101, 102, 190
Schink, B., 133, 186, 292, 297
Scrable, H., 343
Semenov, A., 45
Senin, I., 364
Severin, F., 305, 323, 340, 356, 366
Severin, S., 82–84, 345

Severina, I., 253, 254, 357, 366
Shidlovsky, K., 358
Shuvalov, V., 43
Sigman, D., 169
Skou, J., 5, 283
Skulachev, M., 356, 366, 376
Slater, E., 4, 102, 104
Smith, R., 232, 235, 255, 345, 346
Sohal, R., 340
Sommer, S., 322
Sotnikova, L., 363
Spirin, A., 18
Spivak, I., 361
Starkov, A., 312
Stock, M., 163–166, 172, 174, 236
Stoeckenius, W., 139, 140, 144, 145
Stuchlikova, E., 373, 374
Sulston, J., 318
Sumper, M., 263
Szent-Györgyi, A., 4, 5, 17

T
Tanaka, K., 260
Teather, R., 219
Temkin, M., 16
Tikhonova, G., 130
Tokuda, H., 277, 279
Trauble, H., 263
Trissl, H., 47
Tubbs, P., 216

U
Unemoto, T., 277, 279, 287
Unwin, N., 141, 143

V
Verkhovsky, M., 115, 116, 279, 280
Vinci Leonardo da, 3
Vinogradov, A., 27, 96, 168
Vysokikh, M., 346, 357, 358, 366

W
Walker, J., 5, 160, 163
Wallace, R., 327, 328, 329, 376
Wang, X., 5, 319
Waterbury, J., 203
Watson, J., 4
Weindruch, R., 374, 375
Weismann, A., 327–329, 331, 332, 376
West, I., 219, 266

W (*cont.*)
Wikström, M., 110, 115, 116
Woldegiorgis, G., 224

Y
Yaguzhinsky, L., 345
Yoshida, M., 172, 173
Yoshikawa, S., 112

Z
Zamyatnin, A., 419
Zassenhaus, H., 339
Zinovkin, R., 362, 363
Zorov, D., 258, 260, 262, 315

Subject Index

A

Acetyl-CoA, 74, 75, 214, 276
Acetyl phosphate, 77
Acidophilic bacteria, 34
Aconitase, 312, 313, 315, 316, 323
Adenylate cyclase, 233
Adenosine diphosphate (ADP), 168–170, 172, 174, 175
    inhibition of F_oF_1-ATPase, 180
Adenosine monophosphate (AMP), 15, 16, 183, 184
Adenosine triphosphatase (ATPase)
    Ca^{2+}, 27, 180, 301
    Cl^-, 301, 302
    classification, 176
        type E_1E_2, 180
        type F_oF_1, 177
        type V_oV_1, 179
    H^+, 21–23, 152, 176–185, 212, 221, 251, 268, 269, 288, 297, 301
        in anaerobic bacteria, 177, 182
        in plasmalemma, 183
    K^+, 5, 182, 209, 301
    Na^+, 5, 282, 283, 288, 293
    Na^+/K^+, 27, 154, 180, 283, 284, 301
Adenosine triphosphate (ATP), 14, 15
    as convertible energetic currency, 11, 15, 22, 185, 252, 265, 301
Adenosine triphosphate synthase (ATP-synthase)
    in chloroplasts, 159, 161, 183, 268
    mechanism, 172, 198, 221, 226, 268
    mitochondrial, 160, 161, 163, 166
    Na^+, 27, 293
    stoichiometry H^+/ATP, 165, 174–176, 180, 186
    3D structure, 162–164

    subunit composition, 180, 198
Adrenal gland, 329
Aging, 305, 326, 333, 339, 342, 344, 355
Alkaline phosphatase, 282
Alkaliphilic bacteria, 297
Alternative oxidase (AOX), 121, 122, 337
Amino acid transport, 218
Antenna light-harvesting, 22, 34, 252
Antimycin, 108, 246, 357
Antiporter
    ATP/ADP, 221
    carnitine/acyl carnitine, 216–218
    dicarboxylates, 213, 248, 249
    glutamate/aspartate, 233, 235, 248
    HPO_4^{2-}/malate^{2-}, 213
    K^+/H^+, 218, 249, 250
    Na^+/H^+, 218, 266–268, 280, 288, 289, 300
    sucrose/H^+, 212
Apoptosis, 318, 320, 322
Arsenate, 14, 228
Ascorbate, 60, 357
Atractylate, 221, 224, 238
Attractant, 151, 152, 196, 251, 311
Autophagy, 315
Axial ligand, 38, 52, 63, 64, 104, 109, 111

B

Bacteriochlorophyll, 32, 34–37, 39, 41, 42, 44, 48, 54
    dimer, 34, 35, 38, 42, 44, 53
Bacteriopheophytin, 34, 35, 37, 42, 44
Bacteriorhodopsin
    3D structure, 142
    conformational changes, 143, 148, 149
    dark adaptation, 145
    electrogenic stages, 146

B (*cont.*)
    generation of $\Delta\bar{\mu}H^+$, 139, 141
    in attractant response, 152
    photocycle, 143–145, 147–150, 153
    quantum yield, 23
    schiff base, 140–144, 147–150
    two-dimensional crystals, 139, 143
Bilayer membrane, 32, 88, 222, 227, 299, 357, 377
Binuclear center, 110, 113, 114, 116, 121, 123, 125, 126
Bioenergetics
    definition, 3
    evolutionary aspects, 13–15
    relation to other biological sciences, 5–7, 18
Biotin, 173, 275
Bongkrekic acid, 221, 224
British anti-lewisite, 108
Brown fat, 232–234, 236–239, 246
Buffers of $\Delta\bar{\mu}$ H^+, 11, 265

C
Calcium ion
    accumulation in mitochondria, 208, 211
    activation of phospholipase A_2, 250
    as secondary messenger, 250
Capsaicin, 96, 97, 341
Carboxyatractyloside, 221, 224, 238
Cardiolipin, 362, 365, 366, 369
Carnitine, 214–217
Carotenoids, 33, 43, 140
Caspase, 319
Catalase, 248, 341
Chemiosmotic theory, 5
Chloride transport, 67
m-Chlorocarbonyl cyanide phenylhydrazone (CCCP), 290
Chlorophyll, 22–25, 31, 32, 54, 56–58, 62, 63, 87, 151, 252, 253
Chlorophyll dimer, 56, 61, 63
Chloroplast, 55, 58, 62, 66, 132, 253
    buffering of $\Delta\bar{\mu}$ H^+, 67
    outer membrane, 204
    photoredox chain, 52, 53, 55, 57, 67, 151
    rotation, 203, 204
    structure, 56
Cholesterol, 204, 264
Chromaffin granules, 212
Chromatophore, 23, 32, 35, 45, 47, 51–53, 106, 169, 184–187, 190, 268, 269

Citrate, 186, 213, 275, 276, 287, 313
Cobalamine (vitamin B_{12}), 280, 281
Coenzyme A (CoA), 17, 18, 84
Coenzyme B (CoB), 135
Coenzyme F_{420}, 96, 135–137, 401
Coenzyme M (CoM), 134
Coenzyme Q (CoQ)
    in reaction center complex, 42, 43, 45, 48, 49, 51
    in respiratory chain, 90, 92, 102, 104, 123, 130, 190
Complex b_{c1}. *See* Complex III
Complex $b_6 f$, 56–58, 60, 63, 64, 66, 105, 108, 131
Complex I, 92–97, 308
    3D structure, 99, 101, 102
    bacterial, 93–95
    electron transport, 99
    generation of $\Delta\bar{\mu}$ H^+, 97
    inhibitors, 96, 97, 341
    mitochondrial, 99, 102
    redox centers, 90, 94
    stoichiometry $H^+/\bar{e}$, 98, 99
    subunit composition, 93–96
Complex II. *See* Succinate dehydrogenase
Complex III ($CoQH_2$:cytochrome *c*- oxidoreductase)
    3D structure, 103, 104
    in bacterial photoredox chain, 106
    in respiratory chain, 104
    inhibitors, 108, 312
    redox centers, 104
    subunit composition, 104, 105
Complex IV. *See* Cytochrome *c* oxidase
    Copper, 24, 56–59, 85, 90, 98, 110–115, 121, 123, 125, 126
Coupling ion, 27, 119, 285, 297, 300
Creatine phosphate, 14, 108, 217
Cyanide, 121, 125, 131, 132, 152, 240, 245–247
Cyanobacteria, 55, 131
Cyclic adenosine monophosphate (cAMP), 233, 251
Cyclic guanosine monophosphate (cGMP), 154
Cyclosis, 203, 204
Cytochrome
    a, 109
    b, 35, 49, 51, 56, 63, 64, 90, 103–106, 136, 323, 339
    b_5, 246, 264, 265
    b_{559}, 61, 62
    c, 109

Subject Index 429

c_1, 49, 51, 52, 90, 102–106, 108, 109
d, 109
definition, 24
f, 56, 59, 63–65
o, 52
P-450, 246, 247
tetraheme c-type, 37–41, 44, 47, 49, 51
Cytochrome c oxidase, 88, 106, 108, 110, 113, 115, 120, 123, 125, 128, 131, 337
   3D structure, 112
   aa_3-type, 123, 125
   cbb_3-type, 123, 128
   generation of $\Delta\bar{\mu}H^+$, 115
   inhibitors, 116
   redox centers, 110
   stoichiometry $H^+/\bar{e}$, 115, 116
   subunit composition, 110–112

D

Decarboxylase
   glutaconyl-CoA, 276
   methylmalonyl-CoA, 276, 277, 292
   Na^+-motive, 275
   oxaloacetate, 275, 276
Deoxyribonucleic acid (DNA), 18
Devescovinid, 204, 205
N,N-Dicyclohexylcarbodiimide (DCCD), 152, 167, 171, 177, 180, 181, 202, 203
Diethylstilbestrol, 167, 180, 181
2,4-p-Dinitrophenol, 238, 259
Diuron, 60, 259

E

Electron magnetic resonance (EPR), 94, 95
Endoplasmic reticulum, 6, 27, 178, 183, 246, 264, 301, 308
Endosome, 178, 179, 183
N-Ethylmaleimide, 180, 282

F

Factor F_o
   inhibition by DCCD, 171
   sensitivity to oligomycin, 171
   structure, 161
   subunit composition, 160, 161, 165, 171, 198
Factor F_1
   3D structure, 162, 164
   ATP hydrolysis, 167–169
   catalytic centers, 167, 168, 177
   inhibition by ADP, 169
   nucleotides binding, 167, 168, 172, 179
   subunit composition, 177
   synthesis of bound ATP, 169
   unisite catalysis, 167
Factor F_6, 161
Fatty acid
   as uncouplers in thermogenesis, 234
   covalent binding to proteins, 38, 234
   dicarboxylic, 201, 233
   β-oxidation, 91, 216, 217, 233, 235
   transport, 214, 217, 263, 345
Fatty acylcarnitine, 214–216
Fatty acyl-CoA, 91, 215, 233, 263
Fenton's reaction, 325
Fermentation, 3, 20, 79–82, 132–134, 182, 276
Ferredoxin, 53, 57, 58–60, 66, 96, 278
Ferredoxin-$NADP^+$ reductase (FNR), 60, 66
Ferricyanide, 95
Flagellin, 197, 198
Flagellum, 195–201, 204, 205, 290, 291
   bacterial, 195–197, 205
   eukaryotic, 195, 205
Flavin adenine dinucleotide (FAD), 18, 54, 58, 120, 125, 246, 278–280
Flavin mononucleotide (FMN), 90, 94, 95, 99, 101, 278–280
Formate, 81, 119, 132–134, 276, 281, 293
Formate dehydrogenase, 81, 281, 293
Free respiration, 231, 232, 240, 245–248
Fumarate, 75, 119, 126–128, 312
Fumarate reductase, 126, 127

G

Generator of $\Delta\bar{\mu}H^+\bar{e}$
   primary, 287
   secondary, 176–179, 181, 282, 283
   stoichiometry $H^+/\bar{e}$, 98, 115, 116, 124, 125
Glaucoma, 360, 364, 371, 378
Glucose, 25, 26, 71, 73–75, 80, 81, 119, 127, 133, 184, 212, 302
Glutathione, 309, 341, 366
Glyceraldehyde phosphate dehydrogenase, 76
Glycerol, 38, 40, 71, 233, 248, 250
Glycolysis, 19, 21
Gramicidin, 218
Green photosynthetic bacteria, 22, 53
Guanosine diphosphate (GDP), 77, 154, 233
Guanosine triphosphate (GTP), 72, 77, 154, 221, 222, 233

H
Halophilic bacteria, 288, 297, 300
Halorhodopsin, 149, 150, 152, 153, 299, 301
Heme
  bH, 35, 36, 49, 64, 106, 312, 377
  bL, 49–51, 64, 106, 312
  chlorin, 125
  high-spin, 111, 116, 125
  low-spin, 111, 116, 125
Hemoglobin, 111
2-Heptyl-4-hydroxyquinoline N-oxide (HQNO), 66, 280
Hexokinase, 73, 250
Hibernation, 236
Hydrogenase, 96, 136, 137, 276
Hydroxylamine, 167, 180, 181
Hypoxia, 324–326

K
α-Ketoglutarate, 11, 75, 77, 78, 122, 222, 249
Krebs cycle, 73–75, 77, 91, 127, 128, 186, 190, 213, 250, 312, 323

L
β-Lactamase, 227
Lactate, 20, 80, 81, 83, 84, 119, 132, 212, 246, 276
Leader peptidase, 225
Lipase, 233
Lipoic acid, 77
Lipophilic ions, 136
Lipoprotein, 198
Liposome, 149, 219, 357
Lysophospholipids, 250
Lysosome, 6, 178, 179, 183, 317

M
Macroergic bond, 14
Malate, 119, 127, 128, 213, 240, 248, 249
Malate–aspartate–glutamate shuttle, 248
Malate dehydrogenase, 249
Melatonin, 342
Melibiose, 287
Melittin, 227
Membrane, 101, 149, 161, 180, 188, 227, 252, 300
Menadione (Vitamin K_3), 247
Menaquinone (MQ), 40
Methane, 133–135, 280, 281, 293, 294

Methanofuran, 281, 293
Methanogenesis, 134, 135
Methanophenazine, 136, 137
Methionine, 373
Methylamine, 134, 135
Methylmalonyl-CoA, 276, 277, 292
Mild uncoupling, 312, 367
Mitochondrion
  biogenesis, 226
  Ca^{2+} transport, 27
  filamentous, 254, 256, 258, 260
  giant, 255
  inner membrane, 211, 216, 232, 246, 251
  junction, 256, 258, 260, 261
  lateral transport, 261, 264
  motility, 224
  origin, 224, 226
  outer membrane, 256, 260
  shape, 254, 258
  staining by rhodamine, 161, 249, 250, 255, 259
Mitoptosis, 315, 317, 322
Molybdenum, 124, 282
Monensin, 290
Motility, 195, 196, 202, 203, 205, 224, 254, 267, 289–291
  of bacteria, 197, 203
  of protozoa, 203, 205
  of spermatozoa, 195
Mucidin, 108
Myxothiazol, 108, 240, 312

N
Na^+/K^+-gradient as buffer for $\Delta\bar\mu H^+$, 53, 265
Na^+-cycle, 297, 300
Na^+-motive force(s), 10, 11
Na^+-motor, 290, 291
NADH dehydrogenase, 96, 120, 278
Necrosis, 320–323, 325, 328, 358, 360
Nicotinamide adenine dinucleotide (NAD(H)), 73
Nicotinamide adenine dinucleotide phosphate (NADP(H)), 18
Nigericin, 218, 221, 249
Nitrate
  as inhibitor of V_0V_1-ATPase, 180
  as terminal electron acceptor, 124, 127, 135
  reductase, 124, 126
Nitrite, 92, 124, 131
p-Nitrophenol, 146

Subject Index 431

Nonheme FeS proteins, 24, 35
  in complex I, 94–96
  in complex III, 104–106
  in noncyclic photoredox chain, 56
  in photosystem I, 58, 59
Nonheme iron
  in alternative oxidase, 121
  in photosystem II
  in reaction center complex
Noradrenaline, 233, 236, 238
Nuclear magnetic resonance
            (NMR), 164, 165
Nucleoside diphosphate kinase, 222
Nucleus, 6, 33, 93, 104, 161, 255, 315, 345
Nystatin, 204

O

Obesity, 236
Oligomycin, 160, 167, 170, 171, 180,
            185, 238
Organoptosis, 324, 328
Osmoregulation, 176, 181
Osmotic work, 11, 66, 182, 207, 213, 218, 287,
            297, 300, 302
Osteoporosis, 330, 360, 369, 374, 378
Ouabain, 284, 301
Oxaloacetate, 249, 275, 276

P

Palmitoyl carnitine, 216
Peptidoglycan, 200
Peroxisome, 6, 248, 308, 316
o-Phenanthroline, 47, 62
Phenazine methosulfate, 60
Phenoptosis, 322–324, 326, 328–330, 333,
            376, 377
Pheophytin, 55, 61
Phoborhodopsin, 151, 152
Phosphatidylinositol, 250
Phosphodiesterase, 154
Phosphoenolpyruvate, 74, 77, 78, 207, 302
3-Phosphoglyceraldehyde, 78
2-Phosphoglycerate, 77, 78
Phospholipase, 250
Photophosphorylation, 61
Photoreceptor, 151, 153, 154, 364
Photoredox chain
  cyclic, 52, 57, 67, 151
      in chloroplasts and cyanobacteria, 59,
            299
      in purple bacteria, 24, 34, 56
  noncyclic, 53–55, 57, 67, 151, 190

  in chloroplasts and cyanobacteria, 59,
            151
  in green sulfur bacteria, 54, 55
Photosystem I
  3D structure, 58
  electron transport, 60
  generation of $\Delta\bar{\mu}$, 60, 61
  localization, 67, 68
  subunit composition, 58, 59, 61
Photosystem II
  3D structure, 61, 62
  electron transport, 62, 63
  generation of $\Delta\bar{\mu}$, 63
  localization, 67, 68
  subunit composition, 61, 62
Piericidin, 96
Plasmalemma, 6, 27, 179–183, 207, 247,
            302, 309
Plastocyanin, 51, 54, 56–60, 65, 66,
            108, 131
Plastoquinone (PQ), 24, 55, 57, 60, 108, 132,
            309, 355, 366, 377
Potassium-transporting system, 209
Progeria, 329, 330, 362, 364
Propionate, 64, 65, 104, 132, 276, 292
Propionyl-CoA, 276, 292
Protein
  binding to membrane, 38
  inhibitor F_1, 161
  kinase, 233, 252, 320, 338
  p53, 343, 365, 368
  p66Shc, 337, 338, 341
  Rieske, 51, 52, 90, 102–106, 108
  transport through membrane, 21, 227
Protein phosphatase, 252
Proteoliposome, 45, 47, 48, 60, 106, 107, 146,
            181, 184, 186, 187, 219, 265
Protometer, 152, 153, 251
Proton motive force (Δp), 10, 88
Proton pump, 22, 97–99, 102, 116, 123, 125,
            126, 129, 139, 145, 184, 189
Protonophore, 186, 203, 235, 277, 282, 290,
            297, 366
Purple membrane, 139–141, 143,
            146, 253
Purple photosynthetic bacteria, 52, 53
Pyrophosphate inorganic (PPi), 15, 159, 183,
            184, 186, 268, 269
Pyrophosphate synthase, 52, 159, 183–186
Pyruvate, 73, 74, 77–81, 122, 127, 233,
            275, 276
Pyruvate dehydrogenase
            complex, 81
Pyruvate formate lyase, 81, 276

Q

Q-cycle, 51, 55, 56, 63, 66, 90, 96, 106, 299, 312
Quinol oxidase
  ba_3-type, 123
  bd-type, 125, 126, 128, 129, 131
  bo-type, 125, 126
Quinone, 277, 278

R

Ratio P/O, 174, 237, 238, 240, 246
Reaction center complex (bacterial)
  3D structure, 37
  electron transport, 60, 61
  generation of $\Delta \bar{\mu} H^+$, 33, 101, 253
  redox centers, 61, 63, 94, 110
  subunit composition, 36–37
Reactive oxygen species (ROS), vi, 101, 305, 377
Redox loop (Mitchell loop), 298, 299
Regulation of pH in cytoplasm, 288
Repellent, 151, 152, 196, 251
Respiratory chain, 87
Respiratory control, 122, 238, 312
Reticulum mitochondriale, 254, 259
Retinal, 149
Reverse electron transport, 11, 67, 130, 191, 312, 340, 341
Rhodamine, 249, 259
Rhodoquinone (RQ), 127
Rhodopsin
  cGMP cascade, 154
  generation of $\Delta\Psi$, 45, 48, 49, 297, 301
  photocycle, 143–145, 147–150, 153
Riboflavin (Vitamin B_2), 278, 279
Ribonucleic acid (RNA), 18, 183, 306, 309
  messenger, 160, 233
  ribosomal, 18
Ribozyme, 18
Rotenone, 96, 97, 240, 246, 259, 312

S

Sarcopenia, 330, 333, 370, 374, 375, 378
Secretory vesicles, 178, 179
Sensory rhodopsin, 151, 152
SkQ, 355, 357, 358, 362, 364–368, 372, 374, 376, 378

Skulachev ion, 345
Special pair. See Bacteriochlorophyll dimer
Spermatozoid, 195, 255
Steroid hormones, 246
Stoichiometry $H^+/\bar{e}$, 98, 115, 116, 124–126, 136, 137
Stoichiometry H^+/ATP, 180, 186
Submitochondrial particles, 187
Succinate, 75, 91, 119, 126–128, 132, 172, 174, 209, 211, 276, 292, 312
Succinate dehydrogenase, 91, 126, 312
Succinyl-CoA, 392
Sulfite, 169
Symporter
  Cl^-, H^+, 150
  H^+, metabolite, 176, 288
  $H_2PO_4^-, H^+$, 175, 213
  K^+, H^+, 209, 268
  lactate, H^+, 212
  lactose, H^+, 212, 218–220, 223
  Na^+, metabolite, 287, 288, 300

T

Taxis, 152, 196, 197, 251, 265
Tetrahydromethanopterin (MPT), 134
Tetraphenylborate, 345
Tetraphenylphosphonium, 345, 356, 358, 360
Thermogenesis, 122, 231, 232, 234, 236, 237, 240, 241, 246, 247
Thermophilic bacteria, 99, 285
Thiamine pyrophosphate, 18, 77
Thiosulfate, 131
Thylakoid, 24, 51, 55, 58, 60, 61, 65, 67, 68, 87, 106, 107, 131, 169, 210, 252, 253, 268, 269
Thymus involution, 360, 369, 375, 378
Thyroid hormones, 248, 324
Tonoplast, 178–180, 184, 207, 289
Transducin, 154
Transhydrogenase, 11, 52, 187–190, 299
Transmembrane difference in electric potential ($\Delta\Psi$)
  conversion into Δ pH, 249, 250, 268
  definition, 9
  in thylakoid, 67, 250, 268
  in transmembrane protein transport, 226
Transmembrane difference in H^+ concentration (ΔpH), 9

Subject Index

Transmembrane electrochemical potential difference in H^+ ions ($\Delta\bar{\mu} H^+$)
  as convertible energetic currency, 265
  buffering, 265
  definition, 207
  transmission, 252, 253
Transmembrane electrochemical potential difference in Na^+ ions ($\Delta\bar{\mu} Na^+$), 12, 135, 290
Tungsten, 282

U
Ubiquinone. *See* Coenzyme Q
UCP, 232–236, 238, 239
Uncoupler, 131, 172, 196, 216, 219, 360, 366
Uncoupling
  in brown fat, 232, 236, 238
  in muscles, 237–239
  in plants, 239
Uniport of K^+
  in bacteria, 213
  in chloroplasts, 269
  in mitochondria, 269
Uveitis, 364

V
Valinomycin, 177, 181, 208, 218, 221, 287
Vanadate, 167, 180, 181, 282, 301
Vitamin A, 22, 309
Vitamin B_1, 18
Vitamin B_{12}, 18
Vitamin E, 344, 345
Vitamin K_1, 53, 56–58, 98
Vitamin K_3, 96, 108, 247

W
Water-oxidizing complex (WOC), 57, 62

X
Xenobiotics, 126, 246, 247

Z
Zinc, 34, 108, 110

Index of Organisms

A
Acetobacterium woodii, 293
Acidaminococcus fermentans, 276
Acidithiobacillus (Thiobacillus) ferrooxidans, 129
Aplysia californica, 301
Arabidopsis thaliana, 330, 331
Arum sp., 240
Ascaris sp., 127, 128
Azotobacter vinelandii, 126, 128

B
Bacillus firmus, 291
Bacillus pseudofirmus FTU, 291
Bacillus sp. YN-1, 291
Bacillus subtilis, 196, 228
Bacillus thuringiensis, 301
Bdellovibrio bacteriovorus, 196
Blastochloris (Rhodopseudomonas) viridis, 37–44, 46–51

C
Caenorhabditis elegans, 337, 338
Candida albicans, 260
Ceriodaphnia affinis, 358, 368
Chara sp., 203
Chlamydomonas reinhardtii, 64
Chloroflexus sp., 55
Clostridium symbiosum, 276

E
Endomyces magnusii, 260
Enterococcus hirae, 159, 180, 212, 282, 297

Escherichia coli, 94, 124, 126, 218, 287, 297, 311

F
Florometra serratissima, 413, 414
Fusobacterium nucleatum, 276

H
Haemophilus influenzae, 228, 278
Halobacterium halobium, 139, 287
Halobacterium salinarium, 139, 141, 149, 151, 152, 251, 253, 267, 301
Helicobacter pylori, 96
Heterocephalus glaber, 340
Homo sapiens, 376, 377

I
Ilyobacter tartaricus, 164, 165, 176, 293

K
Klebsiella aerogenes, 275, 276
Klebsiella pneumoniae, 276, 278, 287

L
Lactococcus casei, 159
Limonium vulgaris, 302

M
Malonomonas rubra, 277
Melipona bicolor, 413, 414
Methanobacterium thermoautotrophicum, 209

M (*cont.*)
Methanosarcina barkeri, 136
Methanosarcina mazei, 285
Mixotricha paradoxa, 204, 205
Moorella thermoacetica, 285
Mycobacterium phlei, 246, 287

N
Natronomonas pharaonis, 150
Neisseria gonorrhoeae, 278
Neisseria meningitidis, 278
Neurospora crassa, 94
Nitella sp., 203
Nitrobacter sp., 131
Nothobranchius furzeri, 335, 358, 368
Nothobranchius guentheri, 335
Nothobranchius kuhntae, 335
Nothobranchius rachovii, 335

O
Oncorhynchus keta, 329

P
Paracoccus denitrificans, 107, 110–112, 123, 124, 126
Peptococcus aerogenes, 276
Phormidium uncinatum, 202, 203, 253, 267
Podospora anserina, 337, 358
Polytoma papillatum, 255
Polytomella agilis, 255
Porphyromonas gingivalis, 278
Propionigenium modestum, 276, 292, 293, 297, 300
Pseudomonas aeruginosa, 278

R
Rhodospirillum rubrum, 35, 36, 51, 107, 184, 185, 187, 189, 190, 196, 268
Rickettsia prowazekii, 224

S
Saccharomyces cerevisiae, 166
Salinibacter ruber, 151
Salmonella typhimurium, 197, 276, 287, 297
Schizosaccharomyces pombe, 413
Spiroplasma sp., 203
Symplocarpus renifolius, 241
Synechococcus sp., 203, 204
Syntrophus gentianae, 186

T
Tenebrio molitor, 214
Thermotoga maritima, 285
Thermus thermophilus, 99, 101

V
Veillonella alcalescens, 276
Vibrio alginolyticus, 277–279, 287, 289–291, 297, 300
Vibrio cholerae, 278, 291
Vibrio costicola, 277
Vibrio harveyi, 267, 278
Vibrio parahaemolyticus, 290

Y
Yarrowia lipolytica, 93, 101, 102
Yersinia pestis, 278